History of Physics

The Springer book series *History of Physics* publishes scholarly yet widely accessible books on all aspects of the history of physics. These cover the history and evolution of ideas and techniques, pioneers and their contributions, institutional history, as well as the interactions between physics research and society. Also included in the scope of the series are key historical works that are published or translated for the first time, or republished with annotation and analysis.

As a whole, the series helps to demonstrate the key role of physics in shaping the modern world, as well as revealing the often meandering path that led to our current understanding of physics and the cosmos. It upholds the notion expressed by Gerald Holton that "science should treasure its history, that historical scholarship should treasure science, and that the full understanding of each is deficient without the other." The series welcomes equally works by historians of science and contributions from practicing physicists.

These books are aimed primarily at researchers and students in the sciences, history of science, and science studies; but they also provide stimulating reading for philosophers, sociologists and a broader public eager to discover how physics research – and the laws of physics themselves – came to be what they are today.

All publications in the series are peer reviewed. Titles are published as both print- and eBooks. Proposals for publication should be submitted to Dr. Angela Lahee (angela.lahee@springer.com) or one of the series editors.

Galina Weinstein

Einstein's Legacy

From General Relativity to Black Hole Mysteries

Springer

Galina Weinstein
The Efi Arazi School of Computer Science
Reichman University
Herzliya, Israel

ISSN 2730-7549 ISSN 2730-7557 (electronic)
History of Physics
ISBN 978-3-031-73571-4 ISBN 978-3-031-73572-1 (eBook)
https://doi.org/10.1007/978-3-031-73572-1

This Springer imprint is published by the registered company Springer Nature Switzerland AG
The registered company address is: Gewerbestrasse 11, 6330 Cham, Switzerland

Preface

Einstein's theory of general relativity is widely acknowledged as one of the most transformative breakthroughs in the history of physics. It fundamentally challenged existing paradigms and extended the boundaries of our understanding, revealing a novel and profound conception of spacetime and gravity. Part I of this new book, though loosely based on preprints I previously uploaded to *ArXiv*, has been extensively rewritten and offers a fresh perspective on Einstein's path to the general theory of relativity. Additionally, this part delves into the intricate network of interactions between Einstein and his collaborators, who worked alongside him after 1915 to solve the field equations of general relativity and further develop the theory. I then discuss the pioneering techniques critical in overcoming the mathematical challenges inherent in solving Einstein's field equations, laying the groundwork for the theoretical framework of classical black hole physics. This exploration underscores the ingenuity of early mathematicians and physicists, highlighting the progressive advancements that have shaped our understanding of black holes within classical general relativity.

Part II of this book addresses quantum gravity. It provides a comprehensive historical analysis, starting with black hole thermodynamics, delving into contemporary advancements in quantum gravity, encompassing quantum computers, black hole information theory, AdS/CFT correspondence, ER = EPR conjecture, and recent experimental research on wormholes. The unifying thread throughout Part II is the exploration of various aspects of modern black hole research. This part is based on several preprints I initially uploaded to *ArXiv*. However, I have completely rewritten and revised these preprints for this book, incorporating new insights and integrating them into a cohesive narrative for Part II of this book.

This book addresses historians, philosophers of physics, and physicists interested in learning about how general relativity was invented and the developments that followed afterward in classical and quantum black hole physics.

Finally, I want to make three general comments:

1. In writing this book, I have chosen to forgo direct quotations from primary sources to maintain a cohesive narrative, allowing the content to reflect my understanding and synthesis of the subject matter. I hope this approach provides readers with a clear and engaging journey through the fascinating world of general relativity envisioned by Einstein and those who interacted with him and developed his theory.
2. To preserve the reader's engagement and clarity of comprehension, I have seamlessly woven numerous intricate derivations and technical explanations into the footnotes to clarify complex subjects. This delicate balance ensures the narrative remains unencumbered while providing the necessary depth for those who seek it. Thus, the body of the book primarily features the fundamental equations and theoretical analyses, allowing for a thorough yet accessible exploration of the subject matter. Through these derivations and explanations, I endeavor to convey the relentless efforts and ingenious creativity of the luminaries of general relativity and quantum gravity. Each equation in their seminal works is a testament to days of labor, profound contemplation, a quest for solutions, and boundless creativity. By illuminating the rigorous toil behind these equations, I aim to offer a deeper perspective on the laborious work and challenges that culminate in brilliant solutions or equations, each a glimpse of inventive genius.
3. Einstein's seminal contributions, particularly those articulated in his original manuscripts, are characterized by a certain fluidity in notation—including, but not limited to, crossed-out terms, bracketed insertions reflective of momentary hesitations or reconsiderations, and occasionally terms that, with the benefit of hindsight, may be deemed excessive. While not central to the theoretical constructs underpinning this book, these historical nuances offer invaluable insights into the evolutionary trajectory of Einstein's thoughts and the process underlying the genesis of his revolutionary ideas. Readers with a keen interest in engaging directly with the unadulterated originality of Einstein's formulations, replete with the aforementioned notational peculiarities, are encouraged to consult the *Collected Papers of Albert Einstein* ([CPAE4], [CPAE6], [CPAE7]). This comprehensive compilation serves as a veritable treasure trove for scholars and enthusiasts, offering an unparalleled window into Einstein's intellectual odyssey. Within these pages, one may find the exactitude of Einstein's original equations, presented with all their historical and notational idiosyncrasies, providing a profound context for the discussions in this book.

I want to express my sincere gratitude to my supervisors, Prof. John Stachel and the late Prof. Mara Beller, for their guidance and support during my doctoral studies and afterward. Their mentorship has been invaluable in shaping my academic journey.

I am profoundly grateful to Prof. Peter Galison, whose openness and commitment to diversity allowed me to join the esteemed team, the History, Philosophy, Culture (HPC), the Next Generation Event Horizon Telescope (ngEHT). In a world where acceptance can be challenging, his willingness to support me has been instrumental.

I would also like to extend my gratitude to the Harvard Working Group, the HPC, the ngEHT, and the Black Hole Initiative for fostering an environment that, while not directly involved in the creation of this book, provided a source of inspiration and support that has been invaluable to my research.

I am deeply thankful to Prof. Gil Kalai for allowing me to continue my research endeavors when such chances were scarce. Despite the challenges, I value the platform provided to contribute to and grow within the academic community.

As someone with Asperger's Syndrome, my unique perspective has greatly influenced my approach to writing this book. Despite facing significant obstacles, I acknowledge my determination and effort in completing this work within a year.

I extend my gratitude to all those in the broader scientific community whose work has inspired and motivated me, even if we have not directly interacted.

Finally, I would like to thank the entire Springer team, particularly Dr. Angela Lahee, Dr. Marina Forlizzi, Ravi Vengadachalam, Femina Joshi Arulthas, and others whose support was invaluable in bringing this project to life. Your expertise and guidance have made this experience both smooth and rewarding.

The research and writing of this book were supported by the European Research Council (ERC) under the European Union's Horizon 2020 research and innovation program (grant agreement No. 834735). I am grateful for this funding, which made this work possible.

Herzliya, Israel — Galina Weinstein

Contents

Part I
Classical General Relativity

Chapter 1
Journey Through Relativity 1912–1916: From Covariance to *Entwurf* and Back Again

1.1 Introduction

Albert Einstein's legacy is adorned with myths and anecdotes; one of the most enduring myths is his struggle with mathematics. This notion, however, is as much a product of Einstein's renowned sense of humor as it is a misinterpretation of his prodigious intellect. Einstein was famed not only for his scientific genius but also for his playful spirit. He delighted in humor, often engaging in light-hearted pranks and witty remarks. This facet of his personality undoubtedly contributed to the spread of the myth regarding his mathematical abilities. A telling story emerges as illustrated in Alice Calaprice's *Dear Professor Einstein: Albert Einstein's Letters to and from Children*. On January 3, 1943, a young girl named Barbara from Washington, D.C., confided in Einstein about her struggles with mathematics, expressing that she had to exert more effort than her peers. Einstein's response, penned from Princeton on January 7, 1943, was humorous. He reassured Barbara, saying she need not worry about her difficulties with mathematics, for his challenges were greater [Cal]. Such anecdotes, embraced and perpetuated by the public, fostered an image of Einstein as a brilliant physicist with a supposed weakness in mathematics. Yet, this portrayal is far from the truth. Through my extensive study of both published and unpublished works by Einstein, it becomes clear that he was, in fact, an extraordinary mathematician. His proficiency in mathematical principles and inventing and manipulating mathematical constructs were remarkable.

In this and the ensuing chapters, I endeavor to illuminate Einstein's mathematical genius by presenting various derivations and techniques he employed. Through these explorations, I aim to provide a window into Einstein's unparalleled mastery of physics and mathematics. It is important to remember, however, that while we may reverse-engineer and understand his methods, the true brilliance lies in the original creation of these mathematical insights.

In this chapter, I explore Einstein's gradual development of his general theory of relativity between 1912 and 1916. I trace his journey from the early stages of collaboration with Marcel Grossmann to his mature general relativity.

G. Weinstein, *Einstein's Legacy*, History of Physics,
https://doi.org/10.1007/978-3-031-73572-1_1

Einstein oscillated between various formulations of his field equations before arriving at the groundbreaking conclusion in November 1915. During the early stages of developing his theory of gravitation, Einstein conducted research focused on static gravitational fields [E1-12, E2-12]. Starting in the summer of 1912, he embarked on a journey to explore the correct form of the field equations for his new theory of gravitation. His quest led him to delve into tensors and the intricacies of tensor calculus. During the winter months of 1912–1913, he diligently recorded his gravitational equations in a small notebook that has since become known as the *Zurich notebook* [E3-12]. Throughout this pivotal period, Einstein received invaluable assistance from his old schoolmate turned mathematics professor, Marcel Grossmann. This collaboration marked the beginning of their partnership, with a second significant collaboration occurring a year later when they jointly authored the *Entwurf* paper [EG13].[1]

In the *Zurich notebook*, we witness Einstein's tireless search for gravitational field equations. In the notebook, Einstein presents equations with minimal explanations and derivations. Consequently, comprehending the methodology behind his equations in the *Zurich notebook* requires extrapolation and educated guesses. Many equations recorded in Einstein's *Zurich Notebook* would appear in his later published works. Moreover, a noteworthy precursor to the field equations Einstein later formulated in 1915 can be traced back to equations proposed in 1912 in the *Zurich Notebook*, three years before the completion of the general theory of relativity.[2]

---

[1] During his time at the Zurich Polytechnic, Einstein, the student, encountered challenges studying subjects that did not pique his interest. He predominantly devoted his hours to solitary pursuits, embarking on a self-guided journey through physics. This involved conducting experiments and immersing himself in the works of the pioneering figures in the field. Einstein was, in essence, an autodidact. Einstein's approach to his education at the Polytechnic diverged from the conventional path. He refrained from diligently taking notes and, regrettably, missed numerous mandatory classes. Despite having the opportunity to establish a solid mathematical foundation, he never attended the mathematical classes at the Polytechnic taught by luminaries such as Hermann Minkowski, missing the chance to learn under the tutelage of these great mathematicians. Einstein's lack of enthusiasm extended to other courses at the Polytechnic, which he habitually skipped, much like he did with the great mathematicians' classes. Fortunately, before facing the examinations mandated during his four-year course, Einstein turned to the notes of his dedicated friend and classmate, Grossmann. When Einstein needed employment, Grossmann's father intervened and secured him a position at the Patent Office in 1902. In the summer of 1912, Einstein again sought Grossmann's assistance, the man who had previously come to his rescue. Einstein confided in Grossmann about his predicament and asked him to investigate whether there was a suitable geometry to address the questions. The following day, Grossmann returned with the revelation that Riemannian geometry was the appropriate choice. Thus, Einstein, the autodidact, embarked on yet another self-guided educational journey, this time delving into the mathematical tools developed by luminaries such as Bernhard Riemann, Gregorio Ricci-Curbastro, Elwin Bruno Christoffel, and Tullio Levi-Civita [W17].

[2] The ensuing equations recorded in Einstein's *Zurich Notebook* would subsequently be identified within the pages of his later published works: Eq. (1.3) (the line element), Eqs. (1.12) and (1.20) (the contravariant stress-energy tensor), Eqs. (1.18) and (1.21) (the energy-momentum balance in a gravitational field), Eq. (1.26) (the Beltrami second parameter), Eq. (1.27) (Poisson's equation), Eq. (1.28) (the harmonic condition), Eq. (1.29) (the d'Alembertian-like operator in curved spacetime), Eq. (1.30) (Einstein's operator), a version of Eq. (1.34) (a conservation law for the gravitational

A consortium of researchers dedicated approximately a decade to interpreting Einstein's *Zurich Notebook*, culminating in the publication of their insights [JRSNS]. I endeavor to present an interpretation that, while drawing upon the foundational work of these scholars, diverges subtly in its analytical perspective.

Certain expressions are inferred but not explicitly written in the *Zurich Notebook*. Einstein had suppressed expressions in the notebook and wrote them later in his joint *Entwurf* paper with Marcel Grossmann [EG13] and other papers presenting the *Entwurf* theory ([E13, EG14]). The derivation presented in Sect. 1.2 is based on the intermittent equations in Einstein's *Zurich Notebook* [E3-12]. This foundation is augmented by implicit equations, which, while absent from the notebook, are pivotal and intuitively integral to the derivation process. Equations not recorded in the notebook have been sourced from the *Entwurf* paper [EG13], serving to complete the analytical framework.

From 1912 to 1913, Einstein utilized a notational convention for covariant and contravariant derivatives that might seem unconventional today. Specifically, he wrote both types of derivatives with lower indices, denoting them as $\frac{\partial}{\partial x_\nu}$. This notation underscores the evolving nature of mathematical symbols in the early development of relativity theory, reflecting Einstein's efforts to formulate his ideas in a coherent mathematical framework. The same holds for the inverse of the metric tensor: in the *Zurich Notebook* (and in the 1913 *Entwurf* paper [EG13]) $\gamma_{\mu\nu}$ represents the contravariant metric tensor, the inverse of the covariant $g_{\mu\nu}$. I deliberately adhere to the specific notational convention used by Einstein from 1912 through 1913, where he denoted both covariant and contravariant derivatives with lower indices and the inverse of the metric tensor with $\gamma_{\mu\nu}$. This choice *emphasizes the historical evolution* of Einstein's mathematical notation over a critical period from 1912 through 1913 until 1914. By closely following Einstein's notation, I aim to illuminate the developmental trajectory of his theoretical constructs and the corresponding mathematical expressions. This approach offers a window into Einstein's thought processes during these formative years. We can more easily identify where Einstein might have gone astray or where his intuition was correct but limited by the notation or mathematical tools available at the time. *In my analysis, I highlight how Einstein's mathematical expressions from 1912 co-evolved and influenced his later equations and expressions.*

Einstein's contributions within the *Zurich Notebook* are delineated along the specified trajectory. In the first pages of the *Zurich Notebook*, Einstein's pursuit aimed to distill candidates for his gravitational field equations through invariant formulations. Einstein's earlier computations did not yield definitive candidates for the left-hand

field), Eq. (1.36) (the Riemann curvature tensor), Eq. (1.75) (the Ricci tensor), Eq. (1.76) (the Christoffel symbols), Eq. (1.81) (the November tensor), Eqs. (1.40) and (1.85) (the Laplacian-like term), Eq. (1.25) (early geodesic equation), and Eq. (1.97) (the *Entwurf* field equations—the stress-energy pseudo tensor). Equations (1.41) (a perturbation applied to the Minkowski metric) and (1.58) (the box operator in flat space-time acting on the perturbation) are implicitly assumed. Moreover, the linearized forms of Eq. (1.67) [or Eq. (1.68)], Eq. (1.72) [or Eq. (1.73)] bear a notable resemblance to the field equations Einstein later formulated in 1915. While definitive evidence linking these early equations to the inspiration behind his subsequent field equations remains elusive, scholars propose a potential influence.

side of the gravitational field equations. Instead, they facilitated the development of several pivotal techniques, results, and conceptual insights. These contributions proved instrumental as Einstein later integrated the Riemann and Ricci tensors into his theoretical framework, marking a significant advancement in his quest for a coherent and comprehensive description of gravitation. It was not until page $14L$ of his notebook that Einstein commenced a methodical examination of the Riemann curvature tensor. Arguably, he received that tensor from Grossmann [E3-12].

Einstein was remarkably close to his final formulation of the general theory of relativity three years before he officially presented it. In the *Zurich Notebook*, Einstein explored the Ricci tensor as a potential candidate for the gravitational field equations. This exploration indicates that he was already considering what scholars called the *November tensor* and its relationship with the stress-energy tensor [W18]. Despite coming close to the correct formulation, Einstein temporarily rejected the broader covariance of the field equations he would eventually embrace in 1915.

In the spring of 1913, Einstein made a fateful decision. He discarded what he later claimed to be the fundamentally correct gravitational field equation, which he had meticulously worked on in the final pages of the *Zurich notebook*. Instead, he adopted the non-covariant *Entwurf* field equations of 1913 in collaboration with Grossmann [EG13]. Einstein favored the *Entwurf* theory, which appeared more aligned with the Newtonian limit, albeit based on a calculation error that incorrectly suggested it accounted for rotation. This phase of Einstein's work was characterized by what might be interpreted as *parapraxis*—minor calculation errors reflecting his subconscious inclination towards the *Entwurf* theory as a comprehensive framework for gravity, integrating key principles and frameworks like the Newtonian limit (correspondence principle), conservation principle, and rotational dynamics.

Einstein initially formulated the *Entwurf* field equations based on an invariant operation, specifically an operator he had already derived in the *Zurich Notebook*, rather than directly on the Ricci tensor. Using this operator in the *Entwurf* field equations manifests Einstein's effort to formulate the equations of gravitation in a manner closely related to the Newtonian approximation. In his joint *Entwurf* paper with Grossmann, Einstein provided a detailed methodology for deriving an identity that culminated in formulating the *Entwurf* field equations [EG13].

The Ricci tensor was ultimately the correct choice for Einstein's field equations, representing the gravitational tensor due to the stress-energy tensor. Scholars have shown that while Einstein initially distanced himself from the Ricci tensor, his engagement with it on pages $24R-25R$ in the *Zurich Notebook* reveals a deep, ongoing effort to reconcile it with his mathematical insights [E3-12, JRSNS]. I show in this chapter that Einstein's work also involved reconciling vestiges of the November tensor with the 1913 *Entwurf* formalism.

During 1913, Einstein, collaborating with Michele Besso, faced challenges with the *Entwurf* vacuum field equations, particularly in accurately predicting the advance of Mercury's perihelion. Their initial calculations significantly deviated from the known astronomical data. The *Entwurf* theory's non-generally covariant field equations, justified by the hole argument, were insufficient. Its failure to accurately predict Mercury's perihelion advance and alignment with rotational aspects (Mach's principle) in 1915 marked a critical juncture. However, Einstein's response did not entirely discard the *Entwurf* theory. Instead, he refined it and incorporated its elements into a new framework. This adaptive process, evident in Einstein's November 1915 submissions, utilized aspects of the 1914 *Entwurf* theory [E2-14] within the emerging structure of general relativity, particularly adapting it for unimodular transformations.

Post-1915, Einstein moved away from unimodular coordinates. This shift was not precipitated by experimental disconfirmation but emerged from theoretical reasoning. Revisiting the 1914 *Entwurf* theory in 1916 and integrating it with David Hilbert's axiomatic approach, Einstein derived the generally covariant field equations [E4-16]. In his work, *Foundations of Physics*, Hilbert proposes an interconnection between Einstein's general relativity and Gustav Mie's electrodynamics within an axiomatic framework [Hil15-2]. Einstein's advancements in 1916 culminated in the contracted Bianchi identity. This led to the principle of energy-momentum conservation, thereby refining and correcting elements from the earlier theory [E4-16]. Revisiting and refining previous ideas, rather than outright rejecting his theory, underscores Einstein's methodological approach in his scientific journey.

In this chapter, I show that Einstein's general theory of relativity development exemplifies a complex, recursive process of theoretical evolution. It highlights the importance of integrating and reevaluating past concepts, correcting errors, and continually refining ideas rather than a linear hypothesis testing and rejection trajectory. This approach underscores the intricate blend of calculation errors, theoretical reasoning, and conceptual innovation that marked Einstein's path to one of history's most profound scientific theories.

1.2 From the Zurich Notebook to the 1913 *Entwurf* Theory: Einstein's Theoretical Journey

1.2.1 The Stress-Energy Tensor and the Geodesic Equation

In 1911, within his textbook, *The Relativity Principle*, Max Laue introduced the second-rank tensor, $T_{\mu\nu}$, which came to be recognized as the stress-energy-momentum tensor, or more succinctly, the stress-energy tensor [Lau]. This tensor is pivotal in the theoretical framework of general relativity, bridging the stress components with the mass sources and the densities of energy and momentum.

Subsequently, in 1912, Einstein provided an extensive overview of the special theory of relativity, wherein he engaged deeply with Laue's formulation of the stress-energy tensor. Einstein recapitulated Laue's derivations through his manuscript and underscored the necessity of assigning a symmetric tensor $T_{\mu\nu}$ to every conceivable material process under scrutiny [CPAE4], Doc 1. *This assertion posits that the right-hand side of the field equations in Einstein's gravitational theory should invariably be represented by the tensor* $T_{\mu\nu}$. Further elaboration on the formulation of $T_{\mu\nu}$, particularly in the context of pressureless dust, was articulated by Einstein in his

Zurich Notebook.[3] As I show in this section, Einstein derived the equations of motion for a material particle in a metric field and generalized these to a pressureless dust cloud in a gravitational field. The derivation process uses the principle of least action and the Euler-Lagrange equations to derive momentum and force density expressions in a gravitational field.

I will begin my analysis with page $5R$ of the *Zurich Notebook*. Einstein started with the variation of the action [E3-12][4]:

$$\delta \int ds = 0, \quad \text{or:} \quad \delta \int L dt = 0, \quad \text{with the Lagrangian: } L = -m\frac{ds}{dt}. \tag{1.1}$$

At the top of page $5R$, he introduced the line element as [E3-12][5]:

$$(ds)^2 = g_{11}(dx)^2 + \cdots + g_{44}(dt)^2. \tag{1.2}$$

On page 39L, Einstein wrote the most general form of the spacetime line element (1.2) [E3-12][6]:

THE LINE ELEMENT:

$$\boxed{ds^2 = g_{\mu\nu} dx_\mu dx_\nu.} \tag{1.3}$$

---

[3] Dust refers to a perfect fluid with no pressure or internal stresses, characterized by its rest mass density alone. The stress-energy tensor for dust is a specific case of the energy-momentum tensor, where the stress-energy tensor takes a simplified form due to the assumption that the pressure (p) is zero and the only non-zero component is the energy density. It describes a collection of particles that do not exert pressure on each other.

[4] The expression for the Lagrangian of the system L describes the dynamics of a particle of mass m moving through spacetime, with ds denoting the infinitesimal spacetime interval.

[5] The line element defines the metric tensor $g_{\mu\nu}$ and dx_μ are the differential changes in the spacetime coordinates. Equation (1.2) implies a sum over certain terms of the metric tensor.

[6] This equation represents the line element in general relativity. It describes the infinitesimal interval of spacetime between two events. ds^2 is the spacetime interval squared. It generalizes distance to spacetime, incorporating both spatial distances and temporal durations. dx_μ and dx_ν are differentials of the spacetime coordinates, where μ and ν range over the spacetime dimensions (4 for time, and 1, 2, 3 for the three spatial dimensions in a four-dimensional spacetime). The line element can be seen as a generalization of the Pythagorean theorem to curved spacetime. In flat Euclidean space, $ds^2 = dx^2 + dy^2 + dz^2$ gives the distance squared between two points. When we incorporate time $-c^2dt^2$ and allow for a curved spacetime, the equation in the tensorial formulation is (1.3), with the metric tensor $g_{\mu\nu}$ reflecting the influence of gravity on spacetime geometry. The interval ds^2 is invariant under coordinate transformations. This means that no matter what coordinate system we use, the calculated interval ds^2 between two events remains the same. Using the metric (1.2) [or Eq. (1.3)], we can express ds in terms of the spacetime coordinates and their derivatives, which allows us to write down the Lagrangian L explicitly in terms of $g_{\mu\nu}$ and the velocities $\frac{dx_\mu}{dt}$. The explicit dependence of L on the metric and the velocities reflects the spacetime geometry's influence on the particle's dynamics.

The final step on page $5R$ involved deriving the equations of motion for a material point in a static gravitational field, i.e., the Euler-Lagrange equations from the condition (1.1) that the variation of the action is zero [E3-12]:

$$\text{EQUATIONS OF MOTION:} \quad \frac{d}{dt}\left(\frac{\partial L}{\partial \dot{x}}\right) - \frac{\partial L}{\partial x} = 0. \tag{1.4}$$

Equations (1.1) and (1.2) result in three expressions for force (F_x), energy (E), and angular momentum in the x-direction (J_x) of a material point[7]:

The first expression represents an equation for the force of a material point[8]:

$$\text{FORCE:} \quad F_x = -\frac{1}{2} m \sum_{\mu\nu} \frac{\partial g_{\mu\nu}}{\partial x_1} \frac{dx_\mu}{ds} \frac{dx_\nu}{dt}. \tag{1.5}$$

The second and third expressions represent the energy and angular momentum of the material point [E3-12, EG13][9]:

$$\text{ENERGY:} \quad E = m\left(g_{41}\frac{dx_1}{ds} + g_{42}\frac{dx_2}{ds} + g_{43}\frac{dx_3}{ds} + g_{44}\frac{dx_4}{ds}\right). \tag{1.6}$$

$$\text{ANGULAR MOMENTUM:} \quad \frac{\partial L}{\partial \dot{x}} = J_x = -\frac{m}{\frac{ds}{dt}}\left(g_{11}\dot{x}_1 + g_{12}\dot{x}_2 + g_{13}\dot{x}_3 + g_{14}\right). \tag{1.7}$$

The above two equations lead to the specific case of the force density, energy density, and momentum density in the x-direction.

Einstein introduced the proper volume V_0 [E3-12]:

$$V = V_0 \frac{ds}{dt} \frac{1}{\sqrt{-g}}, \tag{1.8}$$

and the rest mass density or the proper density ρ_0:

$$\rho_0 = \frac{m}{V}, \tag{1.9}$$

[7] These equations involve the metric tensor $g_{\mu\nu}$ and its derivatives and components of the metric tensor, and the derivatives of the coordinates x_μ and x_ν with respect to ds. Further, $\dot{x}_i$ denotes the derivative of x_i with respect to time t. The equations for energy and momentum indicate how the presence of mass-energy curves spacetime and affects these quantities. $\frac{ds}{dt}$ represents the derivative of the proper time s (or the spacetime interval along a path) with respect to the coordinate time t.

[8] The force experienced by a material point is derived from the change in the metric tensor $g_{\mu\nu}$ with respect to the spatial coordinates. This force equation accounts for the metric tensor and its derivative affecting the material point's trajectory.

[9] These equations are formulated similarly to the force equation by how the material point's motion through spacetime coordinates influences its energy and angular momentum, integrating the effects of the metric tensor (spacetime curvature). In the *Zurich Notebook*, Einstein rendered the equation for the energy implicit, electing not to inscribe it explicitly.

to relate the material point's properties to a volume of space, adjusting for relativistic effects through $\frac{ds}{dt}$ and the metric determinant $\sqrt{-g}$.[10] By plugging the expressions for V [Eq. (1.8)] and ρ_0 [Eq. (1.9)] into the initial equations for F_x [Eq. (1.5)], E [Eq. (1.6)], and J_x [Eq. (1.7)], Einstein derived the force, energy, and momentum per volume equations [E3-12] [EG13][11]:

$$\frac{F_x}{V} = -\frac{1}{2}\rho_0\sqrt{-g}\sum_{\mu\nu}\frac{\partial g_{\mu\nu}}{\partial x_1}\frac{dx_\mu}{ds}\frac{dx_\nu}{ds}. \tag{1.10}$$

$$\frac{EJ}{V} = -\rho_0\sqrt{-g}\sum_{\nu} g_{\mu\nu}\frac{dx_\nu}{ds}\frac{dx_4}{ds}. \tag{1.11}$$

Equations (1.10) and (1.11) lead to the simplest form of the contravariant stress-energy tensor of the material process for pressureless dust, analogous to the source mass-density ρ in the Poisson equation [E3-12][12]:

THE CONTRAVARIANT STRESS- ENERGY TENSOR FOR PRESSURELESS DUST:

$$\boxed{T_{\mu\nu} = \rho_0\frac{dx_\mu}{ds}\frac{dx_\nu}{ds}.} \tag{1.12}$$

[10] This allows for the consistent description of the physical quantities across different frames of reference. This consistency is necessary because, in relativity, measurements of distances, times, and volumes can vary depending on the observer's state of motion.

[11] These equations express how F_x, E, and J_x are distributed within a given spacetime volume, factoring in the curvature effects ($g_{\mu\nu}$). For energy E, $\mu = 4$, and for momentum J_x in the x-direction, $\mu = 1$.

[12] Einstein wrote the contravariant stress-energy tensor for pressureless dust as a tensor density $\mathcal{T}_{i\kappa}$. He transformed this to a tensor $T_{\mu\nu}$. ρ_0 is the mass density of the dust in its rest frame, $\frac{dx_\mu}{ds}$ and $\frac{dx_\nu}{ds}$ are the components of the four-velocity of the dust particles, and s is the proper time along the world line of the particles. This form follows from considering the dust's mass-energy content moving through spacetime. Later, Einstein revisited this equation many times. For instance, Eq. (3.56) in Sect. 3.2.3. The energy-momentum tensor for dust is a simplified model that describes a collection of particles that move under the influence of gravity but do not interact with each other through any other force (i.e., they exert no pressure and have no internal stresses). This simplicity allows for a straightforward representation of energy and momentum in spacetime. However, real physical systems may involve pressures, internal stresses, and energy flows. To account for these complexities, the general form of the stress-energy tensor $T_{\mu\nu}$ includes terms for energy density, pressure, and shear stresses. It describes the complete state of matter-energy content, including mass and any form of energy and momentum flux. Based on Einstein's 1916 tensor notation, we can reformulate Eq. (1.12) as Eq. (3.56): $T^{\mu\nu} = \rho_0\frac{dx^\mu}{ds}\frac{dx^\nu}{ds}$.

Substituting Eq. (1.12) into Eqs. (1.10) and (1.11) leads to [E3-12, EG13][13]:

$$\text{FORCE DENSITY:} \quad \frac{F_x}{V} = -\frac{1}{2}\sqrt{-g}\sum_{\mu\nu}\frac{\partial g_{\mu\nu}}{\partial x_1}T_{\mu\nu}. \tag{1.13}$$

$$\text{ENERGY DENSITY:} \quad \frac{E}{V} = -\sqrt{-g}\sum_{\nu} g_{4\nu}T_{\mu\nu}. \tag{1.14}$$

$$\text{MOMENTUM DENSITY:} \quad \frac{J_x}{V} = -\sqrt{-g}\sum_{\nu} g_{1\nu}T_{\mu\nu}. \tag{1.15}$$

Einstein expanded the scope from one component to two general expressions[14]:

$$\text{THE ENERGY- MOMENTUM DENSITY:} \quad \boxed{-\sqrt{-g}\sum_{\nu} g_{\sigma\mu}T_{\mu\nu}.} \tag{1.16}$$

$$\text{THE GRAVITATIONAL FOUR- FORCE DENSITY:} \quad \boxed{-\frac{1}{2}\sqrt{-g}\sum_{\mu\nu}\frac{\partial g_{\mu\nu}}{\partial x_\sigma}T_{\mu\nu}.} \tag{1.17}$$

Einstein set the divergence of the energy-momentum density (1.16) equal to the four-force density (1.17). He combined these two equations into one expression, representing *energy and momentum conservation in a gravitational field* [E3-12]:

THE ENERGY- MOMENTUM BALANCE IN A GRAVITATIONAL FIELD:

$$\boxed{\sum_{\mu\nu}\frac{\partial}{\partial x_\nu}(\sqrt{-g}g_{\sigma\mu}T_{\mu\nu}) - \frac{1}{2}\sum_{\mu\nu}\sqrt{-g}\frac{\partial g_{\mu\nu}}{\partial x_\sigma}T_{\mu\nu} = 0.} \tag{1.18}$$

---

[13] These equations relate the components of $T_{\mu\nu}$ for the dust to the gravitational field described by the metric tensor $g_{\mu\nu}$. Equation (1.10) ($\frac{F_x}{V}$) expresses the force per volume acting on the dust due to the gravitational field. It represents how the curvature of spacetime (represented by $g_{\mu\nu}$ and its derivatives) influences the motion of dust. Equation (1.11) ($\frac{E}{V}$) gives the energy per volume of the dust, showing how the energy content is affected by the gravitational field. The third equation ($\frac{J_x}{V}$) describes the angular momentum in the x-direction per volume of the dust, influenced by spacetime curvature.

[14] He combined Eqs. (1.14) and (1.15) to obtain the stress-energy tensor adjusted by the metric tensor and the square root of the determinant of $g_{\mu\nu}$, reflecting the flow of energy and momentum in spacetime.

The expression (1.18) is a general expression that applies not just to pressureless dust but to any matter or energy distribution described by the stress-energy tensor (1.12).

Einstein then introduced the contravariant tensor density as [E3-12]:

$$\sqrt{-g}T_{\mu\nu} = \Theta_{\mu\nu}. \tag{1.19}$$

By Eq. (1.12), $\Theta_{\mu\nu}$ for a cloud of pressureless dust is given by [E3-12]:

THE CONTRAVARIANT STRESS- ENERGY TENSOR DENSITY FOR PRESSURELESS DUST:

$$\boxed{\Theta_{\mu\nu} = \rho_0 \frac{dx_\mu}{ds}\frac{dx_\nu}{ds}.} \tag{1.20}$$

Einstein then wrote the equivalent form of Eq. (1.18), the energy-momentum balance in a gravitational field [E3-12]:

THE ENERGY- MOMENTUM BALANCE IN A GRAVITATIONAL FIELD:

$$\boxed{\sum_{\mu\nu} \frac{\partial}{\partial x_\nu}\left(g_{\sigma\mu}\Theta_{\mu\nu}\right) - \frac{1}{2}\sum_{\mu\nu} \frac{\partial g_{\mu\nu}}{\partial x_\sigma}\Theta_{\mu\nu} = 0.} \tag{1.21}$$

Like Eq. (1.18), the underlying principle of this expression, or its foundational structure, applies more broadly than just to pressureless dust.

On page 41*R*, Einstein employed the Newtonian equations of motion to analyze the behavior of a particle that, while not influenced by external forces, is restricted to moving along a curved surface. He began with the x-component of the Newtonian equation of motion for a particle. This particle has a mass m, and is constrained to move along the surface defined by $f(x, y, z)$ [E3-12]:

$$\frac{d^2x}{dt^2} - \frac{\lambda}{m}\frac{\partial f}{\partial x} = 0. \tag{1.22}$$

The left-hand side of this equation represents the normal force that confines the particle's motion to the surface defined by $f = 0$, with λ representing the force's magnitude. By replacing the time variable t in Eq. (1.22) with the length of a trajectory s on the surface, Einstein derived the subsequent equation:

$$\frac{d^2x}{ds^2} - \lambda'\frac{\partial f}{\partial x} = 0. \tag{1.23}$$

Through this analysis, Einstein elucidated that the path delineated by the particle within the constraints of the surface constitutes a geodesic (that the particle's trajectory is a geodesic line on the curved surface defined by $f = 0$). He showed that geodesics, defined as curves representing the shortest distance between two points within a given surface, are derived and understood by applying the variational principle (1.1).

On page $43LB$, Einstein integrated the energy-momentum balance in a gravitational field [Eq. (1.18)] from page $5R$ [E3-12]:

THE ENERGY-MOMENTUM BALANCE IN A GRAVITATIONAL FIELD $\sqrt{-g} = 1$:

$$\boxed{\frac{\partial}{\partial x_n}(g_{m\nu} T_{\nu n}) - \frac{1}{2}\frac{\partial g_{\mu\nu}}{\partial x_m} T_{\mu\nu} = 0,} \tag{1.24}$$

incorporating the stress-energy tensor specific to pressureless dust, Eq. (1.12). This derivation led to the following equation:

EARLY GEODESIC EQUATION:

$$\frac{d}{ds}\left(g_{m\nu}\frac{dx_\nu}{ds}\right) - \frac{1}{2}\frac{\partial g_{\mu\nu}}{\partial x_m}\frac{dx_\mu}{ds}\frac{dx_\nu}{ds} = 0. \tag{1.25}$$

Einstein's equation (1.25) is an early form of what would later evolve into the geodesic equation. This equation tries to convey how spacetime's curvature influences objects' paths. It has components that are very similar to the final form of the geodesic Eq. (1.367) (see Sect. 1.7.1). Despite its conceptual proximity, this formulation exhibits significant disparities from the geodesic equation. The early Eq. (1.25) attempts to describe the motion using the metric tensor $g_{\mu\nu}$ and its derivatives directly. The later geodesic Eq. (1.367) uses the Christoffel symbols, which are derived from the first and second derivatives of the metric tensor. The Christoffel symbols provide a more convenient and compact way to describe the geodesic equation.

1.2.2 Einstein's Exploration of Invariants

Subsequently, Einstein searched for potential candidates for the left-hand side of the field equations. On pages $6L$, $6R$, and $7L$ he constructed an expression that is mathematically equivalent to the *Laplace-Beltrami operator* $\Delta\phi$ acting on a scalar field [E3-12]:

THE BELTRAMI SECOND PARAMETER:

$$\frac{1}{\sqrt{-g}}\sum_{\mu\nu}\frac{\partial}{\partial x_\mu}\left(\sqrt{-g}\gamma_{\mu\nu}\frac{\partial\phi}{\partial x_\nu}\right). \tag{1.26}$$

In the mathematical part of the *Entwurf* paper, Marcel Grossmann writes this equation and designates it as the *second parameter of Beltrami* [EG13]. Therefore, it stands to reason that Grossmann was likely the source from which Einstein acquired this expression. It generalizes the notion of the Laplacian, ∇^2 or Δ, to curved space, acting on the scalar function ϕ. The Laplacian is fundamental in Newtonian gravity, where it appears in *Poisson's equation* for the gravitational potential.[15]

POISSON'S EQUATION:

$$\boxed{\nabla^2\Phi = 4\pi G\rho.} \tag{1.27}$$

Einstein applied a harmonic condition on the contravariant tensor $\gamma_{\sigma\nu}$ (which is related to harmonic coordinates) [E3-12][16]:

THE HARMONIC CONDITION:

$$\boxed{\sum_{\mu} \frac{\partial \gamma_{\mu\nu}}{\partial x_{\mu}} = 0,} \tag{1.28}$$

to simplify the derivative operations in the Beltrami second parameter, while certain terms are simplified. The harmonic condition simplifies the geometry-related terms in the Beltrami second parameter, making the equation's overall structure more tractable.

By exploring the Beltrami second parameter, Einstein aimed to understand how classical notions of fields and potentials could be generalized within the relativistic framework. Einstein's exploration of such invariant mathematical constructs provided him with analytical tools to tackle the complexities of his gravitational theory. The Beltrami second parameter could offer insights into how scalar fields and their dynamics are described in a geometrically curved context. It allowed for a deeper understanding of how gravitational fields, described by the metric tensor, could be related to the sources of those fields, such as mass and energy density.

In the subsequent sections of his *Zurich Notebook*, Einstein's mathematical explorations exhibit a marked increase in sophistication without explicit engagement with the Riemann curvature tensor. Einstein's calculations reflect a deepening mathematical inquiry, potentially relevant to the formulation and refinement of the field equations.

On page $12L$, Einstein replaced $\gamma_{\mu\nu}\frac{\partial\phi}{\partial x_{\nu}}$ [in the Beltrami second parameter (1.26)] with $\gamma_{\alpha\beta}\frac{\partial\gamma_{\mu\nu}}{\partial x_{\beta}}$. This led him to write a variant of the Beltrami second parameter, acting on the components of a tensor rather than a scalar function, which complicates it due to the need to consider the tensorial nature of $\gamma_{\mu\nu}$ and its derivatives [E3-12]:

---

[15] In Newtonian gravity, the gravitational field is described by the gravitational potential Φ, which satisfies the Poisson equation:∇^2 is the Laplace operator, G is the gravitational constant, and ρ is the mass density. This equation states that the Laplacian of the gravitational potential is proportional to the local mass density. The Laplacian operator: $\Delta \equiv \nabla^2 = \frac{\partial^2}{\partial x^2} + \frac{\partial^2}{\partial y^2} + \frac{\partial^2}{\partial z^2}$.

[16] Or the general harmonic condition: the condition for a metric to be harmonic. Refer to footnote 25 for further information on this condition.

THE D'ALEMBERTIAN- LIKE OPERATOR IN CURVED SPACETIME:

$$\boxed{\sum_{\alpha\beta} \frac{1}{\sqrt{-g}} \frac{\partial}{\partial x_\alpha} \left(\gamma_{\alpha\beta} \sqrt{-g} \frac{\partial \gamma_{\mu\nu}}{\partial x_\beta} \right).} \tag{1.29}$$

This equation is structurally analogous to applying the d'Alembertian operator in curved spacetime on the metric tensor components $\gamma_{\mu\nu}$. For future reference, I call this equation the *d'Alembertian-like operator in curved spacetime*. This equation has elements in common with the operations involved in the Beltrami second parameter. However, Eq. (1.29) is not the Beltrami second parameter but one of Einstein's early exploratory attempts to formulate candidates for the left-hand side of his field equations.[17]

During this period, Einstein extensively explored different mathematical structures and operations. He was grappling with mathematically grasping the dynamics of spacetime curvature. His exploration included various tensorial equations, of which the metric tensor and its derivatives were central components.

For $\sqrt{-g} = 1$, Eq. (1.29) is presented as [E3-12]:

[17] Olivier Darrigol refers to Eq. (1.29) as a simple generalization of the d'Alembertian operator [Dar]. The Laplace-Beltrami operator is the generalization of the Laplace operator to curved spaces, and the d'Alembertian operator is the generalization of the wave operator to curved spacetime. Although the operator (1.29) looks superficially similar to the structure of the two above-mentioned operators, it involves the inverse of the metric tensor. The expression (1.29) can be viewed as a generalization of the d'Alembertian operator but applied to a rank-2 tensor. From a modern perspective, it is more closely related to the Lichnerowicz Laplacian (or the Laplace-de Rham) operator, which acts on tensor fields. Equation (1.29) is a second-order differential operator acting on the components of a rank-2 tensor field (accounting for the metric via the metric determinant $\sqrt{-g}$), just like how the Lichnerowicz Laplacian operator acts on perturbations of the metric tensor. However, this similarity reflects a more modern perspective, particularly in differential geometry and general relativity. In contemporary physics and mathematics, there is a greater emphasis on understanding how differential operators extend to different fields—scalars, vectors, tensors—especially in curved spacetime. This modern approach often involves looking at operators like the Lichnerowicz Laplacian, which is more sophisticated. The Lichnerowicz Laplacian, or Laplace-de Rham operator, is a natural generalization of the Laplace-Beltrami operator to act on tensor fields, which is why Einstein's equation (1.29) is more akin to it from this viewpoint. Thus, Einstein's operator (1.29) can be seen as a precursor of these more general operators. Since Eq. (1.29) does not include the curvature terms, it is a simpler version of the Lichnerowicz Laplacian because it generalizes the action of the Laplacian to tensors rather than scalars. Einstein substituted $\gamma_{\mu\nu} \frac{\partial \phi}{\partial x_\nu}$ in the Beltrami second parameter (1.26) with $\gamma_{\alpha\beta} \frac{\partial \gamma_{\mu\nu}}{\partial x_\beta}$, a maneuver that underscores his intuitive approach to deriving novel formulations. This maneuver does not strictly generalize the Beltrami second parameter but rather represents an ingenious step on Einstein's part. To highlight this, I describe Eq. (1.29) as a *d'Alembertian-like operator in curved spacetime*. This naming implicitly acknowledges that the operator (1.29) shares structural similarities with the d'Alembertian while being adapted to the specific context of tensor fields.

EINSTEIN'S OPERATOR:

$$\boxed{\sum_{\alpha\beta} \frac{\partial}{\partial x_\alpha}\left(\gamma_{\alpha\beta}\frac{\partial \gamma_{\mu\nu}}{\partial x_\beta}\right).} \tag{1.30}$$

For future reference, *I call this equation Einstein's operator.*[18]

Einstein adopted a convention where $g_{\mu\nu}$ denoted the covariant metric tensor and $\gamma_{\mu\nu}$ represents the contravariant metric tensor. This simplified notation by using $\frac{\partial}{\partial x_\mu}$ consistently, an approach that could be seen as a notational "trick" to streamline expressions, similar in spirit to the convenience offered by Einstein much later in [E1-16] by his summation convention.

On page 13*R* of the *Zurich Notebook*, Einstein applied *the product rule for partial derivatives*[19]:

THE PRODUCT RULE:

$$\frac{\partial}{\partial x_\nu}(uv) = \frac{\partial v}{\partial x_\nu}u + v\frac{\partial u}{\partial x_\nu}, \tag{1.31}$$

to the operator (1.30). According to the product rule[20]:

$$\frac{\partial}{\partial x_\alpha}\left(\gamma_{\alpha\beta}\frac{\partial \gamma_{\mu\nu}}{\partial x_\beta}\right) = \frac{\partial \gamma_{\alpha\beta}}{\partial x_\alpha}\frac{\partial \gamma_{\mu\nu}}{\partial x_\beta} + \boxed{\gamma_{\alpha\beta}\frac{\partial^2 \gamma_{\mu\nu}}{\partial x_\alpha \partial x_\beta}.} \tag{1.32}$$

Einstein assumed $\sqrt{-g} = 1$ in the energy-momentum balance in a gravitational field (1.21), and used it to set the third term of Eq. (1.32) equal to the contravariant stress-energy tensor[21]:

$$\gamma_{\alpha\beta}\frac{\partial^2 \gamma_{\mu\nu}}{\partial x_\alpha \partial x_\beta} = \Theta_{\mu\nu}. \tag{1.33}$$

The dominant second-order spatial derivatives in Eq. (1.33) resemble the classical Laplacian action on the metric components in the limit of a weak field and static conditions. Einstein also wrote the harmonic condition (1.28). Thus, he wanted to show that the field Eq. (1.33) could demonstrate compatibility with energy-momentum conservation under this condition.

---

[18] Scholars have called Eq. (1.30) the *core operator* [JRSNS]. I refer to it as *Einstein's operator* because Einstein called it an *operator* [EG13].

[19] To apply the product rule for partial derivatives, we identify u and v, then compute $\frac{\partial u}{\partial x_\nu}$, and compute $\frac{\partial v}{\partial x_\nu}$. Finally, apply the Eq. (1.31).

[20] Here, the derivative is applied to each factor in turn, consistent with the product rule's directive to take the derivative of each part of the product separately and then sum the results.

[21] The left-hand side of Eq. (1.33) involves the second-order partial derivatives of the contravariant metric tensor $\gamma_{\mu\nu}$ with respect to two different coordinates x_α and x_β. It measures how $\gamma_{\mu\nu}$ changes as both x_α and x_β change, a mixed second derivative if $\alpha \neq \beta$, or a pure second derivative if $\alpha = \beta$.

Einstein then replaced the contravariant stress-energy tensor $\Theta_{\mu\nu}$ in the energy-momentum balance in a gravitational field (1.21) with the left-hand side of Eq. (1.33). This led to a conservation law for the gravitational field [E3-12]:

A CONSERVATION LAW FOR THE GRAVITATIONAL FIELD:

$$\sum_{\nu\sigma} \frac{\partial}{\partial x_\sigma} g_{\rho\nu} \boxed{\gamma_{\alpha\beta} \frac{\partial^2 \gamma_{\mu\nu}}{\partial x_\alpha \partial x_\beta}} - \frac{1}{2} \sum_{\mu\nu} \frac{\partial g_{\mu\nu}}{\partial x_\rho} \boxed{\gamma_{\alpha\beta} \frac{\partial^2 \gamma_{\mu\nu}}{\partial x_\alpha \partial x_\beta}}, \quad \text{where: } \sqrt{-g} = 1. \tag{1.34}$$

Summarizing the discussion until now, in the first pages of the *Zurich Notebook*, Einstein was grappling with the challenge of finding candidates for field equations. Einstein's pursuit aimed to distill candidates for his gravitational field equations through invariant formulations. His earlier computations did not yield definitive candidates for the left-hand side of the gravitational field equations. Instead, they facilitated the development of several pivotal techniques, results, and conceptual insights. These contributions proved instrumental as Einstein later integrated the Riemann and Ricci tensors into his theoretical framework, marking a significant advancement in his quest for a coherent and comprehensive description of gravitation.

1.2.3 Einstein's Exploration of the Ricci Tensor

It was not until page 14L of his notebook that Einstein commenced a methodical examination of the *Riemann curvature tensor*. He wrote a form of the Riemann curvature tensor [E3-12][22]:

THE RIEMANN CURVATURE TENSOR (GROSSMANN'S TENSOR):

$$(i\kappa, lm) = \frac{1}{2}\left(\frac{\partial^2 g_{im}}{\partial x_\kappa \partial x_l} + \frac{\partial^2 g_{\kappa l}}{\partial x_i \partial x_m} - \frac{\partial^2 g_{il}}{\partial x_\kappa \partial x_m} - \frac{\partial^2 g_{\kappa m}}{\partial x_i \partial x_l}\right) + \sum_{\rho\sigma} \gamma_{\rho\sigma} \left[\begin{bmatrix} im \\ \sigma \end{bmatrix}\begin{bmatrix} \kappa l \\ \rho \end{bmatrix} - \begin{bmatrix} il \\ \sigma \end{bmatrix}\begin{bmatrix} \kappa m \\ \rho \end{bmatrix}\right]. \tag{1.36}$$

At the outset of page 14L, Einstein wrote the Christoffel symbols of the first kind [E3-12]:

[22] The Riemann curvature tensor is a fourth-rank tensor, expressed in terms of the Christoffel symbols, which are functions of the metric tensor and its derivatives. The standard expression for the Riemann curvature tensor is:

$$R^\rho{}_{\sigma\mu\nu} = \frac{\partial \Gamma^\rho_{\nu\sigma}}{\partial x_\mu} - \frac{\partial \Gamma^\rho_{\mu\sigma}}{\partial x_\nu} + \sum_\lambda \left(\Gamma^\rho_{\mu\lambda}\Gamma^\lambda_{\nu\sigma} - \Gamma^\rho_{\nu\lambda}\Gamma^\lambda_{\mu\sigma}\right). \tag{1.35}$$

THE CHRISTOFFEL SYMBOLS OF THE FIRST KIND:

$$\boxed{\begin{bmatrix} \mu\nu \\ l \end{bmatrix} = \frac{1}{2}\left(\frac{\partial g_{\mu l}}{\partial x_\nu} + \frac{\partial g_{\nu l}}{\partial x_\mu} - \frac{\partial g_{\mu\nu}}{\partial x_l}\right).} \tag{1.37}$$

and made a notable annotation, referring to Grossmann's tensor on the left, juxtaposed with an inscription of the Riemann curvature tensor (1.36) on the right [E3-12]. This indicated Grossmann's pivotal role in acquainting Einstein with the Riemann tensor, signifying an important phase in Einstein's journey toward articulating covariant field equations.[23]

Einstein contracted the Riemann curvature tensor simplifying it to specific terms within the *Ricci tensor* [E3-12]:

THE RICCI TENSOR:

$$R_{im} = \boxed{\frac{1}{2}\gamma_{\kappa l}\left(\frac{\partial^2 g_{im}}{\partial x_\kappa \partial x_l} + \frac{\partial^2 g_{\kappa l}}{\partial x_i \partial x_m} - \frac{\partial^2 g_{il}}{\partial x_\kappa \partial x_m} - \frac{\partial^2 g_{\kappa m}}{\partial x_i \partial x_l}\right)} + \tag{1.38}$$
$$\gamma_{\kappa l}\sum_{\rho\sigma}\gamma_{\rho\sigma}\left(\begin{bmatrix} im \\ \sigma \end{bmatrix}\begin{bmatrix} \kappa l \\ \rho \end{bmatrix} - \begin{bmatrix} il \\ \sigma \end{bmatrix}\begin{bmatrix} \kappa m \\ \rho \end{bmatrix}\right).$$

At the bottom of page 14*L*, Einstein wrote the following condition [E3-12]:

$$\sum_\kappa\left(\frac{\partial^2 g_{\kappa\kappa}}{\partial x_i \partial x_m} - \frac{\partial^2 g_{i\kappa}}{\partial x_\kappa \partial x_m} - \frac{\partial^2 g_{m\kappa}}{\partial x_\kappa \partial x_i}\right) = 0. \tag{1.39}$$

It appears that Einstein sought to apply this specific constraint to the Ricci tensor. When this restriction is applied, accompanied by the appropriate allocation of indices, the second, third, and fourth terms delineated within the boxed terms in the Ricci tensor (1.38) are simplified to a single term, which bears resemblance to a Laplacian operator:

THE LAPLACIAN- LIKE TERM:

$$\boxed{\frac{1}{2}\gamma_{\kappa l}\frac{\partial^2 g_{im}}{\partial x_\kappa \partial x_l}.} \tag{1.40}$$

I therefore call this term the *Laplacian-like term* [see Eq. (1.85)]. As I will explain later, Einstein failed to see that this term reduces to the Laplacian and the Poisson equations.

---

[23] Scholars have revealed that Einstein employed on page 14*L* the notation introduced by Elwin Bruno Christoffel [Chr]. This finding underscores the historical reliance on Christoffel's mathematical formalism in Einstein's *Zurich Notebook*. Grossmann communicated this formalism to Einstein [JRSNS].

The systematic engagement with the Ricci tensor, as evidenced on page 14L and subsequent pages, underscored a new chapter in Einstein's relentless pursuit.

On page 19R, Einstein started with the following assumption [E13]:

THE FLAT (MINKOWSKI) SPACETIME METRIC + A DEVIATION:

$$\boxed{\text{METRIC TENSOR} = \gamma_{\kappa\kappa} + g_{i\kappa}.} \tag{1.41}$$

He wrote what he called a *restriction* [E3-12][24]:

THE LINEARIZED HARMONIC RESTRICTION:

$$\boxed{\sum_{\kappa} \gamma_{\kappa\kappa} \left(\frac{\partial g_{i\kappa}}{\partial x_{\kappa}} - \frac{1}{2} \frac{\partial g_{\kappa\kappa}}{\partial x_i} \right) = 0,} \tag{1.44}$$

Einstein conjectured that the restriction (1.44) could perhaps be decomposed into two separate restrictions [E3-12]:

---

24 Alternatively, this equation can be written as follows:

$$\sum_{\kappa} \gamma_{\kappa\kappa} \left(2\frac{\partial g_{i\kappa}}{\partial x_{\kappa}} - \frac{\partial g_{\kappa\kappa}}{\partial x_i} \right) = 0. \tag{1.42}$$

$g_{i\kappa}$ in Eq. (1.41) is a small deviation on the flat Minkowski spacetime metric $\gamma_{\kappa\kappa}$. I adhere to Einstein's original notation in addressing these equations as closely as possible. Wherever possible, I provide clarifications to resolve any ambiguity that arises from Einstein's notational inconsistency. This strategy is adopted to preserve the integrity of the original mathematical framework while enhancing the clarity and comprehension of the notebook for the reader. $\gamma_{i\kappa}$ represents the flat spacetime metric:

$$\boxed{\gamma_{i\kappa} = \begin{cases} 1 & \text{if } i = \kappa = 0, \\ -1 & \text{if } i = \kappa \neq 0, \\ 0 & \text{if } i \neq \kappa. \end{cases}} \tag{1.43}$$

$\gamma_{i\kappa} \equiv$ the Kronecker delta $\delta_{i\kappa}$. $\delta_{i\kappa}$ is defined as equal to 1 when $i = \kappa$ and equal to 0 when $i \neq \kappa$. Since Einstein specifically refers to the diagonal elements of the metric, he uses $\gamma_{\kappa\kappa}$. Thus, in Eq. (1.41), $\gamma_{\kappa\kappa}$ (i.e., the Kronecker delta) represents the flat Minkowski metric. In Eq. (1.44), Einstein also uses the Kronecker delta. However, in this equation, $\gamma_{\kappa\kappa}$ expresses conditions on $g_{i\kappa}$.

(1) *The harmonic restriction in the context of linearized gravity*[25]:

THE HARMONIC CONDITION IN LINEARIZED GRAVITY:

$$\sum_{\kappa} \gamma_{\kappa\kappa} \left(\frac{\partial g_{i\kappa}}{\partial x_{\kappa}} \right) = 0. \tag{1.45}$$

(2) *A restriction on the trace of the weak-field metric*:

THE TRACE OF THE METRIC RESTRICTION:

$$\sum_{\kappa} \gamma_{\kappa\kappa} g_{\kappa\kappa} = \text{const.} \tag{1.46}$$

Einstein then wrote a linearized field equation for gravitation [E3-12]:

FIRST LINEARIZED FIELD EQUATION:

$$\boxed{\sum \gamma_{\kappa\kappa} \frac{\partial^2 g_{im}}{\partial x_{\kappa}^2} = \kappa T_{im}.} \tag{1.47}$$

On the right-hand side, we have κ, the constant related to the gravitational constant:

$$\kappa = \frac{8\pi G}{c^4}. \tag{1.48}$$

g_{im} is the deviation (in the covariant form) [see Eq. (1.41)],

$$\text{and:} \quad T_{im} = g_{i\mu} g_{m\nu} \Theta_{\mu\nu}, \tag{1.49}$$

expresses the stress-energy tensor in the covariant form[26]

[25] The harmonic condition imposes a further constraint on the metric tensor and its derivatives. It restricts how the metric tensor components can vary for spacetime coordinates and contributes to the consistency of the field equations with energy-momentum conservation. Equation (1.45) plays a role in simplifying the field Eq. (1.57). The formulation presented in Eq. (1.45) articulates the condition as representing the metric tensor's deviation. This distinction is important, as delineated in Eq. (1.44), positioning the condition outlined in (1.45) divergent from the canonical form of the harmonic condition. Scholars have referred to this specific condition as the *Hertz restriction*, a nomenclature derived from a correspondence in which Einstein later mentioned it in a letter to Paul Hertz [JRSNS] ([CPAE8], Doc. 111). The term *Hertz restriction* is justified because the deviation is consistently written as a metric tensor within Einstein's equation formulations of Eq. (1.44), and the condition (1.45). However, Einstein's consistent use of condition (1.45) and its variants [e.g., Eq. (1.83)] throughout the *Zurich Notebook* akin to the harmonic condition—or more accurately, a harmonic restriction—warrants the continued use of the terminology *harmonic restriction* to accurately reflect the nature of the Eq. (1.45).

[26] $\Theta_{\mu\nu}$ represents the stress-energy tensor in the contravariant form defined by Eq. (1.20).

The ensuing equation on page 19*R* has the form [E3-12]:

$$\sum_{\kappa im} \gamma_{\kappa\kappa} \frac{\partial^2 g_{im}}{\partial x_\kappa^2} \frac{\partial g_{im}}{\partial x_\sigma} = \sum_{\kappa im} \gamma_{\kappa\kappa} \left[\text{TERM 1} - \text{TERM 2} \right]. \tag{1.50}$$

Terms 1 – 2 are represented by:

$$\sum_{\kappa im} \gamma_{\kappa\kappa} \left[\frac{\partial}{\partial x_\kappa} \left(\frac{\partial g_{im}}{\partial x_\kappa} \frac{\partial g_{im}}{\partial x_\sigma} \right) - \frac{1}{2} \frac{\partial}{\partial x_\sigma} \left(\frac{\partial^2 g_{im}}{\partial x_\kappa} \right) \right]. \tag{1.51}$$

Two steps lead to TERM 1 - TERM 2:
(1) TERM 1 inside the sum is derived by applying the product rule of differentiation (1.31) to a product of two functions, where those two functions are:

$$u = \frac{\partial g_{im}}{\partial x_\kappa}, \quad v = \frac{\partial g_{im}}{\partial x_\sigma}. \tag{1.52}$$

$$\frac{\partial}{\partial x_\kappa} (u \cdot v) = \boxed{\frac{\partial}{\partial x_\kappa} \left(\frac{\partial g_{im}}{\partial x_\kappa} \cdot \frac{\partial g_{im}}{\partial x_\sigma} \right)} \quad \text{and other terms, which are ignored.} \tag{1.53}$$

(2) TERM 2 inside the sum ostensibly represents the corresponding right-hand side of the Eq. (1.47); upon closer examination, its derivation from the original Eq. (1.47) appears incongruent. Furthermore, the divergence second term in Eq. (1.51) exhibits notable similarities to the second term delineated in Eq. (1.34). Intriguingly, the Eq. (1.50) suggests that the restrictions previously mentioned were not explicitly imposed upon the field Eq. (1.47). Instead, a substitution was effected, mirroring the methodology previously employed within the notebook, particularly on page 13*R*, as explicitly outlined in Eq. (1.34). Put differently, Einstein substituted the stress-energy tensor in Eq. (1.47) with the second term in the conservation law for the gravitational field [Eq. (1.34)].

Subsequently, Einstein articulated beneath the divergence Eq. (1.50) that the principles of energy and momentum are applied within certain approximations, and the constraints applied [E3-12].

However, Einstein failed to see that the Laplacian-like term (1.40) reduces to the Laplacian and the Poisson equation. This is because in the case of flat spacetime, where the Minkowski metric describes the spacetime, and assuming Cartesian coordinates for simplicity, the metric tensor components g_{im} are constant. In flat spacetime, the metric tensor g_{im} is constant and does not change from point to point, which means its derivatives with respect to spacetime coordinates x_κ are zero: $\frac{\partial g_{im}}{\partial x_\kappa} = 0$. Therefore, in flat spacetime, the Laplacian-like term (1.40) would be zero because the second derivatives of the metric tensor components with respect to the coordinates would also be zero:

$$\sum \gamma_{\kappa\kappa} \frac{\partial^2 g_{im}}{\partial x_\kappa^2} = 0. \tag{1.54}$$

This equation involves second derivatives and summation over indices similar to operations in computing a Laplacian. However, it does not directly represent the Laplacian of a function in flat spacetime due to its specific dependence on the metric tensor's second derivatives, which would be zero in the context of flat spacetime. Therefore, Eq. (1.47) does not yield a Laplacian operator on its own. In other words, the metric tensor components are constant in flat spacetime, characterized by the Minkowski metric $\gamma_{\kappa\kappa}$ (with diagonal components $-1, 1, 1, 1$). This situation is specific to the idealized condition of perfectly flat spacetime, where no curvature or gravitational fields influence the metric tensor's form. The Laplacian operator ∇^2, when applied to a scalar field in flat spacetime, is defined by:

$$\nabla^2 \varphi = \frac{\partial}{\partial x_\mu} \frac{\partial}{\partial x_\nu} \varphi. \tag{1.55}$$

This operator *is not zero by definition*; it describes how the function φ changes in the second order with respect to the coordinates. The Laplacian can reveal, for example, the curvature or the spread of a scalar field in space.

The problem here lies in the contrast between the behavior of the metric tensor in weak gravitational fields and perfectly flat spacetime. In weak gravitational fields, the metric tensor deviates slightly from flatness due to gravitational potential, whereas in perfectly flat spacetime, the metric tensor remains constant. This is relevant to how Einstein conceptualized spacetime in the context of his gravitational theories, particularly how the metric tensor changes in weak gravitational fields.

In his 1912 static gravitational theory, Einstein sought to understand how gravity affects the geometry of space. One of his key insights was replacing the gravitational potential with the variable speed of light. This shift in perspective led him to formulate the metric tensor, which describes the geometry of spacetime to reflect the effects of gravity [E1-12, E2-12].

In the special relativity limit, where gravitational effects are negligible, the metric tensor is diagonal with components $(-1, -1, -1, c^2)$, where c is the speed of light. Here, the three spatial components are equal to -1, representing a flat Euclidean space, while the time component is equal to c^2, reflecting the nature of time in special relativity. Einstein extended this understanding to weak static gravitational fields, such as those produced by a stationary mass. In these scenarios, he observed that the metric tensor degenerated similarly to special relativity, but the component $g_{44} = c^2$ is a function of the spatial coordinates (x, y, z). This implies that in weak gravitational fields, space remains approximately Euclidean. Therefore, Einstein's intuition was that in the weak-field approximation, the spatial metric of a weak gravitational field should be flat.

In other words, Einstein assumed that if both the Minkowski metric and the metric describing a static gravitational field have spatial components (i.e., the 1, 1, 1 components) that are the same, one might consider treating the static field metric as a flat metric within the given context of deriving the Newtonian limit. By utilizing the weak-field approximation and assuming a diagonal metric tensor with components $(-1, -1, -1)$, Einstein could derive Newtonian results [W18]. In the context of the Laplacian-like term (1.54), this method was problematic.[27]

For the Minkowski metric, since the metric tensor components are constant, their second derivatives with respect to spacetime coordinates are zero. Therefore, the Laplacian-like term, represented by Eq. (1.54), evaluates to zero in flat spacetime. However, in the case of a static gravitational field, while the spatial components of the metric may appear flat, the temporal component typically varies as a function of spatial coordinates due to the influence of gravity. Consequently, the metric tensor is not constant, and its second derivatives with respect to spacetime coordinates are generally non-zero. Therefore, the Laplacian-like term does not necessarily evaluate to zero for the static field metric. On the other hand, the Laplacian operator, represented by Eq. (1.55), is defined based on the second derivatives of a scalar field with respect to spacetime coordinates. This operator is applicable both in flat spacetime and in the presence of a static gravitational field. In the latter case, where the metric tensor is not constant, the Laplacian operator represents the variations in the scalar field due to the gravitational potential. Thus, treating the metric describing a static gravitational field as a flat metric within the context of deriving the Newtonian limit means that the Laplacian-like term, as represented by Eq. (1.54), would evaluate to zero because it assumes constant metric tensor components. Consequently, Einstein could not obtain the Poisson equation under this simplification, which relates the gravitational potential to the mass distribution via the Laplacian operator.

Eventually, Einstein, upon encountering a lack of success in obtaining the Newtonian approximation, discarded the field Eq. (1.47) and instead wrote new

[27] In the context of Einstein's gravitational field equations, while the spatial components may appear similar to those of the Minkowski metric, the varying temporal component indicates that spacetime is not flat but rather influenced by a gravitational field. So, while we can draw parallels between the spatial components of the two metrics, it is necessary to consider the overall structure of spacetime and the influence of gravity, as reflected in the temporal component of the metric, when interpreting them. In the framework of special and general relativity, space and time are intrinsically intertwined within the fabric of spacetime. This means that we cannot separate the temporal component from the spatial components of the metric tensor because they collectively describe the geometry of spacetime. The Minkowski metric, which describes flat spacetime in special relativity, describes this unified view of spacetime by incorporating temporal and spatial components.

Similarly, in the context of a static gravitational field, while the spatial components of the metric may appear flat, the temporal component reflects the influence of gravity, tying together the dynamics of spacetime. However, in the context of the Newtonian limit, where gravitational effects are weak and velocities are much smaller than the speed of light, it is natural to believe that the unified view of spacetime described by special and general relativity does not hold. In the Newtonian limit, the gravitational effects are described by Newtonian gravity, which treats space and time separately and assumes a flat, Euclidean space. Einstein reasoned that while the unified view of spacetime described by relativity remains valid in the full theory, in the Newtonian limit, one can treat space and time separately, and the temporal component of the metric tensor becomes less relevant.

field equations that, while bearing some similarities, were distinct from the previous ones [E3-12][28]:

SECOND LINEARIZED FIELD EQUATION:

$$\boxed{\square g_{i\kappa} = \kappa T_{i\kappa},} \qquad \text{where:} \tag{1.57}$$

THE BOX OPERATOR IN FLAT SPACE-TIME ACTING ON THE DEVIATION $g_{i\kappa} = \kappa T_{i\kappa}$:

$$\boxed{\square g_{i\kappa} = \frac{\partial^2 g_{i\kappa}}{\partial x^2} + \frac{\partial^2 g_{i\kappa}}{\partial y^2} + \frac{\partial^2 g_{i\kappa}}{\partial z^2} - \frac{1}{c^2}\frac{\partial^2 g_{i\kappa}}{\partial t^2} = \kappa T_{i\kappa}.} \tag{1.58}$$

Einstein did not write Eqs. (1.41) and (1.58) in the *Zurich Notebook*. Though not delineated verbatim in the notebook, Einstein undoubtedly anticipated these equations, as demonstrated by the logical progression and interrelation of the concepts and calculations detailed therein. In deriving the field Eq. (1.57), it becomes evident that Einstein presupposed Eq. (1.41). The above wave-like Eq. (1.58) is necessary for understanding the underlying framework of Eq. (1.57): Einstein applied the d'Alembertian operator $\square$ to the deviation $g_{i\kappa}$ in a flat spacetime $\gamma_{\kappa\kappa}$ in Eq. (1.41). After his 1913 lecture in Vienna, Einstein formalized the Eq. (1.58) in a paper published a few months later, drawing upon the foundational concepts presented during the lecture [refer to Eq. (1.157) in Sect. 1.3.4].

Einstein was motivated to show that his theory would converge to the Newton-Poisson equation (1.27) under the appropriate conditions. By setting the sources to zero in Eq. (1.57), Einstein studied the behavior of $\frac{\partial g_{i\kappa}}{\partial x_\kappa}$ in the absence of external sources. He imposed the harmonic condition (1.45) on the Eq. (1.57) with no sources. I will proceed with the derivation, illustrating the steps that Einstein likely followed. He applied the harmonic condition to the metric component g_{ik}. He took the second derivative of this equation with respect to x_i using the box operator $\square$:

$$\square\left(\sum_\kappa \gamma_{\kappa\kappa}\left(\frac{\partial g_{i\kappa}}{\partial x_\kappa}\right)\right) = \square(0). \tag{1.59}$$

[28] The d'Alembertian operator $\square$ that measures the second derivatives of the covariant metric tensor components, a generalization of the Laplacian operator in flat spacetime. The d'Alembert operator, in a flat Minkowski spacetime, is defined as:

D'ALEMBERT OPERATOR IN FLAT SPACE-TIME:

$$\square = -\frac{1}{c^2}\frac{\partial^2}{\partial t^2} + \frac{\partial^2}{\partial x^2} + \frac{\partial^2}{\partial y^2} + \frac{\partial^2}{\partial z^2}, \tag{1.56}$$

where c is the speed of light, t is the time coordinate, and x, y, z are the spatial coordinates in a flat Minkowski spacetime. The operator $\square$ combines the second-order time derivative with the second-order spatial derivatives.

We end up with the following result[29]:

$$\sum_{\kappa} (\Box\gamma_{\kappa\kappa}) \frac{\partial g_{i\kappa}}{\partial x_{\kappa}} = \Box\gamma_{\kappa\kappa} \left(\frac{\partial g_{i\kappa}}{\partial x_{\kappa}} \right) = 0. \tag{1.62}$$

We assume that $\gamma_{\kappa\kappa} = 1$ [behaves as a Kronecker delta], and the equation simplifies to:

$$\Box \left(\frac{\partial g_{i\kappa}}{\partial x_{\kappa}} \right) = 0. \tag{1.63}$$

This equation implies a divergence and absence of matter, thus, energy-momentum conservation for pressureless dust:

ENERGY- MOMENTUM CONSERVATION FOR PRESSURELESS DUST:

$$\frac{\partial T_{im}}{\partial x_{\kappa}} = 0. \tag{1.64}$$

However, if $\gamma_{\kappa\kappa}$ equates 1 [we take into consideration Eq. (1.63)], Eq. (1.57) becomes[30]: $\Box g_{\kappa\kappa} = \kappa T =$ const. To solve this problem, Einstein modified the Eq. (1.57) to incorporate an additional term on the right-hand side involving the trace of the stress-energy tensor.

---

[29] The right-hand side of (1.59) is zero because $\Box(0) = 0$:

$$\Box \left(\sum_{\kappa} \gamma_{\kappa\kappa} \left(\frac{\partial g_{i\kappa}}{\partial x_{\kappa}} \right) \right) = 0. \tag{1.60}$$

The d'Alembertian operator distributes across the sum:

$$\Box \left(\sum_{\kappa} \gamma_{\kappa\kappa} \left(\frac{\partial g_{i\kappa}}{\partial x_{\kappa}} \right) \right) = \sum_{\kappa} \Box \left(\gamma_{\kappa\kappa} \frac{\partial g_{i\kappa}}{\partial x_{\kappa}} \right) = \sum_{\kappa} (\Box\gamma_{\kappa\kappa}) \frac{\partial g_{i\kappa}}{\partial x_{\kappa}} = 0. \tag{1.61}$$

[30] Given $g_{\kappa\kappa}$ is a constant, it suggests that the stress-energy tensor is T. For a cloud of pressureless dust, the stress-energy tensor is given by: $T_{ik} = \rho_0 u_i u_k$. ρ_0 is the mass density, and u_i is the four-velocity of the dust. The trace of the stress-energy tensor T is given by $T = T_k^k$, which represents the sum of the diagonal components of the tensor: $T = \rho_0 u_k u^k$.

On page 20*L*, Einstein wrote another field equation [E3-12][31]:

THIRD LINEARIZED FIELD EQUATION:

$$\boxed{\Box(g_{ik}) = \kappa\left(T_{i\kappa} - \frac{1}{4}\delta_{i\kappa}T\right).} \tag{1.66}$$

However, again, Einstein discarded Eq. (1.66) due to their incompatibility with the condition presented in Eq. (1.46).

This setback prompted Einstein to *revise Eq.* (1.57) again, introducing an additional term involving the trace of the metric on the left-hand side. The correction made to Eq. (1.57) demonstrates a shift in Einstein's approach, as he had previously introduced a term involving the trace of the stress-energy tensor on the right-hand side in Eq. (1.66) [W18].

The new linearized field equations introduced by Einstein on page 20*L* take the following form [E3-12][32]:

FOURTH LINEARIZED FIELD EQUATION:

$$\boxed{\Box\left(g_{ik} - \frac{1}{2}\delta_{ij}U\right) = \kappa T_{ik}.} \tag{1.68}$$

[31] Einstein wrote Eq. (1.66) without using the Kronecker delta:

$$\Box(g_{ik}) = \rho_0 \frac{dx_i}{d\tau}\frac{dx_\kappa}{d\tau} - \frac{1}{4}\rho_0 \frac{dx_\kappa}{d\tau}\frac{dx_\kappa}{d\tau}. \tag{1.65}$$

The right-hand side of the Eq. (1.66) involves terms related to the stress-energy tensor and its trace—the first term on the right accounts for the energy-momentum. The additional term on the right is introduced specifically to account for the contribution of the trace of the stress-energy tensor to the gravitational field. We need to convert the contravariant stress-energy tensor (and its trace) for dust on the right-hand side of Eq. (1.65) into its covariant form, see Eq. (1.49).

[32] The first equation is presented as:

$$\Delta(g_{11} - \frac{1}{2}U) = T_{11}. \tag{1.67}$$

This linearized field equation represents a relationship involving the Laplacian operator Δ, the metric component g_{11}, the scalar field - the gravitational potential U, and the component of the energy-momentum tensor T_{11}. Here are the second, third, and fourth equations [E3-12]: $\Delta g_{12} = T_{12}$, $\Delta g_{13} = T_{13}$, $\Delta g_{14} = T_{14}$. These equations are similar to the first one but involve the metric components g_{12}, g_{13}, and g_{14}, and the components of the energy-momentum tensor T_{12}, T_{13}, and T_{14}. They represent other relationships derived under the same conditions.
Scholars have noted a remarkable similarity between Eq. (1.68) and Einstein's tensor (1.426) (see Sect. 1.7.2). They have observed that the left-hand side of the equation under consideration resembles the linearized harmonic restriction (1.44) [JRSNS].

Taking the divergence $\frac{\partial}{\partial x_\kappa}$ of both sides of Eq. (1.68), we get[33]:

$$\frac{\partial}{\partial x_\kappa}\left[\Box\left(g_{ik}-\frac{1}{2}\delta_{ij}U\right)\right]=\Box\left[\frac{\partial}{\partial x_\kappa}\left(g_{ik}-\frac{1}{2}\delta_{ij}U\right)\right]=\frac{\partial T_{ik}}{\partial x_\kappa}. \tag{1.69}$$

In the Eq. (1.69), if the expression within the brackets is meant to represent a field configuration satisfying the linearized harmonic restriction (1.44), then taking the divergence of the whole Eq. (1.68) and applying this condition would result in zero:

$$\Box\left[\frac{\partial}{\partial x_\kappa}\left(g_{ik}-\frac{1}{2}\delta_{ij}U\right)\right]=\frac{\partial T_{ik}}{\partial x_\kappa}=0. \tag{1.70}$$

This outcome is because the linearized harmonic condition (1.44) enforces that the spacetime coordinates are chosen so that the d'Alembertian operator acting on the related field quantities leads to zero, simplifying the analysis. When the divergence of the energy-momentum tensor is zero, it implies the conservation of energy and momentum.

On page 20L, there is a conspicuous absence of both Eqs. (1.69) and (1.70), which suggests a lack of engagement with these formulations. Moreover, no evidence indicates that Einstein contemplated applying the linearized harmonic restriction (1.44) to the field Eq. (1.68). Despite the seemingly intuitive step of applying this restriction to derive Eqs. (1.69) and (1.70), page 20L does not indicate such analytical progression. This omission could plausibly suggest that Einstein did not consider this analytical maneuver. Analysis of Einstein's work in the *Zurich Notebook* demonstrates instances where Einstein overlooked evident analytical steps that might now appear as apparent.

Instead, Einstein tried to take the divergence $\frac{\partial}{\partial x_\sigma}$ of the left-hand side of Eq. (1.68) and substituted T_{ik} with terms that represent conservation for the gravitational field. He wrote an equation of the following format: DIVERGENCE OF LEFT- HAND SIDE OF EQ. (1.68) = TERMS THAT REPRESENT CONSERVATION FOR THE GRAVITATIONAL FIELD. Mirroring his approach with Eq. (1.50), Einstein endeavored to apply a similar methodology here. He commented that this equation can be represented in an extended form. Thus, it can be expressed in a more detailed and expanded mathematical form by incorporating more variables and applying them to more general situations beyond the linearized case Einstein discussed. However, Einstein then simplified the left-hand side of the equation. This led him to (1.68): [E3-12][34]:

---

[33] Equation (1.69) implies the d'Alembertian operator can be distributed over the inner components of the expression, demonstrating its distributive property over addition and subtraction, as well as its commutative property with partial derivatives under suitable conditions (see footnote 29).

[34] Einstein simplified the left-hand side of the Eq. (1.67) by noticing that the Laplacian of U is $-2\Delta U$. Finally, he recognized that:

$$-2\Delta U=\sum T_{\kappa\kappa}=\sum T. \tag{1.71}$$

FIFTH LINEARIZED FIELD EQUATION:

$$\boxed{\Box(g_{ij}) = T_{ij} - \frac{1}{2}\delta_{ij}\left(\sum T\right).} \tag{1.73}$$

Equation (1.73) and the standard form of Einstein's field equations in the covariant form, Eq. (1.290), share a similar form. In hindsight, the linearized Eq. (1.73) represents a simplified version of the Einstein field equations.

In the *Newtonian limit*, Einstein aimed to show that the linearized field Eqs. (1.68) and (1.73) would yield an equation akin to the *Poisson equation* for the gravitational potential U[35]: However, on page 21*L*, Einstein ultimately *abandoned* the Eq. (1.73) for the same reason he had previously discarded the other linearized field equations. The exploration into the properties of various mathematical expressions represented a deliberate effort by Einstein to identify expressions that would satisfy the criteria for Newtonian approximation. However, the dynamic interaction and complex intertwining of the harmonic condition with the Laplacian term catalyzed his pursuit of a novel commencement.

Einstein titled page 22*R Grossmann*. He derived a tensor, representing the gravitational field in terms of the Christoffel symbols and their derivatives. Initially, Einstein considered specifying a covariant differential tensor of the second rank and second-order for the *Ricci tensor* on page 22*R* [E3-12]:

THE RICCI TENSOR:

$$\boxed{T_{il} = \sum_{\kappa l} \frac{\partial}{\partial x_l}\left(\Gamma^{\kappa}_{i\kappa}\right) - \frac{\partial}{\partial x_\kappa}\left(\Gamma^{\kappa}_{il}\right) - \left(\Gamma^{\lambda}_{i\kappa}\right)\left(\Gamma^{\kappa}_{\lambda l}\right) - \left(\Gamma^{\lambda}_{il}\right)\left(\Gamma^{\kappa}_{\lambda\kappa}\right).} \tag{1.75}$$

This equation represents the Laplace operator applied to the scalar field U, equal to the sum of the diagonal components of the tensor $T_{\kappa\kappa}$ which is also equal to the sum T, the trace of $T_{\kappa\kappa}$ over the index κ. Einstein plugged Eq. (1.71) into Eq. (1.67) and obtained for Eq. (1.67) the equation:

$$\Delta g_{11} = T_{11} - \frac{1}{2}\sum T_{\kappa\kappa}. \tag{1.72}$$

.

[35] He aimed to show that Eq. (1.67) [see footnote 32] and (1.71) [see footnote 34] yield:

$$\text{THE POISON EQUATION:} \quad \Delta U = 4\pi G\rho. \tag{1.74}$$

where Δ is the Laplacian operator, U is the gravitational potential, G is the gravitational constant, and ρ is the mass density.

Γ^{κ}_{il} represent the *Christoffel symbols of the second kind*[36]:

THE CHRISTOFFEL SYMBOLS OF THE SECOND KIND:

$$\boxed{\Gamma^{\kappa}_{il} = \frac{1}{2}\gamma_{\kappa\lambda}\left(\frac{\partial g_{\lambda l}}{\partial x_i} + \frac{\partial g_{i\lambda}}{\partial x_l} - \frac{\partial g_{il}}{\partial x_\lambda}\right).} \tag{1.76}$$

Let us derive Einstein's gravitational tensor from the Ricci tensor. We start with the original Ricci tensor as given in Eq. (1.75). We use the relation from page 22R [E3-12][37]:

$$\frac{\partial \log \sqrt{-g}}{\partial x_i} = \Gamma^{i\kappa}_{\kappa}, \tag{1.77}$$

to rewrite some of the Christoffel symbols.

Now, using Eq. (1.77), we can express the first and second terms in Eq. (1.75), T_{il}, differently:

$$\frac{\partial}{\partial x_l}\left(\Gamma^{\kappa}_{i\kappa}\right) - \frac{\partial}{\partial x_\kappa}\left(\Gamma^{\kappa}_{il}\right) = \boxed{\frac{\partial}{\partial x_l}\left(\frac{\partial \log \sqrt{-g}}{\partial x_i}\right) - \frac{\partial}{\partial x_\kappa}\left(\Gamma^{\kappa}_{il}\right).} \tag{1.78}$$

Using Eq. (1.77), we can express the third and fourth terms in Eq. (1.75), T_{il}, differently:

$$\left(\Gamma^{\lambda}_{i\kappa}\right)\left(\Gamma^{\kappa}_{\lambda l}\right) - \left(\Gamma^{\lambda}_{il}\right)\left(\Gamma^{\kappa}_{\lambda\kappa}\right) = \boxed{\left(\Gamma^{\lambda}_{i\kappa}\right)\left(\Gamma^{\kappa}_{\lambda l}\right) - \left(\Gamma^{\lambda}_{il}\right)\left(\frac{\partial \log \sqrt{-g}}{\partial x_i}\right).} \tag{1.79}$$

Next, we substitute Eqs. (1.78) and (1.79) into the expression for T_{il} [Eq. (1.75)] and simplify:

$$T_{il} = \sum_{\kappa l} \frac{\partial}{\partial x_l}\left(\frac{\partial \log \sqrt{-g}}{\partial x_i}\right) - \boxed{\frac{\partial}{\partial x_\kappa}\left(\Gamma^{\kappa}_{il}\right) - \left(\Gamma^{\lambda}_{i\kappa}\right)\left(\Gamma^{\kappa}_{\lambda l}\right)} - \left(\Gamma^{\lambda}_{il}\right)\left(\frac{\partial \log \sqrt{-g}}{\partial x_i}\right). \tag{1.80}$$

Einstein employed unimodular transformations, where $\sqrt{-g} = 1$ (thus, the logarithm of $\sqrt{-g}$ approaches zero). This leaves the Ricci tensor with the second and

[36] The Christoffel symbols describe the connection between covariant derivatives and the contravariant metric tensor in curved spacetime. In these expressions: i, l, and κ are spacetime indices. $\gamma_{\kappa\lambda}$ represents the components of the contravariant metric tensor. $g_{\lambda l}$, $g_{i\lambda}$, and g_{il} are components of the metric tensor. $\frac{\partial}{\partial x^l}\left(\frac{\partial g_{\lambda l}}{\partial x_i}\right)$, $\frac{\partial}{\partial x_l}\left(\frac{\partial g_{i\lambda}}{\partial x_l}\right)$, and $\frac{\partial}{\partial x_\lambda}\left(\frac{\partial g_{il}}{\partial x_\lambda}\right)$ are the partial derivatives of the metric tensor with respect to spacetime coordinates.

[37] Equation (1.77) relates the partial derivative of the logarithm of the square root of the determinant of the metric tensor with the Christoffel symbols ($\Gamma^{i\kappa}_{\kappa}$). The presence of the square root of the determinant of the metric tensor $\sqrt{-g}$ in the expression (1.77) is related to unimodular transformations: transformations (either of coordinates or of the metric tensor) that preserve the absolute value of the determinant of the metric tensor.

third terms in Eq. (1.80). These two terms constitute Einstein's gravitational tensor [E3-12][38]:

THE NOVEMBER TENSOR:

$$\boxed{T^{x}_{il} = \frac{\partial}{\partial x_{\kappa}} \left(\Gamma^{\kappa}_{il}\right) - \left(\Gamma^{\lambda}_{i\kappa}\right)\left(\Gamma^{\kappa}_{\lambda l}\right).} \tag{1.81}$$

Given the Christoffel symbols of the second kind (1.76), Einstein took the divergence of the Christoffel symbols and wrote the first term of the November tensor [E3-12][39]:

THE EXPANDED FORM OF THE DIVERGENCE OF A CHRISTOFFEL SYMBOL:

$$\frac{\partial}{\partial x_{\kappa}} \left(\Gamma^{\kappa}_{il}\right) = \frac{1}{2}\frac{\partial}{\partial x_{\kappa}} \left(\gamma_{\kappa\alpha}\left(\frac{\partial g_{i\alpha}}{\partial x_l} + \frac{\partial g_{l\alpha}}{\partial x_i} - \frac{\partial g_{il}}{\partial x_{\alpha}}\right)\right). \tag{1.82}$$

Einstein later abandoned what was retrospectively termed the November tensor approach. However, he thoroughly examined the November tensor in the *Zurich Notebook* before doing so. Let us briefly analyze this.

Einstein imposed the harmonic condition (1.28) [E3-12]:

$$\sum_{\kappa} \frac{\partial \gamma_{\kappa\alpha}}{\partial x_{\kappa}} = 0, \tag{1.83}$$

on the November tensor (1.81). However, the condition (1.83) can simplify the computation of the November tensor by affecting terms involving the divergence of the Christoffel symbols (1.82). Still, it leaves the curvature-indicating products of Christoffel symbols $\left(\Gamma^{\lambda}_{i\kappa}\right)\left(\Gamma^{\kappa}_{\lambda l}\right)$ intact.[40]

After applying condition (1.83), Einstein got a more streamlined equation [E3-12]:

$$\boxed{-\frac{1}{2}\sum \gamma_{\kappa\alpha}\frac{\partial^2 g_{il}}{\partial x_{\alpha} x_{\kappa}}} - \frac{1}{2}\sum\left(\frac{\partial \gamma_{\kappa\alpha}}{\partial x_l}\frac{\partial g_{i\alpha}}{\partial x_{\kappa}} + \frac{\partial \gamma_{\kappa\alpha}}{\partial x_i}\frac{\partial g_{l\alpha}}{\partial x_{\kappa}}\right). \tag{1.84}$$

---

[38] Scholars coined the term the *November tensor* to describe Eq. (1.81) [JRSNS]. The reason it is referred to in this manner is that when we set T^{x}_{il} equal to the stress-energy tensor, multiplied by κ [the constant related to the gravitational constant, Eq. (1.48)], we obtain the field Eq. (1.244) of Einstein's first paper on general relativity of November 4, 1915.

[39] This operation involves taking the partial derivative of the Christoffel symbols (1.76) with respect to the coordinate x_{κ} and expanding this derivative.

[40] The harmonic condition (1.83) primarily simplifies expressions involving derivatives of the metric tensor and, by extension, can affect terms involving the divergence of Christoffel symbols. However, it does not directly simplify the products of Christoffel symbols. Applying the harmonic condition (1.83), the divergence of the Christoffel symbols (1.82) is directly impacted. The condition (1.83) does not eliminate the products of Christoffel symbols $\left(\Gamma^{\lambda}_{i\kappa}\right)\left(\Gamma^{\kappa}_{\lambda l}\right)$ which are terms that are indispensable for the description of gravitational interactions and curvature described by the November tensor.

Neglecting the other terms involving derivatives of the metric tensor in the first approximation, Einstein thought that the first term in Eq. (1.84):

THE LAPLACIAN- LIKE TERM:

$$\frac{1}{2}\sum \gamma_{\kappa\alpha}\frac{\partial^2 g_{il}}{\partial x_\alpha x_\kappa}, \tag{1.85}$$

could lead to a form analogous to the Poisson equation in Newtonian gravity.

This underscores Einstein's initial approach to extrapolating the Newtonian limit from the Ricci tensor, which diverged significantly from his subsequent comprehension in four distinct facets:

1. In 1912, Einstein employed the harmonic condition (1.28) [or Eq. (1.83)] alongside the constraints imposed by unimodular transformations to truncate the Ricci tensor. This led to a formulation wherein the initial term resembled the Laplacian operator [Eq. (1.84)], albeit with supplementary terms disregarded under the Newtonian approximation. Subsequent reflection made Einstein realize in 1915 that adherence to the condition (1.28) was not a prerequisite for deriving the Laplacian-like term [Eq. (1.85)] (refer to footnote 122 of Sect. 1.6.1). Moreover, only after an additional year of contemplation did Einstein recognize the significance of the Eq. (1.28), or harmonic condition, in the context of gravitational waves, an insight elaborated upon in the discussions of Sect. 3.2.1, Chap. 3. The trajectory of intellectual comprehension regarding Eq. (1.28) by Einstein during the year 1912 suggests an incomplete mastery over its conceptual and practical implications. Even though Einstein recorded formulations in his notebook that ostensibly aligned with the correct structure of his later equations from 1915, this act alone does not substantiate a comprehensive understanding of the appropriate application of the equations. This observation underlines a subtle distinction between the mere transcription of equations that appear formally accurate and the deeper cognitive grasp of their underlying principles and operational utility.

2. As said earlier, in the case of flat spacetime, the metric tensor components $g_{il} = 1$. This means the derivatives of $g_{il} = 1$ with respect to x_α and x_β are zero. Therefore, in flat spacetime, the Laplacian-like term [Eq. (1.85)] would be zero because the second derivatives of the metric tensor components with respect to the coordinates would also be zero:

$$\sum \frac{1}{2}\gamma_{\alpha\beta}\frac{\partial^2 g_{il}}{\partial x_\alpha x_\beta} = 0. \tag{1.86}$$

This equation is similar to operations involved in computing a Laplacian. Recall that the Laplacian operator ∇^2 is not zero by definition, and the static metric is not constant. Thus, Eq. (1.84) did not yield a Laplacian operator.

3. In the nascent phase of his explorations, delineated within the *Zurich Notebook* in 1912, Einstein formulated tensors and drafted truncated expressions derived from the November tensor. His strategy to obtain the Newtonian limit in 1912 can be seen as an intuitive search for equations that would reduce to the familiar Poisson equation of Newtonian gravity under weak static gravitational fields and slow-moving sources. This involved examining the structure of his equations and identifying terms that would dominate or vanish in these limits. Truncating higher-order terms and focusing on those that resemble the Laplacian operator acting on the gravitational potential was a part of this search. During this period, Einstein had yet to recognize the potential for utilizing the methodology leading to the Poisson equation (see derivation at the end of Sect. 1.6.1).[41]

4. The derivation of Eq. (1.84) from the November tensor T_{il}^{x} [Eq. (1.81)] by imposing the harmonic condition (1.28) to arrive at the Newtonian limit suggests a form of *coordinate restriction*. Based on Einstein's derivations in the *Zurich Notebook*, and his inscription on page 19*R* regarding Eq. (1.44) that serves as a *restriction*, it becomes apparent that he was not fully acquainted with the potential to employ Eq. (1.28) as a coordinate condition. This discrepancy reveals that, although Einstein wrote Eq. (1.28) [and Eqs. (1.44) and (1.45)], he only comprehended their true significance and applicability as coordinate conditions years later. This realization would only materialize in 1915–1916. Condition (1.28) arbitrarily restricts the tensor's physical applicability and specifies the regime (weak gravitational fields) where T_{il}^{x} must reproduce the well-established classical results. The condition (1.28) is not aimed at simplifying the T_{il}^{x} or choosing a particular coordinate system for ease of calculations (without altering the physical content of T_{il}^{x}). Einstein used condition (1.28) as a validation step that any candidate for the left-hand side of his field equations undergoes to guarantee that it reduces to the Newtonian limit.[42]

---

[41] How can Eq. (1.85) lead to the Newtonian limit? Consider a weak gravitational field and focus on the 00 (44) component in a static field. The operation (1.85) on $g_{\mu\nu}$ or on the Newtonian potential Φ [Eq. (1.283), refer to footnote 123 in Sect. 1.6.1] in the weak-field limit is akin to taking the Laplacian of the metric deviations. This leads to an equation that, for the 00 component, can be analogous to the Poisson equation for the gravitational potential in the limit of a static field, thus bridging the gap between Einstein's November tensor and Poisson's equation in the Newtonian limit (see explanation at the end of Sect. 1.6.1).

[42] Einstein's method in formulating the left-hand side of his field equations, as revealed in his *Zurich Notebook*, significantly deviates from his later practices. Scholars agree that Einstein applied *coordinate restrictions* to simplify broadly covariant equations into forms that could serve as the fundamental equations of his gravitational theory [JRSNS]. These restrictions were not merely tools for selecting convenient coordinate systems (as coordinate conditions are understood) but defined the theory's equations. Einstein used coordinate restrictions to reconcile his field equations with the principles of energy-momentum conservation and to guarantee that they reduce to Newtonian equations in the limit of weak static fields.

On the next page, 23*L*, Einstein opted to pursue an alternative trajectory. He considered a mathematical construct:[E3-12][43]:

$$\text{THE } \theta_{i\kappa\lambda} \text{ TENSOR:} \quad \boxed{\theta_{i\kappa\lambda} = \frac{1}{2}\left(\frac{\partial g_{i\kappa}}{\partial x_\lambda} + \frac{\partial g_{\kappa\lambda}}{\partial x_i} + \frac{\partial g_{\lambda i}}{\partial x_\kappa}\right).} \tag{1.87}$$

Einstein called the expression (1.87) the $\theta_{ik\lambda}$ tensor, and he used it to relate certain terms in the November tensor to derivatives of the metric tensor [E3-12]:

$$T_{il}^{x} = \sum_{\kappa l} \frac{\partial}{\partial x_\kappa}\gamma_{\kappa\alpha}\left(\theta_{il\alpha} - \frac{\partial g_{il}}{\partial x_\alpha}\right) - \gamma_{\lambda\alpha}\gamma_{\kappa\beta}\left(\theta_{i\kappa\alpha} - \frac{\partial g_{i\kappa}}{\partial x_\alpha}\right)\left(\theta_{i\lambda\beta} - \frac{\partial g_{l\lambda}}{\partial x_\beta}\right). \tag{1.88}$$

He first used condition (1.28) and applied expression (1.87) as a coordinate restriction [he set $\theta_{il\alpha} = 0$, i.e., used Eq. (1.95)] to eliminate various terms from the November tensor. This led to a truncated form of the November tensor on the bottom of page 23*L* [E3-12]:

$$\sum_{\kappa\alpha} \gamma_{\kappa\alpha} \frac{\partial^2 g_{il}}{\partial x_\kappa \partial x_\alpha} + \boxed{\sum_{\rho\alpha} \gamma_{\rho\alpha}\gamma_{\kappa\beta} \frac{\partial g_{i\kappa}}{\partial x_\alpha}\frac{\partial g_{l\rho}}{\partial x_\beta}.} \tag{1.89}$$

For future reference, I call the second term in Eq. (1.89) the *November coupling term*:

THE NOVEMBER COUPLING TERM:

$$\boxed{\sum_{\rho\alpha} \gamma_{\rho\alpha}\gamma_{\kappa\beta} \frac{\partial g_{i\kappa}}{\partial x_\alpha}\frac{\partial g_{l\rho}}{\partial x_\beta}.} \tag{1.90}$$

Subsequently, Einstein truncated Eq. (1.88) by applying just one restriction, Eq. (1.95). By applying this restriction, he derived the negative of expression (1.89), discarded the harmonic condition (1.28), and asserted that it was unnecessary [E3-12].

Einstein was on the cusp of discovering his 1915 general relativity. Despite his efforts, he faced difficulties in achieving his goals: consistency with energy-momentum conservation and recovering the Newtonian limit in the five linearized field Eqs. (1.47), (1.57), (1.66), (1.68), (1.73), and the November tensor (1.81).

[43] Scholars called Eq. (1.87) the *theta condition* [JRSNS]. In Einstein's calculations dealing with Eq. (1.81), unwanted terms involving second-order derivatives of the metric could arise. The expression (1.87) is used to manipulate and simplify these terms by canceling out unwanted contributions. The expression is constructed to transform as a tensor under coordinate transformations. The Christoffel-like expression (1.87) is defined similarly to the Christoffel symbols (1.76). It involves partial derivatives of the metric components $g_{i\kappa}$ with respect to the spacetime coordinates x_λ, x_i, and, x_κ, and it has a similar form to the expression for $\Gamma^{\alpha}_{\mu\nu}$. While it is not exactly the Christoffel symbol, it closely resembles it. It serves a similar purpose in describing the connection coefficients associated with the metric tensor $g_{i\kappa}$.

On page $42R$, Einstein examined the counterclockwise and clockwise Minkowski matrices in rotating coordinates (rotations around the z-axis) describing spacetime in a rotating frame [E3-12]:

COUNTERCLOCKWISE ROTATION AROUND THE Z- AXIS WITH ANGULAR VELOCITY ω:

$$\begin{pmatrix} -1 & 0 & \omega y \\ 0 & -1 & -\omega x \\ \omega y & -\omega x & 1-\omega^2(x^2+y^2) \end{pmatrix}. \tag{1.91}$$

CLOCKWISE ROTATION AROUND THE Z- AXIS WITH ANGULAR VELOCITY ω:

$$\begin{pmatrix} -1+\omega^2 y^2 & -\omega^2 xy & \omega y \\ -\omega^2 xy & -1+\omega^2 x^2 & -\omega x \\ \omega y & -\omega x & 1 \end{pmatrix}. \tag{1.92}$$

The determinant of matrices (1.91) and (1.92) is 1 and they are invertible. Einstein considered what he called the g *metric*:

THE g METRIC:

$$\begin{pmatrix} -(1+\alpha x_2^2) & \alpha x_1 x_2 & \beta x_2 \\ \alpha x_1 x_2 & -(1+\alpha x_1^2) & -\beta x_1 \\ \beta x_2 & -\beta x_1 & 1 \end{pmatrix}. \tag{1.93}$$

where α and β are integration constants.

Despite the similarity between (1.92) and (1.93), the determinant of Eq. (1.93) is:

$$\det(g) = \alpha x_1^2 + \alpha x_2^2 + \beta^2 x_1^2 + \beta^2 x_2^2 + 1. \tag{1.94}$$

Matrix (1.93) satisfies the coordinate restriction: Eq. (1.87) = 0[44]:

$$\text{THE } \theta_{\iota\kappa\lambda} \text{ CONDITION/RESTRICTION:} \quad \boxed{\theta_{\iota\kappa\lambda} = 0.} \tag{1.95}$$

Setting the determinant of (1.93) to 1 leads to a relationship between the constants α and β. Einstein wrote the condition $\alpha = -\beta^2$ arising from the determinant calculation, Eq. (1.94). By adjusting the constants α and β in the metric (1.93) to make the determinant 1, Einstein obtained its inverse, he called the γ *metric*, which surprisingly resembled the rotation metric (1.91) [E3-12]:

THE γ METRIC:

$$\begin{pmatrix} -1 & 0 & \beta x_2 \\ 0 & -1 & -\beta x_1 \\ \beta x_2 & -\beta x_1 & 1+\alpha(x_1^2+x_2^2) \end{pmatrix}. \tag{1.96}$$

[44] Matrix (1.93) is a solution to condition (1.95) but Einstein did not explicitly write Eq. (1.95) on page $42R$ [JRSNS].

While the metric (1.93) is a solution to condition (1.95), the matrix (1.96) is not a solution to this condition.[45] The metric (1.93) and its inverse (1.96) gave the impression of rotation matrices, which were necessary for Einstein's investigation into how his gravitation theory could account for inertial forces in rotating frames. He tried to understand if the effects experienced in such frames (like the centrifugal force) could be explained as manifestations of spacetime curvature. This was central to his principle of equivalence, which posits that the effects of gravity and acceleration are locally indistinguishable. The condition (1.95) not allowing the γ metric (1.96), but allowing the g metric (1.93), indicated a limitation and a need for further refinement in Einstein's theoretical framework in 1912. The final result highlighted the challenges in fully integrating the behavior of rotating systems within Einstein's nascent gravitational theory. Moreover, it symbolized the challenges faced by Einstein in acknowledging that his November tensor (1.81) could yield the Newtonian limit and rotation.

The core of the challenge resides in the tensor $\theta_{i\kappa\lambda}$ [Eq. (1.87)] bearing a notable similarity to the Christoffel symbols of the second kind. This analogy extends to applying expression (1.87) to the Ricci tensor, which resulted in truncated formulations incorporating a term reminiscent of the Laplacian operator [Eq. (1.85)]. Furthermore, the matrices (1.93) and (1.96) displayed congruence with the matrices of rotational dynamics. These equations, in their entirety, closely paralleled the accurate equations that would eventually form the foundational components of the general theory of relativity. Thus, these pseudo-formulations were inadequate in accurately modeling rotational dynamics.

Due to these difficulties (and the failure to reconcile the November tensor with the Newtonian limit and energy-momentum conservation), Einstein eventually moved on to equations different from the November tensor (1.81) and the associated field equations. His search for candidates for the left-hand side of the field equations culminated on pages 26*L* and 26*R* in deriving an identity, which prefigured the formulation he would later introduce in 1913 in collaboration with Marcel Grossmann [refer to Eqs. (1.135) and (1.137) in Sect. 1.3 and the detailed explanation therein]. This preliminary version underwent a simplification process, resulting in the *Entwurf* field equations (the first and second terms in the square brackets) and the corresponding stress-energy pseudo-tensor (the third and fourth terms in the square brackets) [E3-12]:

[45] We calculate the partial derivatives of the components of the matrix (1.96) with respect to the coordinates x_1, x_2 and x_3 (assuming x_3 is another variable in the system, even if it does not explicitly appear in the components of $g_{i\kappa}$), and then plug these derivatives into the Eq. (1.95) for all possible combinations of i, κ, and λ. Not all evaluated expressions for the condition (1.95) are zero. Specifically, two of the expressions result in αx_1 and αx_2, indicating that the matrix (1.96) is not a solution to the condition (1.95) for all combinations of i, κ, and λ. The presence of non-zero terms in the results means that the matrix does not satisfy the specified coordinate restriction (1.95) across all the possible derivative combinations.

THE *entwurf* FIELD EQUATIONS — THE STRESS-ENERGY PSEUDO TENSOR:

$$\frac{\partial \gamma_{\lambda\rho}}{\partial x_m}\left[\frac{\partial}{\partial x_\alpha}\left(\gamma_{\alpha\beta}\sqrt{-g}\frac{\partial g_{\lambda\rho}}{\partial x_\beta}\right) - \sqrt{-g}\gamma_{\mu\nu}\gamma_{\alpha\beta}\frac{\partial g_{\lambda\mu}}{\partial x_\alpha}\frac{\partial g_{\rho\nu}}{\partial x_\beta}\right.$$
$$\left. - \frac{1}{2}\sqrt{-g}\left(\frac{\partial g_{\sigma\tau}}{\partial x_\lambda}\frac{\partial \gamma_{\sigma\tau}}{\partial x_\rho} - \frac{1}{2}g_{\lambda\rho}\gamma_{\alpha\beta}\frac{\partial g_{\sigma\tau}}{\partial x_\alpha}\frac{\partial \gamma_{\sigma\tau}}{\partial x_\beta}\right)\right]. \tag{1.97}$$

The terms in the square brackets in this equation can be expressed verbally as follows:

THE D'ALEMBERTIAN-LIKE OPERATOR IN CURVED SPACETIME [Eq. (1.29)]–

$$\boxed{\sqrt{-g}\gamma_{\mu\nu}\gamma_{\alpha\beta}\frac{\partial g_{\lambda\mu}}{\partial x_\alpha}\frac{\partial g_{\rho\nu}}{\partial x_\beta}} \text{ — THE STRESS-ENERGY PSEUDO TENSOR.} \tag{1.98}$$

Equation (1.97) focuses on covariant forms of the *Entwurf* field equations and the stress-energy pseudo tensor of the gravitational field.

This was a pivotal moment that marked Einstein's inadvertent commitment to the *Entwurf* framework—a misstep eloquently termed as "falling into the *Entwurf* trap" by John Stachel (private conversation). This phrase encapsulates the critical juncture at which Einstein's pursuit of a theory that approached the Newtonian limit and was compatible with energy-momentum conservation momentarily converged on a path that, in hindsight, deviated from the direction of generally covariant field equations.

However, in the next chapter, I show that the structural similarity between the boxed term in Eq. (1.98) and the November coupling term (1.90) reflects an underlying continuity in Einstein's thought process and mathematical exploration. The November coupling term (1.90), associated with the Ricci tensor, involves the manipulation of the metric tensor and its derivatives. The boxed term in Eq. (1.98) found within Einstein's calculations involving the *Entwurf* equations reflects a deeper exploration of differential geometry. The structural similarity between these expressions, found in different contexts of Einstein's work, reflects the continuity and the evolution of Einstein's mathematical and physical insights as he developed the general theory of relativity.

1.3 The 1913 *Entwurf* Theory

1.3.1 Einstein's Physical Equations

Marcel Grossmann collaborated with Einstein again on the new *Entwurf* theory. Their collaboration resulted in a joint paper titled "Outline [*Entwurf*] of a Generalized Theory of Relativity and a Theory of Gravitation" [EG13]. Einstein was the driving force and significantly influenced the project. He was primarily responsible for physical concepts, developing the theory's foundations, and providing the initial ideas and insights that guided their collaboration. Conversely, Grossmann was crucial in providing mathematical expertise and tools to formulate the theory mathematically. While both Einstein and Grossmann contributed to the collaboration, Einstein's influence and ideas were more prominent in shaping the project's direction. Einstein's authority and vision in physics were significant factors in their work together.

Einstein commenced the joint *Entwurf* paper with a derivation of the equations of motion for a material point in a static gravitational field and the expression for the energy-momentum balance in a gravitational field (1.21). His derivation was based on previous equations he had written on pages $5R$ and $39L$ in his *Zurich Notebook* [E3-12] (see derivation in Sect. 1.2.1).[46]

After this derivation, Einstein embarked on deriving the field equations. In 1912, he initially adopted a top-down approach, starting from the Ricci tensor and attempting to demonstrate how it could lead to the Newtonian limit. However, failing to achieve this, he switched to a bottom-up strategy. He set out to derive the field equations of gravitation, which generalize Newton's theory of gravitation's *Poisson equation* (1.27). The generalization Einstein sought would likely take the form [EG13]:

THE GENERAL FORM OF THE FIELD EQUATIONS:

$$\boxed{\kappa\Theta_{\mu\nu} = \Gamma_{\mu\nu}.} \tag{1.99}$$

Following the Newton-Poisson equation (1.27), Einstein assumed that Eq. (1.99) were of second order. However, he struggled to find a differential expression $\Gamma_{\mu\nu}$ that was both a generalization of $\nabla^2\Phi$ and that behaved as a tensor under arbitrary transformations. Einstein realized that the natural candidate for $\Gamma_{\mu\nu}$ would be the contravariant differential tensor of the second rank - the *Ricci tensor*. However, he had already explored this possibility in the *Zurich Notebook*. He discovered that the Ricci tensor did not simplify to the Newtonian limit in the weak-field limit, as shown in Eq. (1.27). Therefore, Einstein opted for a new approach based on his operator, Eq. (1.101) (see Sect. 1.2.3).

[46] Einstein started with the variation of the action (1.1). Using the line element (1.3), he wrote down the Lagrangian L in terms of $g_{\mu\nu}$ and the velocities $\frac{dx_\mu}{dt}$. Equations (1.1) and (1.3) resulted in three expressions for force (F_x) (1.5), energy (E) (1.6), and angular momentum in the x-direction (J_x) of a material point (1.7). He introduced the proper volume (1.8) and the proper density ρ_0 (1.9). By plugging Eq. (1.8) and (1.9) into Eqs. (1.5), (1.6), and (1.7), Einstein derived the force, energy, and momentum per volume, Eqs. (1.10) and (1.11) (the force density, energy density, and momentum density in the x-direction). Equations (1.10) and (1.11) led to the simplest form of the contravariant stress-energy tensor of the material process for pressureless dust $\Theta_{\mu\nu}$ (1.20). Substituting the contravariant stress-energy tensor for pressureless dust (1.20) into Eqs. (1.10) and (1.11) led to new expressions for the force density (1.13), energy density (1.14), and the momentum density (1.15). Einstein expanded the scope from one component to two general expressions: the energy-momentum density (1.16) and the gravitational four-force density (1.17). Finally, he set the divergence of Eq. (1.16) equal to Eq. (1.17) and combined these two equations into one expression. This led to the energy-momentum balance in a gravitational field (1.21). Refer to the equations in Sect. 1.2.1.

Einstein further imposed the following condition: according to Eq. (1.99), for the energy-momentum balance in a gravitational field (1.21) to be satisfied, $\Gamma_{\mu\nu}$ must correspond to the four-force density (1.17):

$$\boxed{-\frac{1}{2\kappa}\sqrt{-g}\sum_{\mu\nu}\frac{\partial g_{\mu\nu}}{\partial x_\sigma}\Gamma_{\mu\nu}.} \tag{1.100}$$

Upon abandoning the Ricci tensor, Einstein redirected his investigation towards the left-hand side of the field equations by employing an expression (1.30) analogous to the d'Alembertian in curved spacetime (1.29) while positing that the determinant of the metric tensor $\sqrt{-g}=1$. He called this expression (1.30) an *operator*[47]:

EINSTEIN'S OPERATOR:

$$\boxed{\sum_{\alpha\beta}\frac{\partial}{\partial x_\alpha}\left(\gamma_{\alpha\beta}\frac{\partial \gamma_{\mu\nu}}{\partial x_\beta}\right).} \tag{1.101}$$

Einstein noted that Eq. (1.101) could, under the approximation of weak fields, lead to an expression involving the d'Alembert operator for flat spacetime acting on $\gamma_{\mu\nu}$[48]:

THE D'ALEMBERT OPERATOR IN FLAT SPACE- TIME ACTING ON $\gamma_{\mu\nu}$:

$$\boxed{\Box=\frac{\partial^2\gamma_{\mu\nu}}{\partial x_1^2}+\frac{\partial^2\gamma_{\mu\nu}}{\partial x_2^2}+\frac{\partial^2\gamma_{\mu\nu}}{\partial x_3^2}-\frac{1}{c^2}\frac{\partial^2\gamma_{\mu\nu}}{\partial x_4^2}.} \tag{1.102}$$

He substituted his operator (1.101) into the *four-force density* (1.100) [EG13]:

SUM OF PARTIAL DERIVATIVES OF THE METRIC TENSOR =

$$\frac{1}{2}\sum_{\mu\nu}\sqrt{-g}\frac{\partial g_{\mu\nu}}{\partial x_\sigma}\left[\boxed{\text{EINSTEIN'S OPERATOR}}+\text{X}\right]=\text{IDENTITY.} \tag{1.103}$$

X = Additional terms omitted when forming the first approximation.
$\Gamma_{\mu\nu}$ = EINSTEIN'S OPERATOR + X.

[47] Recall from Sect. 1.2.2 that I call it *Einstein's operator*. This operator acts on a tensor field with other derivative terms.

[48] In flat spacetime, the d'Alembertian in curved spacetime reduces to the familiar form of the $\Box$ operator, containing the second time derivative minus the Laplacian (see footnote 28). Thus, under the weak-field approximation, Eq. (1.101) relates to operations resembling the Laplacian.

Einstein wrote that it is important to understand that knowing just one element involved in the identity structure allows for determining the entire identity [EG13].[49] He focused on the electrostatics example and how it relates to the conservation of momentum and the search for an identity that encapsulates this conservation.

Einstein posited that the conservation of momentum in electrostatics can be understood through a specific mathematical identity [EG13][50]:

$$\sum_{\mu}\frac{\partial}{\partial x_{\mu}}\left[\left(\frac{\partial\varphi}{\partial x_{\nu}}\frac{\partial\varphi}{\partial x_{\mu}}\right)-\left(\frac{1}{2}\sum_{\mu}\left(\frac{\partial\varphi}{\partial x_{\mu}}\right)^{2}\right)\delta_{\mu\nu}\right]=\frac{\partial\varphi}{\partial x_{\nu}}\boxed{\sum_{\mu}\frac{\partial^{2}\varphi}{\partial x_{\mu}^{2}}}=-\frac{\partial\varphi}{\partial x_{\nu}}\boxed{\rho}. \tag{1.104}$$

The left-hand side of Eq. (1.104) corresponds to a tensor that resembles a pseudo-tensor. Thus, the Eq. (1.104) takes the following form:

THE DIVERGENCE OF THE PSEUDO- TENSOR =
THE DIVERGENCE OF THE DIFFERENTIAL EQUATIONS. (1.105)

The term $\frac{\partial\varphi}{\partial x_{\nu}}\rho$ can be interpreted as the force density in the electrostatic field, which is derived from the electric field E acting on the charge density ρ.[51] Thus, the right-hand side of Eq. (1.104):

$$\frac{\partial\varphi}{\partial x_{\nu}}\sum_{\mu}\frac{\partial^{2}\varphi}{\partial x_{\mu}^{2}}=-\frac{\partial\varphi}{\partial x_{\nu}}\rho, \tag{1.106}$$

shows that the way the electrostatic potential and the electric field vary in space (and thus exert forces on charges) inherently respects the momentum conservation principle.[52]

---

[49] An identity in this context means a mathematical statement true by the nature of the mathematical structures involved (like the metric tensor and its derivatives) rather than a physical law that governs the behavior of matter and energy.

[50] Identity (1.104) represents a form of momentum conservation in the presence of an electric field (the derivatives of φ with respect to the spatial coordinates x_{ν}, x_{μ} correspond to the components of the electric field). The field's influence is described through the gradients of the electrostatic potential φ (the right-hand side).

[51] The potential φ gradient gives us the electric field E, according to $E = -\nabla\varphi$.

[52] In an electrostatic system, the conservation of momentum asserts that the total mechanical momentum of all charges remains constant in the absence of external forces. This constancy occurs because the internal forces, such as those between charges and between charges and the field, only transfer momentum from one part of the system to another without altering the total amount. The divergence of the electric field, rather than the force field, helps us understand how forces act on charge distributions across space. By integrating the force density over the volume occupied by the charges and the field, we account for how momentum is distributed and conserved. Mathematically, this integration assures that the net change in the total system momentum is zero, provided no external forces exist.

Einstein pointed out that if we know even a single element of identity (1.104), we can determine the entire identity [Eq. (1.110)], which, in turn, helps find the differential equation governing the electrostatic potential φ [EG13]:

$$\nabla^2\varphi = \sum_\mu \frac{\partial^2\varphi}{\partial x_\mu^2} = -\rho. \tag{1.107}$$

[The Laplacian = the charge density].[53]

To derive the identity (1.110), Einstein suggests applying the product rule (1.31) multiple times to this term[54]:

$$\sum_\mu \frac{\partial}{\partial x_\mu}\left(\frac{\partial\varphi}{\partial x_\nu}\frac{\partial\varphi}{\partial x_\mu}\right). \tag{1.109}$$

Einstein's approach is deriving an entire identity through multiple applications of the product rule (1.31) without initially knowing the complete form of the identity [Eq. (1.110)] or Eqs. (1.104) and (1.107). The electrostatics identity (1.104) is rooted in the behaviors of electric fields and potentials, which are inherently scalar, making the product rule directly applicable to their analysis. Upon the first application of the product rule (1.31) to the term (1.109), we consider every possible interaction between the derivatives of φ, akin to examining the first derivatives and their products under differentiation. This step does not yet reveal the full identity or Eq. (1.107) but begins to hint at the structure of interactions within the field. By iteratively applying the product rule, we examine deeper layers of interactions (second derivatives, mixed derivatives, etc.). Each application reveals more about the structure of the field's behavior and its mathematical description. As we continue, patterns emerge—specific combinations of derivatives consistently appear. These patterns suggest conservation laws within the field, hinting at underlying principles like momentum or energy conservation. Through this iterative process, we eventually recognize a complete mathematical structure—an entire identity that represents the conservation law (refer to Eq. (1.105)):

---

[53] This is the Laplacian generalized to multiple dimensions. In a general n-dimensional space with coordinates $x_1, x_2, \ldots x_n$, the Laplacian is:

$$\nabla^2\varphi = \sum_{\mu=1}^{n} \frac{\partial^2\varphi}{\partial x_\mu^2}, \tag{1.108}$$

where each term $\frac{\partial^2\varphi}{\partial x_\mu^2}$ is the second derivative of φ with respect to one of the coordinates.

[54] Applying the product rule (1.31) to each term within the summation (1.109), we differentiate the product of $\frac{\partial\varphi}{\partial x_\nu}$ and $\frac{\partial\varphi}{\partial x_\mu}$.

$$\begin{aligned}&\text{THE DIVERGENCE OF THE PSEUDO-TENSOR} =\\&\text{THE DIVERGENCE OF THE DIFFERENTIAL EQUATIONS} +\\&\text{AN EXPRESSION FOR MOMENTUM CONSERVATION.}\end{aligned} \quad (1.110)$$

Identity (1.110) inherently includes Eq. (1.107), even if we did not know it initially. Einstein's approach highlights a profound connection between physical principles (like momentum conservation) and their mathematical representations.

It may appear intuitive and, indeed, compelling to posit that Einstein would deduce his IDENTITY, as represented by Eq. (1.103) through the application of the product rule (1.31) to the boxed term in the following equation:

$$\sum_{\mu\nu} \sqrt{-g}\, \boxed{\frac{\partial g_{\mu\nu}}{\partial x_\sigma} \text{THE D'ALEMBERTIAN-LIKE OPERATOR IN CURVED SPACETIME}}\,. \quad (1.111)$$

In both cases [Eq. (1.110) and the complete form of the *Entwurf* identity, Eq. (1.135)], the mathematical forms represent conservation laws, albeit in different contexts and frameworks. However, a careful examination of the mathematical context reveals that Einstein and Grossmann eschewed what, in hindsight, might be construed as the ostensibly straightforward path, diverging significantly from the methodological approach that one might anticipate based on Einstein's prior elucidations in the physical part of the *Entwurf* paper. Notice that the two terms from gravitational theory (1.111) and electrostatics (1.109), though structurally similar at a glance, embody fundamentally different physical concepts and mathematical constructions due to the nature of the fields they describe and the theories they belong to. The electrostatics term deals with a scalar field (φ), describing the electric potential, while the gravitational term involves a tensor field ($g_{\mu\nu}$), describing the curvature of spacetime. The electrostatic term operates within the context of flat spacetime, while the *Entwurf* term is fundamentally about describing the curvature of spacetime itself.

1.3.2 Grossmann's Calculations in the **Entwurf** *Collaboration*

Grossmann reasoned that while it was possible to define the Ricci tensor (1.75), it could not be reduced to the Poisson equation of Newton's theory of gravitation in the case of an infinitely weak static gravitational field. This realization raised questions about the connection between the Ricci tensor and the gravitational equations, particularly in arbitrary transformations. The key issue was whether the field equations allowed for arbitrary transformations or only a certain group of transformations (linear transformations). Grossmann was uncertain about this distinction, which led to a lack of clarity regarding the role of the Ricci tensor in the gravitational field equations. Grossmann appeared to embrace certain misconceptions that Einstein also held. Specifically, this is evident in deriving an identity that led to the *Entwurf* field equations.

In the second mathematical part of the *Entwurf* paper, Grossmann explained how the IDENTITY was derived. Grossmann writes that the change is from:

GROSSMANN'S OPERATOR:

$$\boxed{\sum_{\alpha\beta\mu\nu} \frac{\partial g_{\mu\nu}}{\partial x_\sigma} \frac{\partial}{\partial x_\alpha} \left(\gamma_{\alpha\beta} \sqrt{-g} \frac{\partial \gamma_{\mu\nu}}{\partial x_\beta} \right),} \tag{1.112}$$

to the proposed identity [Eqs. (1.135) and (1.137)]. Grossmann's operator (1.112) represents a version of Einstein's expression (1.111). However, it seems that Grossmann proposed the boxed expression in Eq. 1.111 to Einstein, further guiding him through deriving the identity.

The transformation from Eq. (1.112) to the identity involves *integrating by parts* to the terms within the expression (1.112) [EG13]. Grossmann transformed the initial expression (1.112) into a form:

$$\sum_{\alpha\beta\mu\nu} \frac{\partial}{\partial x_\alpha} \left(\sqrt{-g} \gamma_{\alpha\beta} \frac{\partial \gamma_{\mu\nu}}{\partial x_\beta} \frac{\partial g_{\mu\nu}}{\partial x_\sigma} \right) - \sum_{\alpha\beta\mu\nu} \sqrt{-g} \gamma_{\alpha\beta} \frac{\partial \gamma_{\mu\nu}}{\partial x_\beta} \frac{\partial^2 g_{\mu\nu}}{\partial x_\sigma \partial x_\alpha}. \tag{1.113}$$

Subsequently, he employed integration by parts on the second term, continuing in this manner through successive steps until the final identity was deduced.

Integration by parts is the integral counterpart of the product rule, formulated to handle integrals of products of functions. It can be derived from the product rule and is expressed as[55]:

$$\int u dv = uv - \int v du. \tag{1.114}$$

We use *integration by parts* to transfer differentiation from one function to another under an integral, based on the formula (1.114). Grossmann's expression (1.113) [and the identity (1.135) and (1.137)] reflects this principle, albeit in a differential form rather than under an integral, due to the nature of the operations involved in tensor

[55] u and v are functions of some variable, and du and dv are their respective differentials.

calculus and the gravitational theory.[56] Grossmann's manipulation highlights the interplay between differentiation, the metric tensor, its derivatives, and the geometric structure enclosed in $\sqrt{-g}$ and $\gamma_{\alpha\beta}$. The differential form of integrating by parts, particularly in manipulating terms involving the derivatives of tensors in expression (1.112), allowed Grossmann to reorganize the equations so that they could make certain properties more manifest. Specifically, it helped him in isolating divergence terms.

Grossmann mentions a relation he used in the derivation:

THE DERIVATIVE OF THE INVERSE METRIC TENSOR IN TERMS OF THE METRIC TENSOR:

$$\boxed{\frac{\partial \gamma_{\rho\mu}}{\partial x_\nu} = -\sum_{\alpha\beta} \gamma_{\rho\alpha}\gamma_{\mu\beta}\frac{\partial g_{\alpha\beta}}{\partial x_\nu}.} \tag{1.115}$$

This identity (1.115) is used in terms that involve the differentiation of $\gamma_{\mu\nu}$ with respect to spacetime coordinates.[57]

[56] To see how this comes about from the initial expression (1.112), Let us consider an analogy to the integration by parts formula, where differentiation and integration operations are inverse. When we apply the product rule for differentiation (1.31) [which is a forward operation analogous to integration by parts (1.114)], we distribute the differentiation across a product of functions, leading to two terms: one where the differentiation directly applies to the product (as in the first term of the Eq. (1.113)), and another where the differentiation is applied in a manner that resembles the leftover from integration by parts (as in the second term of Eq. (1.113)). Thus, *by integrating by parts in a differential sense*, we rearrange the differentiation across the terms in expression (1.112), resulting in the two terms in the expression (1.113): The first term arises directly by applying the derivative to the entire product inside the parentheses. The second term results from transferring one differentiation onto the metric tensor $g_{\mu\nu}$ term, akin to moving the differentiation in integration by parts and recognizing the role of the second derivative of the metric tensor.

[57] Equation (1.115) is derived by differentiating the identity:

THE METRIC TENSOR TIMES ITS INVERSE IS THE IDENTITY MATRIX:

$$\gamma_{\rho\sigma} g_{\sigma\mu} = \delta^\rho_\mu, \quad \text{where:} \tag{1.116}$$

δ^ρ_μ is the Kronecker delta, which is:

$$\delta^\nu_\mu = \begin{cases} 1 & \text{if } \mu = \nu, \\ 0 & \text{if } \mu \neq \nu, \end{cases} \tag{1.117}$$

and applying the product rule (1.31) and solving for the desired derivative.
In 1914, Einstein wrote this relation as follows [E2-14]:

$$\sum_\sigma g_{\mu\sigma} g^{\nu\sigma} = \delta^\nu_\mu. \tag{1.118}$$

and, according to the multiplication theorem for determinants [E2-14]:

$$\left|\sum_\sigma g_{\mu\sigma} g^{\nu\sigma}\right| = |\delta^\nu_\mu| = 1. \quad \text{From this, it follows:} \quad |g_{\mu\nu}| \cdot |g^{\mu\nu}| = 1. \tag{1.119}$$

Grossmann also applied the metric determinant derivative relation, which expresses the derivative of the $\sqrt{-g}$ in terms of the metric tensor and its inverse, as follows[58]:

THE DERIVATIVE OF THE DETERMINANT OF THE METRIC TENSOR

$$\frac{\partial\sqrt{-g}}{\partial x_\alpha} = -\frac{1}{2\sqrt{-g}}\sum_{\mu\sigma}\gamma_{\mu\sigma}\frac{\partial g_{\mu\sigma}}{\partial x_\alpha}. \tag{1.125}$$

Identity (1.118) states that the metric tensor and its inverse act as a matrix and its inverse, canceling each other to produce the identity matrix. We differentiate the identity (1.116) with respect to x_μ. Differentiating both sides of this identity with respect to x_ν yields zero since the Kronecker delta is a constant with respect to spacetime coordinates, and its derivative with respect to any coordinate is zero.

$$\frac{\partial}{\partial x_\mu}(\gamma_{\rho\sigma}g_{\sigma\mu}) = \frac{\partial}{\partial x_\mu}\delta^\rho_\mu = 0. \tag{1.120}$$

The derivative of a constant with respect to any variable is zero, which is why the right-hand side of Eq. (1.120) $\frac{\partial}{\partial x_\mu}\delta^\rho_\mu = 0$. The left-hand side of Eq. (1.120) involves a product, so we apply the product rule (1.31) to the metric tensor and its inverse:

$$\frac{\partial\gamma_{\rho\sigma}}{\partial x_\nu}g_{\sigma\mu} + \gamma_{\rho\sigma}\frac{\partial g_{\sigma\mu}}{\partial x_\nu} = \frac{\partial}{\partial x_\mu}(\gamma_{\rho\sigma}g_{\sigma\mu}). \tag{1.121}$$

As said above, $\frac{\partial}{\partial x_\mu}(\gamma_{\rho\sigma}g_{\sigma\mu})$ is equal to the derivative of the Kronecker delta with respect to the coordinate which is zero. Applying the product rule (1.31) to the identity that connects the metric tensor with its inverse (1.116) gives:

$$\frac{\partial\gamma_{\rho\sigma}}{\partial x_\nu}g_{\sigma\mu} + \gamma_{\rho\sigma}\frac{\partial g_{\sigma\mu}}{\partial x_\nu} = 0 \,[\text{In Einstein's 1914 notation: } \sum_\sigma\left(\frac{\partial g^{\rho\sigma}}{\partial x^\nu}g_{\sigma\mu} + g^{\rho\sigma}\frac{\partial g_{\sigma\mu}}{\partial x^\nu}\right) = 0]. \tag{1.122}$$

Thus, applying the product rule to the left-hand side of the identity (1.116), we get two terms: the derivative of the inverse metric tensor multiplied by the metric tensor plus the inverse metric tensor multiplied by the derivative of the metric tensor. Since Eq. (1.122) includes the identity (1.118), it equals zero.

We isolate the first term on the left-hand side of Eq. (1.122) and rewrite the equation, focusing on the terms involving derivatives of the metric tensor and its inverse:

$$\frac{\partial\gamma_{\rho\sigma}}{\partial x_\nu}g_{\sigma\mu} = -\gamma_{\rho\sigma}\frac{\partial g_{\sigma\mu}}{\partial x_\nu}. \tag{1.123}$$

Now, we multiply both sides by $\gamma_{\mu\beta}$ (allowing to change indices appropriately), leading to Eq. (1.115).

58 In Einstein's 1914 notation:

$$\frac{\partial\sqrt{-g}}{\partial x^\alpha} = \frac{1}{2\sqrt{-g}}\sum_{\mu\sigma}g^{\mu\sigma}\frac{\partial g_{\mu\sigma}}{\partial x^\alpha}. \tag{1.124}$$

This relation played a key role in deriving the identity [Eqs. (1.135) and (1.137)] from Eq. (1.112) [EG13].[59]

Initially, the straightforward approach when encountering terms involving products of functions or, in this case, tensor components and their derivatives, is to say:

---

[59] To obtain Eq. (1.125), we first write the derivative of a square root function:

DERIVATIVE OF A SQUARE ROOT FUNCTION: $$\frac{d}{dx}\left(x^{\frac{1}{2}}\right)=\frac{1}{2}x^{-\frac{1}{2}}=\frac{1}{2\sqrt{x}}. \tag{1.126}$$

.

[If the argument inside the square root is not simply x but a function of x, say $u(x)$, we would use the chain rule to differentiate. Consider the function $h(x)=\sqrt{u(x)}=(u(x))^{\frac{1}{2}}$. The derivative of $h(x)$ with respect to x would be:

$$\frac{d}{dx}\sqrt{u(x)}=\frac{1}{2}(u(x))^{-\frac{1}{2}}\cdot\frac{du(x)}{dx}=\frac{1}{2\sqrt{u(x)}}\cdot u'(x). \tag{1.127}$$

Here, the chain rule is applied because we are differentiating a composition of functions: the square root function and the inner function $u(x)$.]

We then write a straightforward application of Eq. (1.126) to the determinant of the metric tensor. In other words, applying the chain rule (1.127) to the determinant of the metric tensor, we obtain:

THE CHAIN RULE APPLIED TO THE DETERMINANT OF THE METRIC TENSOR:

$$\frac{\partial\sqrt{-g}}{\partial x_\alpha}=\frac{1}{2\sqrt{-g}}\frac{\partial(-g)}{\partial x_\alpha}. \tag{1.128}$$

The chain rule applied to the determinant of the metric tensor in Einstein's 1914 notation:

$$\frac{\partial\sqrt{-g}}{\partial x^\alpha}=\frac{1}{2\sqrt{-g}}\frac{\partial(-g)}{\partial x^\alpha}=\frac{1}{2\sqrt{-g}}(-1)\frac{\partial g}{\partial x^\alpha}=-\frac{1}{2\sqrt{-g}}\frac{\partial g}{\partial x^\alpha}. \tag{1.129}$$

.

The constant factor rule in differentiation allows taking out the -1 as a constant factor.

We then write the derivative of the determinant of a matrix A with respect to a parameter t:

THE DERIVATIVE OF THE DETERMINANT OF A MATRIX:

$$\frac{d}{dt}\det(A)=\det(A)\,\mathrm{tr}\left(A^{-1}\frac{dA}{dt}\right). \tag{1.130}$$

Applying Eq. (1.130) to the metric tensor $g_{\mu\sigma}$, where t corresponds to a coordinate x_σ, we get:

$$\frac{d}{dx^\alpha}\det(g)=\det(g)\,\mathrm{tr}\left(g^{-1}\frac{dg}{dx^\alpha}\right). \quad \text{Thus:} \tag{1.131}$$

$$\frac{\partial g}{\partial x^\alpha}=g\,\mathrm{tr}\left(g^{-1}\frac{\partial g}{\partial x^\alpha}\right). \tag{1.132}$$

This equation involves the determinant of the metric tensor g, its inverse g^{-1}, and the derivative of g with respect to a coordinate x^σ. By the definition of the trace of a matrix product, the right-hand side of Eq. (1.132) is:

$$g\,\mathrm{tr}\left(g^{-1}\frac{\partial g}{\partial x^\alpha}\right)=\sum_{\mu\sigma}g^{\mu\sigma}\frac{\partial g_{\mu\sigma}}{\partial x^\alpha}. \tag{1.133}$$

Combining Eqs. (1.132) and (1.133) leads to:

we apply the product rule for differentiation (1.31) to the term inside the derivative in the expression (1.112). The product rule is a fundamental tool in calculus, providing the first step in manipulating expressions to reveal additional structure or simplify calculations. However, upon deeper examination, especially when deriving the *Entwurf* field equations, the analogy to *integrating by parts* becomes apparent. Thus, the actual transformation results from *integrating by parts* in a tensor calculus framework. Grossmann's transition from a *product rule* interpretation to *integration by parts* perspective reflects a deeper understanding of the mathematical principles at play than Einstein's.

Considering the observable discordance between the directives provided by Einstein in the physical exposition of the *Entwurf* paper (see Sect. 1.3.1) and those delineated by Grossmann within the mathematical section, coupled with the most reasonable assumption that Grossmann provided Einstein with the expression (1.112)—there emerges a compelling argument. This scenario strongly suggests that Grossmann, leveraging his mathematical prowess, guided Einstein through deriving the identity [(1.135) and (1.137)]. This act is a testament to Grossmann's invaluable mathematical contributions to Einstein's work.

Finally the original expression (1.112) resulted in a complex *identity* [EG13]:

$$\text{THE } \textit{entwurf} \text{ IDENTITY}: \tag{1.135}$$

$$\sum_{\alpha\beta\tau\rho} \frac{\partial}{\partial x_\sigma}\left(\sqrt{-g}\gamma_{\alpha\beta}\frac{\partial \gamma_{\tau\rho}}{\partial x_\beta}\frac{\partial g_{\tau\rho}}{\partial x_\sigma}\right) - \frac{1}{2}\sum_{\alpha\beta\tau\rho}\frac{\partial}{\partial x_\sigma}\left(\sqrt{-g}\gamma_{\alpha\beta}\frac{\partial \gamma_{\tau\rho}}{\partial x_\alpha}\frac{\partial g_{\tau\rho}}{\partial x_\beta}\right) =$$

$$\frac{1}{2}\sum_{\mu\nu}\sqrt{-g}\frac{\partial g_{\mu\nu}}{\partial x_\sigma}\left[\text{TERMS}\right] \quad \text{where:} \tag{1.136}$$

$$\text{TERMS} = \text{THE D'ALEMBERTIAN-LIKE OPERATOR IN CURVED SPACETIME } [(1.29)] -$$

$$\sum_{\alpha\beta\tau\rho} \gamma_{\alpha\beta} g_{\tau\rho}\frac{\partial \gamma_{\mu\tau}}{\partial x_\alpha}\frac{\partial \gamma_{\nu\rho}}{\partial x_\beta}$$

$$+\frac{1}{2}\sum_{\alpha\beta\tau\rho}\gamma_{\alpha\mu}\gamma_{\beta\tau}\frac{\partial g_{\tau\rho}}{\partial x_\alpha}\frac{\partial \gamma_{\tau\rho}}{\partial x_\beta}$$

$$-\frac{1}{4}\sum_{\alpha\beta\tau\rho}\gamma_{\mu\nu}\gamma_{\alpha\beta}\frac{\partial g_{\tau\rho}}{\partial x_\alpha}\frac{\partial \gamma_{\tau\rho}}{\partial x_\beta}. \tag{1.137}$$

Evidently, in 1913, Einstein transposed the identity from pages $26L$ and $26R$ of his *Zurich Notebook* into his collaborative work with Marcel Grossmann, incorporating, at most, a few additional clarifications. Based on the previous discussion, it can be

---

THE DERIVATIVE OF THE DETERMINANT:

$$\frac{\partial g}{\partial x^\alpha} = \sum_{\mu\sigma} g^{\mu\sigma}\frac{\partial g_{\mu\sigma}}{\partial x^\alpha}. \tag{1.134}$$

Plugging Eq. (1.134) into the expression (1.129) we get Eq. (1.125).

posited that Grossmann played a pivotal role in guiding Einstein toward the derivation of the identity [(1.135) and (1.137)] that culminated in the formulation of the *Entwurf* field equations, as documented in the final pages of the *Zurich Notebook*. Einstein's persuasion led Grossmann to believe that the Ricci tensor could not reproduce the Newtonian limit. Consequently, Grossmann assisted Einstein in deriving the identity [(1.135) and (1.137)] necessary for the *Entwurf* field equations.

1.3.3 Einstein Derives the Field Equations

Einstein defined $\Delta_{\mu\nu}$, a second-rank tensor derived from the metric tensor $\gamma_{\mu\nu}$ by differential operations and $\theta_{\mu\nu}$, a *contravariant stress-energy pseudo tensor* of the gravitational field γ.

The *contravariant* form of the $\Delta_{\mu\nu}(\gamma)$ (which relates to $\gamma_{\alpha\beta}$) is obtained from the first two terms in the square brackets in Eq. (1.137) [EG13]:

$$\Delta_{\mu\nu}(\gamma) = \text{THE D'ALEMBERTIAN- LIKE OPERATOR IN CURVED SPACETIME} - \sum_{\alpha\beta\tau\rho} \gamma_{\alpha\beta} g_{\tau\rho} \frac{\partial \gamma_{\mu\tau}}{\partial x_\alpha} \frac{\partial \gamma_{\nu\rho}}{\partial x_\beta}. \tag{1.138}$$

The third and fourth terms in the square brackets in Eq. (1.137) are the *contravariant form of the pseudo-tensor* $\theta_{\mu\nu}$ [EG13]:

$$\text{THE CONTRAVARIANT FORM OF THE STRESS- ENERGY PSEUDO- TENSOR:}$$
$$-2\kappa\theta_{\mu\nu} = \sum_{\alpha\beta\tau\rho} \left(\gamma_{\alpha\mu}\gamma_{\beta\nu} \frac{\partial g_{\tau\rho}}{\partial x_\alpha} \frac{\partial \gamma_{\tau\rho}}{\partial x_\beta} - \frac{1}{2} \gamma_{\mu\nu}\gamma_{\alpha\beta} \frac{\partial g_{\tau\rho}}{\partial x_\alpha} \frac{\partial \gamma_{\tau\rho}}{\partial x_\beta} \right). \tag{1.139}$$

After obtaining the contravariant forms for Eqs. (1.138) and (1.139), Einstein formulated the covariant counterparts of these equations, which appeared to be similar in the form [refer to Eq. (1.97), which Einstein wrote on page $26R$ in the *Zurich Notebook*; see Sect. 1.2.3]. The covariant form of $D_{\mu\nu}(g)$ is represented by:

$$D_{\mu\nu}(g) = \text{THE D'ALEMBERTIAN- LIKE OPERATOR IN CURVED SPACETIME} - \sum_{\alpha\beta\tau\rho} \gamma_{\alpha\beta} \gamma_{\tau\rho} \frac{\partial g_{\mu\tau}}{\partial x_\alpha} \frac{\partial g_{\nu\rho}}{\partial x_\beta}. \tag{1.140}$$

The first term on the right-hand side of Eqs. (1.138) and (1.140) is Eq. (1.29), which Einstein wrote on page $12R$ in the *Zurich Notebook* (see Sect. 1.2.2):

THE D'ALEMBERTIAN- LIKE OPERATOR IN CURVED SPACETIME:

$$\boxed{\frac{1}{\sqrt{-g}}\sum_{\alpha\beta}\frac{\partial}{\partial x_\alpha}\left(\gamma_{\alpha\beta}\sqrt{-g}\frac{\partial g_{\mu\nu}}{\partial x_\beta}\right).} \tag{1.141}$$

The second term on the right-hand side in Eqs. (1.138) and (1.140) is a term that bears similarity to what I referred to in Sect. 1.2.3 as the November coupling term, Eq. (1.90), which is related to the weak field approximation[60]:

A TERM SIMILAR TO THE NOVEMBER COUPLING TERM:

$$\boxed{\sum_{\alpha\beta\tau\rho}\gamma_{\alpha\beta}\gamma_{\tau\rho}\frac{\partial g_{\mu\tau}}{\partial x_\alpha}\frac{\partial g_{\nu\rho}}{\partial x_\beta}.} \tag{1.142}$$

Equation (1.90) was part of Einstein's early efforts in 1912 to formulate a truncated November tensor, while the Eq. (1.142) comes from the *Entwurf* field equations. However, both equations' overall structure and operation types (metrics and derivatives products) are similar. This demonstrates that Einstein never fully relinquished the November tensor. Instead, he continually endeavored to integrate elements of it into his evolving theoretical framework.

Subsequently, Einstein wrote an equation representing $t_{\mu\nu}$, the *covariant form of the stress-energy pseudo-tensor* $t_{\mu\nu}$ for the gravitational field:

THE COVARIANT FORM OF THE STRESS- ENERGY PSEUDO- TENSOR:

$$\boxed{-2\kappa t_{\mu\nu}=\sum_{\alpha\beta\tau\rho}\left(\frac{\partial g_{\tau\rho}}{\partial x_\mu}\frac{\partial\gamma_{\tau\rho}}{\partial x_\nu}-\frac{1}{2}g_{\mu\nu}\gamma_{\alpha\beta}\frac{\partial g_{\tau\rho}}{\partial x_\alpha}\frac{\partial\gamma_{\tau\rho}}{\partial x_\beta}\right).} \tag{1.143}$$

Equations (1.138), (1.139), (1.140), (1.143), facilitate the identity, Eq. (1.137):

$$\sum_{\mu\nu}\frac{\partial}{\partial x_\nu}\left(\sqrt{-g}\,\boxed{\text{FIRST TERM}}\right)=\frac{1}{2}\sum_{\mu\nu}\sqrt{-g}\frac{\partial}{\partial x_\sigma}\boxed{\text{SECOND TERM}}, \quad \text{where:} \tag{1.144}$$

THE CONTRAVARIANT FIRST TERM:

$$g_{\sigma\mu}\kappa\theta_{\mu\nu}.$$

THE COVARIANT FIRST TERM: (1.145)

$$\gamma_{\mu\nu}\kappa t_{\mu\sigma}.$$

THE CONTRAVARIANT SECOND TERM:

$$g_{\mu\nu}\left(-\Delta_{\mu\nu}(\gamma)+\kappa\theta_{\mu\nu}\right). \tag{1.146}$$

[60] The second term is nonlinear. This means considering effects beyond the linear approximation. It is relevant if the gravitational field is still weak but not weak enough to ignore all nonlinear effects.

THE COVARIANT SECOND TERM:

$$\gamma_{\mu\nu}\left(-D_{\mu\nu}(g)+\kappa t_{\mu\nu}\right). \tag{1.147}$$

Next, Einstein combined Eqs. (1.138), (1.139), (1.17), and Eq. (1.21) into *the conservation law for the gravitational field*, Eq. (1.135):

$$\sum_{\mu\nu}\frac{\partial}{\partial x_\nu}\left(\sqrt{-g}g_{\sigma\mu}\kappa\theta_{\mu\nu}\right)-\frac{1}{2}\sum_{\mu\nu}\sqrt{-g}\frac{\partial g_{\mu\nu}}{\partial x_\sigma}\kappa\theta_{\mu\nu}=-\frac{1}{2}\sqrt{-g}\sum_{\mu\nu}\frac{\partial g_{\mu\nu}}{\partial x_\sigma}\Delta_{\mu\nu}(\gamma), \tag{1.148}$$

THE CONSERVATION LAW FOR THE GRAVITATIONAL FIELD:

$$\text{or: }\boxed{\sum_{\mu\nu}\frac{\partial}{\partial x_\nu}\left(\sqrt{-g}g_{\sigma\mu}\kappa\theta_{\mu\nu}\right)+\frac{1}{2}\sum_{\mu\nu}\sqrt{-g}\frac{\partial g_{\mu\nu}}{\partial x_\sigma}\left(\Delta_{\mu\nu}(\gamma)-\kappa\theta_{\mu\nu}\right)=0.} \tag{1.149}$$

Equation (1.149) represents a general form of an identity [Eqs. (1.135) and (1.137)]. In this sense, Eq. (1.149) expresses a *conservation law for the gravitational field*. This conservation law must be an IDENTITY because it is formulated purely in terms of the metric tensor and its derivatives, without involving matter or non-gravitational fields (i.e., without the stress-energy tensor). The next step was writing the field equations. In essence, Eq. (1.149) provides the expression for the gravitational tensor $\Gamma_{\mu\nu}$, the right-hand side of Eq. (1.99):

$$\Gamma_{\mu\nu}=\Delta_{\mu\nu}(\gamma)-\kappa\theta_{\mu\nu}=\kappa\Theta_{\mu\nu} \tag{1.150}$$

Hence, the *contravariant* Entwurf *field equations* take the form:

$$\Delta_{\mu\nu}(\gamma)=\kappa(\Theta_{\mu\nu}+\theta_{\mu\nu}). \tag{1.151}$$

Similarly, using Eq. (1.147), Einstein wrote the *covariant* form with sources of the *Entwurf* field Eqs. (1.151) [EG13][61]:

THE ENTWURF FIELD EQUATIONS:

$$\boxed{D_{\mu\nu}(g)=\kappa(T_{\mu\nu}+t_{\mu\nu}).} \tag{1.152}$$

The covariant form of the Entwurf *vacuum field equations*, which describe the gravitational field in the absence of matter and energy, is expressed as follows:

[61] The *Entwurf* field equations describe how spacetime curvature (as represented by the operator $D_{\mu\nu}(g)$ acting on the metric g) relates to the distribution of matter and energy. In a 4-dimensional spacetime, the symmetric rank-2 tensors used in the *Entwurf* field equations have 10 independent components due to their symmetry. This symmetry property means that the components are equal when their indices are exchanged [$D_{\mu\nu}(g)=D_{\nu\mu}(g)$]. Therefore, the number of independent equations is reduced to 10.

THE ENTWURF VACUUM FIELD EQUATIONS:

$$\boxed{D_{\mu\nu}(g) = 0.} \tag{1.153}$$

We plug Eq. (1.140) into Eq. (1.152) to replace $D_{\mu\nu}(g)$ on the left-hand side, and we get:

$$\boxed{\begin{array}{r}\text{THE D'ALEMBERTIAN- LIKE OPERATOR IN CURVED SPACETIME}-\\ \text{A TERM SIMILAR TO THE NOVEMBER COUPLING TERM} = -\kappa(T_{\mu\nu} + t_{\mu\nu}).\end{array}} \tag{1.154}$$

Thus, this equation integrates both invariants and remnants of the truncated November tensor, reflecting the evolution of Einstein's gravitational theory from the *Zurich Notebook* to the *Entwurf* theory.

The *Entwurf* field equations are covariant under a restricted group of transformations, specifically linear transformations, rather than fully generally covariant. The absence of general covariance in the *Entwurf* theory can be seen when trying to transform the equations under a general coordinate transformation. The $D_{\nu\mu}(g)$ operator, which is part of the gravitational field equations, as defined in the *Entwurf* theory, will not maintain its form under such transformations.

1.3.4 The Vienna Lecture

In September of 1913, during the eighty-fifth assembly of the Congress of German Natural Scientists and Physicians in Vienna, Einstein delivered a lecture entitled "On the Present State of the Problem of Gravitation," which elucidated his nascent *Entwurf* theory. This exposition precipitated a vigorous debate among his contemporaries, who were divided in their reception of his ideas. The proceedings of this lecture and the ensuing discussions were later published in the December issue of the *Physikalische Zeitschrift* [E13].

Within the ambit of this lecture, Einstein revisited the gravitational equations he had previously delineated in his 1913 *Entwurf* paper [EG13]. Despite the anticipatory nature of this revision, it largely reiterated the foundational premises of his earlier formulation, offering scant novelty in terms of the conceptual underpinnings of the field equations themselves. Notwithstanding, a methodological evolution was evident in how Einstein approached his gravitational equations. Einstein wrote the energy-momentum balance in a gravitational field as follows [EG14][62]:

[62] $\mathcal{T}_{\sigma\nu}$ represents the stress-energy tensor of matter in the gravitational field [refer to Eq. (1.19) in Sect. 1.2.1]. See Eq. (1.18).

THE ENERGY-MOMENTUM BALANCE IN A GRAVITATIONAL FIELD:

$$\sum_{\nu} \frac{\partial \mathcal{T}_{\sigma\nu}}{\partial x_{\nu}} = \boxed{\frac{1}{2} \sum_{\mu\nu\tau} \frac{\partial g_{\mu\nu}}{\partial x_{\sigma}} \gamma_{\mu\tau}} \mathcal{T}_{\tau\nu}. \tag{1.155}$$

Unlike his previous formulations of the energy-momentum balance in a gravitational field in the *Zurich Notebook* and the *Entwurf* paper [EG13], Einstein included the gradient of the metric tensor $\frac{1}{2} \sum_{\mu\nu\tau} \frac{\partial g_{\mu\nu}}{\partial x_{\sigma}} \gamma_{\mu\tau}$ on the right-hand side of Eq. (1.155). This finally led him in 1914 to choose the gradient of the metric tensor (1.213), see Sect. 1.5.2), to represent the gravitational field's components in his *Entwurf* theory.

Einstein subsequently initiated a linearization methodology applied to the *Entwurf* field equations, a strategic approach aimed at transforming these inherently complex and nonlinear equations into a simplified, linear variant.[63]

Let us examine the linearization method applied to the *Entwurf* field equations in [E13]. By employing linearization to the *Entwurf* field equations and integrating a sequence of assumptions, a significant divergence from the unsuccessful attempts documented in the *Zurich Notebook* is observed. Specifically, in 1913, this methodological approach facilitated the achievement of the Newtonian limit. Einstein started with:

THE FLAT (MINKOWSKI) SPACETIME METRIC + A DEVIATION:

$$\boxed{g_{\mu\nu} = \delta_{\mu\nu} + g^{*}_{\mu\nu}.} \tag{1.156}$$

The $g^{*}_{\mu\nu}$ represents the slight deviations from the flat Minkowski spacetime due to weak gravitational fields [refer to Eq. (1.41), Sect. 1.2.2].[64] The result is a set of linearized field equations, where the complex non-linear terms of the *Entwurf* field Eq. (1.152) which involve products and higher powers of $g^{*}_{\mu\nu}$ are not present.[65] Einstein interpreted this approximation as valid when the gravitational field is weak and spacetime is close to being flat.

---

[63] It is imperative to acknowledge Einstein's prior utilization of this method, as documented in the *Zurich Notebook* [Eq. (1.41) (The flat Minkowski spacetime metric + a deviation), (1.57) (the second linearized field equation), and (1.58) (the box operator in flat space-time acting on the deviation $g_{i\kappa} = \kappa T_{i\kappa}$), refer to Sect. 1.2.2]. Recall that in examining the equations in the *Zurich Notebook* written a year before the paper [E13], it becomes evident that Eqs. (1.41) and (1.58), central to Einstein's derivation in the *Zurich Notebook*, were not explicitly stated but can be inferred with a high degree of certainty. This inference is supported by a comprehensive analysis of the explicitly presented equations [e.g., Eq. (1.44) (the linearized harmonic restriction), (1.45) (the harmonic condition in linearized gravity), (1.47) (the first linearized field equation), and Eq. (1.57)].

[64] In this approximation, terms involving products and higher powers of the perturbations $g^{*}_{\mu\nu}$ are neglected. This is analogous to truncating the Taylor series after the first-order term, assuming that the higher-order terms are negligibly small due to the smallness of $g^{*}_{\mu\nu}$.

[65] Einstein started with the *Entwurf* field Eq. (1.152) [the operator (1.152) with $\sqrt{-g} = 1$ because in flat spacetime, the determinant of the metric tensor simplifies to -1, i.e., Eq. (1.101)]. He linearized these field equations around a background flat Minkowski spacetime, leading to a wave-like Eq. (1.157).

The *Entwurf* field Eq. (1.152) are simplified in the weak-field approximation and reduced to a wave-like equation [E13][66]:

THE D'ALEMBERT OPERATOR IN FLAT SPACE-TIME ACTING ON $g^*_{\mu\nu} = \kappa T_{\mu\nu}$:

$$\Box g^*_{\mu\nu} = \frac{\partial^2 g^*_{\mu\nu}}{\partial x^2} + \frac{\partial^2 g^*_{\mu\nu}}{\partial y^2} + \frac{\partial^2 g^*_{\mu\nu}}{\partial z^2} - \frac{1}{c^2}\frac{\partial^2 g^*_{\mu\nu}}{\partial t^2} = \kappa T_{\mu\nu}. \tag{1.157}$$

One might legitimately inquire whether Einstein had conceived of gravitational waves as early as 1913, assuming he pursued the above path.[67] Einstein made the explicit connection between the field equations and gravitational waves, particularly through using the d'Alembertian operator in a linearized form of the field equations, in 1916 after the initial formulation of his 1915 general relativity.[68] *However, in 1913, instead of following this approach to gravitational waves, Einstein made a series of assumptions that led him to the Newtonian limit* [E13]:

(1) A static gravitational field.[69]
(2) The metric deviation $g^*_{\mu\nu}$ goes to zero at infinity.[70]
(3) A pressureless and static cloud of dust.[71]
(4) All components of the metric tensor $g^*_{\mu\nu}$ other than the time-time component ($\mu = \nu = 4$) are negligible ($g^*_{\mu\nu} = 0$).[72]

Einstein's assumptions collectively simplify the problem and lead to the derivation of the Poisson equation for g^*_{44}, which describes how the gravitational potential is

[66] Einstein calculated $\Box g^*_{\mu\nu}$, i.e., applied the time and spatial derivatives to $g^*_{\mu\nu}$. Thus, he applied the d'Alembertian operator $\Box$ to the perturbation $g^*_{\mu\nu}$ in a flat spacetime $\delta_{\mu\nu}$ in Eq. (1.156).

[67] It is important to note that the d'Alembertian $\Box$ does not necessarily imply gravitational waves. The d'Alembertian is a differential operator used in physics to describe a wide range of wave phenomena and field equations, not exclusively gravitational waves. The d'Alembertian operator is one of the mathematical tools used in physics, including studying electromagnetic waves.

[68] When the d'Alembertian operator is applied to the perturbed metric $g^*_{\mu\nu}$ under the linear approximation, and the results are incorporated into the field equations, one obtains *a wave-like* equation for the perturbations. Equation (1.157) describes how these perturbations, which can be interpreted as gravitational waves, propagate through spacetime. In other words, the perturbations $g^*_{\mu\nu}$ in the spacetime metric propagate wave-likely, similar to how electromagnetic waves propagate in Maxwell's theory. This equation resembles Eq. (3.28) [see footnote 13 in Chap. 3].

[69] The field is unchanging over time, eliminating the need to consider time-dependent variations. This assumption simplifies the time derivatives in the equations.

[70] The spacetime approaches a flat Minkowski spacetime. Spacetime remained asymptotically flat, and the deviations had a limited range.

[71] This dust distribution serves as the source of gravity. Einstein reduced the stress-energy tensor $T_{\mu\nu}$ to primarily the mass density component, aligning with his *static field assumption*.

[72] Einstein simplified the metric to a diagonal form to approximate a static metric where only the time-time component is nonzero, equal to c^2 ($c^2 \neq 1$). By simplifying the metric, he greatly reduced the number of terms and equations in the field equations. The dominant term is the gravitational potential $g^*_{\mu\nu}$, directly related to the Newtonian gravitational potential.

related to the mass density ρ_0 in a static, weak gravitational field. These assumptions allowed Einstein to focus on the important features of a static gravitational field.

With Einstein's above specific set of assumptions, his wave Eq. (1.157) reduces to focus only on the time component [E13][73]:

$$\boxed{\Box g^*_{\mu\nu} = -\frac{1}{c^2}\frac{\partial^2 g^*_{44}}{\partial t^2} = \kappa T_{44}.} \tag{1.158}$$

By focusing only on g^*_{44} and neglecting the other components and off-diagonal terms, Einstein concentrated on the most significant aspect of the gravitational field for a static scenario: how g^*_{44} behaves under the influence of a static cloud of dust.[74]

According to the assumptions mentioned above, Einstein identified T_{44} with the Poisson equation's mass density ρ_0 (refer to assumption 3), and assumed that the spacetime is asymptotically flat (i.e., that at spatial infinity, the metric deviations $g^*_{\mu\nu}$ approach a Minkowski metric $\delta_{\mu\nu}$, refer to assumption 2). In this limit, the wave operator $\Box$ reduces to a spatial Laplacian operator as the time derivatives become negligible compared to spatial derivatives. With these assumptions, $\Box g^*_{\mu\nu}$ becomes Δg^*_{44} and the Eq. (1.158) simplifies to an equation resembling Einstein's field equation from his static gravitational fields theory [E2-12], i.e., a form of Poisson's equation for g^*_{44} [E13]:[75]

[73] In other words, this reduction implies that, under Einstein's assumptions, the gravitational field's spatial variations and dynamical aspects are not considered. Instead, the focus is on how the gravitational potential (related to g^*_{44}) is influenced by the presence of mass in a static way.

[74] In this equation, $\Box g^*_{\mu\nu}$ simplifies to just the second time derivative of the g^*_{44} component, reflecting the static nature of the field and the focus on the time-time component of the metric tensor g^*_{44}.

[75] In his 1912 theory of static gravitational fields, Einstein formulated a nonlinear field equation, represented by the following equation [E2-12]:

THE NONLINEAR STATIC FIELD EQUATION:

$$\boxed{c\Delta c - \frac{1}{2}(\nabla c)^2 = \kappa c^2 \sigma.} \tag{1.159}$$

This equation shows that the Laplacian of the gravitational potential (Δc) is related to both the energy density associated with matter ($c^2\sigma$) and the energy density associated with the gravitational field itself [$\frac{1}{2\kappa}(\nabla c)^2$]. Let us move the energy density term in this equation involving the gradient of the static gravitational potential c to the right-hand side:

$$c\Delta c = \frac{1}{2}(\nabla c)^2 + \kappa c^2 \sigma. \tag{1.160}$$

Equation (1.160) shows that this non-linear Eq. (1.160) shares some structural similarity with the Poisson equation. However, in Eq. (1.160), the second term $\frac{1}{2}(\nabla c)^2$ is non-linear, which distinguishes it from the linear Poisson equation. The Laplacian term equals a linear source term in a Poisson equation. In Einstein's equation (1.160), the presence of the non-linear term $\frac{1}{2}(\nabla c)^2$ introduces interactions between different components of the gravitational potential c through its gradient, making the equation non-linear.

A FORM OF POISSON'S EQUATION:

$$\boxed{\Delta g^*_{44} = \kappa c^2 \rho_0.} \tag{1.161}$$

For components other than g^*_{44} (i.e., μ, ν not equal to 4), the equation states $\Delta g^*_{\mu\nu} = T_{\mu\nu} = 0$.[76] The solution to the Poisson equation represents the metric $g^*_{\mu\nu}$ in this static, weak gravitational field.

In the paper [E13], Einstein articulated a derivation [of Eqs. (1.158) and (1.161)], a methodology introduced during his lecture in Vienna in 1913. In a discussion following the Vienna lecture, Max Born asked Einstein about the speed at which gravity's influence spreads according to his *Entwurf* theory. It did not seem obvious to Born that gravitational effects would travel at the speed of light; the relationship seemed to him quite complicated. Einstein responded that writing the equations for cases where the perturbations introduced to the field are extremely small is surprisingly straightforward [refer to Eq. (1.156)]. In such situations, the metric tensor ($g_{\mu\nu}$) deviates only slightly from what it would be without the disturbance ($g^*_{\mu\nu}$), and these small disturbances ($g^*_{\mu\nu}$) move through the field at the speed of light [CPAE4], Doc 18.

In 1913, Einstein's primary goal was to reconcile his emerging *Entwurf* theory with the well-established Newtonian physics, particularly in the so-called Newtonian limit. The limited covariance of the *Entwurf* field equations was sufficient for this purpose. The focus on achieving consistency with the Newtonian limit and the partial success of the *Entwurf* theory have initially hindered Einstein's development of fully covariant field equations. As evident from the derivation presented earlier, Einstein *did not fully* explore gravitational waves in his *Entwurf* theory. Furthermore, the full implications of Eq. (1.157) in the context of gravitational waves were not fully explored by Einstein at this time. It was only after completing the general theory of relativity in 1915–1916 that Einstein started to explore the implications of his theory for gravitational waves [refer to Eq. (3.4) and the equations after that in Sect. 3.2.2, Chap. 3]. However, the mathematical groundwork laid in 1913 [Eqs. (1.156), (1.157), (1.158)] was foundational for the later discovery of gravitational waves, even though this was not Einstein's primary goal at the time. Understanding gravitational waves required a more complete and refined formulation of general relativity, which Einstein had not yet fully developed. Moreover, the underlying field equations must be covariant to describe gravitational waves accurately. This is because the properties of these waves should not depend on the observer's frame of reference. Einstein's field equations, in their final form, are covariant and form the foundation for predicting and understanding gravitational waves. Nevertheless, the response mentioned above to Born elucidates that Einstein acknowledged the implication of his gravitational theory, which hints toward the propagation of undulations at the velocity of light.

---

[76] This condition means that for metric tensor components other than the time-time component, the Laplacian of $g^*_{\mu\nu}$ is zero, implying no gravitational potential associated with these components. Similarly, $T_{\mu\nu} = 0$ means that the stress-energy tensor components ($T_{\mu\nu}$) associated with these spacetime points are zero. In other words, those points have no matter or energy density.

1.4 Two Tests of the *Entwurf* Theory

1.4.1 The First Test: Mercury's Perihelion Advance

In June 1913, Einstein's closest friend, Michele Besso, visited him in Zurich. Besso, a proficient engineer, had a deep knowledge of physics and mathematics.[77] Together, Einstein and Besso attempted to solve the *Entwurf* vacuum field equations to determine Mercury's perihelion advance in the Sun's static field. Besso visited Einstein again in August 1913. During this time, Einstein and Besso continued to work on the project. Their joint work consists of the *Einstein-Besso manuscript* [CPAE4], Doc 14, which contains calculations with little explanation.

Let us concentrate on Einstein and Besso's calculations. The derivation starts with assuming a weak gravitational field far from the Sun, which the Minkowski flat metric of special relativity can approximate. The metric components are represented by $g_{\mu\nu}$, where μ and ν are indices. Einstein kicks things off with the calculations in the *Einstein-Besso manuscript*. He uses the form of Poisson's Eq. (1.161) [where he writes g_{44} for g^*_{44}] to find the metric field of the Sun to first-order approximation when considering a spherically symmetric, static gravitational field produced by the Sun [W17, CPAE4], Doc 14:

$$g_{44} = c^2\left(1 - \frac{A}{r}\right). \tag{1.162}$$

A represents a correction factor in the metric component g_{44} that considers the gravitational effects in a spherically symmetric gravitational field of the Sun:

$$A = \left(1 + \frac{\alpha}{r}\right). \tag{1.163}$$

The term $\frac{\alpha}{r}$ in the metric represents the deviation from flatness due to the Sun's gravity. α in this term is represented by[78]:

$$\boxed{\alpha = \frac{\kappa M}{4\pi} = \frac{2GM}{c^2}.} \tag{1.164}$$

The spatial components g_{11}, g_{22} and g_{33} are flat, reflecting a static, spherically symmetric gravitational field without directly incorporating the effects of spacetime curvature on spatial dimensions. This approach is somewhat aligned with a Newtonian perspective, focusing on gravitational potential affecting time but not fully accounting for the curvature of space itself.

[77] In 1926, Einstein wrote that Besso had extraordinary knowledge of pure science [CPAE15], Doc. 436.

[78] κ is Einstein's gravitational constant [Eq. (1.48)], G is the Newtonian gravitational constant, M is the mass of the Sun, and c is the speed of light.

To account for more accurate results, Einstein calculates the second-order metric for the gravitational field of the Sun through successive approximations, leading to a more complex metric field[79]:

$$g_{44} = c^2 \left(1 - \frac{A}{r} + \frac{5}{8}\frac{A^2}{r^2}\right). \tag{1.165}$$

This is the metric field of the static Sun, as obtained through the *Entwurf* theory, including second-order corrections. Now, Besso handles the calculations in the *Einstein-Besso manuscript.* The motion of a planet (in this case, Mercury) within this gravitational field is described using equations of motion for a mass point, which include the following three equations [CPAE4], Doc 14[80]:

$$\text{Conservation of momentum: } 2\dot{f} = \dot{\varphi} r^2 = B\frac{ds}{dt}. \tag{1.166}$$

$$\text{Energy: } E = g_{44}\frac{dt}{ds} \quad \text{and:} \quad \text{Angular momentum: } J_x = \frac{\dot{x}}{\frac{ds}{dt}}. \tag{1.167}$$

Besso reduced Einstein's equation (1.6) for the energy and (1.7) for the angular momentum in the x direction to equations for a mass point moving slowly in the static field of the Sun.[81] He derived an angular momentum conservation equation called the *area law*. However, Besso's *area law* does not directly represent the planet's motion under the influence of gravity. This law differs from Kepler's second law because the rate of change of the area swept out by the radius vector from the Sun to Mercury ($\frac{ds}{dt}$) is not constant in the *Entwurf* theory:

---

[79] The second-order metric is derived by applying the *Entwurf* vacuum field equations. Einstein substitutes the first-order metric [Eq. (1.162)] into the *Entwurf* vacuum field Eq. (1.153)). By substituting the first-order metric into these field equations, Einstein derives the second-order metric [Eq. (1.165)]. This second-order metric more accurately describes the Sun's gravitational field in the *Entwurf* theory.

[80] This equation describes the relationship between the rate of change of the area swept out per unit time $\dot{f}$ and $\frac{ds}{dt}$. It represents the conservation of angular momentum in polar coordinates for a planet orbiting the Sun. The angular momentum is conserved as the planet moves in its orbit. Equation (1.166) equates two different expressions $2\dot{f}$ and $\dot{\varphi} r^2$ that describe angular momentum. The area constant B is introduced on the right-hand side, specifically in the expression $\frac{ds}{dt}$ as a proportionality constant to relate the angular velocity $\dot{\varphi}$ and the squared radial coordinate r^2 on the left-hand side to the rate of change of $\frac{ds}{dt}$, i.e., the velocity component in the fourth coordinate direction (representing time). $\dot{f}$ represents the area speed, i.e., the rate of change of the area swept out by the radius vector from the Sun to the planet per unit of time.

[81] Besso employed the Eqs. (1.6) and (1.7) delineated in Einstein's *Entwurf* paper [EG13] to resolve the Mercury perihelion problem. Besso's energy equation involves g_{44} tying the energy to the temporal component of the metric tensor [see Eq. (1.167)]. Einstein's energy Eq. (1.6) aggregates contributions from all spatial and temporal components. Besso's angular momentum equation is a simpler relationship between velocity and spacetime curvature than Einstein's equation (1.7), without explicitly invoking the metric tensor's spatial components [see Eq. (1.167)]. Refer to Sect. 1.2.1 for Einstein's equations.

BESSO'S AREA LAW:

$$\boxed{\frac{dB}{dt} = \frac{d}{dt}\left(\frac{\dot{\varphi}r^2}{\frac{ds}{dt}}\right) = 0.} \tag{1.168}$$

Besso's *area law* Eq. (1.168) is a key component of Besso's calculations of the precession of Mercury's perihelion.

Besso then combined the first Eq. (1.167), E, with the conservation of momentum, Eq. (1.166), and derived the resulting equation [CPAE4], Doc 14:

$$E\dot{f} = Bg_{44}. \tag{1.169}$$

He inserted Eq. (1.165) into Eq. (1.169), and derived for Mercury moving in the Sun's gravitational field g_{44} (1.165)[82]:

$$E(\dot{\varphi}r^2) = Bg_{44} = Bc^2\left(1 - \frac{A}{r} + \frac{5}{8}\frac{A^2}{r^2}\right). \tag{1.170}$$

Einstein and Besso wrestled with the last set of equations across several pages in the *Einstein-Besso manuscript*, ultimately reaching a rather disheartening outcome. When Eq. (1.170) is integrated over the entire orbit (from perihelion to aphelion), it yields the accumulated change in the angle φ, which corresponds to the precession of Mercury's perihelion[83]:

$$\int_{r_1}^{r_2} d\varphi = \pi\left(1 + \frac{5}{8}\frac{A}{a(1-e^2)}\right). \tag{1.171}$$

From Eq. (1.171), we get:

$$\varphi(r_2) - \varphi(r_1) = \pi\left(1 + \frac{5}{8}\frac{A}{a(1-e^2)}\right) \approx \pi + \pi\frac{5}{8}\frac{A}{a(1-e^2)}. \tag{1.172}$$

---

[82] The left-hand side of the equation, $E(\dot{\varphi}r^2)$, represents the energy of an orbiting planet (i.e., Mercury) in polar coordinates. It is the product of the angular velocity $\dot{\varphi}$, the square of the radial distance (r^2), and Mercury's energy (E). On the right-hand side of the equation, we have the constants B and c^2 and the parameter A, defined by Eq. (1.163). These constants and corrections account for the effects of gravity on Mercury's energy. Equation (1.170) relates E [Eq. (1.167)], φ [Eq. (1.166)], r and B [defined in Eq. (1.166)], g_{44}, c^2 and A [Eq. (1.163)].

[83] This equation describes how the angle φ changes as the planet Mercury moves along its elliptical orbit. Specifically, $\int_{r_1}^{r_2} d\varphi$ represents the integral of the angle φ as Mercury moves along its elliptical orbit from perihelion (r_1) and aphelion (r_2). In other words, the angle swept out by the radius vector from the Sun to Mercury as the planet orbits the Sun between its perihelion and aphelion.

Focusing only on the second term for a half-orbit:

$$\pi \frac{5}{8} \frac{A}{a(1-e^2)}, \tag{1.173}$$

and doubling this for the full orbit, we get:

$$2\pi \frac{5}{8} \frac{A}{a(1-e^2)} = \pi \frac{5}{4} \frac{A}{a(1-e^2)}. \tag{1.174}$$

The calculation focuses on determining the orbit of Mercury and accounting for the additional influences beyond what Newtonian mechanics would typically predict. Equation (1.174) represents the precession of Mercury's perihelion, giving us the amount by which the perihelion of Mercury advances in its orbit per revolution according to the *Entwurf* theory.

Einstein and his close friend Besso, in 1913, grappled with the *Entwurf* vacuum field equations to discern the perihelion advance of Mercury. This feat encountered miscalculations and subsequent corrections, resulting far from the observed discrepancy. According to the *Entwurf* theory, the field of a static Sun produced an advance of Mercury's perihelion per revolution, Eq. (1.174), corresponding to $18''$. This value falls short of explaining the observed value of approximately $43''$.[84] Einstein did not acknowledge the $18''$ disappointing outcome until he revisited his calculations in the autumn of 1915 (as referenced in Sect. 1.5.3).

1.4.2 The Second Test: Rotation

In 1913, Einstein revisited and embarked again on the mission to solve the issues with the rotation metric.[85] This time, he approached the problem with new insights and his developing theoretical framework of the *Entwurf* theory. Now, Einstein believed

[84] However, Einstein's initial result showed an exaggerated figure of $1821''$. This discrepancy was noted in the *Einstein-Besso manuscript*, where Einstein verified the calculation [CPAE4], Doc. 14. It turned out that Einstein's result was exaggerated by 100. This was a significant deviation, considering that in Newton's theory, the Sun causes no perihelion motion at all. In contrast, the *Entwurf* theory suggested a perihelion motion more than three times the observed total motion. Eventually, Einstein and Besso identified a numerical error in their calculation—a mistaken value for the Sun's mass, overestimated by a factor of 10. Since perihelion motion is proportional to the square of this mass, the error is magnified to a factor of 100. The erroneously high figure of $1821''$ was a result of Einstein's computational error, while the $18''$ result was incorrect due to the *Entwurf* theory itself. If the *Entwurf* theory had predicted $1800''$, it would have been decisively refuted. It is suggested in [Jan-21], that while the corrected value of $18''$ was significantly lower than the initial miscalculation, it was still less than half of the observed total motion of Mercury's perihelion, $43''$. This shortfall was disappointing but not immediately seen as a fatal flaw of the *Entwurf* theory.

[85] Refer to Sect. 1.2.3 for a detailed explanation of Einstein's difficulties with the rotation metric.

he had fully resolved the matter, significantly boosting his confidence in his new theory.

In the *Einstein-Besso manuscript*, specifically on pages 41 − 42, Einstein explored whether the Minkowski metric, when applied in rotating coordinates, would align with the solutions of his *Entwurf* vacuum field equations. To determine this, he started by incorporating the components of the rotation metric into the *Entwurf* vacuum field equations, limiting his calculations to a first-order approximation. This step was essential for assessing if the metric field of a rotating system could be considered a valid solution within the framework of the *Entwurf* theory. However, during this process, Einstein made several errors while substituting the first-order components of the rotation metric into the *Entwurf* vacuum field equations. These mistakes inadvertently led him to the conclusion he was seeking. He specifically questioned whether the solution for the g_{44} component of the metric tensor, as derived from the *Entwurf* field equations, matched the g_{44} component that one would get by transforming the Minkowski metric from an inertial system to a rotating system. In the manuscript, Einstein concluded affirmatively, indicating that the two g_{44} values were the same. Alongside this solution, he noted its correctness [CPAE4], Doc. 14. This instance highlights a critical moment in Einstein's journey toward his general theory of relativity, where his initial assumptions and calculations, though eventually proven to be flawed, played a significant role in his ongoing theoretical developments.

Einstein performed a series of calculations to bridge his understanding of general relativity (as per the *Entwurf* theory at this stage) with Mach's principle. He was trying to see if the effects of a rotating universe, as embodied by a rotating metric, could be derived from his *Entwurf* field equations. The analysis concludes that, with some approximation methods, the rotating metric is a valid solution to the *Entwurf* equations, giving some credence to the notions inspired by Mach's principle.

Intriguingly, the rotation challenges and the anomaly in Mercury's perihelion advance were interrelated. For diminutive angular velocities, a rotating system's metric field largely mirrors the Sun's metric field—the metric examined in the 1913 *Einstein-Besso manuscript* to deduce Mercury's perihelion advance [W18]. Einstein wrote down the covariant and contravariant components of the metric tensor for the rotating spacetime. He wanted to determine if the rotating metric was a solution to the *Entwurf* vacuum field equations. He used an approximation method similar to the one he used for the perihelion advance of Mercury. By plugging in the first-order approximation of $g_{\mu\nu}$ into the *Entwurf* vacuum field equations (1.153), Einstein aimed to see if this metric described a vacuum spacetime.

Einstein focused on the $\mu = \nu = 4$ component of the *Entwurf* operator $D_{44}(g)$. This led him to the following equation[86]:

$$D_{44}(g) = -\Delta g_{44} - 8\omega^2. \tag{1.175}$$

[86] See derivation in [Jan-99].

He considered t_{44}, the 44-component of the stress-energy pseudo-tensor of the gravitational field. He believed the result for κt_{44} was:

$$\kappa t_{44} = -4\omega^2. \tag{1.176}$$

Contrary to Einstein's expectations, the actual outcome is as follows[87]:

$$\kappa t_{44} = -\omega^2. \tag{1.179}$$

Hence, t_{44}, in the rotating system is proportional to ω^2, where recall that ω is the angular velocity of the system. However, at this preliminary stage, Einstein did not recognize this miscalculation in the equation of the stress-energy pseudo tensor. Plugging Eqs. (1.175) and (1.176) into the *Entwurf* field equation:

$$D_{44}(g) + \kappa t_{44} = 0, \quad \text{Einstein obtained:} \tag{1.180}$$

$$D_{44}(g) + \kappa t_{44} = -\Delta g_{44} - 8\omega^2 + 4\omega^2 = 0. \quad \text{In other words:} \tag{1.181}$$

$$\Delta g_{44} = -4\omega^2. \tag{1.182}$$

This equation indicates a sign error on Einstein's part. Einstein mistakenly omitted the minus sign in Eq. (1.176) and wrote $+4\omega^2$ instead of $-4\omega^2$. Δg_{44} is a term that arises when the $D_{44}(g)$ operator is applied to the metric. So, from Einstein's perspective, the $-8\omega^2$ term comes from applying the $D_{44}(g)$ operator to the metric. κt_{44} contributes an additional $+4\omega^2$. When we sum up the ω^2 contributions, we get a net contribution of $-4\omega^2$.

Einstein, still unaware of the mathematical error he had committed, then proposed a solution for g_{44} in the form of a term proportional to the squared distance from the axis of rotation:

$$C\omega^2(x^2 + y^2), \quad \text{where } C \text{ is a constant.} \tag{1.183}$$

In two-dimensional polar coordinates (r, θ), r^2 is given by $r^2 = (x^2 + y^2)$. Hence:

$$\Delta g_{44} = C\omega^2 \Delta r^2. \tag{1.184}$$

[87] Equation (1.143) [Sect. 1.3.3] represents $t_{\mu\nu}$. For $\mu = \nu = 4$, this equation takes the form:

$$\kappa t_{44} = \frac{1}{2} \sum_{\alpha\beta\tau\rho} \left(\frac{\partial g_{\tau\rho}}{\partial x_4} \frac{\partial \gamma_{\tau\rho}}{\partial x_4} - \frac{1}{2} g_{44} \gamma_{\alpha\beta} \frac{\partial g_{\tau\rho}}{\partial x_\alpha} \frac{\partial \gamma_{\tau\rho}}{\partial x_\beta} \right). \tag{1.177}$$

The first term in Eq. (1.177) does not contribute to second-order terms proportional to ω^2 because it involves derivatives with respect to a spatial coordinate x_4. The second term yields:

$$-\frac{1}{2} \left(\frac{1}{2} g_{44} \gamma_{22} \frac{\partial g_{14}}{\partial x_2} \frac{\partial \gamma_{14}}{\partial x_2} \right) \to -\frac{1}{4} (4\omega)^2. \tag{1.178}$$

$\Delta r^2 = 4$, and Eq. (1.184) becomes[88]:

$$\Delta g_{44} = 4C\omega^2. \tag{1.188}$$

Equating Eqs. (1.188)–(1.182), results in: $C = -1$. This ultimately led Einstein directly to the desired outcome he had been seeking:

$$g_{44} = 1 - \omega^2(x^2 + y^2). \tag{1.189}$$

This equation represents the time-time component of the metric for the rotating system. This result matches what one would obtain through a direct transformation from a non-rotating to a rotating frame. Given the squared terms in x and y, it shows the influence of rotation, particularly as one moves further out radially from the center of rotation [CPAE4], Doc. 14, [W17].

1.4.3 Einstein's Tensor Approach Versus Nordström's Scalar Theory

In 1912, Gunnar Nordström introduced an alternative to Einstein's *Entwurf* theory. Nordström sought a scalar theory where the speed of light remained constant while still adhering to the principle of relativity and maintaining the equivalence of gravitational and inertial mass. His theory was more closely aligned with the framework of special relativity, maintaining its fundamental postulates [Nor-12]. In his quest to formulate a theory of gravity, Einstein introduced a more complex structure. Instead of a single scalar gravitational potential used in Nordström's theory and earlier attempts by Einstein, the *Entwurf* theory utilized a metric tensor field with ten independent components (see Sect. 1.3.3). This tensor described the gravitational field in a more complex and comprehensive manner. The speed of light could vary depending on the gravitational field, which is a departure from the constancy of light speed in special

88 Δr^2 in polar coordinates is given by:

$$\Delta r^2 = \frac{1}{r}\frac{\partial}{\partial r}\left(r\frac{\partial(r^2)}{\partial r}\right) + \frac{1}{r^2}\frac{\partial^2(r^2)}{\partial\theta^2}. \tag{1.185}$$

Now, we calculate the derivatives:

$$\frac{\partial(r^2)}{\partial r} = 2r, \quad \frac{\partial^2(r^2)}{\partial\theta^2} = 0. \tag{1.186}$$

We substitute these derivatives back into the expression (1.185):

$$\Delta r^2 = \frac{1}{r}\frac{\partial}{\partial r}(r \cdot 2r) + \frac{1}{r^2} \cdot 0 = \frac{1}{r}\frac{\partial}{\partial r}(2r^2) = \frac{1}{r} \cdot 4r = 4. \tag{1.187}$$

relativity. This variation is not in the sense of changing the local speed of light but in how light's path is influenced by the curvature of spacetime caused by gravity.

Nordström's theory did not predict light rays bending in a gravitational field. The *Entwurf* theory, on the other hand, anticipated a light deflection in the Sun's gravitational field of $0.83''$, a prediction later proven inaccurate. Einstein's 1915 general relativity theory doubled this prediction to $1.7''$. In 1914, Erwin Freundlich proposed an expedition to test Einstein's light-bending prediction. However, the expedition was never realized because of World War I, sparing Einstein a potential experimental refutation since he was unaware that the $0.83''$ prediction was erroneous. Both the *Entwurf* and Nordström's theories failed to predict the perihelion motion of Mercury correctly. By 1913, both conflicting theories failed to explain the observed perihelion motion value of $43''$. Consequently, before 1915, the *Entwurf* theory lacked empirical backing, and choosing between the two theories—*Entwurf* or Nordström's—was empirically unfeasible.

Einstein had reservations about Nordström's theory for neglecting rotation, even though the *Entwurf* theory had its inadequacies regarding rotation. Despite Einstein's criticism, Nordström continued developing his theory, believing that alternative approaches to gravitation were needed. Nordström revised his theory after discussions with Einstein, leading to a more natural and simpler theory than the *Entwurf* theory [Nor-13-1, Nor-13-2]. It still faced limitations, like not fully aligning with Mach's principle and not predicting the bending of light rays near the sun. Einstein considered Nordström's theory further and even formulated a version combining their ideas. He developed postulates for this gravitational theory and explored equations of motion in a scalar gravitational field [E13].

Einstein collaborated with Adriaan Fokker from Leiden to develop a theory that strictly obeyed general covariance, reformulating Nordström's theory using tools from the *Entwurf* theory. Their work suggested the importance of the Riemann-Christoffel tensor in a gravitational theory. In his *Entwurf* theory with Grossmann, Einstein struggled to develop a theory of gravitation that was consistent with the observed phenomena and generally covariant. However, in their joint work, Einstein and Fokker reformulated Nordström's theory within a generally covariant framework [EF].

This was an ironic twist, as Einstein struggled to achieve general covariance in his gravitational theory. The effort to bring general covariance to a competitor's theory and the success in doing so highlighted both Einstein's intellectual generosity and his deep commitment to pursuing a more complete understanding of gravity. However, Einstein's work with Fokker was important for a few reasons. First, it showed that a generally covariant formulation of gravitational theory was indeed possible, addressing a significant theoretical challenge. Second, it demonstrated that different theories of gravity (like Nordström's) could be incorporated into this framework. Finally, it paved the way for Einstein's further development of general relativity, where he would eventually formulate a theory of gravitation that was both generally covariant and consistent with experimental data, such as the bending of light by gravity.

By 1915, Einstein had resolved his theory's difficulties with general covariance. The irony continued with Nordström's reaction to Einstein's 1915 theory. Nordström, a proponent of his scalar theory of gravity, recognized the significance and potential of Einstein's general relativity. Consequently, he renounced his scalar theory and began working within the framework of general relativity.[89]

The interactions between Einstein and Nordström illustrate a vital aspect of scientific advancement: collaboration and competition often go hand in hand, driving progress and leading to breakthroughs. Einstein's ability to contribute significantly to Nordström's theory and Nordström's openness to embracing Einstein's later advancements underscore the collaborative nature of scientific discovery.

1.5 The 1914 *Entwurf* Theory

1.5.1 The Mature Einstein-Grossmann Entwurf *Theory*

In 1914, in a joint paper with Grossmann, "Covariance Properties of the Field Equations of the Theory of Gravitation Based on the General Theory of Relativity," Einstein refined the 1913 *Entwurf* theory, demonstrating the validity of its field equations, particularly in adapted coordinate systems [EG14]. Einstein and Grossmann commenced the refinement by considering the energy-momentum balance in a gravitational field (1.155) [EG14]. $\mathcal{T}_{\sigma\nu}$ represents the energy-momentum tensor density of matter in the gravitational field [refer to Eq. (1.19) in Sect. 1.2.1][90]:

$$\mathcal{T}_{\sigma\nu} = \sum_{\mu} \sqrt{-g}\gamma_{\sigma\mu}T_{\mu\nu} = \sum_{\mu} \sqrt{-g}g_{\sigma\mu}\Theta_{\mu\nu}. \tag{1.190}$$

Einstein and Grossmann further represented $t_{\sigma\nu}$, the stress-energy pseudo-tensor of the gravitational field, by an equation of a similar form:

$$\boldsymbol{t}_{\sigma\nu} - \sum_{\mu} \sqrt{-g}\gamma_{\sigma\mu}t_{\mu\nu} - \sum_{\mu} \sqrt{-g}g_{\sigma\mu}\vartheta_{\mu\nu}. \tag{1.191}$$

---

[89] I provide further elaboration on this topic in Sect. 2.3, Chap. 2 and Sect. 3.2.3, Chap. 3.

[90] Einstein first integrated equations previously outlined in the *Zurich Notebook* and the *Entwurf* paper. This led to the derivation of an equation that represents the stress-energy pseudo tensor of the gravitational field (1.193) and to the *Entwurf* Eq. (1.194). This process was initiated by addressing the energy-momentum balance in a gravitational field (1.155) (see Sect. 1.3.4). Subsequently, he extended the Eq. (1.19) (refer to Sect. 1.2.1) to Eq. (1.190) and defined the stress-energy pseudo tensor density by an equation of similar mathematical form (1.191). Einstein attempted to amalgamate equations characterized by general covariance—such as those representing the energy-momentum balance in a gravitational field and the stress-energy tensor—with the *Entwurf* field Eqs. (1.151) and (1.152), which conspicuously lacked this property of general covariance.

They wrote the equation for the stress-energy pseudo-tensor of the gravitational field $\boldsymbol{t}_{\sigma\nu}$[91]:

THE STRESS-ENERGY PSEUDO-TENSOR OF THE GRAVITATIONAL FIELD:

$$-2\kappa \boldsymbol{t}_{\sigma\nu} = \sqrt{-g}\left(\sum_{\beta\rho\tau} \gamma_{\beta\nu} \frac{\partial g_{\rho\tau}}{\partial x_\sigma} \frac{\partial \gamma_{\rho\tau}}{\partial x_\beta} - \frac{1}{2} \sum_{\alpha\beta\rho\tau} \delta_{\sigma\nu} \gamma_{\alpha\beta} \frac{\partial g_{\rho\tau}}{\partial x_\alpha} \frac{\partial \gamma_{\rho\tau}}{\partial x_\beta}\right). \tag{1.193}$$

Using Eq. (1.191), the *Entwurf* equations for the stress-energy pseudo-tensor (1.139) and (1.143) lead to the derivation of Eq. (1.193) [EG14].

Einstein and Grossmann integrated Eqs. (1.151), (1.152) (the contravariant and covariant forms of the *Entwurf* field equations) into a single set of field equations. The result they obtained was a less general form of the *Entwurf* field equations [EG14]:

A LESS GENERAL FORM OF THE ENTWURF FIELD EQUATIONS:

$$\sum_{\alpha\beta\mu} \frac{\partial}{\partial x_\alpha}\left(\sqrt{-g}\, \gamma_{\alpha\beta} g_{\sigma\mu} \frac{\partial g_{\mu\nu}}{\partial x_\beta}\right) = -\kappa(\mathcal{T}_{\sigma\nu} + t_{\sigma\nu}). \tag{1.194}$$

The above Eq. (1.194) is largely expressed in terms of covariant tensors, particularly on the right-hand side, where $\mathcal{T}_{\sigma\nu} + t_{\sigma\nu}$ are in covariant form. Thus, Einstein and Grossmann integrated Eqs. (1.151), (1.152) into a field equation that was predominantly in a covariant format.[92]

Einstein aimed to show that his *Entwurf* field equations (1.194) maintain energy-momentum conservation in adapted coordinate systems.[93] He proceeded by indicating that he conducted a differentiation with respect to x_ν (he took the partial derivative for x_ν) on both sides of the Eq. (1.194). Concurrently, he considered the conservation of energy-momentum[94]:

---

[91] The Kronecker delta $\delta_{\sigma\nu}$ is used in the second summation term. When $\sigma = \nu$, then $\delta_{\sigma\nu} = 1$, and the second term in Eq. (1.193) simplifies to:

$$-\frac{1}{2} \sum_{\alpha\beta\rho\tau} 1 \cdot \gamma_{\alpha\beta} \frac{\partial g_{\rho\tau}}{\partial x_\alpha} \frac{\partial \gamma_{\rho\tau}}{\partial x_\beta}. \tag{1.192}$$

When σ and ν are not equal, $\delta_{\sigma\nu}$ equals 0, and this term disappears from the sum. So, $\delta_{\sigma\nu}$ acts as a switch that selects certain terms in the equation based on whether σ and ν are equal. When they are equal, it selects the term; when they are not equal, it sets the term to zero.

[92] While the *Entwurf* field equations could be expressed using different tensor components (covariant, contravariant, or mixed), the critical aspect was their lack of general covariance, i.e., they did not maintain their form under all possible changes of coordinates. This distinction is necessary for understanding Einstein's *Entwurf* field equations.

[93] Although the paper was co-authored with Grossmann, the concepts discussed predominantly reflect Einstein's intellectual contributions and were developed primarily under his direction. Consequently, for this discussion, references to these ideas will be attributed specifically to Einstein.

[94] This equation represents the conservation of energy-momentum, which states that the covariant divergence of the stress-energy-momentum tensor density and the stress-energy pseudo tensor

THE ENERGY- MOMENTUM CONSERVATION:

$$\boxed{\sum_{\nu} \frac{\partial}{\partial x_{\nu}} (\mathcal{T}_{\sigma\nu} + t_{\sigma\nu}) = 0.} \tag{1.195}$$

Following the differentiation of Eq. (1.194) with respect to x_{ν} and taking into consideration Eq. (1.195), Einstein derived the following relations [EG14][95]:

THE B RESTRICTION TO ADAPTED COORDINATES:

$$\boxed{B_{\sigma} = \sum_{\alpha\beta\mu\nu} \frac{\partial^2}{\partial x_{\nu} x_{\alpha}} \left(\sqrt{-g}\, \gamma_{\alpha\beta} g_{\sigma\mu} \frac{\partial g_{\mu\nu}}{\partial x_{\beta}} \right) = 0.} \tag{1.196}$$

The B_{σ} differential conditions are nothing but coordinate restrictions because they specify certain relationships that must hold between the second partial derivatives of the metric components for the spacetime coordinates x_{α} and x_{ν}. These restrictions impose additional constraints on the metric tensor $g_{\mu\nu}$ and ensure that the *Entwurf* theory's solutions conform to these restrictions. B_{σ} means that Eq. (1.194) is modified and adapted to satisfy the coordinate restrictions B_{σ}.

In their 1914 joint publication, Einstein and Grossmann introduced a variational formalism as the foundational framework for deriving the *Entwurf* field equations. According to Einstein and Grossmann, this approach substantiated that the *Entwurf* field Eqs. (1.152) emerge from applying the variational principle.

They first showed that the *Entwurf* field Eq. (1.194) is equivalent to [EG14]:

$$\int \left(\delta \mathcal{L} - 2\kappa \sum_{\mu\nu} \sqrt{-g} T_{\mu\nu} \delta\gamma_{\mu\nu} \right) d\tau = 0. \tag{1.197}$$

where the Lagrangian density $\mathcal{L}$ is given by:

$$\mathcal{L} = \frac{1}{2} \sqrt{-g} \sum_{\alpha\beta\rho\tau} \gamma_{\alpha\beta} \frac{\partial g_{\rho\tau}}{\partial x_{\alpha}} \frac{\partial \gamma_{\rho\tau}}{\partial x_{\beta}}. \tag{1.198}$$

---

density is zero. $\mathcal{T}_{\sigma\nu}$ and $t_{\sigma\nu}$ are used to account for the total energy and momentum in the system. In simple terms, Eq. (1.195) tells us how energy-momentum changes in response to the curvature of spacetime. In the absence of gravity, the conservation of the energy-momentum tensor alone ($\frac{\partial \mathcal{T}_{\sigma\nu}}{\partial x_{\nu}} = 0$) suffices. In a generally covariant theory, the vanishing divergence of the energy-momentum tensor is a direct mathematical consequence of the field equations rather than an additional requirement. Yet, in his attempts to build a bridge between the conservation laws and his gravitational field equations, Einstein found apparent inconsistencies when his field equations were not generally covariant. This led him to look for compensating terms to warrant conservation, affecting the mathematical form of his proposed *Entwurf* field equations.

95 B_{σ} is a set of four differential conditions for the metric tensor $g_{\mu\nu}$.

Einstein and Grossmann employed the variational principle[96]:

THE ACTION:

$$J = \int \mathcal{L} d\tau. \tag{1.199}$$

They calculated the Lagrangian density's variation $\delta\mathcal{L}$.

The first step was calculating the variation of $\sqrt{-g}$:

$$\delta\sqrt{-g} = -\frac{1}{2}\sum_{\mu\nu}\sqrt{-g}g_{\mu\nu}\delta g_{\mu\nu}. \tag{1.200}$$

The second step involved calculating the variation of the derivative of the metric tensor and its inverse: $\delta\left(\frac{\partial g_{\tau\rho}}{\partial x_\alpha}\right)$ and $\delta\left(\frac{\partial \gamma_{\tau\rho}}{\partial x_\beta}\right)$.

The third step: After finding these variations in steps one and two, Einstein and Grossmann expanded the variations of the metric derivatives by applying tensorial manipulations (e.g., integration by parts).

The fourth step: They incorporated the expressions found in step three into the Lagrangian density variation $\delta\mathcal{L}$ and thus into δJ. The δJ then includes contributions from each varied component. To find J, they integrated $\delta\mathcal{L}$ over τ and applied the principle of least action $\delta J = 0$:

$$\delta J = \int \delta\mathcal{L}\, d\tau = 0. \tag{1.201}$$

The fifth step: Einstein and Grossmann showed that $\delta\mathcal{L}$ = Eqs. (1.140) and (1.141) (the d'Alembertian-like operator in curved spacetime) + (1.143) (the covariant form of the stress-energy pseudo-tensor). This led them to rewrite Eq. (1.201) representing the covariant form of the *Entwrurf* field equations [EG14]:

$$\delta J = \int \sqrt{-g}\sum_{\mu\nu}\left[D_{\mu\nu}(g) = \kappa(T_{\mu\nu} + t_{\mu\nu})\right]\delta\gamma_{\mu\nu} d\tau = 0. \tag{1.202}$$

The sixth step: Finally, they concluded that since $\delta\gamma_{\mu\nu}$ is mutually independent, then this condition allows the *Entwurf* field equations in the covariant form (1.152) to be derived from a Lagrangian density via the variational principle [EG14].

Einstein and Grossmann extended their gratitude to the mathematician Paul Bernays within their paper, acknowledging his pivotal recommendation to employ the variational principle, which significantly streamlined their theoretical framework [EG14].

---

96 The action J is defined in terms of the integral of the Lagrangian density $\mathcal{L}$ over τ.

1.5.2 *The Foundation of the* Entwurf *Theory*

In 1914, Einstein authored a review paper titled "The Formal Foundation of the General Theory of Relativity," where he presented a fully developed and comprehensive articulation of the *Entwurf* theory [E2-14]. This publication marked a pivotal moment as Einstein transitioned from collaborative endeavors with Marcel Grossmann to establishing himself as the sole contributor to the advanced formulation of the general theory of relativity.

Einstein first established the equations describing how material processes, the motion of matter, occur in a gravitational field, which is considered as given. These equations aimed to describe how gravity affects the behavior of physical systems. The material-process equations were designed to be covariant under arbitrary transformations of the spacetime coordinates. Einstein wrote Eq. (1.155), the energy-momentum balance in a gravitational field (see Sect. 1.3.4). In this formulation, Einstein maintained the structural integrity of Eq. (1.155) but modified the notation employed. Specifically, whereas the inverse metric tensor was previously denoted by $\gamma_{\mu\tau}$, he opted for $g^{\mu\tau}$ to represent this tensor [E2-14]:

THE ENERGY- MOMENTUM BALANCE IN A GRAVITATIONAL FIELD:

$$\boxed{\sum_{\nu} \frac{\partial \mathcal{T}_{\tau}^{\nu}}{\partial x_{\nu}} = \frac{1}{2} \sum_{\mu\tau\nu} g^{\tau\mu} \frac{\partial g_{\mu\nu}}{\partial x_{\sigma}} \mathcal{T}_{\tau}^{\nu}.} \tag{1.203}$$

The choice made by Einstein to include the gradient of the metric tensor $\frac{\partial g_{\mu\nu}}{\partial x_{\sigma}}$ on the right-hand side of Eq. (1.203) was a key insight in the development of his theory of gravitation. It led him to choose the metric tensor gradient [$\Gamma^{\tau}_{\nu\sigma}$ (1.213)] to represent the gravitational field's components in the [E2-14] *Entwurf* theory.

Furthermore, Einstein enhanced the representation by offering a more precise depiction of the tensor density. The stress-energy tensor of continuously distributed masses neglecting the effects of surface forces $\mathcal{T}_{\tau}^{\nu}$ is represented by [E2-14][97]:

$$\mathcal{T}_{\tau}^{\nu} = \rho_0 \sqrt{-g} \frac{dx_{\nu}}{ds} \sum_{\mu} g_{\tau\mu} \frac{dx_{\mu}}{ds}. \tag{1.204}$$

These two equations combine the equations that Einstein wrote in the *Zurich Notebook*, in his *Entwurf* paper with Grossmann, and his paper based on the Vienna lecture.[98] Einstein revisited and revised his formulation in his *Zurich Notebook* and his *Entwurf* theory on the motion of material points. He elucidated the motion of

97 When surface forces are ignored, $\mathcal{T}_{\tau}^{\nu}$ describes the bulk properties of the continuous mass distribution, including its energy density, momentum density, and stress. This simplification is used to focus on $\mathcal{T}_{\tau}^{\nu}$ without considering the details of how it interacts with external boundaries and surfaces. Einstein wrote Eq. (1.204) in 1913 [EG13] [refer to Eq. (1.10) in Sect. 1.2.1, footnote 12.].

98 Equation (1.204) combines equations (1.14) (energy density) and (1.15) (momentum density), which Einstein wrote in 1912–1913 (see Sect. 1.2.1). Einstein had already formulated Eq. (1.203)

material points for the presence and absence of gravitational fields. The foundational principle is that the trajectory of a material point in a gravitational field follows a geodesic line within a four-dimensional manifold. This is expressed through a series of equations that incorporate the Christoffel symbols, which are pivotal in defining the motion of objects within the curved spacetime. Einstein followed his previous derivations. The geodesic equation is derived from the principle of least action—or, more specifically, from the variation of the integral of the line element (ds) [E2-14][99]:

$$\sum_{\mu} g_{\sigma} \frac{d^2 x_{\mu}}{ds^2} + \sum_{\mu\nu} \Gamma_{\mu\nu\sigma} \frac{dx_{\mu}}{ds} \frac{dx_{\nu}}{ds} = 0, \tag{1.207}$$

incorporating the Christoffel symbols of the first kind:

$$\Gamma_{\mu\nu\sigma} = \frac{1}{2} \left(\frac{\partial g_{\mu\sigma}}{\partial x_{\nu}} + \frac{\partial g_{\nu\sigma}}{\partial x_{\mu}} - \frac{\partial g_{\mu\nu}}{\partial x_{\sigma}} \right). \tag{1.208}$$

Utilizing the Christoffel symbols of the second kind:

$$\Gamma^{\tau}\mu\nu = \sum_{\sigma} g^{\sigma\tau} \Gamma_{\mu\nu\sigma}, \tag{1.209}$$

Einstein reformulated the geodesic equation (1.207) in its most comprehensive form [E2-14][100]:

---

in his Vienna lecture paper [E13] and Eq. (1.204) on page $5R$ in his *Zurich Notebook* [E3-12]. This Eq. (1.204) also appears in his 1913 joint paper with Grossmann [refer to Eq. (1.11) in Sect. 1.2.1].

99 The variational principle states that for a free particle, the actual path taken between two points is the one for which the variation (δ) of the integral of ds along the path is zero, expressed as Eq. (1.1). The spacetime interval ds is given by Eq. (1.3) (see Sect. 1.2.1). The action (S) related to this motion, which is to be extremized, can be written in terms of the Lagrangian (L) that depends on ds, as:

$$S = \int L ds = \int \sqrt{g_{\mu\nu} \frac{dx_{\mu}}{d\tau} \frac{dx_{\nu}}{d\tau}} d\tau \tag{1.205}$$

where τ is the proper time for timelike geodesics. The variation of S with respect to the coordinates x_{μ} leads to the Euler-Lagrange equations:

$$\frac{d}{d\tau} \left(\frac{\partial L}{\partial (\frac{dx_{\mu}}{d\tau})} \right) - \frac{\partial L}{\partial x_{\mu}} = 0 \tag{1.206}$$

When these equations are applied to the Lagrangian derived from the metric tensor $g_{\mu\nu}$, they yield the geodesic equation.

100 $\frac{d^2 x_{\tau}}{ds^2}$ represents the second derivative of the coordinate x_{τ} with respect to the proper time s along the geodesic. This describes how the coordinate x_{τ} changes as the particle moves on a geodesic through spacetime. $\left(\frac{dx_{\mu}}{ds}\right)\left(\frac{dx_{\nu}}{ds}\right)$ represents the 4-velocity of the particle. Equation (1.210) relates the curvature of spacetime (described by the Christoffel symbols Γ) to the motion of particles.

THE GEODESIC EQUATION:

$$\boxed{\frac{d^2x_\tau}{ds^2} + \sum_{\mu\nu} \Gamma^\tau{}_{\mu\nu} \frac{dx_\mu}{ds}\frac{dx_\nu}{ds} = 0.} \tag{1.210}$$

In its most comprehensive form, the geodesic equation reveals the inherent relation between the geometry of spacetime and the motion of material points within it. Through this equation, Einstein draws a parallel to the classical law of inertia [E2-14].

After establishing the equations for material processes, Einstein formulated the field equations of gravitation. However, these field equations were not generally covariant; they did not maintain their form under all possible transformations of the spacetime coordinates. This limitation was considered a significant drawback because general covariance was indispensable for a fully relativistic theory of gravity. During his intense and tumultuous journey toward the formulation of his field equations, Einstein reached a point where he temporarily convinced himself that generally covariant field equations were not a feasible requirement for the theory of gravity he was developing.

Despite their non-covariant nature, Einstein's adherence to the *Entwurf* field equations was significantly influenced by an argument he devised known as the *hole argument*. This reasoning played a crucial role in his initial resistance to adopting fully covariant gravitational field equations, as it ostensibly demonstrated that covariance could lead to physical inconsistencies in the theory of gravitation. During 1913–1914, Einstein presented several versions of the *hole argument* against general covariance, implying that generally covariant field equations were impermissible [E2-14]. In its initial form, Einstein's hole argument could be seen as a theoretical maneuver to address the non-uniqueness of solutions in generally covariant field equations. However, it is important to note that the hole argument was an important step in Einstein's journey toward the final formulation of the general theory of relativity. In the end, Einstein moved away from the hole argument and its constraints.

Here is the 1914 version of the hole argument [E2-14]: Suppose $g_{\mu\nu}$ represents the metric tensor in the coordinate system K, and $g'_{\mu\nu}$ represents it in another coordinate system K'. Einstein imagined a certain region of spacetime Σ (a hole) where we do not know the exact values of the metric tensor. Outside this hole, $g_{\mu\nu}$ [the field $G(x)$] and $g'_{\mu\nu}$ [the field $G'(x')$] describe the same gravitational field, i.e., they are just different coordinate representations of the same physical reality. Inside the hole, however, they might differ.

Einstein's clever insight or maneuver trick was the following: he proposed swapping the coordinates for the values inside the hole. He took the expressions for the gravitational field in K' (described by $g'_{\mu\nu}$) but evaluated them at the coordinate points of K. So, even though it is the field from K', it is being described in the coordinates of K. Mathematically, this involves replacing x' (coordinates of K') with x (coordinates of K) for the values of $g'_{\mu\nu}$ inside the hole. This results in a new field, which we can denote as $G'(x)$. The maneuver implies that there can be two different solutions [$G(x)$ and $G'(x)$] for the gravitational field outside the hole, even though they are based on the same coordinate x in the region of the hole. This raised a concern: if general covariance were true, and the field equations were generally covariant, then there would not be a unique solution for the metric inside the hole.

The hole restricts the usable coordinate systems to only adapted ones. These adapted systems justify the restricted covariance of Einstein's *Entwurf* equations and explain why fully generally covariant equations could be problematic: they allow too much freedom and ambiguity in their solutions. By 1914, Einstein demonstrated the compatibility of his hole argument with the constraints of adapted coordinate systems. This framework justified the constrained covariance of the field equations and Einstein's *Entwurf* gravitational tensor [Eq. (1.217)] and provided a rationale for generally rejecting covariant field equations [E2-14, W17, W18]. The hole argument was a challenge that Einstein posed against the principle of general covariance in his developing *Entwurf* theory. It demonstrated, to Einstein's thinking at the time, that if the field equations were generally covariant, then they could admit multiple, distinct solutions (different metric tensors) that represented the same physical situation inside of a given region (the hole), but different situations outside that region.

One could posit that Einstein's continued defense of the *Entwurf* theory had shades of dogmatism. But not quite. While Einstein was deeply committed to testing his theories, trying to salvage the *Entwurf* theory, he leaned on the crutches of adapted coordinate systems and the hole argument. In 1914, the hole argument provided Einstein with a refuge against the constraints of general covariance and allowed him to delve into the intricacies of the *Entwurf* theory. While ensnared by the clutches of the *Entwurf* theory, Einstein authored the comprehensive review article dedicated to this theory, "The Formal Foundation of the General Theory of Relativity" [E2-14]. In this article, he articulated numerous formulations that would later influence and recur in his 1915–1916 general theory of relativity (this is discussed in Sects. 1.6 and 1.7).

In the above-mentioned review article, Einstein employed a variational procedure to derive the *Entwurf* field equations. Recall that he had previously utilized this methodology in collaboration with Grossmann [EG14]. Einstein started with the action (1.199), as represented by [E2-14][101]:

$$\boxed{J = \int L\sqrt{-g}\, d\tau.} \tag{1.211}$$

He proposed a specific Lagrangian based on derivatives of the metric tensor:

THE LAGRANGIAN:

$$\boxed{L = -\sum_{\mu\rho\tau\nu} g^{\tau\nu}\Gamma^{\rho}_{\mu\tau}\Gamma^{\mu}_{\rho\nu}.} \tag{1.212}$$

[101] Recall that the integral includes the Lagrangian density of the gravitational field, $\mathcal{L} = L\sqrt{-g}$.

This Lagrangian is not represented in terms of the Christoffel symbols. $\Gamma^{\tau}_{\nu\sigma}$ in Eq. (1.212) represents the components of the gravitational field, which are defined as the gradient of the metric tensor[102]:

THE ENTWURF COMPONENTS OF THE GRAVITATIONAL FIELD:

$$\boxed{\Gamma^{\tau}_{\nu\sigma} = \frac{1}{2}\sum_{\mu} g^{\tau\mu}\frac{\partial g_{\mu\nu}}{\partial x_{\sigma}}.} \tag{1.213}$$

Plugging these components into the Lagrangian (1.212), Einstein derived the *Entwurf* Lagrangian[103]:

EINSTEIN'S ENTWURF LAGRANGIAN:

$$L = \frac{1}{4}\sum_{\alpha\beta\tau\nu} g^{\alpha\beta}\frac{\partial g_{\tau\nu}}{\partial x_{\alpha}}\frac{\partial g^{\tau\nu}}{\partial x_{\beta}}. \tag{1.215}$$

Einstein articulated a theorem. According to this theorem, the gravitational field, represented by the metric tensor $g_{\mu\nu}$, is varied by an infinitely small amount. When we vary this field, we replace $g_{\mu\nu}$ with $g_{\mu\nu} + \delta g_{\mu\nu}$. Doing this, we examine how $g_{\mu\nu}$ affects the action J under small perturbations. Einstein required the variation $\delta g_{\mu\nu}$ to vanish at the boundaries of a region Σ (at the boundaries of the hole). With this variation, not only the metric tensor but also the Lagrangian $L\sqrt{-g}$ and the action J become $\mathcal{L} + \delta\mathcal{L}$ and $J + \delta J$, respectively. The action J represents the integral of the Lagrangian density over spacetime. Then, the first variation of the action δJ: $\Delta(\delta J) = 0$ is stationary (zero) regardless of the specifics of $\delta g_{\mu\nu}$. However, this holds provided the coordinate systems are *adapted coordinate systems* and for any *justified transformations* (substitutions between adapted coordinate systems). Setting $\delta J = 0$ leads to the *Entwurf* field equations valid in adapted coordinate systems. The theorem states that δJ is invariant under the above variations and within these adapted coordinate systems. This means that the variation in the action does not depend on the particular choice of coordinates as long as they are adapted coordinates. Einstein argued that this invariance was necessary for demonstrating that the *Entwurf* field equations derived from the action principle were covariant (in adapted coordinate systems) [E2-14].

---

[102] Recall that Einstein chose to include the gradient of the metric tensor on the right-hand side of the energy-momentum balance (1.203) [EG14]. This led him to choose the metric tensor gradient to represent the gravitational field's components in his *Entwurf* theory.

[103] Substituting Eq. (1.213) into Eq. (1.212) we get the scalar:

$$L = -\frac{1}{4}\sum_{\mu\rho\tau\nu\alpha\beta} g^{\tau\nu}g^{\rho\alpha}g^{\mu\beta}\frac{\partial g_{\alpha\mu}}{\partial x_{\tau}}\frac{\partial g_{\beta\nu}}{\partial x_{\rho}}. \tag{1.214}$$

Plugging the derivative of the inverse metric tensor in terms of the metric tensor, [Eq. (1.115) see Sect. 1.3.2] into Eq. (1.214), simplifies the Lagrangian to the *Entwurf* Lagrangian (1.215).

Einstein proved his theorem.[104] He varied the action J, given by the integral of the Lagrangian density $\mathcal{L} = L\sqrt{-g}$ by infinitesimally changing the metric tensor $g_{\mu\nu}$. The first variation of the action δJ must vanish for any small $\delta g_{\mu\nu}$, given that these disappear at the hole's boundary Σ. By analyzing δJ, Einstein deduced the form of the *Entwurf* gravitational tensor $\mathcal{G}_{\mu\nu}$. He began with the Lagrangian, which depends on the contravariant fundamental metric tensor $g^{\mu\nu}$ and its first derivatives, $g_\sigma^{\mu\nu}$.[105] The variation of the action, as described in [E2-14], is expressed as follows[106]:

$$\delta J = \int d\tau \sum_{\mu\nu} \delta g^{\mu\nu} \left(\frac{\partial L\sqrt{-g}}{\partial g^{\mu\nu}} - \sum_\sigma \frac{\partial}{\partial x_\sigma} \left(\frac{\partial L\sqrt{-g}}{\partial g_\sigma^{\mu\nu}} \right) \right). \tag{1.216}$$

Einstein defined the Euler-Lagrange equations in the parentheses of Eq. (1.216) as the *Entwurf* gravitational tensor $\mathcal{G}_{\mu\nu}$ [E2-14]:

THE ENTWURF GRAVITATIONAL TENSOR:

$$\boxed{\mathcal{G}_{\mu\nu} = \frac{\partial L\sqrt{-g}}{\partial g^{\mu\nu}} - \sum_\sigma \frac{\partial}{\partial x_\sigma} \left(\frac{\partial L\sqrt{-g}}{\partial g_\sigma^{\mu\nu}} \right).} \tag{1.217}$$

Plugging Eq. (1.217) into Eq. (1.216) gives:

$$\delta J = \int d\tau \sum_{\mu\nu} \delta g^{\mu\nu} \mathcal{G}_{\mu\nu}. \tag{1.218}$$

Einstein identified $\mathcal{G}_{\mu\nu}$ (which later became the Einstein tensor; see discussion in Sect. 1.7.3). He concluded that $\delta g_{\mu\nu} \mathcal{G}_{\mu\nu}$ under the integral (1.218) was invariant under coordinate transformations in the adapted system[107]:

$$\frac{1}{\sqrt{-g}} \sum \delta g^{\mu\nu} \mathcal{G}_{\mu\nu}. \tag{1.219}$$

Einstein argued that this expression (1.219) was invariant because dividing a scalar density by $\sqrt{-g}$ converts it into a true tensor. Hence, by showing that the quotient (1.219) was an invariant, Einstein deduced that the ratio:

---

[104] I will discuss Einstein's proof in the context of comparing Einstein's 1914 *Entwurf* theory and his later general relativity in Sect. 1.7.3.

[105] Recall that the expression $g_\sigma^{\mu\nu} \equiv \frac{\partial g^{\mu\nu}}{\partial x_\sigma}$ represents the components of the metric tensor's derivative with respect to the coordinate x_σ.

[106] This equation represents the variation of the action for the metric tensor components $\delta g^{\mu\nu}$, corresponding to the Euler-Lagrange equations for the *Entwurf* Lagrangian density $\mathcal{L}$.

[107] Einstein argued that since $\sqrt{-g}d\tau$ is a scalar (i.e., the invariant volume element over spacetime), the integrand multiplied by the differential $d\tau$ must also form a scalar. The integral in Eq. (1.218) divided by $\sqrt{-g}$ remains invariant (the action J must be a scalar quantity or an invariant under transformations in adapted coordinate systems.

$$\frac{\mathcal{G}_{\mu\nu}}{\sqrt{-g}}, \tag{1.220}$$

must be a covariant tensor and vanish in all adapted coordinate systems [E2-14].[108]

A key point in Einstein's theorem was the restriction to adapted coordinate systems. In these systems, δJ is invariant, implying that the changes in the action due to variations in $g_{\mu\nu}$ do not depend on the specific choice of coordinates as long as these coordinates are adapted to the gravitational field [E2-14]. This was a significant step toward demonstrating the covariant nature of the field equations within these coordinate systems. However, it fell short of the general covariance later achieved in general relativity.

Using equations (1.211), (1.217), and (1.218), Einstein derived S_σ^ν:

$$S_\sigma^\nu = \sum_{\mu\tau}\left(g^{\nu\tau}\frac{\partial L\sqrt{-g}}{\partial g^{\sigma\tau}} + g_\mu^{\nu\tau}\frac{\partial L\sqrt{-g}}{\partial g_\mu^{\sigma\tau}} + \frac{1}{2}\delta_\sigma^\nu L\sqrt{-g} - \frac{1}{2}g_\sigma^{\mu\tau}\frac{\partial L\sqrt{-g}}{\partial g_\nu^{\mu\tau}}\right). \tag{1.221}$$

He stressed that the *Entwurf* gravitational tensor [Eq. (1.217)] was covariant only in adapted coordinate systems. Thus, he imposed the restriction [E2-14]:

THE B RESTRICTION TO ADAPTED COORDINATES:

$$\boxed{B_\sigma = \sum_{\alpha\sigma\nu}\frac{\partial^2}{\partial x_\sigma \partial x_\alpha}\left(g^{\nu\alpha}\frac{\partial L\sqrt{-g}}{\partial g_\sigma^{\mu\nu}}\right) = 0.} \tag{1.222}$$

Plugging the *Entwurf* components of the gravitational field (1.213) into Eq. (1.196) (the other form of the B restriction to adapted coordinates), we obtain another form of the B_σ restriction:

$$\boxed{\sum_{\alpha\beta}\frac{\partial^2}{\partial x^\alpha \partial x^\alpha}\left(\sqrt{-g}g^{\alpha\beta}\Gamma^\nu_{\beta\sigma}\right) = 0.} \tag{1.223}$$

Einstein's *Entwurf* field equations prior to imposing the restriction B_σ are the following [E2-14][109]:

THE ENTWURF FIELD EQUATIONS:

$$\boxed{\mathcal{G}_{\mu\nu} = -\kappa \mathcal{T}_\sigma^\nu.} \tag{1.224}$$

[108] If a tensor vanishes (i.e., all its components are zero) in one coordinate system, it must vanish in all coordinate systems. This is because the properties of tensors are independent of the coordinate system used to describe them.

[109] Equation (1.224) involves the correlation between Einstein's *Entwurf* gravitational tensor density Eq. (1.217) $\mathcal{G}_{\mu\nu}$ and the stress-energy tensor density $\mathcal{T}_\sigma^\nu$.

We can fully contract (sum over repeated indices) the *Entwurf* field Eq. (1.224):

$$\sum_{\tau\alpha}\left(-g^{\nu\tau}\frac{\partial L\sqrt{-g}}{\partial g^{\sigma\tau}}-g_{\alpha}^{\nu\tau}\frac{\partial L\sqrt{-g}}{\partial g_{\alpha}^{\sigma\tau}}\right)+\sum_{\alpha\tau}\frac{\partial}{\partial x_{\alpha}}\left(g^{\nu\tau}\frac{\partial L\sqrt{-g}}{\partial g_{\alpha}^{\sigma\tau}}\right)=-\kappa T_{\sigma}^{\nu}. \tag{1.225}$$

We impose the condition B_σ [Eq. (1.222)] on Eq. (1.225) by taking the divergence of both sides of this equation.[110] This simplification is represented as:

$$\text{THE SECOND TERM: } \sum_{\nu}\frac{\partial^2}{\partial x_\alpha \partial x_\alpha}\left(g^{\nu\tau}\frac{\partial L\sqrt{-g}}{\partial g_{\alpha}^{\sigma\tau}}\right)=0. \tag{1.226}$$

We then obtain the equation:

$$\boxed{\sum_{\nu}\frac{\partial}{\partial x_{\alpha}}\left(T_{\sigma}^{\nu}+\frac{1}{\kappa}\sum_{\tau\alpha}\left(-g^{\nu\tau}\frac{\partial L\sqrt{-g}}{\partial g^{\sigma\tau}}-g_{\alpha}^{\nu\tau}\frac{\partial L\sqrt{-g}}{\partial g_{\alpha}^{\sigma\tau}}\right)\right)=0.} \tag{1.227}$$

We can rewrite Eq. (1.227) by introducing t_{σ}^{ν}:

THE STRESS- ENERGY PSEUDO- TENSOR OF THE GRAVITATIONAL FIELD:

$$\boxed{-\mathbf{t}_{\alpha}^{\tau}=-\frac{1}{\kappa}\sum_{\tau\alpha}\left(g^{\nu\tau}\frac{\partial L\sqrt{-g}}{\partial g^{\sigma\tau}}+g_{\alpha}^{\nu\tau}\frac{\partial L\sqrt{-g}}{\partial g_{\alpha}^{\sigma\tau}}\right).} \tag{1.228}$$

With the newly introduced pseudo tensor, we can bring Eq. (1.227) into the form[111]:

THE CONSERVATION OF ENERGY- MOMENTUM:

$$\boxed{\sum_{\nu}\frac{\partial}{\partial x_{\alpha}}\left(T_{\sigma}^{\nu}+t_{\sigma}^{\nu}\right)=0.} \tag{1.229}$$

Einstein replaced Eqs. (1.225) and (1.227) with an equation similar to Eq. (1.194) [E2-14][112]:

---

[110] To do this, we apply the divergence operator $\frac{\partial}{\partial x_\alpha}$ to each term on both sides of the equation. The condition B_σ simplifies the second term on the left-hand side of the equation, making it equal to zero.

[111] The energy-momentum conservation Eq. (1.229) represents the result of taking the divergence of both sides of the Eq. (1.225) and imposing the condition $B_\sigma = 0$ on the left-hand side. The first term on the left-hand side of Eq. (1.229) ($\sum_\nu \partial x_\alpha T_\sigma^\nu$) represents the divergence of the stress-energy tensor. The second term ($\sum_\nu \partial x_\alpha t_\sigma^\nu$) represents the divergence of the stress-energy pseudo tensor.

[112] We start with Eqs. (1.225) and (1.227). The derivatives of L the *Entwurf* Lagrangian [Eq. (1.215)] with respect to $g^{\sigma\tau}$ and its spatial derivatives $\frac{\partial g^{\sigma\tau}}{\partial x_\alpha}$ are computed. In the *Entwurf* Lagrangian [Eq. (1.215)], variations produce terms that involve sums of products of metric components and derivatives. The variation of the Lagrangian involves terms that relate to the gravitational field components $\Gamma_{\nu\sigma}^{\tau}$ [Eq. (1.213)], leading to field equations that incorporate the $\Gamma_{\nu\sigma}^{\tau}$ explicitly.

$$\boxed{\sum_{\alpha\beta} \frac{\partial}{\partial x^{\alpha}} \left(\sqrt{-g} g^{\alpha\beta} \Gamma^{\nu}_{\beta\sigma}\right) = -\kappa(\mathcal{T}^{\nu}_{\sigma} + \mathbf{t}^{\nu}_{\sigma}),} \tag{1.230}$$

where, $\mathcal{T}^{\nu}_{\sigma}$ is the stress-energy tensor density, and t^{ν}_{σ} is the stress-energy pseudo-tensor density of the gravitational field [E2-14]. Recall that this equation is a combination of the 1913 *Entwurf* field Eqs. (1.151) and (1.152).

Einstein plugged Eq. (1.212) into the third and fourth terms on the right-hand side of Eq. (1.221) and obtained an expression for the stress-energy pseudo tensor density[113]:

THE STRESS- ENERGY PSEUDO- TENSOR OF THE GRAVITATIONAL FIELD:

$$\boxed{t^{\nu}_{\sigma} = \frac{\sqrt{-g}}{\kappa} \sum_{\mu\rho\tau\tau'} \left(g^{\nu\tau} \Gamma^{\rho}_{\mu\sigma} \Gamma^{\mu}_{\rho\tau} - \frac{1}{2} \delta^{\nu}_{\sigma} g^{\tau\tau'} \Gamma^{\rho}_{\mu\tau} \Gamma^{\mu}_{\rho\tau'}\right).} \tag{1.232}$$

In a letter dated October 12, 1915, and addressed to Lorentz, Einstein wrote the third and fourth terms on the right-hand side of Eq. (1.221) as a stress-energy pseudo-tensor [CPAE8], Doc. 129:

THE STRESS- ENERGY PSEUDO- TENSOR OF THE GRAVITATIONAL FIELD:

$$\boxed{t^{\nu}_{\sigma} = \frac{1}{2\kappa} \left(\delta^{\nu}_{\sigma} L \sqrt{-g} - \sum_{\mu\tau} g^{\mu\nu}_{\sigma} \frac{\partial L \sqrt{-g}}{\partial g^{\mu\tau}_{\nu}}\right).} \tag{1.233}$$

The pseudo-tensor is not unique; its exact form can differ depending on the definition. Still, its divergence should always yield a conservation law for the combined energy-momentum of matter and the gravitational field. This is essential for the principle of conservation of energy-momentum (1.229).

Einstein stated that the *Entwurf* gravitational tensor $\mathcal{G}_{\mu\nu}$ [Eq. (1.217)] follows the same equation form as Eq. (1.203) (the energy and momentum conservation in a gravitational field) [E2-14][114]:

---

[113] The third term of Eq. (1.221) assumes the variation of the determinant of the metric tensor:

$$\delta \sqrt{-g} = \frac{1}{2} \sqrt{-g} g^{\mu\nu} \delta g_{\mu\nu} \tag{1.231}$$

where $g^{\mu\nu}$ is the inverse of the metric tensor $g_{\mu\nu}$, and $\delta g_{\mu\nu}$ is the variation of the metric tensor components.

[114] The second term on the left-hand side of the Eq. (1.234) involves the components of the gravitational field $\Gamma^{\tau}_{\nu\sigma}$, which are defined as the gradient of the metric tensor according to Eq. (1.213). This term involves the derivatives of the metric tensor $g^{\mu\nu}$ with respect to spacetime coordinates and the *Entwurf* gravitational tensor $\mathcal{G}$. The general format of this equation is familiar from page 13*R* of the *Zurich Notebook*. Refer to Eq. (1.34), Sect. 1.2.2.

A CONSERVATION LAW FOR THE GRAVITATIONAL FIELD:

$$\sum_{\nu\tau} \frac{\partial g^{\tau\nu}}{\partial x_\nu} \mathcal{G}_{\sigma\tau} + \frac{1}{2} \sum_{\mu\nu} \frac{\partial g^{\mu\nu}}{\partial x_\sigma} \mathcal{G}_{\mu\nu} = 0. \tag{1.234}$$

With the condition (1.222) and Eq. (1.217), Einstein rewrote Eq. (1.234) as a new coordinate condition $C_\sigma = 0$ intended to restrict the choice of coordinates to those that satisfy the condition [E2-14, CPAE8], Doc. 129:

$$C_\sigma = \sum_\nu \frac{\partial S_\sigma^\nu}{\partial x_\nu} - B_\sigma = 0. \tag{1.235}$$

$C_\sigma = 0$ is a coordinate restriction rather than a consequence of the *Entwurf* field equations. S_σ^ν is given by Eq. (1.221). Einstein believed that when this condition was satisfied, the form of the theory was simplified and made valid for adapted coordinate systems [E2-14]. He emphasized that satisfying his *Entwurf* field Eqs. (1.224) necessitated fulfilling *two* conditions (or restrictions) $B_\sigma = 0$ and:

$$S_\sigma^\nu \equiv 0. \tag{1.236}$$

Thus, if $S_\sigma^\nu = 0$ in Eq. (1.235), it implies that B_σ must also be zero for the equation to be satisfied.

Einstein believed (incorrectly, as it soon turned out) that enforcing certain mathematical restrictions, like the condition (1.222), would resolve his problems: it led to a restricted form of the field equations that were covariant only under certain transformations, i.e., the adapted ones and energy-momentum conservation was satisfied for adapted coordinate systems.

1.5.3 Einstein's Recognition and Revision

During the spring-summer of 1915, the *Entwurf* theory came under intense scrutiny from Einstein's peers. They argued that Einstein's inability to produce generally covariant field equations stemmed from inherent flaws in his gravitational theory. Yet Einstein felt his approach was justifiable, chiefly due to the theorem he articulated in his *Entwurf* paper; see the discussion in Sect. 1.5.2.

Between March and May of 1915, Tulio Levi-Civita engaged in a correspondence with Einstein, expressing his concerns about Einstein's theorem and the use of adapted coordinate systems in his theoretical framework. Levi-Civita demonstrated that there were adapted coordinate systems for which the tensor (1.220) did not vanish [CPAE8], Doc. 60, 64, 66. Initially, Einstein maintained his stance steadfastly, adhering to the remnants of his *Entwurf* theory. There was a period when Einstein, influenced by Levi-Civita's insightful letters, regained complete confidence in his adapted coordinates and *Entwurf* theory, becoming even more convinced of its correctness [CPAE8], Doc. 71.

In July 1915, Einstein shifted his commitment to the *Entwurf* theory. His correspondence with Arnold Sommerfeld, who contemplated the publication of a second edition of [LMES], highlights this ambivalence. Einstein advised Sommerfeld against including his 1914 review paper on the *Entwurf* theory [E2-14], arguing that none of the current presentations were complete. Concurrently, Einstein disclosed his intentions to author a concise book to elucidate the principles of relativity theory, focusing on introducing the general theory from its inception [CPAE8], Doc. 96. This letter underscores Einstein's evolving perspective during this period—he maintained an engagement with the *Entwurf* theory. Yet, his dedication to it was waning, marking a pivotal moment in developing his theoretical framework.

Einstein delivered six lectures at Göttingen on his *Entwurf* theory. Invited by David Hilbert, Einstein's lectures, which commenced on June 28 and concluded with his return to Berlin on July 5, 1915, aimed to present his theoretical advancements. In his correspondence with Heinrich Zangger, approximately two to three weeks afterward, Einstein revealed his perception of having successfully convinced Hilbert of the theory's validity. He further noted his persuasion of Arnold Sommerfeld, while Max Planck and Max von Laue remained ostensibly detached from the theory [CPAE8], Doc. 101. Concurrently, in his letter to Wander Johannes de Haas and Geertruida de Haas, Einstein expressed satisfaction over his apparent success in convincing Hilbert and Felix Klein of his theory's merits [CPAE8], Doc. 110.

This brief period during the summer of 1915 in Göttingen appeared to offer a resurgence in accepting the *Entwurf* theory, bolstering Einstein's confidence in its consistency. However, the seeming revival of the *Entwurf* theory was ephemeral. Despite the initial resurgence, the foundational issues previously identified during Einstein's spring correspondence with Levi-Civita began to resurface following Einstein's subsequent excursion to Zurich and his return to Berlin in September 1915 [CPAE8], Doc. 111.

By October 9, 1915, Einstein found himself at a critical juncture in the evolution of his theoretical framework, particularly in dialogue with Paul Hertz. Hertz had raised objections to Einstein's use of adapted coordinate systems, presenting critiques that Einstein struggled to grasp fully. In response, Einstein queried Hertz in a letter, arguing that the law of gravitation could not be generally covariant, and sought Hertz's agreement or refutation of this position. Einstein's inquiry to Hertz underscores his continued, albeit waning, attachment to the *Entwurf* theory and indicates the emergence of significant doubts concerning its validity. His emotional tone in the correspondence, described as both nervous and cynical, further reflects the tension he experienced during this period of theoretical uncertainty.

As Einstein reported to his friends in the summer of 1915, he was under the impression that his lectures at Göttingen had convincingly articulated the merits of his *Entwurf* theory to David Hilbert. Contrary to Einstein's belief, Hilbert identified a significant error in Einstein's 1914 review paper on the theory. In 1916, Einstein explicitly recognized and acknowledged the error previously pointed out by Hilbert [CPAE8], Doc. 207. Recall that Levi-Civita also communicated critiques of Einstein's

Entwurf 1914 theory to Einstein. This collaborative critique among contemporaries was instrumental in refining Einstein's propositions, ultimately contributing to the robustness and general acceptance of his general theory of relativity.

In the autumn of 1915, the problems in the *Entwurf* theory led to the gradual decline of the theory's prominence when Einstein reevaluated the calculations in the *Einstein-Besso manuscript*. In September 1915, he wrote to Erwin Freundlich, revealing a critical contradiction he had discovered in his *Entwurf* theory. He informed Freundlich that he had identified a miscalculation. He discovered an arithmetic error, leading to the realization that the metric field describing rotation was not, in fact, a solution to his *Entwurf* field Eq. (1.180). This realization seemed to reinforce Einstein's earlier discovery that his predicted value for Mercury's perihelion motion, a mere $18''$, was also flawed. In his letter dated September 30, 1915, Einstein acknowledged to Freundlich that the shortcomings in his 1913 solution for Mercury's perihelion motion were likely due to the same computational errors [CPAE8], Doc. 123. The following day, Einstein wrote to Otto Naumann, revisiting his calculations on pages $41 - 42$ of the *Einstein-Besso manuscript*. He concluded that the metric field for a rotating system did not align with the solutions of the *Entwurf* field equations and that the solution obtained differed from what would emerge from directly transforming the Minkowski metric to a rotating coordinate system [CPAE8], Doc. 124. By re-evaluating the calculations of pages $41 - 42$ of the *Einstein-Besso manuscript*, Einstein demonstrated that the *Entwurf* field equations he had previously formulated were not satisfied [CPAE8], Doc. 124. In 1913, Einstein thought that κt_{44} had the following relation to ω^2: $\kappa t_{44} = (-)4\omega^2$ [Eq. (1.176), Sect. 1.4.2]. Einstein identified his error and subsequently documented Eq. (1.179) in place of Eq. (1.176). He then aimed to guarantee that Eq. (1.180) held for a rotating system. His new calculation yielded:

$$D_{44}(g) + \kappa t_{44} = -\Delta g_{44} - 2\omega^2 - \omega^2 = 0. \tag{1.237}$$

In 1913, Einstein held the belief that $\Delta g_{44} = -4\omega^2$ (as indicated in Eq. (1.182), Sect. 1.4.2). However, in 1915, he realized that his initial assumption regarding Δg_{44} was incorrect, and it differed from his subsequent findings:

$$\Delta g_{44} = -3\omega^2. \tag{1.238}$$

He proposed a solution for g_{44} in the form of a term proportional to the squared distance from the axis of rotation, Eq. (1.183) [Sect. 1.4.2]. By equating Eqs. (1.188) and (1.238), we can determine the value of the constant C:

$$\Delta g_{44} = 4C\omega^2 = -3\omega^2. \tag{1.239}$$

Consequently, Einstein found that the constant C has a value of $-\frac{3}{4}$. In other words, C is not equal to 1. This marks the culmination of Einstein's calculation, where he provides a solution for the metric component g_{44} for a rotating system:

$$g_{44} = 1 - \frac{3}{4}\omega^2(x^2 + y^2). \tag{1.240}$$

Einstein's above-derived solution for the g_{44} component from the *Entwurf* field equations differed from the one achieved by directly transforming the Minkowski metric into the rotating coordinate system, i.e., Eq. (1.189) [CPAE8], Doc. 123, 124.

In October 1915, Einstein realized that his Lagrangian function (1.212) did not solely lead to the *Entwurf* field equations [where $\Gamma^{\tau}_{\nu\sigma}$ is given by Eq. (1.213)]. He gradually recognized that defining $\Gamma^{\tau}_{\nu\sigma}$ in the Lagrangian as the gradient of the metric, as specified by Eq. (1.213), results in the *Entwurf* field equations. By redefining $\Gamma^{\tau}_{\nu\sigma}$ as the Christoffel symbols of the second kind [Eq. (1.245)], the same Lagrangian would naturally yield covariant field equations that are not restricted to adapted coordinate systems. This realization led him to understand that his insistence on covariance with respect to adapted coordinate systems was a flawed approach and that general covariance was necessary in his theory. Einstein recognized that he had selected an unsuitable Lagrangian function (1.215). This meant to Einstein in the autumn of 1915 that covariance with respect to adapted coordinate systems was misguided, prompting Einstein to require general covariance [CPAE8], Doc. 153, 177. It is important to note that the Lagrangian in the 1915 general relativity looks structurally similar to Eq. (1.212). However, the two underlying meanings and implications are significantly different due to the definitions of the connection terms. In the 1915 Lagrangian, Γ is the Christoffel symbol derived from the metric tensor [Eq. (1.245)].

Upon revising his understanding and correcting the errors previously embedded in his approach, Einstein sought to update Hilbert on these corrections. On November 7, 1915, he communicated these changes to Hilbert, enclosing a manuscript outlining the revised field equations and corrections to his earlier *Entwurf* 1914 review paper, particularly representing the nascent formulation of the general theory of relativity [CPAE8], Doc. 136. This exchange marked a significant turning point, as it came after Einstein recognized the fallacious aspects of his previous methodologies approximately four weeks earlier. The swift amendments to his theoretical approach were acknowledged later by Hilbert in the context of solving the precession of Mercury's perihelion[CPAE8], Doc. 149. However, this moment also signified the onset of an intellectual competition between Einstein and Hilbert, which I will briefly discuss in Sect. 1.7.2.

1.6 The 1915 General Relativity

1.6.1 Returning to General Covariance

In November 1915, after much reflection, Einstein distanced himself from the *Entwurf* field equations. He omitted any reference to the hole argument in his November 1915 theory, realizing its objective was not achieved. This argument neither protected the *Entwurf* theory from criticism nor from being debunked through observation and experiment. In his review article of 1914, Einstein further developed his hole argument. He showed that two different solutions, $G(x)$ and $G'(x)$, can coexist outside the hole (in spacetime), but inside the hole, they coincide. This suggests that, generally, covariant field equations alone cannot uniquely determine what happens inside the hole, as the choice of coordinate systems matters. After dissociating himself from the field equations of the *Entwurf* theory, Einstein grasped that there are no preferred structures or features, such as the hole in spacetime. By November 1915, and upon returning to general covariance, Einstein recognized that no preferred structures in spacetime were independent of the metric tensor field in general relativity. He realized that spacetime is not predefined but emerges from the gravitational field. The theory has no pre-existing spacetime framework until the gravitational field equations are solved. He understood that until a specific metric tensor field is defined, there is no well-defined spacetime.

Einstein's lengthy and intricate journey from 1912 culminated with revisiting his original concept—the 1912 November tensor. Instead of discarding his 1914 *Entwurf* theory, Einstein refined it in the first relativity paper he submitted on November 4, 1915, to the *Proceedings of the Royal Prussian Academy of Sciences, Berlin*. His 1916 review article, *The Foundation of the General Theory of Relativity* [E1-16] and subsequent paper, "Hamilton's Principle and the General Theory of Relativity," further elaborated this refinement [E4-16]. Analyzing the *Entwurf* theory, one can find that it already encompasses the foundational mathematical structure of general relativity, except for its field equations. Einstein's November submissions heavily drew from the 1914 *Entwurf* theory's formalism. Given that Einstein did not start from scratch, he swiftly revised the *Entwurf* theory into the new general theory of relativity, beginning this transformation in early October 1915 and completing the first November 4, 1915 paper within a single month.

So, what refinements did Einstein introduce? He proposed that his gravitational tensor be invariant under unimodular transformations.[115] In the *Entwurf* theory, Einstein introduced tensor densities, which are like tensors but have an extra factor of $\sqrt{-g}$. Considering only unimodular transformations, this factor disappears from many fundamental equations, simplifying calculations [E1-15]. By 1915, when Einstein postulated $\sqrt{-g} = 1$, he could ignore the effects of volume scaling on his equations by focusing on unimodular transformations. The equations retained their form under unimodular transformations, simplifying the mathematics and eliminating inconsistencies.

---

[115] Einstein recognized that the square root of the determinant of the metric tensor, $\sqrt{-g}$, equals 1. By postulating $\sqrt{-g} = 1$, Einstein restricted his attention to transformations with a determinant of 1. By considering only unimodular transformations, Einstein sought a middle ground between full general covariance and the simpler special relativistic invariance. Under unimodular transformations, the determinant of the metric tensor remains constant, and, in Einstein's first November paper [E1-15], it is equal to -1. Proposing that the gravitational tensor be invariant under unimodular transformations simplifies the calculations. However, while this approach makes Einstein's calculations more straightforward, it also constrains the generality of the field equations in his first November paper.

The main issue with the *Entwurf* theory was that its field equations were not generally covariant, i.e., that the field equations did not take the same form in all coordinate systems. General covariance is a property that Einstein believed a proper theory of gravitation should possess. But, at the time, he thought that achieving such covariance would lead to equations that did not conserve energy and momentum or lead to the Newtonian limit. Thus, he initially settled for the restricted covariance of the *Entwurf* theory. The main difference of the new November 4, 2015 paper compared to the previous 1914 *Entwurf* theory lies in the field equations.

In the first November 1915 paper, Einstein focused on the Ricci tensor, derived from the Riemann-Christoffel tensor (1.36) (see Sect. 1.2.2). By contracting the Riemann-Christoffel tensor, we get the Ricci tensor, a second-rank tensor (1.75) (Sect. 1.2.3):

THE RICCI TENSOR:

$$\{\rho\mu,\lambda\nu\} = \sum_{\rho} g^{\mu\rho}\left(\boxed{\frac{\partial}{\partial x_{\rho}}\Gamma^{\rho}_{\nu\sigma} - \Gamma^{\rho}_{\nu\lambda}\Gamma^{\lambda}_{\rho\sigma}} + \boxed{\Gamma^{\rho}_{\rho\lambda}\Gamma^{\lambda}_{\nu\sigma} - \frac{\partial}{\partial x_{\nu}}\Gamma^{\rho}_{\rho\sigma}}\right). \tag{1.241}$$

Equation (1.241) takes the form [E1-15][116]:

$$\{\rho\mu,\lambda\nu\} = G_{\mu\nu} = R_{\mu\nu} + S_{\mu\nu}, \tag{1.242}$$

Recall that Einstein's *Zurich Notebook* from 1912 indicates that he was already working on the decomposition of $G_{\mu\nu}$ into the Ricci tensor $R_{\mu\nu}$ and the other part, $S_{\mu\nu}$. Remember that Einstein had already worked with transformations that kept the determinant of g equal to 1 in 1912 [E3-12] (see Sect. 1.2.3). Under such transformations, all the quantities $G_{\mu\nu}$, $R_{\mu\nu}$ and $S_{\mu\nu}$ behaved as tensors. Einstein recognized that the two parts of $S_{\mu\nu}$ were tied to the derivative of the logarithm of $\sqrt{-g}$. When he postulated $\sqrt{g} = 1$, this term went to zero, leaving $R_{\mu\nu}$ as the primary curvature measure. The Ricci tensor simplifies to $R_{\mu\nu}$ in this particular instance. Once again, Einstein had performed this derivation three years earlier [E3-12] [See the Sect. 1.2.3 Eqs. (1.80) and (1.81).] Thus, if $S_{\mu\nu} = 0$, the Eq. (1.242) simplifies to the following form:

$$G_{\mu\nu} = R_{\mu\nu} = \frac{\partial}{\partial x_{\rho}}\Gamma^{\rho}_{\nu\sigma} - \Gamma^{\rho}_{\nu\lambda}\Gamma^{\lambda}_{\rho\sigma}. \tag{1.243}$$

Einstein postulated that $R_{\mu\nu}$ was his new gravitational tensor and was directly related to the stress-energy tensor $T_{\mu\nu}$. His new field equations gave this relationship:

[116] In this equation, $R_{\mu\nu}$ and $S_{\mu\nu}$ represent components of the $G_{\mu\nu}$ tensor. Specifically, $R_{\mu\nu}$ is determined by the first and second terms of the Eq. (1.80) [see Sect. 1.2.3], while $S_{\mu\nu}$ is defined by this equation's third and fourth terms. Einstein grasped that the two parts of $G_{\mu\nu}$ had distinct physical meanings [E1-15].

FIRST NOVEMBER 1915 GRAVITATIONAL FIELD EQUATION:

$$\boxed{R_{\mu\nu} = -\kappa T_{\mu\nu}.} \tag{1.244}$$

The constraint $\sqrt{g} = 1$ meant that the field equations were not generally covariant. Recall that general covariance is the principle that the laws of physics should be the same in all coordinate systems. Einstein realized this was a limitation and actively worked to generalize his theory further [E1-15].

In the *Entwurf* theory, Einstein initially expressed the components of the gravitational field in terms of the gradient of the metric tensor, as described in Eq. (1.213). This equation is based on just one term of the three that comprise the full Christoffel symbols, indicating the simplification Einstein employed at that stage. In the November 4, 1915 paper, he wrote the geodesic equation (1.210) from his 1914 *Entwurf* paper and said that the preference for using Christoffel symbols (refer to Eq. (1.76) in Sect. 1.2.3):

THE CHRISTOFFEL SYMBOLS OF THE SECOND KIND:

$$\boxed{\Gamma^{\sigma}_{\mu\nu} = -\frac{1}{2}\sum_{\alpha} g^{\sigma\alpha}\left(\frac{\partial g_{\mu\alpha}}{\partial x_{\nu}} + \frac{\partial g_{\nu\alpha}}{\partial x_{\mu}} - \frac{\partial g_{\mu\nu}}{\partial x_{\alpha}}\right),} \tag{1.245}$$

as the components of the gravitational field is particularly justified because they appear in the fundamental equations of the geodesic line. From a physical perspective, these equations represent the motion of a material point within a gravitational field [E1-15].

In the *Zurich notebook*, Einstein had already written the Riemann-Christoffel tensor using the Christoffel symbols of the second kind. However, he gave it up in favor of the *Entwurf* theory, see Sect. 1.2.3. Einstein needed to go through the *Entwurf* theory to understand the total value of the Christoffel symbols in his theory. In 1915, Einstein used the Christoffel symbols to represent the connection coefficients. The Christoffel symbols define the covariant derivative, geodesic equations, and the Riemann curvature tensor. This reflects how Einstein's understanding and the mathematical framework he used evolved. In the *Entwurf* theory, Einstein formulated a theory of gravitation without resorting to the full machinery of Riemannian geometry, which is why there were differences in the connection used. As Einstein continued his work in November 1915, he moved towards using the full Christoffel symbols and the Riemann curvature tensor in his final general relativity formulation, recognizing Riemannian geometry's significance in describing curved spacetime.

Einstein clarified that he and Grossmann had initially explored the November tensor three years before his 1915 general relativity. Still, at that time, he mistakenly believed that these field equations failed to converge with Newton's approximation. It was only in November 1915 that Einstein recognized this as an error. He acknowledged that arriving at the generally covariant field equations was comparatively straightforward. However, the real challenge, and indeed a highly demanding one, was in understanding how these equations extended the Newtonian Poisson's equation and adhered to the principles of energy-momentum conservation. Einstein reflected that the breakthrough in resolving issues with the Newtonian limit and energy-momentum conservation came with his insight that the Christoffel symbols [Eqs. (1.245) or (1.76), not the metric gradient [Eq. (1.213)], should be viewed as representing the gravitational field components. His earlier oversight regarding the significance of the Christoffel symbols had hindered his understanding of the problem, leading him into the complexities of the *Entwurf* field equations. With this newfound understanding, Einstein could derive the field equations of general relativity that successfully led to the Newton-Poisson equation [CPAE8], Doc. 153.[117]

Einstein incorporated into the energy-momentum balance in a gravitational field [Eq. (1.203)] the Christoffel symbols (1.245):

THE ENERGY- MOMENTUM BALANCE WITH CHRISTOFFEL SYMBOLS:

$$\sum_{\alpha} \frac{\partial T_{\sigma}^{\alpha}}{\partial x_{\alpha}} = -\sum_{\alpha\beta} \Gamma_{\sigma\beta}^{\alpha} T_{\alpha}^{\beta}. \tag{1.246}$$

He then wrote his field equations in the following form:

THE FIRST NOVEMBER FIELD EQUATIONS:

$$\boxed{\sum_{\alpha} \frac{\partial \Gamma_{\mu\nu}^{\alpha}}{\partial x_{\alpha}} + \sum_{\alpha\beta} \Gamma_{\mu\beta}^{\alpha} \Gamma_{\nu\alpha}^{\beta} = -\kappa T_{\mu\nu}.} \tag{1.247}$$

Mirroring his method in 1914, Einstein again used a variational approach. He demonstrated that his newly derived field equations are consistent with energy-momentum conservation:

First, he started with Eq. (1.197): the variation of the integral of a Lagrangian density $\mathcal{L}$ modified by a coupling between the metric tensor $g^{\mu\nu}$ and the stress-energy tensor $T_{\mu\nu}$ [E1-15]:

$$\delta J = \delta \left[\int \left(\mathcal{L} - \kappa \sum_{\mu\nu} g^{\mu\nu} T_{\mu\nu} \right) d\tau \right]. \tag{1.248}$$

The variation δ signifies the infinitesimal changes in the metric tensor while keeping the stress-energy tensor fixed. Einstein wrote that the Lagrangian density $\mathcal{L}$ (1.212) is given by [E1-15]:

---

[117] The underlying reasons for Einstein's three-year divergence from the November tensor, during which he grappled with a theory that, in hindsight, appears to have been a diversion, as shown in this chapter, are far more intricate than they seem at first glance. The complexity behind Einstein's departure from the November tensor and his three-year journey through non-covariant field equations before reverting to the November tensor and Ricci tensors of 1912 is multi-faceted, as I demonstrate in this book.

THE LAGRANGIAN:

$$\boxed{\mathcal{L} = \sum_{\sigma\tau\alpha\beta} g^{\sigma\tau}\Gamma^{\alpha}_{\alpha\beta}\Gamma^{\beta}_{\tau\alpha}.} \tag{1.249}$$

This is Eq. (1.212). However, unlike Eq. (1.212), the Lagrangian (1.249) involves the metric tensor $g^{\sigma\tau}$, and Christoffel symbols $\Gamma^{\alpha}_{\alpha\beta}$ and $\Gamma^{\beta}_{\tau\alpha}$.

From these two equations, Einstein derived the Euler-Lagrange equation, the representation of his field equations when derived from a variational principle with the Lagrangian $\mathcal{L}$ [E1-15][118]:

$$\sum_{\alpha} \frac{\partial}{\partial x_{\alpha}} \left(\frac{\partial \mathcal{L}}{\partial g^{\mu\nu}_{\alpha}} \right) - \frac{\partial \mathcal{L}}{\partial g^{\mu\nu}} = -\kappa T_{\mu\nu}. \tag{1.252}$$

Einstein then delved into the structure of the energy-momentum content of the gravitational field. He contracted Eq. (1.252) and obtained [E1-15]:

$$\boxed{\sum_{\alpha\mu\nu} \frac{\partial}{\partial x_{\alpha}} \left(g^{\mu\nu}_{\sigma} \frac{\partial \mathcal{L}}{\partial g^{\mu\nu}_{\alpha}} \right) - \frac{\partial \mathcal{L}}{\partial x_{\sigma}}} = -\kappa \sum_{\mu\nu} T_{\mu\nu} g^{\mu\nu}_{\sigma}. \tag{1.253}$$

He defined the stress-energy pseudo-tensor of the gravitational field [E1-15]:

THE STRESS- ENERGY PSEUDO- TENSOR OF THE GRAVITATIONAL FIELD:

$$\boxed{-2\kappa t^{\alpha}_{\sigma} = \delta^{\alpha}_{\sigma}\mathcal{L} - \sum_{\mu\nu} g^{\mu\nu}_{\sigma} \frac{\partial \mathcal{L}}{\partial g^{\mu\nu}_{\alpha}}.} \tag{1.254}$$

This is Eq. (1.233); see Sect. 1.5.2.

If we take the divergence of both sides of the Eq. (1.254) with respect to the index α, we get:

$$2\kappa \sum_{\alpha} \frac{\partial}{\partial x_{\alpha}} (t^{\alpha}_{\sigma}) = \boxed{\sum_{\alpha\mu\nu} \frac{\partial}{\partial x_{\alpha}} \left(g^{\mu\nu}_{\sigma} \frac{\partial \mathcal{L}}{\partial g^{\mu\nu}_{\alpha}} \right) - \frac{\partial \mathcal{L}}{\partial x_{\sigma}}.} \tag{1.255}$$

[118] Einstein performed the variation of the action integral J given in Eq. (1.248), and thus varied the metric $g^{\mu\nu}$ in both the gravitational Lagrangian $\mathcal{L}$ and the matter interaction term $\kappa \sum_{\mu\nu} g^{\mu\nu} T_{\mu\nu}$. The result of this is:

$$\delta\mathcal{L} - \kappa\delta \sum_{\mu\nu} g^{\mu\nu} T_{\mu\nu}. \tag{1.250}$$

The extremum condition of the action leads to the Euler-Lagrange equations:

$$\sum_{\alpha} \frac{\partial}{\partial x_{\alpha}} \left(\frac{\partial \mathcal{L}}{\partial g^{\mu\nu}_{\alpha}} \right) - \frac{\partial \mathcal{L}}{\partial g^{\mu\nu}} = 0. \tag{1.251}$$

When considering the additional term involving the stress-energy tensor $T_{\mu\nu}$, this equation modifies to Eq. (1.252).

Plugging Eq. (1.255) into Eq. (1.253) brings it to the following form:

$$\sum_{\alpha} \frac{\partial}{\partial x_\alpha}(t_\sigma^\alpha) = \boxed{\frac{1}{2} \sum_{\mu\nu} \frac{\partial g^{\mu\nu}}{\partial x_\sigma} T_{\mu\nu}.} \tag{1.256}$$

Now, Einstein reformulated his *Entwurf* [Eq. (1.203)] energy-momentum balance in a gravitational field as follows [E2-14][119]:

THE ENERGY- MOMENTUM BALANCE IN A GRAVITATIONAL FIELD:

$$\boxed{\sum_{\lambda} \frac{\partial T_\sigma^\lambda}{\partial x_\lambda} = -\frac{1}{2} \sum_{\mu\nu} \frac{\partial g^{\mu\nu}}{\partial x_\sigma} T_{\mu\nu}.} \tag{1.257}$$

Inserting Eqs. (1.257) into (1.256) leads to the total energy-momentum conservation in a gravitational field [E1-15][120]:

THE ENERGY- MOMENTUM CONSERVATION:

$$\boxed{\sum_{\nu} \frac{\partial}{\partial x_\nu}(T_\sigma^\nu + t_\sigma^\nu) = 0.} \tag{1.258}$$

In Einstein's November 4, 1915, general theory of relativity, energy-momentum conservation is derived without the adapted coordinates constraints on the coordinate system. However, Einstein still restricted the covariance of his field equations to the subclass unimodular transformations. Unlike the adapted coordinates, the restriction to unimodular coordinates was Einstein's attempt to simplify the formulation of his gravitational theory, avoiding certain complexities and achieving a conservation law for energy and momentum.

After demonstrating his field equations satisfy the principle of energy-momentum conservation, Einstein dedicated himself to deriving the weak-field limit. First, Einstein contracted the field Eq. (1.247): he multiplied the field Eq. (1.247) by the inverse metric $g^{\mu\nu}$ and then summed over both indices μ and ν to obtain a scalar equation [E1-15]:

---

119 Notice the similarity of the second term of Eq. (1.257) to the gravitational four-force density (1.17) for $\sqrt{-g} = 1$ (refer to Sect. 1.2.1). In other words, for $\sqrt{-g} = 1$, the second terms of both Eqs. (1.257) and (1.18) have a similar structure, focusing on the interaction between the metric tensor (or its inverse) and the stress-energy tensor $T_{\mu\nu}$. It is paramount to acknowledge the sequence in which the theoretical constructs were developed; importantly, the Eq. (1.18) was inscribed within Einstein's *Zurich Notebook* before Einstein established the gravitational field components as the gradient of the metric tensor. This chronological precedence underscores the intuitive grasp Einstein had in 1912 of the underlying principles of unimodular transformations ($\sqrt{-g} = 1$).

120 This expression includes a summation over all spacetime indices ν, indicating that the energy-momentum conservation holds in all spacetime directions.

$$\sum_{\alpha\beta}\frac{\partial^2 g^{\alpha\beta}}{\partial x_\alpha \partial x_\beta}-\sum_{\sigma\tau\alpha\beta} g^{\sigma\tau}\Gamma^{\alpha}_{\sigma\beta}\Gamma^{\beta}_{\tau\alpha}+\sum_{\alpha\beta}\frac{\partial}{\partial x_\alpha}\left(g^{\alpha\beta}\frac{\partial\log\sqrt{-g}}{\partial x_\beta}\right)=-\kappa T. \quad (1.259)$$

Second, Einstein multiplied the field Eq. (1.247) by the inverse metric $g^{\mu\nu}$ and summed only over index ν to obtain another form of the contracted field equations:

$$\sum_{\alpha\nu}\frac{\partial}{\partial x_\alpha}\left(g^{\nu\lambda}\Gamma^{\alpha}_{\mu\nu}\right)-\boxed{\sum_{\alpha\beta\nu} g^{\nu\beta}\Gamma^{\alpha}_{\nu\mu}\Gamma^{\lambda}_{\beta\alpha}}=-\kappa T^{\tau}_{\mu}. \quad (1.260)$$

Einstein then plugged the left-hand side of the stress-energy pseudo-tensor [Eq. (1.232)]:

$$t^{\nu}_{\sigma}-\frac{1}{2}\delta^{\nu}_{\sigma}\sum_{\mu\nu\alpha\beta} g^{\mu\nu}\Gamma^{\alpha}_{\mu\beta}\Gamma^{\beta}_{\nu\alpha}=\boxed{-\sum_{\mu\nu\alpha} g^{\mu\nu}\Gamma^{\alpha}_{\mu\sigma}\Gamma^{\nu}_{\nu\alpha},} \quad (1.261)$$

into Eq. (1.260):

$$\sum_{\alpha\nu}\frac{\partial}{\partial x_\alpha}\left(g^{\nu\lambda}\Gamma^{\alpha}_{\mu\nu}\right)-\frac{1}{2}\delta^{\lambda}_{\mu}\sum_{\mu\nu\alpha\beta} g^{\mu\nu}\Gamma^{\alpha}_{\mu\beta}\Gamma^{\beta}_{\nu\alpha}=-\kappa\left(T^{\lambda}_{\mu}+t^{\lambda}_{\mu}\right). \quad (1.262)$$

Now, Einstein took the divergence of Eq. (1.262) with respect to the spacetime coordinate x_μ:

$$\frac{\partial}{\partial x_\mu}\left(\boxed{\sum_{\alpha\nu}\frac{\partial}{\partial x_\alpha}\left(g^{\nu\lambda}\Gamma^{\alpha}_{\mu\nu}\right)}-\frac{1}{2}\delta^{\lambda}_{\mu}\sum_{\mu\nu\alpha\beta} g^{\mu\nu}\Gamma^{\alpha}_{\mu\beta}\Gamma^{\beta}_{\nu\alpha}\right)=-\kappa\frac{\partial}{\partial x_\mu}\left(T^{\lambda}_{\mu}+t^{\lambda}_{\mu}\right). \quad (1.263)$$

The energy-momentum conservation (1.258) leads to the right-hand side of Eq. (1.263) to be zero. The left-hand side of Eq. (1.263) is rearranged to[121]:

[121] The derivative of the metric tensor in terms of the Christoffel symbols is usually expressed as:

$$\frac{\partial g^{\beta\lambda}}{\partial x^\alpha}=-g^{\beta\mu}g^{\lambda\nu}\frac{\partial g_{\mu\nu}}{\partial x^\alpha}=-g^{\beta\mu}g^{\lambda\nu}\left(\Gamma^{\sigma}_{\mu\alpha}g_{\nu\sigma}+\Gamma^{\sigma}_{\nu\alpha}g_{\mu\sigma}\right). \quad (1.264)$$

If we simplify and re-examine this structure, we obtain the following identity involving Christoffel symbols:

$$g_{\nu\beta}\frac{\partial g^{\nu\lambda}}{\partial x^\alpha}=g^{\mu\lambda}\Gamma^{\nu}_{\mu\alpha}g_{\nu\beta}. \quad (1.265)$$

If both sides of the identity (1.265) are multiplied by the inverse metric tensor $g^{\beta\kappa}$ to so-called cancel $g_{\nu\beta}$, we would get:

$$g^{\beta\kappa}g_{\nu\beta}\frac{\partial g^{\nu\lambda}}{\partial x^\alpha}=g^{\beta\kappa}g^{\mu\lambda}\Gamma^{\nu}_{\mu\alpha}g_{\nu\beta}. \quad (1.266)$$

This simplifies to:

$$\delta^{\kappa}_{\nu}\frac{\partial g^{\nu\lambda}}{\partial x^\alpha}=g^{\mu\lambda}\Gamma^{\nu}_{\mu\alpha}\delta^{\kappa}_{\nu}. \quad (1.267)$$

$$\frac{\partial}{\partial x_\mu}\left(\boxed{\sum_{\alpha\beta}\frac{\partial^2 g^{\alpha\beta}}{\partial x_\alpha \partial x_\beta}}\right) - \frac{1}{2}\delta^\lambda_\mu \sum_{\mu\nu\alpha\beta}\frac{\partial}{\partial x_\mu}\left(g^{\mu\nu}\Gamma^\alpha_{\mu\beta}\Gamma^\beta_{\nu\alpha}\right) = 0. \tag{1.269}$$

The Eq. (1.269) simplifies to the following equation involving the second derivatives of the metric and derivatives of the Christoffel symbols [E1-15]:

$$\frac{\partial}{\partial x_\mu}\left(\sum_{\alpha\beta}\frac{\partial^2 g^{\alpha\beta}}{\partial x_\alpha \partial x_\beta} - \sum_{\sigma\tau\alpha\beta} g^{\sigma\tau}\Gamma^\alpha_{\beta\sigma}\Gamma^\beta_{\alpha\tau}\right) = 0. \tag{1.270}$$

The above equation implies that the entire expression enclosed within the parentheses should equal zero:

$$\sum_{\alpha\beta}\frac{\partial^2 g^{\alpha\beta}}{\partial x_\alpha \partial x_\beta} - \sum_{\sigma\tau\alpha\beta} g^{\sigma\tau}\Gamma^\alpha_{\beta\sigma}\Gamma^\beta_{\alpha\tau} = 0. \tag{1.271}$$

Plugging Eq. (1.271) into Eq. (1.259), it reduces to the following form:

$$\sum_{\alpha\beta}\frac{\partial}{\partial x_\alpha}\left(g^{\alpha\beta}\frac{\partial \log\sqrt{-g}}{\partial x_\beta}\right) = -\kappa T. \tag{1.272}$$

If $\sqrt{-g} = 1$, this equation implies that $-kT = 0$. Einstein concluded that according to this equation, it would be impossible to choose a coordinate system in which $\sqrt{-g} = 1$ [E1-15].

Einstein wrote the first term of the Eq. (1.271) while disregarding the second term [E1-15]:

$$\sum_{\alpha\beta}\frac{\partial^2 g^{\alpha\beta}}{\partial x_\alpha \partial x_\beta} = 0. \tag{1.273}$$

He contended that Eq. (1.273) serves as a first-order approximation of Eq. (1.271). He asserted that, when considering the first approximation of (1.273), we possess the liberty to impose the following condition [E1-15]:

Recall that δ^κ_ν is the Kronecker delta, which reduces the equation by replacing ν with κ everywhere it appears. This leads to:

$$\frac{\partial g^{\kappa\lambda}}{\partial x^\alpha} = g^{\mu\lambda}\Gamma^\kappa_{\mu\alpha}. \tag{1.268}$$

We must match the indices correctly and ensure that all differentiation terms are consistent. Converting the indices appropriately and plugging this equation into the first term (the boxed term) on the left-hand side of Eq. (1.263) gives the first term (the boxed term) on the left-hand side of Eq. (1.269).

$$\boxed{\sum_{\beta}\left(\frac{\partial g^{\alpha\beta}}{\partial x_{\alpha}}\right) = 0.} \tag{1.274}$$

This is the familiar harmonic condition, Eq. (1.28) [see Sect. 1.2.2], which Einstein resurrected from his 1912 *Zurich notebook*, rescuing it from obscurity. If we take the derivative of this equation for x_{ν} we obtain Eq. (1.273).

Einstein then derived the equation:

$$\frac{1}{2}\sum_{\alpha}\frac{\partial^2 g_{\mu\nu}}{\partial x_{\alpha}^2} = \kappa T_{\mu\nu}, \tag{1.275}$$

from his field Eq. (1.247).[122] The left-hand side of Eq. (1.275) [Eq. (1.280), see footnote 122.] is the Laplacian-like term [see Eq. (1.85) in Sect. 1.2.3.]

From Eq. (1.275), Einstein derived the Newton-Poisson equation as a first-order approximation [E1-15].[123]

---

[122] We begin with the field equation:

$$\sum_{\alpha}\frac{\partial \Gamma^{\alpha}_{\mu\nu}}{\partial x_{\alpha}} = -\kappa T_{\mu\nu}, \tag{1.276}$$

and expand the Christoffel symbols $\Gamma^{\alpha}_{\mu\nu}$ in terms of their components. Starting with Eq. (1.245), we will first work on the first term involving: $\frac{\partial g_{\beta\mu}}{\partial x_{\nu}}$. We expand the first term:

$$\frac{1}{2}g^{\alpha\beta}\frac{\partial g_{\beta\mu}}{\partial x_{\nu}} = \frac{\partial g_{\beta\mu}}{\partial x^{\nu}}. \tag{1.277}$$

Now, we expand the second term:

$$\frac{1}{2}g^{\alpha\beta}\frac{\partial g_{\beta\nu}}{\partial x_{\mu}} = \frac{\partial g_{\beta\nu}}{\partial x^{\mu}}. \tag{1.278}$$

Next, we rewrite the Christoffel symbols with these expanded terms:

$$\Gamma^{\alpha}_{\mu\nu} = \frac{1}{2}\frac{\partial g_{\beta\mu}}{\partial x^{\nu}} + \frac{1}{2}\frac{\partial g_{\beta\nu}}{\partial x^{\mu}} - \boxed{\frac{1}{2}\frac{\partial g_{\mu\nu}}{\partial x_{\beta}}.} \tag{1.279}$$

To express the third term as the second partial derivative of the metric tensor components with respect to x_{α}, we take the derivative of the third term:

$$\frac{1}{2}\sum_{\alpha}\frac{\partial^2 g_{\mu\nu}}{\partial x_{\alpha}^2}. \tag{1.280}$$

[123] This is my reconstruction, which is based on an analysis of a few papers authored by Einstein, for example, see [E1-16]. In Einstein's later derivations, he considered the line element Eq. (1.3) (refer to Sect. 1.2.1). For a stationary observer (whose spatial coordinates do not change, $dx_{\mu} = 0$ for $\mu = 1, 2, 3$). The line element simplifies to:

In an Addendum to the November 4, 1915 paper, submitted on November 11, 1915, Einstein showed that the Eq. (1.272) cannot be satisfied for all coordinate systems.[124] When the divergence term in Eq. (1.272) goes to zero, then this leads directly to generally covariant field equations.[125] The field equations become coordinate-independent, i.e., they have the same form in all coordinate systems [E2-15]:

THE GENERALLY COVARIANT FIELD EQUATIONS:

$$\boxed{G_{\mu\nu} = \kappa T_{\mu\nu}.} \tag{1.287}$$

$$ds^2 = g_{44}dx_4^2. \tag{1.281}$$

The g_{44} component of the metric tensor thus directly affects the measurement of time intervals in the presence of a gravitational field. In flat Minkowski spacetime (absence of gravity), $g_{44} = -1$, and the line element reduces to $ds^2 = -dt^2$. In a weak gravitational field and nearly flat spacetime, the metric can be approximated as:

$$g_{44} = -\left(1 + \frac{2\Phi}{c^2}\right), \quad \text{which implies that:} \tag{1.282}$$

$$g_{44} \approx \frac{2\Phi}{c^2}. \tag{1.283}$$

In the weak field approximation, the Eq. (1.275) becomes Eq. (1.161):

$$\frac{1}{2}\nabla^2 g_{44} = \kappa\rho c^2. \tag{1.284}$$

The right-hand side of Eq. (1.275) is proportional to the energy density, which can be related to the mass density ρ by $T_{44} \approx \rho c^2$ (since $T_{\mu\nu}$ in the Newtonian limit is dominated by the mass-energy component). This is consistent with the pressureless dust approximation, where the mass density ρ directly influences the gravitational field, but the pressure and other forms of stress-energy are negligible. Equation (1.284) is analogous to the Poisson equation in Newtonian gravity (1.27), suggesting that g_{44} is related to the gravitational potential Φ. We set: $\kappa = \frac{8\pi G}{c^4}$ [Eq. (1.48)], and by direct comparison of the structure of the two Eqs. (1.284) and (1.27):

$$\frac{1}{2}\nabla^2 g_{44} = \frac{8\pi G}{c^4}\rho c^2, \qquad \nabla^2\Phi = 4\pi G\rho, \tag{1.285}$$

we see that g_{44} is related to the gravitational potential Φ by Eq. (1.283):

$$\Phi = \frac{c^2}{2}g_{44}. \tag{1.286}$$

[124] That is because if $\sqrt{-g} = 1$ in a coordinate system, then $\log 1 = 0$, and the right-hand side of the equation becomes $\kappa T = 0$. This would imply that the trace of the stress-energy tensor $\sum_\sigma T_\sigma^\sigma$ must vanish in that coordinate system. Hence, Eq. (1.272) is a kind of coordinate condition that cannot be satisfied in all coordinate systems. Einstein then wrote the full field equation (in the presence of matter and energy).

[125] This simplification means that any additional terms dependent on the derivatives of $\sqrt{-g}$ disappear, leaving the $G_{\mu\nu}$ unaffected by changes in the determinant of the metric tensor.

In a vacuum (where $T_{\mu\nu} = 0$), this reduces to:

THE GENERALLY COVARIANT VACUUM FIELD EQUATIONS:

$$\boxed{G_{\mu\nu} = 0.} \tag{1.288}$$

Einstein concluded that if $\sqrt{-g} \neq 1$, then the $S_{\mu\nu}$ term retains its presence within Eq. (1.242), which subsequently evolves into Eqs. (1.287) and (1.288). Conversely, reinstating the condition $\sqrt{-g} = 1$ redirects the analysis to the November 4, 1915 field Eqs. (1.243) and (1.244). In other words, Einstein could choose a coordinate system where $\sqrt{-g} = 1$ holds. Such a choice simplified his calculations. However, it did not restrict the covariance of the field Eq. (1.287) [E2-15]. In this coordinate system, the vacuum version of Eq. (1.247) is[126]:

THE VACUUM FIELD EQUATIONS (THE SECOND NOVEMBER FIELD EQUATIONS):

$$\boxed{R_{\mu\nu} = \sum_{\alpha} \frac{\partial \Gamma^{\alpha}_{\mu\nu}}{\partial x_{\alpha}} + \sum_{\alpha\beta} \Gamma^{\alpha}_{\mu\beta} \Gamma^{\beta}_{\nu\alpha} = 0, \quad \sqrt{-g} = 1.} \tag{1.289}$$

1.6.2 *Broadening the Covariance of the Field Equations*

As his work progressed, Einstein discovered that the full, unrestricted theory naturally warranted energy-momentum conservation, provided he included the trace of the energy-momentum tensor T in his field equations. This realization led him to the generally covariant field equations [E4-15][127]:

THE FINAL FORM OF EINSTEIN'S 1915 FIELD EQUATIONS:

$$\boxed{G_{\mu\nu} = -\kappa \left(T_{\mu\nu} - \frac{1}{2} g_{\mu\nu} T \right).} \tag{1.290}$$

$G_{\mu\nu}$ is the Ricci tensor. Equation (1.290) are generally covariant.

[126] Even if, in a particular calculation, we choose $\sqrt{-g} = 1$, this is just a local property of the coordinates we have chosen for convenience. One can always transform to another coordinate system where $\sqrt{-g} \neq 1$. The physical predictions of Einstein's theory remain invariant under such transformations. The equations' covariance means that the equations' form remains unchanged under any coordinate transformation.

[127] As mentioned in Sect. 1.2.3, Einstein had previously formulated a first-order approximation version of Eq. (1.73) in his *Zurich notebook* of 1912 that bears a striking resemblance in form to his field Eq. (1.290) of 1915. Whether Einstein retained this knowledge from 1912 or had completely forgotten about it, one cannot deny that his act of formulating such a field equation in 1912 indicates his thought process.

The trace of the stress-energy tensor $T_{\mu\nu}$ is represented by[128]:

$$\boxed{T = \sum_{\rho\sigma} g^{\rho\sigma} T_{\rho\sigma} = \sum_{\sigma} T^{\sigma}_{\sigma}.} \tag{1.291}$$

Einstein imposed the unimodular condition $\sqrt{-g} = 1$, which led him to a revised form of the field Eq. (1.247) [and the vacuum field Eq. (1.289)] [E4-15]:

THE THIRD NOVEMBER FIELD EQUATIONS:

$$\boxed{R_{\mu\nu} = \sum_{\lambda} \frac{\partial \Gamma^{\lambda}_{\mu\nu}}{\partial x_{\lambda}} + \sum_{\rho\lambda} \Gamma^{\lambda}_{\mu\rho} \Gamma^{\rho}_{\nu\lambda} = -\kappa\left(T_{\mu\nu} - \frac{1}{2} g_{\mu\nu} T\right), \quad \sqrt{-g} = 1.} \tag{1.292}$$

He obtained the contracted form of Eq. (1.292) [E4-15][129]:

$$\sum_{\alpha\beta} \frac{\partial^2 g^{\alpha\beta}}{\partial x_{\alpha} \partial x_{\beta}} - \sum_{\sigma\tau\alpha\beta} g^{\sigma\tau} \Gamma^{\alpha}_{\sigma\beta} \Gamma^{\beta}_{\tau\alpha} + \sum_{\alpha\beta} \frac{\partial}{\partial x_{\alpha}} \left(g^{\alpha\beta} \frac{\partial \log(\sqrt{-g})}{\partial x_{\beta}} \right) = \kappa T. \tag{1.293}$$

Since $\sqrt{-g} = 1$, the third term on the left-hand side of Eq. (1.293) drops out, leaving us with the following equation:

$$\sum_{\alpha\beta} \frac{\partial^2 g^{\alpha\beta}}{\partial x_{\alpha} \partial x_{\beta}} - \sum_{\sigma\tau\alpha\beta} g^{\sigma\tau} \Gamma^{\alpha}_{\sigma\beta} \Gamma^{\beta}_{\tau\alpha} = \kappa T. \tag{1.294}$$

Using Eq. (1.232), Einstein rewrote the second term as[130]:

$$\kappa t = \sum_{\sigma\tau\alpha\beta} g^{\sigma\tau} \Gamma^{\alpha}_{\sigma\beta} \Gamma^{\beta}_{\tau\alpha}. \tag{1.295}$$

---

[128] The left-hand side $\sum_{\rho\sigma} g^{\rho\sigma} T_{\rho\sigma}$ uses the indices ρ and σ to sum over all components of the stress-energy tensor $T_{\mu\nu}$ with the metric tensor $g^{\rho\sigma}$ contracting the indices. This notation reflects Einstein's use of the summation symbol $\sum$ at the time. It is worth noting that Einstein had not yet introduced the Einstein summation convention, which he would invent less than a year later. With the summation convention, repeated indices imply summation. The second part $\sum_{\sigma} T^{\sigma}_{\sigma}$ is a simplified way to express the same trace by summing over the diagonal components of $T_{\mu\nu}$ where σ appears as both indices. It emphasizes that we are summing only over the components where μ and ν are equal, which correspond to the diagonal elements of the tensor.

[129] Einstein contracted the indices μ and ν, summing over them. He rearranged the terms to separate the Christoffel symbols and the metric tensor terms.

[130] The rationale behind equating the first term in Eq. (1.232) to κT is a first approximation of Einstein's field equations (1.292).

Substituting Eq. (1.295) into Eq. (1.294), gives:

$$\sum_{\alpha\beta} \frac{\partial^2 g^{\alpha\beta}}{\partial x_\alpha \partial x_\beta} - \kappa t = \kappa T. \tag{1.296}$$

The first term on the left-hand side of Eq. (1.296) involves second-order partial derivatives of the metric tensor components $g^{\alpha\beta}$ with respect to the coordinates x_α and x_β. This means that if this term is zero in Eq. (1.296), then the condition:

$$\kappa(t + T) = 0, \tag{1.297}$$

is fulfilled. Taking the divergence of both sides of (1.297) satisfies the energy-momentum conservation (1.258).

1.6.3 Calculating Mercury's Perihelion Advance

In 1915, Einstein correctly explained the anomalous perihelion precession of Mercury [E3-15]. He started with his vacuum field Eq. (1.289) and imposed the determinant condition from his November 11, 1915 paper, "On the General Theory of Relativity (Addendum)" [E2-15]. which implies that the determinant of the metric tensor equals -1 [E3-15]:

THE DETERMINANT OF THE METRIC CONDITION:

$$\boxed{|g_{\mu\nu}| = -1.} \tag{1.298}$$

Equation (1.298) represents the condition for unimodular coordinates[131]:

$$\boxed{\sqrt{-g} = 1.} \tag{1.299}$$

The condition (1.299) was necessary for Einstein's calculation of Mercury's perihelion advance. When Einstein worked on explaining the perihelion of Mercury, he was dealing with a very specific and complex problem. The orbit of Mercury deviates slightly from what Newtonian mechanics predicts, a discrepancy that needs a precise explanation. In this context, Einstein used the unimodular condition (1.299) as part of an approximate method to simplify the problem. This approach allowed him to derive a solution that could explain the observed perihelion shift without delving into a more comprehensive solution. By focusing on an approximate solution, Einstein acknowledged that his approach was a stepping stone toward a more complete under-

131 $|g_{\mu\nu}|$ represents the absolute value of the determinant of the metric tensor. If the determinant of the metric tensor $g_{\mu\nu}$ is equal to -1, then taking the square root of the negative determinant would give 1. This can be expressed as: $\sqrt{|g_{\mu\nu}|} = \sqrt{-(-1)} = 1$.

standing. Guided by this belief, Einstein clarified that he planned to focus solely on deriving a solution for now, without delving into whether it is the sole possible option [E3-15]. By not questioning the uniqueness of his solution at that time, Einstein was accepting that his explanation, though valid and successful in explaining Mercury's orbit, might be one of several possible solutions within the broader framework of his general relativity.

Einstein put forward four criteria for the Sun's gravitational field [E3-15]:

(1) All components of $g_{\mu\nu}$ are independent of x_4 (time independence). This implies a static field.
(2) $g_{\rho 4} = g_{4\rho} = 0$ for $\rho = 1, 2, 3$. This criterion simplifies the metric by eliminating terms that mix spatial and temporal components.
(3) Spherical symmetry.
(4) The solution vanishes at infinity. This criterion guarantees that the spacetime metric approaches the flat Minkowski metric at great distances from the mass. Under these conditions and to the first approximation, Einstein found the metric [E3-15][132]:

THE METRIC IN THE FIRST APPROXIMATION:

$$\boxed{g_{\rho\sigma} = -\delta_{\rho\sigma} + \alpha\left(\frac{\partial^2 r}{\partial x_\rho \partial x_\sigma} - \frac{\delta_{\rho\sigma}}{r}\right) = -\delta_{\rho\sigma} - \alpha\frac{x_\rho x_\sigma}{r^3}, \quad g_{44} = 1 - \frac{\alpha}{r}.} \tag{1.300}$$

At this juncture, I will diverge to analyze this equation for future reference in the subsequent chapters. The right-hand side of the metric given in Einstein's first-order approximate solution (1.300) represents a linearized approximation of the gravitational field around a spherically symmetric mass, such as the Sun, in the weak-field limit[133]:

Einstein transformed the left-hand side terms in Eq. (1.300):

---

[132] r is the radial distance from the Sun, and α is a parameter defined by Eq. (1.164)). The first term on the right-hand side, $-\delta_{\rho\sigma}$, represents the Minkowski metric component, indicating a flat spacetime background without gravitational effects. The second term on the right-hand side, $\alpha\frac{x_\rho x_\sigma}{r^3}$, contributes to the metric tensor components that involve the constant α and the product of two coordinates x_ρ and x_σ, divided by the cube of the radial distance from the Sun. r denotes a radial coordinate, particularly in spherically symmetric spacetimes. $\frac{\partial^2 r}{\partial x_\rho \partial x_\sigma}$ is a second-order spatial derivative of the radial coordinate. In his 1915 papers, Einstein derived terms with second derivatives in the weak field regime [see, for instance, Eq. (1.273)] [E1-15]. Indeed, second derivatives of the metric components naturally appear in the weak field approximation. Finally, the third term on the left-hand side, the term $\frac{\delta_{\rho\sigma}}{r}$: in the diagonal components of the metric tensor, this term would introduce a scaling factor that varies inversely with r. In the off-diagonal components (where $\rho \neq \sigma$), this term would contribute nothing since $\delta_{\rho\sigma}$ would be zero.

[133] Refer to Eq. (1.41) in Sect. 1.2.1:

$$g_{\rho\sigma} = -\delta_{\rho\sigma} + \text{A DEVIATION}. \tag{1.301}$$

$$-\delta_{\rho\sigma}+\alpha\left(\frac{\partial^2 r}{\partial x_\rho \partial x_\sigma}-\frac{\delta_{\rho\sigma}}{r}\right), \tag{1.302}$$

into the right-hand side terms[134]:

---

[134] The transformation from Eqs. (1.302) to (1.314) is performed as follows:

$$\text{We start by finding } \frac{\partial r}{\partial x}: \quad \frac{\partial r}{\partial x}=\frac{\partial}{\partial x}\sqrt{x^2+y^2+z^2}. \tag{1.303}$$

Given the function (1.315), using the chain rule:

$$\frac{\partial r}{\partial x}=\frac{\partial r}{\partial u}\cdot\frac{\partial u}{\partial x}, \text{ where:} \tag{1.304}$$

$$u=x^2+y^2+z^2. \tag{1.305}$$

$$\frac{\partial r}{\partial u}=\frac{1}{2\sqrt{u}} \quad \text{since } r=\sqrt{u}. \tag{1.306}$$

$$\frac{\partial u}{\partial x}=2x, \tag{1.307}$$

since u is given by Eq. (1.305) and y, z are constants with respect to x.

Now we combine Eqs. (1.306) and (1.307) and plug them into Eq. (1.304), which involves the chain rule:

$$\frac{\partial r}{\partial x}=\frac{1}{2\sqrt{u}}\cdot 2x=\frac{x}{\sqrt{u}}. \quad \text{We substitute Eqs. (1.305) and (1.315) back:} \tag{1.308}$$

$$\frac{\partial r}{\partial x}=\frac{x}{\sqrt{x^2+y^2+z^2}}. \quad \text{And finally simplify:} \tag{1.309}$$

$$\frac{\partial r}{\partial x}=\frac{x}{r}. \tag{1.310}$$

We can write Eq. (1.310) equivalently as:

$$\frac{\partial r}{\partial x}=\frac{1}{2}\cdot\frac{1}{\sqrt{x^2+y^2+z^2}}\cdot 2x=\frac{x}{\sqrt{x^2+y^2+z^2}}=\frac{x}{r}. \tag{1.311}$$

Then, applying the product rule and plugging Eq. (1.311) into the second derivative of r with respect to x, we calculate the second derivative:

$$\frac{\partial^2 r}{\partial x^2}=\frac{\partial}{\partial x}\left(\frac{x}{r}\right)=\frac{\partial}{\partial x}\left(x\cdot\frac{1}{r}\right)=\frac{\partial x}{\partial x}\cdot\frac{1}{r}+x\cdot\frac{\partial}{\partial x}\left(\frac{1}{r}\right)=\frac{1}{r}+x\cdot\left(-\frac{x}{r^3}\right). \tag{1.312}$$

$$\text{We have calculated that:} \quad \frac{\partial^2 r}{\partial x^2}=\frac{1}{r}+x\cdot\left(-\frac{x}{r^3}\right). \tag{1.313}$$

Inserting Eq. (1.313) into (1.302) becomes: $-\delta_{\rho\sigma}+\alpha\left[\left(\frac{1}{r}+x\cdot\left(-\frac{x}{r^3}\right)\right)-\frac{\delta_{\rho\sigma}}{r}\right]$. In this equation, the second and fourth terms cancel out because $\delta_{\rho\sigma}$ is 1 when $\rho=\sigma$ and 0 otherwise. Therefore, we have:

$$-\delta_{\rho\sigma}+\alpha\left(x\cdot\left(-\frac{x}{r^3}\right)\right). \quad \text{This leads us to Eq. (1.314).}$$

$$-\delta_{\rho\sigma} - \alpha\frac{x_\rho x_\sigma}{r^3}, \tag{1.314}$$

where Einstein defined:

$$r = \sqrt{x^2 + y^2 + z^2}. \tag{1.315}$$

The spatial components of Eq. (1.300):

$$g_{\rho\sigma} = -(\delta_{\rho\sigma} + \alpha\frac{x_\rho x_\sigma}{r^3}), \tag{1.316}$$

can be written in the following form [E1-16][135]:

$$\boxed{g_{11} = -\left(1 + \frac{\alpha}{r}\right).} \tag{1.319}$$

Since the third component g_{44} in Eq. (1.300) is [E1-16][136]:

$$\boxed{g_{44} = \left(1 - \frac{\alpha}{r}\right),} \tag{1.320}$$

Einstein's first-order approximation solution, as shown in (1.300), can be expressed in the following manner [Ea-Jan]:

$$ds^2 = (1 - \frac{\alpha}{r})c^2dt^2 - \sum_{\rho\sigma}(\delta_{\rho\sigma} + \alpha\frac{x_\rho x_\sigma}{r^3})dx_\rho dx_\sigma. \tag{1.321}$$

However, in his perihelion of Mercury paper [E3-15], Einstein did not express his approximate solution as a line element. The line element (1.321) describes a spacetime with a correction term to the flat spacetime metric involving the parameter α

[135] g_{11} represents the component of the metric tensor associated with the radial dimension (i.e., the r coordinate) in spherical coordinates. It describes how distances are measured along the radial direction. g_{11} modifies how lengths and distances are calculated near a massive object. Let us derive Eq. (1.319) from Eq. (1.316). To find g_{11}, we would need to set $\rho = \sigma = 1$ in Eq. (1.316). Assuming that the coordinates are in a standard spherical system (where 1 represents the radial coordinate), $x_\rho = x_\sigma = r$, Eq. (1.316) becomes:

$$-(\delta_{11} + \alpha\frac{r^2}{r^3}). \tag{1.317}$$

Since $\delta_{11} = 1$, (because $\rho = \sigma$), the Eq. (1.317) simplifies to:

$$-(1 + \alpha\frac{1}{r}), \tag{1.318}$$

which is Einstein's equation (1.319). In 1916, using Eq. (1.319), Einstein calculated length contraction [E1-16].

[136] g_{44} in Eq. (1.320) affects the measurement of time intervals. It represents how time dilates in a gravitational field. This is further discussed in Sect. 1.7.1.

and coordinates x_ρ and x_σ. It can be expressed in spherical coordinates [Ea-Jan][137]:

$$\boxed{ds^2 = (1 - \frac{\alpha}{r})dt^2 - (1 + \frac{\alpha}{r})dr^2 - r^2(d\theta^2 + \sin^2\theta d\phi^2).} \tag{1.322}$$

Einstein did not write Eq. (1.322) in his perihelion of Mercury paper because he imposed the determinant condition (1.299).[138]

Einstein combined the metric in the first-order approximation (1.300) and the Christoffel symbols (1.245). This yielded the components of the gravitational field in the first order approximation [E3-15]:

$$\Gamma^\tau_{\rho\sigma} = -\alpha\left(\delta_{\rho\sigma}\frac{x_\tau}{r^2} - \frac{3}{2}\frac{x_\rho x_\sigma x_\tau}{r^5}\right). \tag{1.326}$$

He considered indices $\rho, \sigma, \tau \in 1, 2, 3$. For the gravitational field's time components, Γ^σ_{44}, the following equation applies:

COMPONENTS OF THE GRAVITATIONAL FIELD (FIRST ORDER APPROXIMATION):

$$\Gamma^\sigma_{44} = -\frac{\alpha}{2}\frac{x_\sigma}{r^3}. \tag{1.327}$$

By only considering terms where $\mu = \nu = 4$ on the left-hand side of Eq. (1.289), and inserting the results from Eq. (1.327), Einstein derived [E3-15]:

$$\sum_\sigma \frac{\partial \Gamma^\sigma_{44}}{\partial x_\sigma} = \frac{\alpha^2}{2r^4}. \tag{1.328}$$

137 Einstein wrote the first two terms of this line element in his 1916 gravitational waves paper. Refer to Eq. (3.3) in Sect. 3.2.1 and the discussion in Sect. 3.2.3.

138 Indeed, the metric tensor $g_{\rho\sigma}$, Eq. (1.322):

$$g_{\rho\sigma} = \begin{pmatrix} 1+\frac{\alpha}{r} & 0 & 0 & 0 \\ 0 & -1 & 0 & 0 \\ 0 & 0 & -1 & 0 \\ 0 & 0 & 0 & 1-\frac{\alpha}{r} \end{pmatrix} \tag{1.323}$$

does not satisfy the unimodular condition (1.299). The determinant of a diagonal matrix is the product of Eq. (1.322)'s diagonal elements. So, the determinant g of matrix (1.323) is:

$$g = \left(1 + \frac{\alpha}{r}\right) \times (-1) \times (-1) \times \left(1 - \frac{\alpha}{r}\right) = \frac{\alpha^2}{r^2} - 1. \tag{1.324}$$

The square root of the negative determinant of the metric tensor is:

$$\sqrt{-g} = \sqrt{1 - \frac{\alpha^2}{r^2}}. \tag{1.325}$$

This expression does not simplify to 1. Thus, (1.322) does not satisfy the unimodular condition (1.299).

This led to the second-order approximation:

COMPONENTS OF THE GRAVITATIONAL FIELD (SECOND ORDER APPROXIMATION):

$$\Gamma_{44}^{\sigma} = -\frac{\alpha}{2}\frac{x_\sigma}{r^3}\left(1 - \frac{\alpha}{r}\right). \tag{1.329}$$

After obtaining the necessary gravitational field components, Einstein utilized the geodesic Eq. (1.210) to describe the motion of a point mass (i.e., a planet) in the curved space-time around the Sun [E3-15].

Einstein assumed that the planet's velocity was much less than the speed of light.[139] Using the previously derived gravitational field components Γ_{44}^{ν}, Eq. (1.327), he arrived at the equations of motion [E3-15][140]:

EQUATIONS RESEMBLING NEWTON'S SECOND LAW WITH GRAVITATIONAL POTENTIAL:

$$\begin{aligned} \frac{d^2 x_\nu}{ds^2} &= \Gamma_{44}^{\nu} = -\frac{\alpha}{2}\frac{x_\nu}{r^3}. \quad \text{for } \nu = 1, 2, 3, \\ \frac{d^2 x_4}{ds^2} &= 0. \end{aligned} \tag{1.330}$$

Einstein derived a form resembling Newton's second law, where the potential Φ is: $\Phi = -\frac{\alpha}{2r}$. Suppose we set $s = x_4$ to a first-order approximation. In that case, this simplifies Eq. (1.330) to Newton's second law with gravitational potential[141]:

NEWTON'S SECOND LAW WITH GRAVITATIONAL POTENTIAL:

$$\boxed{\frac{d^2 x_\nu}{dt^2} = \frac{\Phi x_\nu}{r^2}.} \tag{1.331}$$

Einstein then expressed three key concepts [E3-15]:

(1) *The area law* in polar coordinates (i.e., Kepler's second law) B: This states that the area swept out by the planet as it orbits the Sun remains constant[142]:

---

139 Hence, dx_1, dx_2, dx_3 are considered smaller than dx_4. dx_4 refers to the time component, and this assumption is equivalent to saying that the spatial velocities are much smaller than the speed of light.

140 These equations represent how a test particle (or planet) would move under the influence of the gravitational field near the Sun. The equation $\frac{d^2 x_4}{ds^2} = 0$ signifies that there is not any acceleration in the time component.

141 x_ν represents the position of a particle, r represents the distance between the center of mass and the particle, and the equation describes the motion of a particle under the influence of a spherically symmetric gravitational field.

142 r is the radial distance between the planet and the Sun, and $\frac{d\phi}{ds}$ represents the rate of change of the angular coordinate ϕ for the path length s of the planet. B is the area law constant.

EINSTEIN'S AREA LAW:

$$\boxed{B = r^2 \frac{d\phi}{ds}.} \tag{1.332}$$

Consider Besso's area law (1.168). We need additional assumptions to obtain Einstein's area law (1.332). Besso's equation alone does not directly lead to Einstein's equation (1.332). To make that connection, we need more context related to B, r, ϕ, and s. s is related to Einstein's geodesic equation, as it represents the parameterization of the path followed by a particle moving under the influence of gravity, such as a planet orbiting the Sun. Einstein used the geodesic equation [Eq. (1.210)] and the Christoffel symbols [Eq. (1.326)] to calculate the motion of Mercury in a curved spacetime caused by the Sun's gravitational field. In 1915, these Christoffel symbols described the gravitational field components of the static Sun.

In the 1913 *Entwurf* theory, Einstein did not write the geodesic equation [EG13]. It was only after a year, in the 1914 *Entwurf* review paper, "The Formal Foundation of the General Theory of Relativity," that Einstein derived the geodesic Eq. (1.210) in terms of the Christoffel symbols (1.245) [E2-14] (refer to Sect. 1.5.2).

(2) *The energy* E: A combination of kinetic and potential energies[143]:

$$\text{THE ENERGY LAW:} \quad E = \frac{1}{2}u^2 + \Phi. \tag{1.333}$$

In 1913, Besso wrote Einstein's *Entwurf* Eq. (1.7) and the temporal part (the g_{44} component of the metric tensor) of Eq. (1.6) [see Eq. (1.167), Sect. 1.4.1]. Besso's equation (1.167) pertains to the energy in a *static gravitational field*. On the other hand, in 1915, Einstein used a fundamentally different approach to describing energy, a *Newtonian approximation*. He wrote Eq. (1.333) describing the total energy E grounded in the *Newtonian mechanics* framework, wherein the total energy E of a particle is expressed as a sum of its kinetic and potential energy components. Such an equation serves as a bridge between classical mechanics and the relativistic treatment of gravity. The importance of this equation in the context of Mercury's perihelion advance lies in its ability to account for the relativistic corrections to the classical gravitational potential energy. The relativistic correction to the potential Φ leads to a more accurate prediction of Mercury's orbit, including the observed advance of its perihelion, which Newtonian mechanics could not explain alone. The gravitational potential Φ represents the potential energy per unit mass in a gravitational field. In a weak-field approximation, Einstein shows that his equations reduce to a form that resembles Newtonian gravity, with corrections for relativistic effects.

---

[143] This equation represents energy conservation, i.e., the total energy (E) in terms of the kinetic energy $\frac{1}{2}u^2$ and the gravitational potential energy Φ.

(3) *The effective velocity of the planet*[144]:

$$u^2 = \frac{dr^2 + r^2 d\phi^2}{ds^2}. \tag{1.334}$$

Einstein used the geodesic Eq. (1.210) to calculate the planet's motion. He considered the first-order gravitational field components Γ^{σ}_{44} derived previously by Eq. (1.327). He then incorporated second-order effects expressed by Eq. (1.329) [E3-15][145]:

RELATIVSTIC EQUATION OF MOTION:

$$\frac{d^2 x_\nu}{ds^2} = -\frac{\alpha}{2}\frac{x_\nu}{r^3}\left(1 - 2E + \frac{3B^2}{r^2}\right). \tag{1.335}$$

Under these approximations, the resulting equations yield Kepler's laws of planetary motion. This is significant because it ties Einstein's general relativity to classical mechanics, showing that under certain approximations, Einstein's theory reduces to the previously accepted laws of gravity. Einstein's derivation also shows that the *area law* holds even in the second-order approximation.

Equation (1.335) is transformed into two equations [E3-15]:

THE NEWTONIAN MOTION OF A PARTICLE IN A POTENTIAL FIELD Φ + A CORRECTION:

$$\frac{d^2 x_\nu}{ds^2} = -\frac{\partial \Phi}{\partial x_\nu}, \quad \Phi = -\frac{\alpha}{2r}\left(1 + \frac{B^2}{r^2}\right). \tag{1.336}$$

The derivation of Eq. (1.336) from Eq. (1.335) is of significant importance and will be discussed in detail herein. This derivation is important because it allows the incorporation of relativistic effects into the orbit of Mercury, paralleling the procedure followed in the Newtonian framework. The transformation of Eq. (1.335) into Eq. (1.336) involves two primary components: a mathematically ingenious trick and straightforward mathematical manipulations, such as the application of the chain rule and the product rule. While conceptually sophisticated, Einstein accomplished this ingenious derivation within a few days, highlighting its complexity and the proficiency of Einstein.

For $E \neq 0$, rescaling B as $B' = B\sqrt{1-2E}$ and introducing a new parameter $s = \tilde{s}\frac{1}{\sqrt{1-2E}}$, guarantees that the transformation of Eq. (1.335) into the form involving the gradient of the potential (1.336) remains consistent. By implementing these substitutions, Einstein derived the following equation in place of Eq. (1.335) [E3-15]:

[144] This equation describes the planet's effective velocity in polar coordinates (r, ϕ) due to the Sun's gravitational field. ϕ represents the planet's angular position in its orbit around the Sun; the term $r^2 d\phi^2$ accounts for the component of motion along the angular direction, contributing to the planet's overall velocity.

[145] Equation (1.335) describes how the position x_ν of Mercury changes as it moves along its orbit under the influence of the Sun's gravitational field. It uses the components of the four-velocity $\frac{dx_\nu}{ds}$ to represent how the velocity changes along the geodesic path.

$$\frac{d^2x_\nu}{d\tilde{s}^2} = -\frac{\alpha}{2}\frac{x_\nu}{r^3}\left(1 + \frac{3B'^2}{r^2}\right). \tag{1.337}$$

When $E = 0$, the new parameter and the rescaling by $\sqrt{1-2E}$ are superfluous. The Eq. (1.335) directly simplifies to the following form [Ea-Jan]:

$$\frac{d^2x_\nu}{ds^2} = -\frac{\alpha}{2}\frac{x_\nu}{r^3}\left(1 + \frac{3B^2}{r^2}\right). \tag{1.338}$$

However, if E (the total mechanical energy) of a planetary orbit around the sun is zero, it does not represent an elliptical orbit because of the energy law (1.333). For an elliptical orbit, E must be less than zero. This indicates that the orbiting body is gravitationally bound to the central body (e.g., a planet to the sun) and follows an elliptical trajectory. To solve this problem, Einstein employed a mathematical trick by introducing a new parameter and rescaling the area constant B. By this strategic modification, the resulting Eq. (1.337) closely approximated the form of the equation derived under the assumption that $E = 0$. Thus, the mathematical transformation allowed Einstein to simplify Eq. (1.335) to a form consistent with Newtonian mechanics.[146] $\frac{\partial\Phi}{\partial x_\nu}$ [Eq. (1.336)] matches the right-hand side of Eq. (1.337).[147]

[146] We introduce a new parameter $s = \tilde{s}\frac{1}{\sqrt{1-2E}}$ and compute the first and second derivatives with respect to s:

$$\frac{d}{ds} = \frac{d\tilde{s}}{ds}\frac{d}{d\tilde{s}} = \sqrt{1-2E}\frac{d}{d\tilde{s}}, \qquad \frac{d^2}{ds^2} = (\sqrt{1-2E})^2\frac{d^2}{d\tilde{s}^2} = (1-2E)\frac{d^2}{d\tilde{s}^2}. \tag{1.339}$$

We replace $\frac{d^2x_\nu}{ds^2}$ in Eq. (1.335) with $(1-2E)\frac{d^2x_\nu}{d\tilde{s}^2}$:

$$(1-2E)\frac{d^2x_\nu}{d\tilde{s}^2} = -\frac{\alpha}{2}\frac{x_\nu}{r^3}\left(1 - 2E + \frac{3B^2}{r^2}\right), \tag{1.340}$$

and then divide both sides by $1-2E$:

$$\frac{d^2x_\nu}{d\tilde{s}^2} = -\frac{\alpha}{2}\frac{x_\nu}{r^3}\left(\frac{1 - 2E + \frac{3B^2}{r^2}}{1-2E}\right). \tag{1.341}$$

So the equation in terms of $\tilde{s}$ is:

$$\frac{d^2x_\nu}{d\tilde{s}^2} = -\frac{\alpha}{2}\frac{x_\nu}{r^3}\left(1 + \frac{3B^2}{r^2(1-2E)}\right). \tag{1.342}$$

Now we introduce a rescaled parameter, $B' = B\sqrt{1-2E}$. Substituting B' into Eq. (1.342) leads to Eq. (1.337). If we set $E = 0$, the rescaling by $\sqrt{1-2E}$ becomes unnecessary because $\sqrt{1-2E} = 1$. Starting with Eq. (1.335) with $E = 0$, we obtain Eq. (1.338).

[147] Given the second equation of (1.336):

$$\Phi = -\frac{\alpha}{2r}\left(1 + \frac{B^2}{r^2}\right), \quad \text{where } r = \sqrt{x_\mu x^\mu} \quad \text{(assuming summation over } \mu\text{)}. \tag{1.343}$$

We need to compute $\frac{\partial\Phi}{\partial x_\nu}$ and $\frac{\partial\Phi}{\partial r}$. We first compute $\frac{\partial r}{\partial x_\nu}$:

Combining the effective velocity of the planet (1.334) and the Eq. (1.336) leads to [E3-15][148]:

$$\frac{dr^2 + r^2 d\phi^2}{ds^2} = 2E - 2\Phi = 2E + \frac{\alpha}{r} + \frac{\alpha B^2}{r^3}. \tag{1.352}$$

Employing the area law (1.332), Einstein eliminated s from Eq. (1.352) and got a relativistic version of the Newtonian equation that describes planetary orbits [E3-15]:

$$\boxed{\frac{\partial r}{\partial x_\nu}} = \frac{\partial}{\partial x_\nu}\left(x_\mu x^\mu\right)^{1/2} = \frac{1}{2}\left(x_\mu x^\mu\right)^{-1/2} \cdot 2x_\nu = \boxed{\frac{x_\nu}{r}.} \tag{1.344}$$

We will first compute the derivative with respect to r:

$$\frac{\partial \Phi}{\partial r} = \frac{\partial}{\partial r}\left(-\frac{\alpha}{2r}\left(1 + \frac{B^2}{r^2}\right)\right). \tag{1.345}$$

Applying the product rule to Eq. (1.345), we get:

$$\frac{\partial \Phi}{\partial r} = -\frac{\alpha}{2}\left(\frac{\partial}{\partial r}\left(\frac{1}{r}\right)\left(1 + \frac{B^2}{r^2}\right) + \frac{1}{r} \cdot \frac{\partial}{\partial r}\left(1 + \frac{B^2}{r^2}\right)\right). \tag{1.346}$$

Now, we calculate each term in Eq. (1.346) separately:

$$\frac{\partial}{\partial r}\left(\frac{1}{r}\right) = -\frac{1}{r^2}, \quad \frac{\partial}{\partial r}\left(\frac{B^2}{r^2}\right) = -\frac{2B^2}{r^3}, \quad \frac{\partial}{\partial r}(1) = 0 \quad \text{(the derivative of a constant 1 is zero)}. \tag{1.347}$$

Putting the terms from Eq. (1.347) into the expression (1.346), we get:

$$\frac{\partial \Phi}{\partial r} = -\frac{\alpha}{2}\left(-\frac{1}{r^2}\left(1 + \frac{B^2}{r^2}\right) + \frac{1}{r}\left(-\frac{2B^2}{r^3}\right)\right). \tag{1.348}$$

Simplifying further Eq. (1.348), leads to:

$$\boxed{\frac{\partial \Phi}{\partial r} = \frac{\alpha}{2}\left(\frac{1}{r^2} + \frac{3B^2}{r^4}\right).} \tag{1.349}$$

Using the chain rule, we combine Eqs. (1.344) and (1.349):

$$\frac{\partial \Phi}{\partial x_\nu} = \frac{\partial \Phi}{\partial r} \cdot \frac{\partial r}{\partial x_\nu} = \frac{\alpha}{2}\left(\frac{1}{r^2} + \frac{3B^2}{r^4}\right) \cdot \frac{x_\nu}{r}. \tag{1.350}$$

Simplifying each term inside the parentheses of Eq. (1.350), we get:

$$\boxed{\frac{\partial \Phi}{\partial x_\nu} = \frac{\alpha x_\nu}{2r^3}\left(1 + \frac{3B^2}{r^2}\right).} \tag{1.351}$$

This matches the right-hand side of Eq. (1.337).

[148] This equation incorporates on the left-hand side the effective velocity of the planet in the Sun's gravitational field u^2 [Eq. (1.334)], and on the right-hand side E [Eq. (1.333)], B [Eq. (1.332)], and α [Eq. (1.164)].

RELATIVISTIC PLANETARY ORBIT EQUATION:

$$\boxed{\left(\frac{dx}{d\phi}\right)^2 = \frac{2E}{B^2} + \frac{\alpha}{B^2}x - x^2 + \alpha x^3.} \tag{1.353}$$

Integrating Eq. (1.353) from the perihelion to the aphelion yields [E3-15][149]:

$$\phi = \pi\left(1 + \frac{3}{2}\frac{\alpha}{a(1-e^2)}\right). \tag{1.354}$$

$$\Delta\phi \equiv \varepsilon = 2\pi - \phi = 2\pi - \pi\left(1 + \frac{3}{2}\frac{\alpha}{a(1-e^2)}\right). \quad \text{Thus:} \tag{1.355}$$

THE PRECESSION PER ORBIT OF MERCURY:

$$\boxed{\varepsilon = \pi\frac{3\alpha}{a(1-e^2)}.} \tag{1.356}$$

Einstein transformed the Eq. (1.356) to [E3-15][150]:

$$\boxed{\varepsilon = 24\pi^3\frac{a^2}{T^2c^2(1-e^2)}.} \tag{1.357}$$

He applied this formula to Mercury and found that it predicts a precession of the orbit of 43$''$ per century. This transformation is an example of how Einstein's general relativity modifies *classical mechanics* to account for the effects of gravity on space-time.[151] Einstein highlighted that his November 18, 1915 paper underscored that Kepler's laws are precisely upheld in his generally covariant theory of general

---

[149] Equation (1.354) shows that π represents the angle swept out by the radius vector from perihelion to aphelion during one orbit.

[150] Here, ε represents the angle of precession or the precession per orbit, a is the semi-major axis of the ellipse (of Mercury's orbit), T is the orbital period, and recall that α is a constant determined by the sun's mass [Eq. (1.164)], c is the speed of light, and e is the eccentricity of the ellipse.

[151] Einstein does not provide the exact algebraic steps, which require some work. Let us transform Eq. (1.356) into Eq. (1.357). Einstein hints about introducing T (orbital period in seconds) [E3-15]. Indeed, Einstein's relationship involves the square of the period T^2, as seen in Eq. (1.357). We first need to express α in terms of a, T and c. This involves Kepler's third law for planetary orbits around the Sun:

$$T^2 = \frac{4\pi^2a^3}{GM}. \tag{1.358}$$

Let us use this relation to connect the Eqs. (1.356) and (1.357). We substitute the expression (1.164) for α into Eq. (1.356):

$$\varepsilon = \pi\frac{3\frac{2GM}{c^2}}{a(1-e^2)} = \pi\frac{6GM}{ac^2(1-e^2)}. \tag{1.359}$$

Next, we use Kepler's third law Eq. (1.358) to replace M. Therefore:

relativity, as evidenced in Eqs. (1.332) and (1.357) [E3-15]. In contrast, Besso's *area law* did not align with Kepler's second law, which did not hold under the *Entwurf* theory framework.

Einstein's 1915 derivation of Mercury's precession marked a significant advancement over his 1913 collaborative efforts with Besso, demonstrating profound improvements. In December 1915, Einstein remarked to Besso about being surprised by the occurrence of the $g_{11}...g_{33}$ components in his 1915 calculation of Mercury's perihelion precession [CPAE8], Doc. 162. Einstein highlighted a pivotal element in his derivation when he noted this [W18]. The first-order metric of 1915, Eq. (1.300), presented in Einstein's 1915 Mercury perihelion paper [E3-15], marks a significant advancement over the previous 1913 first-order Eq. (1.162) presented in the *Einstein-Besso Manuscript*. In 1915, Einstein introduced a non-flat component $(-\alpha\frac{x_\rho x_\sigma}{r^3})$ for the spatial parts of the metric tensor (g_{11}, g_{22}, g_{33}). This term represents the effect of mass on the curvature of space itself, not just on the curvature of time. Einstein's surprise, as he expressed in his letter to Besso, likely stemmed from the realization and confirmation that even to first-order approximation, the spatial geometry of the universe is not flat but curved by the presence of mass. This curvature of space is vital to the theory's ability to predict the precession of Mercury's orbit with unprecedented accuracy. The earlier models, including the 1913 version stemming from the *Entwurf* theory, did not fully account for this spatial curvature in their first-order approximations. By incorporating these spatial curvature components into his 1915 theory, Einstein provided a comprehensive explanation for gravitational phenomena, including those previously unexplained (or explained in an unsatisfactory way), like the deflection of light near the Sun and the perihelion precession of Mercury.

In the 1913 calculations with the *Entwurf* theory, Einstein and Besso were working with a framework that was based on a static, spherically symmetric gravitational field, and it focused mainly on the g_{44} component of the metric tensor, which describes how mass affects the temporal component of spacetime. However, this approach did not fully account for the curvature of space itself induced by mass. The calculation led to a prediction of Mercury's perihelion precession closer to the observed value than any Newtonian prediction could provide. However, it was still significantly off from the actual observed value.

In his November 18, 1915, paper on Mercury's perihelion, Einstein discovered that his newly formulated field equations could account for the deflection of a ray of light as it passed by the Sun. Employing the metric to first-order approximation

---

$$M = \frac{4\pi^2 a^3}{GT^2}. \tag{1.360}$$

We substitute Eq. (1.360) into the equation for ε (1.359):

$$\varepsilon = \pi \frac{6G\frac{4\pi^2 a^3}{GT^2}}{ac^2(1-e^2)} = \pi \frac{24\pi^2 a^2}{T^2c^2(1-e^2)}. \tag{1.361}$$

Finally, we simplify the equation and obtain Eq. (1.357).

(1.300), Einstein reported a significant finding. He considered the line element Eq. (1.3) and applied *Huygens' principle* to[152]:

THE LINE ELEMENT FOR LIGHT RAYS:

$$\boxed{ds^2 = \sum_{\mu\nu} g_{\mu\nu} dx^\mu dx^\nu = 0.} \tag{1.362}$$

Einstein asserted that through straightforward calculations, he determined that a light ray passing near the Sun at a distance Δ would experience an angular deflection of magnitude $\frac{2\alpha}{\Delta}$ equivalent to 1.7 arc seconds. This finding contrasted his earlier calculations from 1907 and 1911, which had relied on the equivalence principle, and his 1913 *Entwurf* theory calculations. These earlier calculations had consistently yielded a deflection of half the magnitude, $\frac{\alpha}{\Delta}$ (equivalent to 0.83 arc seconds). Therefore, Einstein concluded that a ray of light passing close to the Sun should undergo a deflection of 1.7 seconds of arc, twice the previously predicted 0.85 seconds of arc [E3-15].

1.7 The 1916 General Relativity

1.7.1 Einstein's 1916 General Theory of Relativity

In March 1916, Einstein submitted a comprehensive review article on his general theory of relativity, titled "The Foundation of the General Theory of Relativity," to the *Annalen der Physik*, which was then published in May of the same year [E1-16].

THE FIELD EQUATIONS AND ENERGY- MOMENTUM CONSERVATION

In this 1916 paper, Einstein introduced the summation convention. He wrote the field Eq. (1.292) using the summation convention where necessary, particularly in the repeated indices within the Christoffel symbols and their derivatives [E1-16].[153]

$$R_{\mu\nu} = \frac{\partial \Gamma^\alpha_{\mu\nu}}{\partial x_\alpha} + \Gamma^\alpha_{\mu\beta}\Gamma^\beta_{\nu\alpha} = -\kappa\left(T_{\mu\nu} - \frac{1}{2}g_{\mu\nu}T\right) \quad \sqrt{-g} = 1. \tag{1.363}$$

---

152 This equation reflects that the proper time along a light ray's path is zero, as light travels at the speed of light c in a vacuum.

153 The indices μ and ν on the left-hand side are free indices and do not imply summation: the first term on the left-hand side $\frac{\partial \Gamma^\alpha_{\mu\nu}}{\partial x^\alpha}$ contains the index α, which is repeated as a subscript and a superscript. This indicates summation over α. So, the summation convention applies to the first term. In the second term $\Gamma^\alpha_{\mu\beta}\Gamma^\beta_{\nu\alpha}$ the indices α and β are repeated as both subscripts and superscripts, implying summation over these indices as well. The right-hand side of the equation contains expressions with indices μ and ν ($T_{\mu\nu}$ and $g_{\mu\nu}$). These indices are free, not repeated, and not subject to summation.

Einstein contracted Eq. (1.363) and then substituted the Eq. (1.232) (the stress-energy pseudo tensor) into the obtained result. The final result is [E1-16]:

$$\frac{\partial}{\partial x_\alpha}\left(g^{\sigma\beta}\Gamma^\alpha_{\mu\beta} - \frac{1}{2}\delta^\sigma_\mu g^{\lambda\beta}\Gamma^\alpha_{\lambda\beta}\right) = -\kappa\left(T^\sigma_\mu + t^\sigma_\mu\right). \tag{1.364}$$

He took the divergence $\frac{\partial}{\partial x_\sigma}$ of Eq. (1.364):

$$\frac{\partial}{\partial x_\sigma}\left(\frac{\partial}{\partial x_\alpha}\left(g^{\sigma\beta}\Gamma^\alpha_{\mu\beta} - \frac{1}{2}\delta^\sigma_\mu g^{\lambda\beta}\Gamma^\alpha_{\lambda\beta}\right)\right) = \frac{\partial}{\partial x_\sigma}\left(-\kappa\left(T^\sigma_\mu + t^\sigma_\mu\right)\right), \tag{1.365}$$

and demonstrated that the left-hand side of Eq. (1.365) is an identity:

$$\frac{\partial^2}{\partial x_\alpha \partial x_\sigma}\left(g^{\sigma\beta}\Gamma^\alpha_{\mu\beta} - \frac{1}{2}\delta^\sigma_\mu g^{\lambda\beta}\Gamma^\alpha_{\lambda\beta}\right) \equiv 0. \tag{1.366}$$

Hence, if the identity in Eq. (1.366) vanishes, the field Eq. (1.363) satisfy the conservation of energy-momentum (1.258) [E1-16]. Equation (1.366) is the precursor to the contracted Bianchi identity (1.476) (see the discussion in Sect. 1.7.3).

THE FIELD EQUATIONS AND THE NEWTONIAN LIMIT

In the same review paper, "The Foundation of the General Theory of Relativity," Einstein finally achieved his long-awaited goal. He demonstrated that when considering the weak-field approximation, the geodesic equation leads to Newton's equations of motion as a first-order approximation, and the field equations lead to the Newton-Poisson equation [Refer to Eq. (1.27) in Sect. 1.2.2]. Einstein rewrote the geodesic Eq. (1.210) using his summation convention [E1-16]:

THE GEODESIC EQUATION:

$$\boxed{\frac{d^2 x_\tau}{ds^2} = \Gamma^\tau_{\mu\nu}\frac{dx_\mu}{ds}\frac{dx_\nu}{ds}.} \tag{1.367}$$

When $\mu = \nu = 4$ (representing the time component of spacetime), we get the following three equations [E1-16]:

(1) The first equation [E1-16][154]:

$$\frac{d^2 x_\tau}{dt^2} = \Gamma^\tau_{44}. \tag{1.369}$$

[154] The geodesic Eq. (1.367) reduces to the geodesic equation for a particle's acceleration along a geodesic, specifically dealing with the (time) 44 component of the Christoffel symbols:

$$\frac{d^2 x_\tau}{ds^2} = \Gamma^\tau_{44}. \tag{1.368}$$

Setting $s = t$ in Eq. (1.368) brings it to the form (1.369).

(2) The second and third equations are the same as Eq. (1.369) but modified to the weak-field approximation by incorporating gravitational effects:

$$\frac{d^2x_\tau}{dt^2} = \begin{cases} -\frac{1}{2}\left(\frac{\partial g_{4\tau}}{\partial x_4} - \frac{\partial g_{4\tau}}{\partial x_4} + \frac{\partial g_{44}}{\partial x_\tau}\right), & \text{for } \tau = 1, 2, 3 \\ -\begin{bmatrix} 44 \\ 4 \end{bmatrix}, & \text{for } \tau = 4 \end{cases} \tag{1.370}$$

The first and second terms $\frac{\partial g_{4\tau}}{\partial x_4}$ cancel each other out, and the Eq. (1.370) simplifies to [E1-16][155]:

$$\boxed{\frac{d^2x_\tau}{dt^2} = -\frac{1}{2}\frac{\partial g_{44}}{\partial x_\tau},} \tag{1.371}$$

Equation (1.371) corresponds to the equation of motion for a material point in the Newtonian theory of gravity, the Eq. (1.331) from Einstein's perihelion of Mercury paper (see Sect. 1.6.3).

Next, Einstein wrote the relationship between the 44 component of the stress-energy tensor (T_{44}) and the density of matter (ρ) in the weak-field approximation [E1-16][156]:

$$T_{44} = \left(\frac{dx_4}{dt}\right)^2 \rho = \rho = T. \tag{1.372}$$

In the Newtonian approximation, Einstein derived Poisson's equation from the field Eq. (1.363). He first provided a general representation for the left-hand side of the Eq. (1.363), where τ can take values 1, 2, 3, or 4 [E1-16]:

$$\frac{\partial}{\partial x_\tau}\Gamma^\tau_{\mu\nu} = \frac{\partial}{\partial x_1}\begin{bmatrix} \mu\nu \\ 1 \end{bmatrix} + \frac{\partial}{\partial x_2}\begin{bmatrix} \mu\nu \\ 2 \end{bmatrix} + \frac{\partial}{\partial x_3}\begin{bmatrix} \mu\nu \\ 3 \end{bmatrix} - \frac{\partial}{\partial x_4}\begin{bmatrix} \mu\nu \\ 4 \end{bmatrix}, \quad \text{for } \tau = 1, 2, 3, 4. \tag{1.373}$$

He focused on the specific component $\mu = \nu = 4$ of Eq. (1.373), representing the calculation of $\frac{\partial^2 g_{44}}{\partial x_1^2}$ and similar terms:

$$\frac{\partial}{\partial x_\tau}\Gamma^\tau_{44} = \boxed{\frac{\partial}{\partial x_1}\begin{bmatrix} 44 \\ 1 \end{bmatrix} + \frac{\partial}{\partial x_2}\begin{bmatrix} 44 \\ 2 \end{bmatrix} + \frac{\partial}{\partial x_3}\begin{bmatrix} 44 \\ 3 \end{bmatrix}} - \frac{\partial}{\partial x_4}\begin{bmatrix} 44 \\ 4 \end{bmatrix}. \tag{1.374}$$

155 $\frac{d^2x_\tau}{dt^2}$ represents the acceleration of a particle with respect to time. The particle is accelerated in the gravitational potential $\frac{g_{44}}{2}$.

156 In this equation $\frac{dx_4}{dt}$ represents the square of the component of the 4-velocity with respect to time and T represents the trace of the stress-energy tensor. The component (T_{44}) of the stress-energy tensor where x_4 corresponds to time, and ρ is the energy density of the dust, is assumed to be constant in the specific case of the weak field approximation.

He obtained an equation that only involves spatial derivatives and removes the time components entirely. This equation involves derivatives of g_{44} for spatial coordinates [E1-16]:

$$\frac{1}{2}\left(\frac{\partial^2 g_{44}}{\partial x_1^2}+\frac{\partial^2 g_{44}}{\partial x_2^2}-\frac{\partial^2 g_{44}}{\partial x_3^2}\right)=-\frac{1}{2}\nabla^2 g_{44}. \tag{1.375}$$

Equation (1.375) represents the final result for the left-hand side of Eq. (1.363), corresponding to the 44 component of the metric tensor. After the above intermediate steps are used to calculate the left-hand side of the field Eq. (1.363), Einstein calculated the right-hand side for the specific component $\mu = \nu = 4$. Plugging Eq. (1.372) into the right-hand side of the field Eq. (1.363) gives:

$$-\kappa\left(\rho-\frac{1}{2}\rho\right). \tag{1.376}$$

So, considering the Newtonian limit, combining (1.375) and (1.376) leads to the Poisson equation [E1-16]:

$$\text{POISSON'S EQUATION:} \quad \boxed{\nabla^2 g_{44} = \kappa\rho.} \tag{1.377}$$

This equation is equivalent to Eq. (1.161) (the speed of light is set to $c = 1$). Eq. (1.377) is the final result, where the Laplacian $\nabla^2 g_{44}$ of the gravitational potential g_{44}, and $\kappa\rho$ is related to the mass density, consistent with the Newtonian theory of gravity.

Einstein successfully derived the Newtonian limit of general relativity, but there was a lingering issue regarding the interpretation of the gravitational potential. In the Newtonian framework, the gravitational potential is treated as an external field acting on particles. This contrasts with the more unified approach of general relativity, where gravitational effects are an intrinsic part of the spacetime curvature as described by the metric tensor. In Einstein's theory, the gravitational potential is interpreted by the metric tensor and the Levi-Civita connection [the Christoffel symbols (1.245)].[157] The curvature still influences particle motion in regions with minimal spacetime curvature, such as the Newtonian limit. However, according to Newtonian theory, in these regions, the gravitational potential is treated as an external field acting on particles, separate from the structure of spacetime itself.

Only later, in 1923, with the development of the affine connection, Élie Joseph Cartan established a more accurate and formal method for calculating the Newtonian limit [Car-23, Car-24]. Unlike the Levi-Civita connection, an affine connection is not necessarily metric-compatible. By allowing the metric and connection to be

[157] In 1915, Einstein formulated general relativity using the Christoffel symbols (the Levi-Civita connection), which are derived from the metric tensor $g_{\mu\nu}$ and its first derivatives. Thus, the Levi-Civita connection is tied directly to the metric tensor. General relativity's foundation rests on the curvature of spacetime described by the Riemann curvature tensor, which is constructed using the Christoffel symbols.

considered independently (relaxing metric compatibility), the affine connection can describe gravitational effects without relying on spacetime curvature. Allowing these to be independent means having a non-flat connection even when the metric is flat, representing gravitational effects without curvature. In the Newtonian limit of general relativity, the spatial metric can become flat (Euclidean), reflecting the absence of curvature or gravitational effects in the sense of general relativity. However, the gravitational potential still affects particle motion, as described by Newton's law of gravitation. The affine connection conveys this by introducing a non-flat connection that governs the inertial motion of particles under gravity, even in the absence of spacetime curvature. The introduction of the affine connection helped resolve the conceptual disparity between Newtonian gravity and Einstein's theory. Introducing the affine connection into the formulation of gravitational theory provided a framework to smoothly transition between the relativistic description of gravity and the classical Newtonian theory. While the Levi-Civita connection was necessary for Einstein in developing his general relativity, it alone could not encapsulate all gravitational effects. Thus, the affine connection represents a significant advancement over relying solely on the Levi-Civita connection.[158]

[158] In 1923, Einstein turned his attention to unifying general relativity with electromagnetism. In his paper, "The Theory of the Affine Field" [E-23], Einstein found the affine connection framework useful. The more general affine connection is not restricted to being derived from the metric and can incorporate both gravitational and electromagnetic potentials. In his paper, Einstein attempts to unify and describe both gravitation and the electromagnetic field using affine connections. However, Einstein's modified field equations (devised in his attempts to unify the gravitational and electromagnetic fields) face several significant challenges and problems. These equations do not provide a model for an electron free of singularities, and they describe the dynamics of the two fields but remain inconsistent with known physical laws. In classical electromagnetism, electric fields are typically generated by electric charges. However, unifying gravitation and electrodynamics in the theory described in Einstein's paper suggests that Einstein's modified field equations imply the existence of electric fields even when no electric charges are present. Thus, Einstein points out that a parameter β in his modified field equations must be set to a very small value. This adjustment guarantees that the predictions of his theory do not include scenarios that are not observed in reality, such as the presence of electric fields without any charges. β acts as a fine-tuning parameter that can be adjusted to account for small discrepancies between the theory and observations.
First, Einstein's new unified field theory can be reduced to the previous form of general relativity and the tensorial formulation of electromagnetism by letting the parameter β vanish.
Second, the factor β can be seen as an *ad-hoc* assumption introduced to tweak the theory to remain consistent with known empirical observations and theoretical principles. It serves as a parameter that could help reconcile the theoretical model with existing physical laws, particularly those of electromagnetism.
Third, Einstein's modified field equations are almost identical to those derived by Hermann Weyl from his theoretical framework. The only difference is the sign of the constant β.
In 1918, in his paper "Gravitation and Electricity," Weyl comprehensively discussed the affine connection. He presented an attempt to unify general relativity and electrodynamics. He integrated concepts from both theories within a unified geometric framework, primarily through affine connections and metrics. In his paper, Weyl discusses the affine connection, central to spacetime geometry in general relativity. He then introduces the additional structure interpreted as the electromagnetic potential and an associated field tensor, representing the electromagnetic field. This integration suggests that the electromagnetic field is described geometrically, similar to how the metric tensor, in general relativity, describes the gravitational field [Wey].

A TENSORIAL FORMULATION OF ELECTROMAGNETISM AND THE ELECTROMAGNETIC STRESS- ENERGY TENSOR

In his 1916 review article, "The Foundation of the General Theory of Relativity," Einstein wrote Maxwell's equations in a covariant form [E1-16]. He used tensor notation, which makes the laws of electromagnetism explicitly covariant. First, Einstein defined the electromagnetic field tensor $F_{\mu\nu}$ in terms of the electromagnetic potential ϕ_μ (which includes the electric potential and the magnetic vector potential)[159]:

$$F_{\mu\nu} = \frac{\partial \phi_\mu}{\partial x_\nu} - \frac{\partial \phi_\nu}{\partial x_\mu}. \tag{1.378}$$

Einstein wrote the two sets of Maxwell's equations in the tensor form [E1-16]:

$$\frac{\partial F_{\rho\sigma}}{\partial x_\tau} + \frac{\partial F_{\sigma\tau}}{\partial x_\rho} + \frac{\partial F_{\tau\rho}}{\partial x_\sigma} = 0. \tag{1.379}$$

Equation (1.379) represents the absence of magnetic monopoles and the fact that magnetic field lines are always closed loops.[160]

Einstein then wrote the covariant form of Gauss's law as [E1-16][161]:

$$\frac{\partial F^{\mu\nu}}{\partial x_\nu} = J^\nu, \tag{1.380}$$

Einstein showed that $F^{\mu\nu}$ is related to the covariant tensor by the metric tensor $g_{\mu\nu}$ in the following way[162]:

$$F^{\mu\nu} = g^{\mu\rho} g^{\nu\sigma} F_{\rho\sigma}. \tag{1.382}$$

[159] This equation expresses the electric and magnetic fields compactly in a unified framework. $F_{\mu\nu}$ concisely represents the electric and magnetic fields in a relativistic (Lorentz-invariant) framework.

[160] In more conventional vector notation, this equation corresponds to the two Maxwell equations $\nabla \cdot \mathbf{B} = 0$ (indicating there are no magnetic monopoles and the divergence of the magnetic field B is zero) and Faraday's law of induction $\nabla \times \mathbf{E} = -\frac{\partial \mathbf{B}}{\partial t}$ (showing how a time-varying magnetic field creates a circulating electric field).

[161] Gauss's law in its non-covariant form: $\nabla \times \mathbf{E} = \frac{\rho}{\epsilon_0}$, where ρ is the charge density. J^ν is the four-current density vector, including charge and current densities. $F^{\mu\nu}$ is the contravariant form of the electromagnetic field tensor associated with $F_{\rho\sigma}$, representing the electromagnetic field in covariant form. Ampère's law (which, in its non-covariant form, is $\nabla \times \mathbf{B} = \mu_0 \mathbf{J} + \mu_0 \varepsilon_0 \frac{\partial \mathbf{E}}{\partial t}$) is included in equation (1.380) as well, since $F^{\mu\nu}$ encompasses both electric and magnetic fields.

[162] Einstein uses the identity:

$$T^{\mu\nu} = \sum_{\rho\sigma} g^{\mu\rho} g^{\nu\sigma} T_{\rho\sigma}. \tag{1.381}$$

Drawing upon his 1914 *Entwurf* theory, Einstein deduced the stress-energy tensor for an electromagnetic field in 1916, utilizing the electromagnetic field tensor $F_{\mu\nu}$ as described in Eq. (1.379) [E1-16][163]:

ELECTROMAGNETIC STRESS- ENERGY TENSOR:

$$\boxed{T_{\mu\nu} = \frac{1}{4} g_{\mu\nu} F_{\alpha\beta} F^{\alpha\beta} - F_{\mu\alpha} F^{\nu\alpha}.} \tag{1.386}$$

A second form of the stress-energy tensor that uses the electromagnetic field tensor [Eq. (1.378)] is the following:

$$T_{\mu\nu} = \frac{1}{4} g_{\mu\nu} \phi_{\alpha\beta} \phi^{\alpha\beta} - \phi_{\mu\alpha} \phi^{\nu\alpha}. \tag{1.387}$$

This form generalizes Eq. (1.386) to a field tensor $\phi_{\mu\nu}$, equivalent to $F_{\mu\nu}$.

LIGHT DEFLECTION NEAR THE SUN

In the same review article, Einstein also derived the bending of light. In 1915, Einstein initially refrained from presenting intricate calculations, providing a succinct overview of the primary derivation of light bending near the Sun (see end of Sect. 1.6.3). However, a few months later, he furnished the comprehensive derivation. In his 1916 review article, "The Foundation of the General Theory of Relativity," Einstein computed the deflection of light rays passing through the Sun's gravitational field, as predicted by his new theory of general relativity [E1-16]. He begins with the geodesic equation for photons [see Eq. (1.367)]:

---

[163] In his 1914 *Entwurf* review article, Einstein wrote the relation [E2-14]:

A MIXED TENSOR TRANSFORMS THE ELECTROMAGNETIC FIELD TENSOR:

$$F^{\mu\nu} = \frac{1}{2} \sum_{\alpha\beta} G^{\mu\nu}_{\alpha\beta} F^{\alpha\beta}. \tag{1.383}$$

Multiplying both sides by $\frac{1}{2} G^{\sigma\tau}_{\mu\nu}$ and summing over μ and ν, we obtain:

$$\frac{1}{2} \sum G^{\sigma\tau}_{\mu\nu} F^{\mu\nu} = \frac{1}{4} \sum_{\alpha\beta\mu\nu} G^{\sigma\tau}_{\mu\nu} G^{\mu\nu}_{\alpha\beta} F^{\alpha\beta}. \tag{1.384}$$

Einstein shows that Eq. (1.384) can be cast into the form:

$$F^{\sigma\tau} = \frac{1}{2} \left(F^{\sigma\tau} - F^{\tau\sigma} \right) = \frac{1}{4} \sum_{\alpha\beta\mu\nu} G^{\sigma\tau}_{\mu\nu} G^{\mu\nu}_{\alpha\beta} F^{\alpha\beta}, \tag{1.385}$$

$F^{\sigma\tau}$ is a new inverse tensor formed via relations (1.383) and (1.384).

In these identities, Einstein used the tensor $G^{\mu\nu}_{\alpha\beta}$ to act as a transformation or projection tensor on the electromagnetic field tensor $F^{\mu\nu}$. The above three identities were part of Einstein's exploration of the connections between gravity and electromagnetism.

THE GEODESIC EQUATION FOR PHOTONS:

$$\boxed{g_{\mu\nu}\frac{dx^{\mu}}{ds}\frac{dx^{\nu}}{ds}=0,} \tag{1.388}$$

and then sets the line element ds^2 (1.3) equal to zero for null geodesics, i.e., along the world lines of light rays [Eq. (1.362)].

Einstein defines the magnitude of the velocity of light γ as experienced by an observer on Earth[164]:

$$\boxed{\gamma=\sqrt{\left(\frac{dx^1}{dx^4}\right)^2+\left(\frac{dx^2}{dx^4}\right)^2+\left(\frac{dx^3}{dx^4}\right)^2},} \tag{1.389}$$

and predicts that light should be deflected when passing near massive objects. He writes Eq. (1.164),[165] and then calculates the light ray's total deflection B as it passes by the massive Sun [E1-16][166]:

$$\boxed{B=\int_{-\infty}^{+\infty}\frac{\partial\gamma}{\partial x^1}dx^2.} \tag{1.390}$$

Next, Einstein employs the approximation for the metric tensor components $g_{\rho\sigma}$, as given in Eq. (1.300), near a massive object. These components describe the curvature of spacetime caused by the mass of the Sun. The specific form of the metric components $g_{\rho\sigma}$ and g_{44} in Eq. (1.300) is derived from the solutions of the generally covariant vacuum field equations (1.289) for the gravitational field of the Sun. Subsequently, Einstein calculates γ [Eq. (1.389)] based on the metric components from Eq. (1.300):

$$\boxed{\gamma=\frac{dx^2}{dx^4}=\sqrt{-\frac{g_{44}}{g_{22}}}=1-\frac{\alpha}{2r}\left(1+\frac{x_2^2}{r^2}\right).} \tag{1.391}$$

[164] γ is calculated in terms of the derivatives of spacetime coordinates with respect to the coordinate representing time dx_4. In this equation, dx^{μ} represents the infinitesimal change in the spacetime coordinates (x^1, x^2, x^3, x^4) along the path of a light ray, where μ ranges from 1 to 3. The expression $\left(\frac{dx^1}{dx^4}\right)^2+\left(\frac{dx^2}{dx^4}\right)^2+\left(\frac{dx^3}{dx^4}\right)^2$ represents the squared components of the velocity vector of light in a vacuum along the spatial coordinates, which are normalized by dividing by dx^4 to obtain a dimensionless quantity. This equation quantifies how the velocity of light changes in a curved spacetime. γ depends on how $g_{\mu\nu}$ varies along the path of the light ray.

[165] Equation (1.164) relates the angle of deflection α of a light ray passing near a massive object (the Sun) to the mass of the object M and the fundamental constants, Newton's gravitational constant G and the speed of light c; see Sect. 1.4.1.

[166] It is an integral of the derivative of γ [Eq. (1.389)] with respect to one of the spatial coordinates (x^1), showing how much the light ray's trajectory is bent due to the curvature of spacetime. dx^2 is the second coordinate component and is a spacetime coordinate dx^{μ}.

This leads to the total deflection B [Eq. (1.390)] of a light ray passing near a massive object (the Sun) using the expression for γ from Eq. (1.391)[167]:

$$B = \frac{\kappa M}{2\pi \Delta} = \frac{4GM}{c^2 \Delta} = \frac{2\alpha}{\Delta}. \tag{1.392}$$

Finally, Einstein obtained the specific value for the deflection B in terms of α [Eq. (1.164)] and Δ, providing a concrete prediction for the angle of deflection of light rays passing by the Sun [E1-16]:

THE DEFLECTION ANGLE:

$$\boxed{B = \frac{2\alpha}{\Delta} = \frac{1.7 \text{ seconds of arc}}{\Delta}.} \tag{1.393}$$

Equation (1.393) for light deflection depends on the metric components [Eq. (1.300)] and the metric variations determined by the vacuum field Eq. (1.289).

In the *Entwurf* theory, the calculated deflection angle B for the gravitational field of the Sun is found to be half the value compared to what is predicted by Einstein's generally covariant 1915–1916 general relativity:

$$B = \int_{-\infty}^{+\infty} \frac{\partial \gamma}{\partial x^1} dx^2 = \frac{2GM}{c^2 \Delta} = \boxed{\frac{\alpha}{\Delta} = \frac{0.85\text{s of arc}}{\Delta}.} \tag{1.394}$$

THE ROTATING DISK

In his 1916 review article, "The Foundation of the General Theory of Relativity," Einstein considered the rotating disk to explore the effects of rotation on the geometry of space. Specifically, he shows how the circumference and radius of the disk are perceived differently in different reference frames due to the effects of special relativity and general relativity. This leads to non-Euclidean geometry in the context of rotating reference frames, highlighting the kinematic effects of rotation on spatial measurements. Einstein illustrated general covariance through the rotating disk thought experiment [E1-16]. This thought experiment, with conceptual roots traced back to Einstein's 1912 writings on the static gravitational theory [E1-12], offers an intuitive understanding of its profound implications. As presented in his 1916 review paper [E1-16] and later in his 1917 book, *Relativity: The Special and the General Theory* [Ein-17], the rotating disk thought experiment bridges the coordinate-dependent structure of special relativity and the coordinate-independent framework of general relativity.

Einstein considers two reference systems: K (a Galilean, non-rotating system)[168] and K' (a system uniformly rotating relative to K.[169] The origins and axes of K and

---

[167] This equation relates the deflection to the mass of the Sun M and the distance of the light ray's closest approach Δ to the Sun.

[168] An inertial frame that is not rotating.

[169] A non-inertial frame of reference. This disk is spinning relative to the inertial frame K.

K' coincide. In both frames, K and K', measurements of a circle's circumference and diameter are taken. In K, measurements are normal. A rod at rest relative to K measures the circumference and diameter, yielding the classical circumference-to-diameter ratio of π. In K', things get interesting. Using a rod at rest relative to K' (the rotating frame), the circumference measurement yields a ratio greater than π. Einstein explains this from the perspective of K: The measuring rod along the circumference undergoes Lorentz contraction. The rod appears shorter, thus requiring more rods to measure the circumference fully, increasing the measured value. The rod along the diameter does not experience this contraction (since it is not moving relative to the disk's center), so the diameter remains normal.[170] In the rotating frame K', when the observer tries to measure the circumference using rods at rest relative to K', they are in constant motion relative to the non-rotating frame K. From the perspective of an observer in the inertial frame K, the rods along the circumference in the rotating frame K' would seem to undergo Lorentz contraction. However, this is not the direct reason why measurements within K' yield a ratio greater than π. The real crux of the matter is the inherent different spatial geometry in K'.

Time dilation is also considered. Two clocks are placed on the disk K'—one at the center and one at the edge. To an observer in K, the clock at the edge runs slower than the clock at the center due to its motion, illustrating time dilation as per special relativity. Two clocks of identical construction will run at different rates if they are situated at different distances from the center of the disk, a result valid from the standpoint of an observer rotating with the disk.

The conclusions from the rotating disk experiment imply that space and time do not have a direct, physical reality independent of the observer's frame of reference. Their properties can change depending on the observer's motion and the presence of gravitational fields. This revolutionary realization suggested that space and time are not merely the stage on which events unfold but are dynamic entities influenced by matter and energy.

The disk thought experiment demonstrates that the traditional Euclidean geometry does not hold in the system K'. Classical geometry is inadequate for describing it. Time in K', like space, cannot be universally defined and is affected by the motion (or gravitational field, by equivalence). Einstein realized that if the laws of physics as we understand them do not hold in a non-inertial frame like K', this suggests a more general principle about the nature of space and time. The breakdown of Euclidean geometry and universal time in K' implies that the traditional, fixed coordinates of space and the universal flow of time are not inherent properties of the universe. Instead, they are constructs that apply in specific conditions (e.g., within inertial frames or away from significant gravitational fields). This realization led Einstein to the principle of general covariance, which posits that all systems of coordinates, on principle, are equally suitable for the description of nature [E1-16]. Einstein's leap was to realize that if all coordinate systems are equally valid, then our descriptions

[170] Lorentz contraction only affects the dimensions parallel to the direction of motion. For an observer in frame K, any object moving relative to them would seem to contract along the direction of its motion.

of nature, including gravity, must be formulated in a way that is independent of any specific coordinate system. Thus, Einstein provided a compelling reason for formulating his general relativity, which uses tensor calculus and generally covariant field equations to describe the laws of physics in any coordinate system [E1-16].

This view was later criticized by Erich Kretschmann. As early as 1917, Kretschmann argued that Einstein's general covariance lacked physical content. Kretschmann claimed that general covariance is a mathematical feature rather than a physical one. He suggested that any theory could be rewritten using tensor calculus in a generally covariant form without changing its testable assertions [Kre]. This critique highlighted a fundamental issue in Einstein's theory. Hence, if any theory could be rewritten using tensor calculus in a generally covariant form, it does not necessarily add physical meaning. Einstein acknowledged that general covariance, in itself, is not a physical principle. However, he argued that the true physical content of general relativity is not just its general covariance but also its specific form of equations and field equations. These equations are not just a rephrasing of older physics in a new language; they provide new predictions (like the bending of light by gravity and gravitational redshift). The use of general covariance in general relativity is significant because it reflects the theory's ability to describe physics in any coordinate system, which is vital for a theory that aims to describe gravity in terms of the curvature of spacetime. This flexibility is not merely mathematical but has physical implications, especially in extreme conditions like near black holes. Kretschmann's critique of general covariance highlighted a key philosophical issue: the distinction between mathematical form and physical content. While general covariance might not imply new physics, Einstein's general covariant formalism did have significant physical implications. His field equations were not merely a covariant rephrasing of Newtonian gravity but represented a fundamentally different understanding of gravity as spacetime curvature.

Due to its circular motion, a rotating frame undergoes continuous acceleration. This acceleration leads to effects like centrifugal force and the Coriolis effect, which are absent in inertial frames. In a small enough region of a gravitational field (like a freely falling elevator), the effects of gravity can be locally negated. In such a region, a local inertial frame can be established where the laws of special relativity are held.

In 1907 and 1911, Einstein articulated a principle of equivalence, which posits that the experience of an observer in a homogeneous gravitational field is indistinguishable from that of an observer undergoing uniform acceleration in a region of space devoid of gravitational fields. This principle implies that an observer in free fall within a gravitational field cannot distinguish their state from that of being in inertial motion without gravity. The principle of equivalence asserts that when conducting any physical experiment, an observer in a static gravitational field will observe certain effects due to gravity; an observer in a region of space free from gravitational fields but uniformly accelerating in a manner opposite and equal to the gravitational acceleration experienced by the first observer will observe identical effects [Ein-07, Ein-11].

In March 1912, Einstein further reformulated and refined the equivalence principle. The new principle states that experiencing acceleration in a static gravitational field (such as free fall) is locally equivalent to being at rest in a space free of static gravitational fields [E2-12]. This idea is exemplified in Einstein's 1907 thought experiment (the happiest thought of his life), where an observer falling freely from the roof of a house does not experience the effects of a gravitational field in their immediate vicinity during the fall. In essence, for the falling observer, the effects of gravity are locally nullified, rendering their state equivalent to being at rest in a gravity-free space. This means that an observer in free fall (such as an astronaut orbiting Earth) will not feel the effects of gravity and will experience weightlessness, even though they are within a gravitational field.

In the disk thought experiment, Einstein considers a rotating reference frame K'. Unlike a freely falling elevator (a local inertial frame), a rotating frame experiences non-uniform motion, as described in the experiment. However, in the rotating disk thought experiment, the effects experienced on the disk can be analogized to gravitational effects if we apply the above weak equivalence principle. The centrifugal force experienced by an observer on the disk mimics a gravitational field that increases with distance from the center. The time dilation on the rotating disk can be interpreted analogously to gravitational time dilation through the lens of the equivalence principle. An observer near the disk's rim would experience time more slowly than at the center. The closer an observer is to the rim of the disk, where the tangential velocity is higher, the greater the effect on the passage of time. This is a manifestation of time dilation due to relative velocity, akin to what is described in special relativity. In a gravitational field, an observer at a higher gravitational potential (further from the mass causing the gravitational field) will experience time passing faster than an observer at a lower gravitational potential (closer to the mass). This effect is known as gravitational time dilation. Einstein proposed that this gravitational time dilation leads to the gravitational redshift effect. Atoms can be considered natural clocks due to the regularity of the light (spectral lines) they emit. The frequency of this emitted light is consistent when observed in a uniform gravitational field. When light is emitted from a lower gravitational potential and moves to a higher one, it loses energy. This energy loss manifests as a decrease in the frequency of light or a redshift. The frequency of light is directly related to the passage of time. If time passes more slowly at a lower gravitational potential (due to time dilation), the frequency of light emitted from that position will be lower when observed at a higher potential. This is because, from the perspective of the higher potential, the clocks (atoms emitting light) at the lower potential are ticking more slowly. Let us incorporate this information into Einstein's equations from 1911 and 1916.

The rate of the clock ν, when moving relative to K but at rest with respect to the disk, is given by the time dilation formula in special relativity [Ein-11]:

$$\nu = \nu_0 \sqrt{1 - \frac{v^2}{c^2}}. \tag{1.395}$$

For small velocities (compared to the speed of light, c), Eq. (1.395) can be approximated as:

$$\nu = \nu_0 \left(1 - \frac{1}{2}\frac{v^2}{c^2}\right). \tag{1.396}$$

The interpretation of these equations is that time dilation results from the disk's rotation in a flat spacetime. This effect is due to the velocity-dependent nature of time in special relativity, extended to include rotational motion. The rotating disk scenario primarily involves velocity-induced time dilation in a non-inertial reference frame without significant gravitational effects.

By substituting v^2 with $\omega^2 r^2$, Einstein wrote the expression (1.396) as:

$$\nu = \nu_0 \left(1 - \frac{1}{c^2}\frac{\omega^2 r^2}{2}\right). \tag{1.397}$$

Equation (1.397) leads to the conclusion that two clocks of identical construction will run at different rates if they are situated at different distances from the disk's center. This result is also valid from the standpoint of the observer rotating with the disk.

Einstein represented the difference in centrifugal force as a potential difference Φ[171]:

$$\Phi = -\frac{\omega^2 r^2}{2}. \tag{1.398}$$

Inserting this result (1.398) into the Eq. (1.397) for ν, we get:

$$\nu = \nu_0 \left(1 + \frac{\Phi}{c^2}\right). \tag{1.399}$$

In his 1916 review article on general relativity, "The Foundation of the General Theory of Relativity," Einstein considered a clock at rest in a static gravitational field [E1-16]. The spatial differentials dx^1, dx^2, and dx^3 are zero for a clock at rest, as the clock is not moving in space. The only non-zero differential is dx^4, representing the time component in a four-dimensional spacetime framework. For the clock period, we set the line element interval ds [Eq. (1.3)] equal to 1 (one tick of the clock), leading to:

$$1 = g_{44}(dx^4)^2. \tag{1.400}$$

Here, g_{44} is the component of the metric tensor related to time.

Rearranging the Eq. (1.400), we get:

$$dx^4 = \frac{1}{\sqrt{g_{44}}}. \tag{1.401}$$

---

[171] Φ is defined as the negative work required to move a unit mass from the clock's position to the center against the centrifugal force.

Using a series expansion for small gravitational potentials, Eq. (1.401) can be approximated as:

$$dx^4 = 1 - \frac{(g_{44} - 1)}{2}. \tag{1.402}$$

The term $(g_{44} - 1)$ is proportional to the gravitational potential, leading to[172]:

$$\boxed{1 - \frac{(g_{44} - 1)}{2} = 1 - \frac{\alpha}{r}.} \tag{1.403}$$

This shows that the gravitational potential affects the rate at which time passes (and hence the clock's rate). From this derivation, Einstein concluded that a clock runs more slowly in a stronger gravitational field or closer to masses. According to this derivation, for a clock at rest in a gravitational field, the time component of the metric tensor g_{44} directly influences the rate at which time passes, leading to gravitational redshift.

The derivation discussed above shows that the rotating disk thought experiment is related to special relativity but introduces complexities beyond the simple scenarios usually considered in special relativity. To analyze the rotating disk fully, Einstein needed to consider the effects of general relativity [Eqs. (1.400) and (1.401)]. While the weak equivalence principle allows us to treat small regions of a gravitational field as equivalent to an accelerating frame (thus applying special relativity locally) [Eqs. (1.395)–(1.399)], this does not fully account for the rotational dynamics of the disk. Einstein's equivalence principle bridges the gap between these two derivations. It states that the effects of acceleration (as experienced by observers on the rotating disk) are locally indistinguishable from those of a gravitational field. In the line element derivation [Eqs. (1.400) and (1.401)], the metric tensor incorporates the influence of gravity on spacetime, thus affecting the rate of clocks in a manner analogous to the velocity-based time dilation seen in the rotating disk scenario [Eqs. (1.395)–(1.399)]. Ultimately, the line element derivation in general relativity provides a more general and comprehensive understanding of gravitational redshift, encompassing both the effects of velocity (as seen in the rotating disk) and gravitational potential on the passage of time.

Similarly, the length contraction (analogous to Lorentz contraction in special relativity) can be viewed in terms of gravitational effects. Let us reverse-engineer Einstein's calculation using Eq. (1.319) to show how the presence of a gravitational field alters the measurement of distances, leading to the conclusion of length contraction. Consider the line element (1.3) again. When a unit-measuring rod is laid parallel to the x-axis, only the dx_1 component (along the x-axis) is non-zero; the other components dx_2, dx_3, dx_4 are zero. Einstein writes the condition $ds^2 = -1$

[172] Einstein related the metric tensor to the gravitational potential [the second equation of Eq. (3.1), refer to footnote 5]. Equation (1.403) is analogous to the component of the metric tensor (3.136) [α is given by Eq. (1.164)], and is similar to Eq. (1.320).

(a space-like interval), which implies the rod has a proper length of 1 unit in this spacetime geometry. Therefore, the line element simplifies to [E1-16]:

$$-1 = g_{11}dx_1^2. \tag{1.404}$$

It represents the spatial distance measured along the x-axis. Substituting g_{11}, the metric component Eq. (1.319), into the simplified line element Eq. (1.404), we get:

$$-1 = -(1 + \frac{\alpha}{r})dx_1^2. \tag{1.405}$$

The final result is the following equation [E1-16][173]:

$$dx_1 \approx 1 - \frac{1}{2}\frac{\alpha}{r}. \tag{1.409}$$

Equation (1.409) shows that in a weak gravitational field (where $\frac{\alpha}{r}$ is small), the effect of the field on the length measured in the direction of the field (x direction) is a slight contraction, as expected from the principles of general relativity. Einstein concluded that in the presence of a gravitational field when a measuring rod is aligned radially (along a radius), it seems slightly shortened compared to the coordinate system it is being measured against. This effect is due to the influence of the gravitational field on the rod [E1-16].

1.7.2 *Hilbert's Foundations of Physics*

On November 20, 1915, Hilbert presented a paper titled "Foundations of Physics" to the Göttingen Academy of Sciences, in which he included his version of the gravitational field Eq. (1.415). Five days later, on November 25, 1915, Einstein

[173] We remove the negative signs on both sides of the Eq. (1.405) and rearrange the equation to isolate dx_1^2 on one side:

$$dx_1^2 = \frac{1}{1 + \frac{\alpha}{r}}. \quad \text{To solve for } dx_1, \text{ we take the square root of both sides:} \tag{1.406}$$

$$dx_1 = \sqrt{\frac{1}{1 + \frac{\alpha}{r}}}. \quad \text{The expression inside the square root can be rewritten as:} \tag{1.407}$$

$$dx_1 = \frac{1}{\sqrt{1 + \frac{\alpha}{r}}}. \tag{1.408}$$

Since we are considering a unit measuring rod and assuming small $\frac{\alpha}{r}$, we can expand this to the first order. For a small $\frac{\alpha}{r}$ the square root function $\sqrt{1 + \frac{\alpha}{r}}$ can be expanded using a Taylor series around $\frac{\alpha}{r} = 0$. The result is Eq. (1.409).

presented his version of the gravitational field Eq. (1.290). There have been claims that Hilbert published field Eq. (1.424) resembling Einstein's field Eq. (1.290), just days before Einstein did. Much debate surrounds whether Hilbert outpaced Einstein in the race to derive the correct final form of the field equations. People have argued that what is noteworthy is that six days earlier, on November 20, 1915, Hilbert derived the same field equations that Einstein had been searching for over an extended period. Some have asserted that the brilliant and highly skilled mathematician elegantly arrived at the correct form of these equations, unlike Einstein's path, which involved various restrictions, conditions, approximations, and challenges.

Scholars in the history of science have dedicated significant effort to scrutinizing Hilbert's manuscripts and publications. They meticulously examined the equations in his handwritten documents, compared them with those in his published works, determined publication dates precisely, and analyzed the consistency between manuscript equations and their counterparts in his published papers. To streamline the discussion, I will summarize the overarching findings from this extensive research [Sta-02][174]: (1) The field equations initially appeared in Einstein's final November paper on the general theory of relativity [E4-15]. (2) Hilbert's proofs of his November 20, 1915 lecture "Foundations of Physics," dated December 6, 1915, do not include the same field equations that Einstein had derived at the end of November 1915. They contain the field Eqs. (1.415) and (1.416) [Hil15-1]. 3) Between December 6, 1915, and March 31, 1916, Hilbert derived the field Eq. (1.424). 4) Scholars have concluded that Hilbert only added the field Eq. (1.424) and revised his paper after having read Einstein's paper of November 25, 1915 [E4-15].

Hilbert's *Foundations of Physics* suggests an interconnection between Einstein's general relativity and Gustav Mie's electrodynamics in an axiomatic framework.[175]

---

[174] Hilbert's presentation on November 20 was only officially published on March 31, 1916 [Hil15-2]. Scholars found that in his published paper, "Foundations of Physics," Hilbert made corrections based on Einstein's final field equations from November 25, 1915 (published on December 2, 1915) [E4-15]. The archival materials housed within the Hilbert collection include a set of proofs dated December 6, 1915, for his November 20 communication [Hil15-1]. These proofs exhibit notable discrepancies compared to the version published on March 31, 1916 [Hil15-2]. The proofs do not include the field Eq. (1.424). Such variations have sparked considerable academic debate regarding the intellectual exchange between Hilbert and Einstein in November 1915. Scholars concluded that Hilbert revised his November 20, 1915, paper sometime between December 1915 and March 1916, incorporating the gravitational field equations in terms of the Einstein tensor [Eq. (1.424)] after reading Einstein's November 25, 1915 paper. In the published version of his paper from March 1916, Hilbert added a reference to Einstein's November 25, 1915 paper, acknowledging that the resulting field equations of gravitation appeared to align with the theory of general relativity established by Einstein in his recent papers [Sta-02].

[175] Gustav Mie developed a unified theory of matter and gravitation in 1912. Mie's theory represents a significant attempt to integrate the principles of electromagnetism with gravitational theory. His theory was built on the idea of an electromagnetic ether. According to Mie, atoms, molecules, and electrons are not separate from this ether but are singularities within it. These singularities represent places where the ether assumes a particular state that can be described as electric charges. Mie proposed that the standard Maxwell equations do not hold due to the intense fields in these singularities, such as inside electrons. He suggested that new equations were needed to explain these environments, pushing beyond the existing theories of electrodynamics and mechanics. Mie's theory aimed

Hilbert ends his paper by saying that a few simple assumptions in two axioms, *Axiom I* and *Axiom II*, are sufficient to develop a comprehensive theory. The theory Hilbert developed through these two axioms fundamentally changes our understanding of space, time, and motion. This transformation is aligned with the ideas presented by Einstein's theory of relativity, which redefined how these concepts are understood in physics. Hilbert believed that the fundamental equations derived from these two axioms would help illuminate complex processes within atoms that were previously poorly understood. In other words, he suggested a unification of electromagnetism and gravitation, offering new insights into the structure and behavior of matter at a microscopic level. Hilbert was also convinced that his fundamental equations would clarify the most fundamental processes previously hidden within the atom; properties and phenomena often seem arbitrary or derived empirically, but he argued that they could theoretically be deduced from fundamental mathematical relationships. Hilbert expressed his ultimate aspiration: that physics could, in principle, be treated as a science akin to geometry, where truths can be derived from basic axioms through logical reasoning. This would be the triumph of the axiomatic method. In this endeavor, Hilbert uses the calculus of variations and invariant theory [Hil15-2].

AXIOM I: Mie's Axiom of the World Function. The law of physical events is determined by a world function H, the following arguments:

$$g_{\mu\nu}, \quad \frac{\partial g_{\mu\nu}}{\partial w_\sigma}, \quad \frac{\partial^2 g_{\mu\nu}}{\partial w_\sigma \partial w_\rho}, \quad q_s, \quad \frac{\partial q_s}{\partial w_\sigma}, \tag{1.410}$$

and the variation of the integral of H[176]:

HILBERT'S ACTION INTEGRAL:

$$\boxed{\int L\sqrt{-g}\,dw, \quad i = 1, 2, 3, 4.} \tag{1.411}$$

Hilbert's approach extensively employed a metric variational principle using a Lagrangian density. He formulated the action integral as Eq. (1.411) [Hil15-2].[177]

to show that gravitational phenomena could be directly derived from the electromagnetic properties of matter. He believed that electric and magnetic fields, electric charges, and electric currents were sufficient to describe all material phenomena, including gravitation. Mie's theory aligned with the special theory of relativity and utilized its mathematical framework. He maintained that his theory was entirely consistent with the principle of relativity and could reproduce known gravitational facts [Mie]. Despite its ambitious scope and formal structure, Mie's theory was critiqued for its lack of tangible physical content and was ultimately untestable with the experimental methods available at the time. However, his ideas significantly influenced Hilbert and contributed to the ongoing discourse around integrating electromagnetism with gravitational theory.

[176] I will substitute H with $\mathcal{L}$ to refer to the Lagrangian for clarity. where $L\sqrt{-g}$ represents the Lagrangian density $\mathcal{L}$.

[177] Hilbert used a metric signature [i.e., the *West Coast* convention $(+, -, -, -)$ rather than the *East Coast* convention $(-, +, ++)$] where the determinant of the metric tensor g is positive. Hence, he used $\sqrt{g}$ rather than $\sqrt{-g}$ and did not need to negate it.

Hilbert illustrated that the Ricci tensor $K_{\mu\nu}$ allows for the following invariant [Hil15-2]:

$$K = g^{\mu\nu} K_{\mu\nu}. \tag{1.412}$$

where $K_{\mu\nu}$ is the Ricci tensor, and K is the Ricci scalar. He demonstrated, as Einstein realized [see Eqs. (1.428) and (1.430)], that the Ricci scalar, K (represented by R in Einstein's writings), is a natural candidate for the gravitational action's Lagrangian, especially since it is invariant under coordinate transformations [Hil15-2].
Hilbert derived the field equations from an action [Hil15-2]:

HILBERT'S ACTION:

$$\boxed{J = \int \mathcal{L} dw,} \tag{1.413}$$

where $\mathcal{L}$ includes terms beyond the Ricci scalar K [Hil15-2]:

UNIFIED LAGRANGIAN:

$$\boxed{\mathcal{L} = K + M.} \tag{1.414}$$

K is given by Eq. (1.412) and it represents the gravitational part of the Lagrangian. M includes terms related to electromagnetic fields and perhaps other forms of matter. In Hilbert's theory, which attempts to unify electromagnetism with gravity, M includes terms that account for these additional interactions.

AXIOM II: (Axiom of General Invariance): The world function H [$\mathcal{L}$] is invariant under any transformation of the world parameters w_s.

Hilbert then formulated a theorem regarding the action J. J is an invariant, i.e., remains unchanged regardless of transformations applied to the world parameters w_s. The action J is stationary, leading to the derivation of Euler-Lagrange equations. In the set of n Euler-Lagrange equations derived, four equations can be derived from the remaining $n-4$. This implies that not all equations are independent; some are consequences of others.

To obtain the field equations, Hilbert used AXIOM I to derive two equations resembling the Euler-Lagrange equations [Hil15-2][178]:

HILBERT'S GRAVITATIONAL FIELD EQUATIONS:

$$[\sqrt{-g}K]_{\mu\nu} \equiv \frac{\partial\sqrt{-g}K}{\partial g^{\mu\nu}} - \boxed{\frac{\partial}{\partial w_\sigma}} \frac{\partial\sqrt{-g}K}{\partial g_\sigma^{\mu\nu}} + \boxed{\frac{\partial^2}{\partial w_\sigma \partial w_\rho}} \frac{\sqrt{-g}\partial K}{\partial g_{\sigma\rho}^{\mu\nu}} = 0, \quad \mu, \nu = 1, 2, 3, 4. \tag{1.415}$$

[178] The first Eq. (1.415) represent gravitational field equations. Hilbert states that four Lagrangian differential equations (1.416) pertain specifically to the four electrodynamical potentials, denoted as q_s.

HILBERT'S GENERALIZED ELECTRODYNAMICAL EQUATIONS:

$$[\sqrt{-g}M]_{\mu\nu} \equiv \frac{\partial\sqrt{-g}M}{\partial q_\mu} - \boxed{\frac{\partial}{\partial w_\sigma}} \frac{\partial\sqrt{-g}M}{\partial q_{\mu\nu}} = 0, \quad \mu = 1, 2, 3, 4. \tag{1.416}$$

Hilbert aimed to demonstrate that the generalized Maxwell's equations (1.416) could be linked to the gravitational Eq. (1.415). This linkage would show how electromagnetism could be integrated within the framework of general relativity and highlight the interconnections between these fundamental forces. Attempting to unify electromagnetism and gravity under a single theoretical framework and demonstrate that these fields manifest the same underlying principles, Hilbert wrote his field Eqs. (1.415) and (1.416) as follows [Hil15-2]:

HILBERT'S FIELD EQUATIONS:

$$\boxed{[\sqrt{-g}K]_{\mu\nu} + \frac{\partial\sqrt{-g}M}{\partial g^{\mu\nu}} = 0.} \tag{1.417}$$

Rearranging Hilbert's gravitational field Eq. (1.415), we get:

$$\frac{\partial\sqrt{-g}K}{\partial g^{\mu\nu}} = \frac{\partial}{\partial w_\sigma}\frac{\partial\sqrt{-g}K}{\partial g^{\mu\nu}_\sigma} - \frac{\partial^2}{\partial w_\sigma w_\rho}\frac{\partial\sqrt{-g}K}{\partial g^{\mu\nu}_{\sigma\rho}}. \tag{1.418}$$

According to Eq. (1.416), the matter-electromagnetic Lagrangian M does not depend on the derivatives of the metric tensor. Thus, using Eq. (1.418), the variation of the action J is represented by:

$$\delta J = \int \delta\sqrt{-g}K\, dw = \int \left(\frac{\partial\sqrt{-g}K}{\partial g^{\mu\nu}}\right)\delta g^{\mu\nu}\, dw. \tag{1.419}$$

AXIOM I implies these equations. In this case, the action (1.413) is relevant, where $\mathcal{L} = K$. This implies a formulation where $K_{\mu\nu} = R_{\mu\nu}$ (using Einstein's notation). The matter-electromagnetic Lagrangian M does not directly produce the terms on the right-hand side of the field Eq. (1.424). Hilbert derived field equations that resembled the final form of Einstein's field equations but within a context where the Lagrangian (1.411) is not solely the Ricci scalar K but involves additional fields and interactions. This equation can be seen as a generalization or a part of a more complex setup of a unified field theory, where $K_{\mu\nu}$ and K represent generalized curvature terms and contributions from other fields integrated into the gravitational action $\mathcal{L} = K$.

The expression inside the integral (1.419) gives the functional derivative of J with respect to $g^{\mu\nu}$:

$$\frac{\delta J}{\delta g^{\mu\nu}} = \frac{\partial\sqrt{-g}K}{\partial g^{\mu\nu}}. \tag{1.420}$$

Setting the functional derivative (1.420) equal to zero should imply the field equations. Specifically, expanding the derivative using the product rule, we get:

$$\frac{\partial \sqrt{-g} K}{\partial g^{\mu\nu}} = \sqrt{-g} \frac{\partial K}{\partial g^{\mu\nu}} + K \frac{\partial \sqrt{-g}}{\partial g^{\mu\nu}}. \tag{1.421}$$

The derivative of the determinant of the metric tensor ($\sqrt{-g}$) with respect to the components of the metric tensor ($g^{\mu\nu}$) involves the determinant of the metric tensor[179]:

$$\frac{\partial \sqrt{-g}}{\partial g^{\mu\nu}} = -\frac{1}{2} \frac{1}{\sqrt{-g}} g g_{\mu\nu} = -\frac{1}{2} \sqrt{-g} g_{\mu\nu}. \tag{1.422}$$

Next, we assume the form of K such that:

$$\frac{\partial K}{\partial g^{\mu\nu}} = K_{\mu\nu}. \tag{1.423}$$

Substituting equations (1.422) and (1.423) back into the expression (1.421), we obtain:

HILBERT'S FIELD EQUATIONS (RESEMBLING EINSTEIN'S FIELD EQUATIONS):

$$\boxed{\left[\sqrt{-g} K\right]_{\mu\nu} = \sqrt{-g} \left(K_{\mu\nu} - \frac{1}{2} g_{\mu\nu} K \right).} \tag{1.424}$$

Thus, AXIOM I and AXIOM II both lead to the field equations (1.424).

In 1924, Hilbert equated the second term on the left-hand side of Eq. (1.417) to [Hil15-1]:

$$-\frac{\partial \sqrt{-g} M}{\partial g^{\mu\nu}} = \frac{1}{\sqrt{-g}} T_{\mu\nu}. \tag{1.425}$$

He substituted this into the field Eq. (1.424) and obtained Einstein's 1915 field equations:

EINSTEIN'S FIELD EQUATIONS:

$$\boxed{R_{\mu\nu} - \frac{1}{2} g_{\mu\nu} R = \kappa T_{\mu\nu}.} \tag{1.426}$$

---

[179] Recall that in 1913, Marcel Grossmann applied the same underlying construct but computed the derivative of the determinant of the metric tensor with respect to the coordinates rather than with respect to the components of the metric tensor; see Eq. (1.125) in Sect. 1.3.2.

(valid in unimodular coordinates):

HILBERT'S 1924 FIELD EQUATIONS:

$$\boxed{\left(K_{\mu\nu} - \frac{1}{2}g_{\mu\nu}K\right) = T_{\mu\nu}.} \tag{1.427}$$

1.7.3 Einstein's Shift Away from Unimodular Coordinates

Einstein's journey to his general theory of relativity was one of multiple detours and changes in approach. His equations during 1915–1916 show various stages in the mathematical formulation of the gravitational theory, especially as it transitioned from the older *Entwurf* theory to the 1916 general relativity [E4-16].

The action integral in the *Entwurf* theory was in the form of Eq. (1.211). L is the Lagrangian density and $d\tau$ is the spacetime volume element. This formulation made it so that the action would be invariant under coordinate transformations. In his 1915 general relativity, Einstein employed the same Lagrangian and derived the field equations through a metric variational principle. However, this treatment was unsatisfactory because it relied on the coordinate condition $\sqrt{-g} = 1$. In his review paper, "The Foundation of the General Theory of Relativity," Einstein apologized for developing covariant field equations with respect to a coordinate system where $\sqrt{-g} = 1$ [E1-16].

Einstein wrote to Moritz Schlick that the fact that the field equations of his theory can be simplified *a posteriori* by choosing the reference system so that the determinant equation $|g_{\mu\nu}| = -1$ is fulfilled, is epistemologically insignificant [CPAE8], Doc. 165. While the simplification involving the unimodular coordinates was a valuable tool for developing the general theory of relativity, it does not necessarily provide deeper insight into the nature of physical reality itself. The simplification was a mathematical convenience for Einstein, allowing him to make more straightforward calculations and manipulations of the field equations of his new general relativity.

However, as explained by Einstein, this does not enhance our understanding of the theory's fundamental principles or the physical phenomena it aims to describe. Einstein differentiates between the tools he used to develop the general theory of relativity in 1915, which describes nature, and the epistemological significance of this description. In 1916, he was wary of over-relying on mathematical crutches such as unimodular coordinates that might obscure the physical intuition and understanding behind his theory. Ultimately, Einstein believed that the physical principles (equivalence, correspondence, conservation, and general covariance) should guide the development of the general theory of relativity rather than letting a specific mathematical formalism dictate our understanding of physical reality.

In the 1916 review paper manuscript, "The Foundation of the General Theory of Relativity," the apology mentioned above can be found in a note on page 40a, which is only partially written, with the bottom half remaining empty.[180] In it, Einstein mentioned that the most general laws governing the gravitational field and matter had been derived, but only for a coordinate system where $\sqrt{-g} = 1$. This approach led to significantly simplifying formulas and calculations without sacrificing the requirement of general covariance. This was achieved by specializing the coordinate system from generally covariant equations. Einstein still pondered whether the field equations could be formulated without assuming $\sqrt{-g} = 1$ to achieve conservation of energy and momentum. He answered yes and eventually found that both conservation principles held. Still, he chose not to include these comprehensive considerations in the 1916 review paper because they did not contribute anything objectively new.

The detailed considerations that Einstein decided not to communicate in the 1916 review paper [E1-16] were contained in an unpublished five-page appendix to part *D* of "The Foundation of the General Theory of Relativity" [E1-16] titled "Appendix: Formulation of the Theory based on a Variational Principle" [CPAE6], Doc. 31. This five-page Appendix forms the core of Einstein's short paper of 1916 titled "Hamilton's Principle and the General Theory of Relativity" [E4-16]. In this paper, Einstein presented the comprehensive considerations as promised in his 1916 review paper, "The Foundation of the General Theory of Relativity" [E1-16]. Through this work [CPAE6], Doc. 31, Einstein successfully relinquished the restrictive unimodular coordinate condition $\sqrt{-g} = 1$, allowing him to treat the variational principle without this restriction [E4-16].

Einstein started his paper "Hamilton's Principle and the General Theory of Relativity" by acknowledging the contributions of Hendrik Antoon Lorentz and David Hilbert, saying they formulated general relativity in a comprehensive form and using a variational principle. He then expressed his goal to do the same and present the fundamental constructs of general relativity clearly and comprehensively, adhering closely to the principle of general relativity itself. He highlighted two main differences in his approach compared to these previous works: unlike Hilbert's presentation, Einstein intended to make as few assumptions as possible about the constitution

---

[180] Einstein's comments were appended as a note on page 40a following the equation of the electromagnetic stress-energy tensor (1.386) [CPAE6], Doc. 31 (see the end of Sect. 1.7.1). If we vary the electromagnetic Lagrangian $\mathcal{L}_{EM}$ with respect to the potential ϕ_μ, this leads to the derivation of $F_{\mu\nu}$ [Eq. (1.378)]. The variation of the action: $\int \mathcal{L}_{EM}\sqrt{-g}d\tau$ with respect to ϕ_μ gives the Euler-Lagrange equations, which lead to Maxwell's equations in general relativity. As seen in Sect. 1.7.1, Einstein did not derive the Eq. (1.378) by varying an electromagnetic Lagrangian with respect to the potential. Einstein defined Eq. (1.378) and then derived Maxwell's equations in general relativity and the electromagnetic stress-energy tensor (1.386) [E1-16]. In deriving the Reissner-Nordström solution, Gunnar Nordström varied the electromagnetic part of the Lagrangian with respect to the potential (see the discussion in Sect. 2.3.3, Chap. 2). Nordström explicitly acknowledged his inspiration from Einstein's paper, "Hamilton's Principle and the General Theory of Relativity" [E4-16], wherein Einstein introduced a variational method (this paper is the topic of this section). In adherence to historical accuracy, I abstain from speculating on undocumented aspects. However, whether Einstein contemplated deriving his electromagnetic tensor in 1916 by a variational method remains an open question.

of matter. This approach was aimed to keep the theory general and applicable across various physical scenarios without being constrained by specific characteristics of matter. Moreover, Einstein asserted that he contrasts his current approach with his earlier work, emphasizing that the choice of coordinate systems was left completely free in this presentation (he was alluding to unimodular coordinates) [E4-16].

Inspired by Hilbert's AXIOM I, Einstein formulated the following framework [E4-16]. The gravitational field is described by the metric tensor, $g_{\mu\nu}$ or its inverse $g^{\mu\nu}$. The matter fields, including the electromagnetic field, are described by functions $q(\rho)$. These are general functions defined over spacetime coordinates. The Lagrangian density $\mathcal{L}$ is defined as follows:

$$\mathcal{L} = \sqrt{-g}R, \tag{1.428}$$

is a function of the metric tensor components $g_{\mu\nu}$, their first and second derivatives with respect to spacetime coordinates x_σ and x_τ, respectively, and the matter fields $q_{(\rho)}$ and their first derivatives [E4-16]:

$$\begin{aligned} &g_{\mu\nu}, \quad g^{\mu\nu}_{\sigma} = \frac{\partial g^{\mu\nu}}{\partial x_\sigma}, \quad g^{\mu\nu}_{\sigma\tau} = \frac{\partial^2 g^{\mu\nu}}{\partial x_\sigma \partial x_\tau}, \\ &q_{(\rho)}, \quad q_{(\rho)\alpha} = \frac{\partial q_{(\rho)}}{\partial x_\alpha}. \end{aligned} \tag{1.429}$$

R in Eq. (1.428) is the Ricci scalar (the contraction of the Ricci tensor):

$$R = g^{\mu\nu} R_{\mu\nu}. \tag{1.430}$$

This is Hilbert's Eq. (1.412).

Einstein then stated the principle of stationary action (or the least action principle) through the variational principle [E4-16]:

$$\delta S = \delta \int \mathcal{L} d\tau = 0. \tag{1.431}$$

Initially, the Lagrangian $\mathcal{L}$ depends on the metric tensor $g_{\mu\nu}$, its first derivatives $g^{\mu\nu}_{\sigma}$, and its second derivatives $g^{\mu\nu}_{\sigma\tau}$, and the matter fields $q_{(\rho)}$, and their derivatives $q_{(\rho)\alpha}$. However, Einstein considered a simplification by assuming that $\mathcal{L}$ does not depend on the second derivatives of the metric tensor $g^{\mu\nu}_{\sigma\tau}$. This assumption simplified the mathematical complexity of the Lagrangian and the derivation of the field equations from the variational principle. By employing partial integration, Einstein transformed the integral of $\mathcal{L}$ over the spacetime volume $d\tau$ [Eq. (1.431)] into a new form involving $\mathcal{G}$ and $\mathcal{G}^*$ [E4-16][181]:

$$S = \int \mathcal{G} d\tau = \int \mathcal{G}^* d\tau + F. \tag{1.434}$$

[181] To remove the second derivatives $g^{\mu\nu}_{\sigma\tau} = \frac{\partial^2 g^{\mu\nu}}{\partial x_\sigma \partial x_\tau}$, we integrate by parts. We start with the integral:

$\int \frac{\partial^2 f}{\partial x_\alpha \partial x_\beta} d\tau$ and rewrite the second derivative. We recognize that we can write this as the divergence of the first derivative of f : $\int \frac{\partial}{\partial x_\beta}\left(\frac{\partial f}{\partial x_\alpha}\right) d\tau$.

By the divergence theorem, the integral of the divergence over a volume is equal to the flux of the vector field through the boundary:

$\int \frac{\partial}{\partial x_\beta}\left(\frac{\partial f}{\partial x_\alpha}\right) d\tau = \left[\frac{\partial f}{\partial x_\alpha}\right]_{\text{boundary}}$. We subtract the term that accounts for the reversed order: $-\int \frac{\partial^2 f}{\partial x_\beta \partial x_\alpha} d\tau$. Combining the boundary term and the remaining integral, we get the final formula:

SECOND- ORDER INTEGRATION BY PARTS FORMULA:

$$\boxed{\int \frac{\partial^2}{\partial x_\alpha \partial x_\beta} f d\tau = \left[\frac{\partial}{\partial x_\beta} f\right]_{\text{boundary}} - \int \frac{\partial^2}{\partial x_\beta \partial x_\alpha} f d\tau.} \tag{1.432}$$

Now, we start with the action integral on the left-hand side of Eq. (1.434) that involves $\mathcal{G}$, dependent on the metric tensor $g^{\mu\nu}$, its first derivatives $g^{\mu\nu}_\sigma$, and its second derivatives $g^{\mu\nu}_{\sigma\tau}$. During the variation of the action, we encounter terms involving the second derivatives $g^{\mu\nu}_{\sigma\tau}$. These terms can be written as:

$\int \frac{\partial \mathcal{G}}{\partial \left(g^{\mu\nu}_{\sigma\tau}\right)} - \delta g^{\mu\nu}_{\sigma\tau} d\tau$. Recognizing that $\delta g^{\mu\nu}_{\sigma\tau} = \frac{\partial}{\partial x_\sigma} \delta g^{\mu\nu}_\tau$, we can rewrite this term as:

$\int \frac{\partial \mathcal{G}}{\partial \left(g^{\mu\nu}_{\sigma\tau}\right)} \frac{\partial}{\partial x_\sigma} \delta g^{\mu\nu}_\tau d\tau$. This highlights the dependence on the second derivatives of the metric tensor.

We generalize this to the integral involving the second derivatives: $\boxed{\int \frac{\partial \mathcal{G}}{\partial \left(g^{\mu\nu}_{\sigma\tau}\right)} g^{\mu\nu}_{\sigma\tau} d\tau.}$

Accordingly, applying the integration by parts formula from Eq. (1.432) to the term $f = g^{\mu\nu}_{\sigma\tau}$, we get:

$$\int \frac{\partial \mathcal{G}}{\partial \left(g^{\mu\nu}_{\sigma\tau}\right)} g^{\mu\nu}_{\sigma\tau} d\tau = \boxed{\left[\frac{\partial \mathcal{G}}{\partial \left(g^{\mu\nu}_{\sigma\tau}\right)} \frac{\partial g^{\mu\nu}}{\partial x_\tau}\right]_{\text{boundary}}} - \int \frac{\partial}{\partial x_\tau}\left(\frac{\partial \mathcal{G}}{\partial \left(g^{\mu\nu}_{\sigma\tau}\right)} \frac{\partial g^{\mu\nu}}{\partial x_\sigma}\right) d\tau. \tag{1.433}$$

The first term on the right-hand side is a boundary term, and the second is a volume term. After partial integration, the new form of the Lagrangian density $\mathcal{G}^*$ will depend on $g^{\mu\nu}$ and $\frac{\partial g^{\mu\nu}}{\partial x_\sigma}$, but not on $\frac{\partial^2 g^{\mu\nu}}{\partial x_\sigma \partial x_\tau}$: $\mathcal{G} = \mathcal{G}^* +$ boundary terms where the boundary terms are given by F. The boundary terms F arise from the integration by parts process and are given by evaluating the terms at the boundary of the integration domain [the boxed term in Eq. (1.433)].

$\mathcal{G}^*$ is a function that depends on the metric tensor $g^{\mu\nu}$, its first derivatives $g_\sigma^{\mu\nu}$, the matter fields $q_{(\rho)}$, and the first derivatives of the matter fields $q_{(\rho)\alpha}$. However, it does not depend on the second derivatives of the metric tensor $g_{\sigma\tau}^{\mu\nu}$.

The integral F over the boundaries is a term that arises from the boundary conditions during integration by parts (see footnote 181).

Einstein applied the variational principle to both sides of Eq. (1.434) and this led to [E4-16][182]:

$$\delta\left(\int \mathcal{G} d\tau\right) = \delta\left(\int \mathcal{G}^* d\tau\right). \tag{1.437}$$

Thus, the new variational principle involves the function $\mathcal{G}^*$ instead of $\mathcal{L}$ and the variation:

$$\delta S = \delta \int \mathcal{G}^* d\tau = 0. \tag{1.438}$$

This modification simplifies the analysis since $\mathcal{G}^*$ is dependent on fewer variables, making the equations easier to handle.

Finally, Einstein derived the field equations for both gravitational and matter fields from the new variational principle (1.438) [E4-16]:

$$\begin{aligned} \frac{\partial}{\partial x_\alpha}\left(\frac{\partial \mathcal{G}^*}{\partial g_\alpha^{\mu\nu}}\right) - \frac{\partial \mathcal{G}^*}{\partial g^{\mu\nu}} &= 0, \\ \frac{\partial}{\partial x_\alpha}\left(\frac{\partial \mathcal{G}^*}{\partial q_{(\rho)\alpha}}\right) - \frac{\partial \mathcal{G}^*}{\partial q_{(\rho)}} &= 0. \end{aligned} \tag{1.439}$$

These are the Euler-Lagrange equations adapted to the fields and their interactions in the theoretical framework defined by Einstein at the beginning.

Recall that $\mathcal{L}$ does not depend on the second derivatives of the metric tensor $g_{\sigma\tau}^{\mu\nu}$. It only depends on the metric tensor $g^{\mu\nu}$, its first derivatives $g_\sigma^{\mu\nu}$, the matter fields $q_{(\rho)}$, and the first derivatives of the matter fields $q_{(\rho)\alpha}$.

---

[182] Applying the variational principle to both sides of the Eq. (1.434), we vary the integral of $\mathcal{G}$ and the sum of the integral of $\mathcal{G}^*$ plus the boundary term F:

$$\delta\left(\int \mathcal{G} d\tau\right) = \delta\left(\int \mathcal{G}^* d\tau + F\right). \tag{1.435}$$

Given the linear nature of the variation operator, we can separate the variations:

$$\delta\left(\int \mathcal{G}^* d\tau + F\right) = \delta\left(\int \mathcal{G}^* d\tau\right) + \delta F. \tag{1.436}$$

We assume that the field quantities (the metric tensor and matter fields) vanish at the boundaries of the integration domain. This implies that the variation of the boundary term F is zero: $\delta F = 0$. Therefore, the equation simplifies to (1.437).

Like Hilbert, Einstein divided the Lagrangian density $\mathcal{L}$ into two parts [E4-16]:

TOTAL LAGRANGIAN:

$$\boxed{\mathcal{L} = \mathcal{G} + \mathcal{M}.} \tag{1.440}$$

One part represents the gravitational field $\mathcal{G}$ and the other is the matter Lagrangian $\mathcal{M}$, which describes the matter content and its interaction with the gravitational field. Equation (1.440) is similar to Hilbert's equation (1.414). The $\mathcal{G}$ part of the Lagrangian density $\mathcal{L}$ includes gravitational fields, and it depends on the metric tensor $g^{\mu\nu}$, its first derivatives $g^{\mu\nu}_{\sigma}$, and its second derivatives $g^{\mu\nu}_{\sigma\tau}$. The matter Lagrangian $\mathcal{M}$ depends on $g^{\mu\nu}$, the matter fields $q_{(\rho)}$ and the first derivatives of the matter field $q_{(\rho)\alpha}$.

Separating the total Lagrangian $\mathcal{L}$ into $\mathcal{G}$ and $\mathcal{M}$ allowed Einstein to rewrite Eq. (1.439) as follows [E4-16][183]:

EINSTEIN'S NEW FIELD EQUATIONS:

$$\boxed{\frac{\partial}{\partial x_\alpha}\left(\frac{\partial \mathcal{G}^*}{\partial g^{\mu\nu}_{\alpha}}\right) - \frac{\partial \mathcal{G}^*}{\partial g^{\mu\nu}} = \frac{\partial \mathcal{M}}{\partial g^{\mu\nu}}.} \tag{1.441}$$

$$\frac{\partial}{\partial x_\alpha}\left(\frac{\partial \mathcal{M}}{\partial q_{(\rho)\alpha}}\right) - \frac{\partial \mathcal{M}}{\partial q_{(\rho)}} = 0. \tag{1.442}$$

Einstein then asserted that the general covariance of the field Eqs. (1.441) and (1.442) follows from the invariance of the following functions (scalar invariants) [E4-16]:

$$\frac{\mathcal{G}}{\sqrt{-g}}, \quad \frac{\mathcal{M}}{\sqrt{-g}}, \tag{1.443}$$

and the scalar invariant (1.428).[184]

Evidently, Hilbert's approach inspired Einstein. Let us compare Hilbert's Euler-Lagrange equations [(1.415) and (1.416)] to Einsein's (1.441) and (1.442). Hilbert's equations involve a more complex structure and additional terms that accommodate higher derivatives of the fields. Hilbert's equation (1.415) includes second derivatives with respect to the coordinates w_σ and w_ρ. Hilbert included second-order derivatives because he explored effects beyond general relativity, incorporating the effects of Mie's theory of the electrodynamics of matter. Einstein's general relativity did not require using Euler-Lagrange equations involving higher-order metric tensor

[183] Recall that $\mathcal{G}^*$ represents the same gravitational Lagrangian as $\mathcal{G}$ under the assumptions that Einstein made during his formulation and derivation.

[184] This means they remain unchanged under any coordinate transformations. The invariance of $\mathcal{G}$ and $\mathcal{M}$ under coordinate transformations implies that the equations formulated with these quantities will be generally covariant, meaning that the form of these equations remains the same regardless of how coordinates are changed. This property is fundamental in general relativity, which posits that the laws of physics are the same in all coordinate systems.

derivatives. Unlike Hilbert, Einstein did not work in 1916 on a unified field theory. Therefore, equations involving higher-order derivatives of the metric tensor extended beyond his general relativity equations.

Moreover, if $\mathcal{M}$ includes derivatives of $q_{(\rho)}$ higher than the first order, the Euler-Lagrange equations (1.442) will no longer be sufficient to derive the correct field equations for the dynamics of $q_{(\rho)}$. These equations must be extended to include terms for higher derivatives, leading to a more general form of Euler-Lagrange equations tailored for higher-order Lagrangians [E4-16]. Thus, while Einstein acknowledged that the Euler-Lagrange equations for $\mathcal{M}$ (1.442) were fundamentally important for a complete description of the dynamics of matter fields within his gravitational theory, he omitted these equations without significantly affecting his theory.

In the five-page appendix to part *D* of his paper, "The Foundation of the General Theory of Relativity," Einstein derived his field equations. He started from Eqs. (1.439) or (1.441) and assumed that since the integral of $\mathcal{G}^*$ is invariant when the variations at the integration boundaries vanish [Eq. (1.438)], the variations $\delta g^{\mu\nu}$ can be freely chosen. Thus, we get [CPAE8], Doc. 31:

$$\delta S = \int \delta g^{\mu\nu} \left[\frac{\partial}{\partial x_\alpha} \left(\frac{\partial \mathcal{G}^*}{\partial g_\alpha^{\mu\nu}} \right) - \frac{\partial \mathcal{G}^*}{\partial g^{\mu\nu}} \right] d\tau \tag{1.444}$$

The right-hand side of Eq. (1.444) depends only linearly on the metric tensor and its first and second derivatives with respect to the coordinates [Eq. (1.429)]. Thus, it follows:

$$\frac{\partial}{\partial x_\alpha} \left(\frac{\partial \mathcal{G}^*}{\partial g_\alpha^{\mu\nu}} \right) - \frac{\partial \mathcal{G}^*}{\partial g^{\mu\nu}} = \sqrt{-g} \left(\alpha R_{\mu\nu} + \beta g_{\mu\nu} g^{\sigma\tau} R_{\sigma\tau} \right). \tag{1.445}$$

$\alpha = 1$ and $\beta = -\frac{1}{2}$ represent constants. $R_{\mu\nu}$ is the Ricci tensor. Substituting R the Ricci scalar [Eq. (1.430)] into Eq. (1.445), we get [CPAE8], Doc. 31:

$$\frac{\partial}{\partial x_\alpha} \left(\frac{\partial \mathcal{G}^*}{\partial g_\alpha^{\mu\nu}} \right) - \frac{\partial \mathcal{G}^*}{\partial g^{\mu\nu}} = \frac{1}{\kappa} \sqrt{-g} \left(R_{\mu\nu} - \frac{1}{2} g_{\mu\nu} R \right). \tag{1.446}$$

From these equations, we can conclude that Einstein's above derivation in the Appendix to his 1916 review paper involved integrating various prior works, especially incorporating elements from his own 1914 *Entwurf* theory and Hilbert's derivation presented in his "Foundation of Physics" paper [Hil15-2] (see Sect. 1.7.2). This integration is particularly evident when comparing equations from different sources. For instance, the Eq. (1.216) from the 1914 *Entwurf* theory corresponds closely to the Eq. (1.444) in Einstein's 1916 Appendix. Similarly, a comparison between the Eq. (1.217) in the 1914 *Entwurf* theory and the left-hand side of the Eq. (1.445) in Einstein's 1916 manuscript reveals notable similarities. Furthermore, a clear relationship emerges when examining Hilbert's derivation and equation as expressed in (1.424) and Einstein's corresponding Eq. (1.446) in the 1916 Appendix. Einstein's adoption of Hilbert's notation, including the designation of the Ricci scalar with the letter K, further underscores the extent to which, in the above derivation, he assimilated elements of Hilbert's work into his theoretical framework.

Next, Einstein derived properties of the function $\mathcal{G}^*$. These were elucidated through considerations of variational principles. The equation $\frac{\mathcal{G}}{\sqrt{-g}}$ [Eq. (1.443)] is invariant under arbitrary transformations. However, $\frac{\mathcal{G}^*}{\sqrt{-g}}$ is invariant under linear substitutions of the coordinates. The starting point is the infinitesimal linear coordinate transformation x_ν, which Einstein had already written in his *Entwurf* review paper of 1914 [E2-14, E4-16][185]:

$$x'_\nu = x_\nu + \Delta x_\nu. \tag{1.447}$$

Einstein derived the transformation laws for the metric tensor components $g^{\mu\nu}$ and their derivatives $g^{\mu\nu}_\sigma$ [E2-14, E4-16][186]:

VARIATION OF THE METRIC TENSOR $\Delta g^{\mu\nu}$:

$$\Delta g^{\mu\nu} = g^{\mu\alpha}\frac{\partial \Delta x_\nu}{\partial x_\alpha} + g^{\nu\alpha}\frac{\partial \Delta x_\mu}{\partial x_\alpha}, \tag{1.448}$$

VARIATION OF THE DERIVATIVE OF THE METRIC TENSOR $\Delta g^{\mu\nu}_\sigma$:

$$\Delta g^{\mu\nu}_\sigma = \frac{\partial \Delta g^{\mu\nu}}{\partial x_\sigma} - g^{\mu\nu}_\alpha \frac{\partial \Delta x_\alpha}{\partial x_\sigma}. \tag{1.449}$$

These transformations of metric tensors under changes in coordinates are specifically linear transformations. According to Eq. (1.438)[187]:

$$\Delta \int \mathcal{G}^* d\tau = 0. \tag{1.452}$$

Einstein then calculated the variation of $\mathcal{G}^*$, $\Delta\mathcal{G}^*$, and:

[185] Equation (1.447) represents how the coordinates x_ν transform under an infinitesimal change Δx_ν. These transformations are localized, i.e., they only affect a limited region of spacetime. Δx_ν vanishes outside some arbitrarily chosen region of spacetime.

[186] Equations (1.448) and (1.449) represent the variations in the metric tensor components due to the coordinate transformations given by $\Delta g^{\mu\nu}$ and $\Delta g^{\mu\nu}_\sigma$.

[187] Given the integral expression:

$$\int \frac{\mathcal{G}^*}{\sqrt{-g}} d\tau = \int \frac{\mathcal{G}^*}{\sqrt{-g}} \sqrt{-g}\, d\tau = \int \mathcal{G}^* d\tau, \tag{1.450}$$

we consider the variation of this integral:

$$\Delta \int \mathcal{G}^* d\tau = \int \Delta\mathcal{G}^* d\tau, \quad \text{since } d\tau \text{ does not vary.} \tag{1.451}$$

.

$$\sqrt{-g}\Delta\left(\frac{\mathcal{G}^*}{\sqrt{-g}}\right). \tag{1.453}$$

Thus, $\mathcal{G}^*$ is treated as a scalar density.[188] Taking into consideration the transformations (1.448) and (1.449), Einstein ensured that physical laws were coordinate-invariant. The final result he obtained is [E4-16]:

$$\sqrt{-g}\Delta\left(\frac{\mathcal{G}^*}{\sqrt{-g}}\right) = S_\sigma^\nu \frac{\partial \Delta x_\sigma}{\partial x_\nu} + 2\frac{\partial \mathcal{G}^*}{\partial g_\alpha^{\mu\sigma}} g^{\mu\nu} \frac{\partial^2 \Delta x_\sigma}{\partial x_\nu \partial x_\alpha}. \tag{1.454}$$

This equation includes terms that account for changes in the metric components $\Delta g^{\mu\nu}$ and $\Delta g_\sigma^{\mu\nu}$ as well as the variation of the coordinate changes Δx_σ with respect to the coordinates x_ν [Eqs. (1.448) and (1.449)].
S_σ^ν is defined as follows [E4-16][189]:

$$S_\sigma^\nu = 2\left(\frac{\partial \mathcal{G}^*}{\partial g^{\mu\sigma}} g^{\mu\nu} + \frac{\partial \mathcal{G}^*}{\partial g_\alpha^{\mu\sigma}} g_\alpha^{\mu\sigma}\right) + \delta_\sigma^\nu \mathcal{G}^* - \frac{\partial \mathcal{G}^*}{\partial g_\nu^{\mu\alpha}} g_\sigma^{\mu\alpha}. \tag{1.457}$$

The $\sqrt{-g}\Delta\left(\frac{\mathcal{G}^*}{\sqrt{-g}}\right) = 0$ in Eq. (1.454) is a stationary action under a variation of $\mathcal{G}^*$.[190]

[188] Einstein multiplied the invariant $\frac{\mathcal{G}^*}{\sqrt{-g}}$ by $\sqrt{-g}$ [see Eq. (1.450)]. Thus, he converted it back into a scalar density. After applying a variation Δ to $\frac{\mathcal{G}^*}{\sqrt{-g}}$, it results in a scalar density, not a scalar. Integrating a scalar density, such as $\mathcal{G}^*$, through τ yields a result that is independent of the choice of coordinates [Eq. (1.450)].

[189] We begin by calculating $\Delta(\mathcal{G}^*)$ using the transformations (1.448) (1.449). To find the variation of $\mathcal{G}^*$, recall that $\mathcal{G}^*$ depends on the metric tensor $g^{\mu\nu}$ and its first derivatives $g_\sigma^{\mu\nu}$:

$$\Delta\mathcal{G}^* = \frac{\partial \mathcal{G}^*}{\partial g^{\mu\nu}} \Delta g^{\mu\nu} + \frac{\partial \mathcal{G}^*}{\partial g_\sigma^{\mu\nu}} \Delta g_\sigma^{\mu\nu}. \tag{1.455}$$

We substitute the variations from Eqs. (1.448) to (1.449) into this expression (1.455) to find the total variation of $\mathcal{G}^*$:

$$\Delta\mathcal{G}^* = \frac{\partial \mathcal{G}^*}{\partial g^{\mu\nu}} \left(g^{\mu\alpha} \frac{\partial \Delta x_\nu}{\partial x_\alpha} + g^{\nu\alpha} \frac{\partial \Delta x_\mu}{\partial x_\alpha}\right) + \frac{\partial \mathcal{G}^*}{\partial g_\sigma^{\mu\nu}} \left(\frac{\partial \Delta g^{\mu\nu}}{\partial x_\sigma} - g_\alpha^{\mu\nu} \frac{\partial \Delta x_\alpha}{\partial x_\sigma}\right). \tag{1.456}$$

The above equations finally lead to Eq. (1.454) where S_σ^ν is given by Eq. (1.457). Thus, these two equations are derived from the transformations (1.448) and (1.449) and the variations.

[190] We recognize that the left-hand side of Eq. (1.454) can be written by using the product rule to expand the left-hand side of Eq. (1.456). Thus, the variation of $\sqrt{-g}\mathcal{G}^*$ involves:

$$\Delta\left(\sqrt{-g}\mathcal{G}^*\right) = \sqrt{-g}\Delta\mathcal{G}^* + \mathcal{G}^*\Delta\sqrt{-g}. \tag{1.458}$$

We substitute the variation of $\sqrt{-g}$:

$$\Delta\sqrt{-g} = -\frac{1}{2}\sqrt{-g}g^{\mu\nu}\Delta g_{\mu\nu} \text{ in Eq. (1.458) and obtain:} \tag{1.459}$$

Thus, according to Eq. (1.457), $\mathcal{G}^*$ satisfies the condition [E4-16][191]:

$$S_\sigma^\nu \equiv 0, \tag{1.467}$$

and the second term in Eq. (1.454) vanishes.[192] Einstein argued that the value of the integral of the action [Eq. (1.434)] over the boundary does not change during the

---

$$\Delta\left(\sqrt{-g}\mathcal{G}^*\right) = \sqrt{-g}\left(\Delta\mathcal{G}^* - \frac{1}{2}\mathcal{G}^* g^{\mu\nu}\Delta g_{\mu\nu}\right). \tag{1.460}$$

Now, we recognize that: $$\frac{\Delta\left(\sqrt{-g}\mathcal{G}^*\right)}{\sqrt{-g}} = \sqrt{-g}\Delta\left(\frac{\mathcal{G}^*}{\sqrt{-g}}\right). \tag{1.461}$$

Thus, we can substitute the left-hand side of Eq. (1.461) with the right-hand side on the left-hand side of Eq. (1.460). We have the final equation:

$$\sqrt{-g}\Delta\left(\frac{\mathcal{G}^*}{\sqrt{-g}}\right) = \Delta\mathcal{G}^* - \frac{1}{2}\mathcal{G}^* g^{\mu\nu}\Delta g^{\mu\nu}. \tag{1.462}$$

To find the field equations, we vary the action (1.452) with respect to $\mathcal{G}^*$:

$$\delta S = \int \delta\left(\sqrt{-g}\mathcal{G}^*\right) d\tau. \tag{1.463}$$

Using the expression (1.460), we have: $$\delta\left(\sqrt{-g}\mathcal{G}^*\right) = \sqrt{-g}\left(\delta\mathcal{G}^* - \frac{1}{2}\mathcal{G}^* g^{\mu\nu}\delta g_{\mu\nu}\right). \tag{1.464}$$

The variational principle states that for the action to be stationary, $\delta S = 0$ for arbitrary variations $\delta\mathcal{G}^*$.

Therefore: $$\int \sqrt{-g}\left(\delta\mathcal{G}^* - \frac{1}{2}\mathcal{G}^* g^{\mu\nu}\delta g_{\mu\nu}\right) d\tau = 0. \tag{1.465}$$

Compare this equation to Eq. (1.197), Sect. 1.5.1. For the integral to be zero for arbitrary variations $\delta\mathcal{G}^*$ and $\delta g_{\mu\nu}$, the integrand itself must vanish.

Dividing through by $\sqrt{-g}$, we get: $$\delta\mathcal{G}^* - \frac{1}{2}\mathcal{G}^* g^{\mu\nu}\delta g_{\mu\nu} = 0. \tag{1.466}$$

[191] This implies that the right-hand side of Eq. (1.454) must be zero. For this to hold for arbitrary variations Δx_σ, each term on the right-hand side must individually vanish. Since $\frac{\partial \Delta x_\sigma}{\partial x_\nu}$ can vary independently, for the term $S_\sigma^\nu \frac{\partial \Delta x_\sigma}{\partial x_\nu}$ to be zero for all variations, Eq. (1.467) must hold.

[192] If Eq. (1.467) holds, then the Eq. (1.454) simplifies to:

$$2\frac{\partial \mathcal{G}^*}{\partial g_\alpha^{\mu\sigma}} g^{\mu\nu}\frac{\partial^2 \Delta x_\sigma}{\partial x_\nu \partial x_\alpha} = 0. \tag{1.468}$$

This reduced form comprises the integrand of Eq. (1.470). Given that this term must vanish for arbitrary variations, this implies that the coefficient of the second term, $\frac{\partial \mathcal{G}^*}{\partial g_\alpha^{\mu\sigma}} g^{\mu\nu}$, must also be zero.

transformation ($\Delta(F) = 0$).[193] He concluded that the invariance of the action under coordinate transformations implies that [E4-16][194]:

$$\Delta\left(\int \mathcal{G}\, d\tau\right) = \Delta\left(\int \mathcal{G}^*\, d\tau\right). \tag{1.469}$$

The variation of the action involves how $\mathcal{G}^*$ changes due to coordinate transformations, which is described by Eq. (1.454).

S_σ^ν were instrumental in deriving the contracted Bianchi identity.
With condition (1.467) and Eq. (1.454), Einstein wrote the integral [E4-16]:

$$\int \frac{\partial \mathcal{G}^*}{\partial g_\alpha^{\mu\sigma}} g^{\mu\nu} \frac{\partial^2 \Delta x_\sigma}{\partial x_\nu \partial x_\alpha} d\tau = 0. \tag{1.470}$$

from which he derived the contracted Bianchi identities [E4-16][195]:

[193] see footnote 182. Recall that Einstein considered transformations (1.447) that only affected a limited region of spacetime and vanished in an infinitesimal neighborhood of the boundary.

[194] Since both $\frac{\mathcal{G}}{\sqrt{-g}}$ and $\sqrt{-g}\tau$ (the infinitesimal proper time interval) are invariants, any variations in these quantities due to coordinate transformations should cancel out when we take the variation on both sides of the equation. If the left-hand side of Eq. (1.469) vanishes ($\Delta\left(\int \mathcal{G}\, d\tau\right)$), the right-hand side also vanishes.

[195] We start with the integral: $\int \frac{\partial^2 f}{\partial x_\alpha \partial x_\beta} g\, d\tau$, transfer one derivative from f to g. Recall from footnote 181 that $\frac{\partial^2 f}{\partial x_\alpha \partial x_\beta}$ can be written as the divergence of a first derivative:

$$\int \frac{\partial}{\partial x_\beta}\left(\frac{\partial f}{\partial x_\alpha} g\right) d\tau - \int \frac{\partial f}{\partial x_\alpha}\frac{\partial g}{\partial x_\beta} d\tau. \text{ By the divergence theorem, the first term (integral) is equal to:} \tag{1.471}$$

$$\int \frac{\partial}{\partial x_\beta}\left(\frac{\partial f}{\partial x_\alpha} g\right) d\tau = \left[\frac{\partial f}{\partial x_\alpha} g\right]_{\text{boundary}}. \quad \text{This gives us (see footnote 181):} \tag{1.472}$$

FIRST- ORDER INTEGRATION BY PARTS WITH SECOND DERIVATIVE TERMS:

$$\int \frac{\partial^2 f}{\partial x_\alpha \partial x_\beta} g\, d\tau = \left[\frac{\partial f}{\partial x_\alpha} g\right]_{\text{boundary}} - \int \frac{\partial f}{\partial x_\beta}\frac{\partial g}{\partial x_\alpha} d\tau. \tag{1.473}$$

To derive the contracted Bianchi identity (1.476) from the Eq. (1.470), Einstein used the formula integration by parts (1.473) twice. 1. *Integration by parts (first time):* first, we consider the term $\frac{\partial^2 \Delta x_\sigma}{\partial x_\nu \partial x_\alpha}$, and use integration by parts to remove one of the derivatives:

$$0 = \int \frac{\partial \mathcal{G}^*}{\partial g_\alpha^{\mu\sigma}} g^{\mu\nu} \frac{\partial^2 \Delta x_\sigma}{\partial x_\nu \partial x_\alpha} d\tau = \boxed{\int \frac{\partial}{\partial x_\nu}\left(\frac{\partial \mathcal{G}^*}{\partial g_\alpha^{\mu\sigma}} g^{\mu\nu}\right)\frac{\partial \Delta x_\sigma}{\partial x_\alpha} d\tau} - \int \frac{\partial}{\partial x_\nu}\left(\frac{\partial \mathcal{G}^*}{\partial g_\alpha^{\mu\sigma}} g^{\mu\nu}\frac{\partial \Delta x_\sigma}{\partial x_\alpha}\right) d\tau. \tag{1.474}$$

The first term (the boxed term) represents the boundary contribution. This term arises from the boundary when the integration by parts formula is applied. The remaining integral term represents the volume contribution with the derivative shifted. By shifting one of the derivatives from Δx_σ

THE CONTRACTED BIANCHI IDENTITY:

$$\boxed{\frac{\partial^2}{\partial x_\nu \partial x_\alpha}\left(g^{\mu\nu}\frac{\partial \mathcal{G}^*}{\partial g_\alpha^{\mu\sigma}}\right) = 0.} \tag{1.476}$$

Einstein derived a strong condition implying conservation properties about the field equations described by $\mathcal{G}^*$ and the metric tensor $g^{\mu\nu}$.

He defined the gravitational pseudo-tensor t_σ^ν in terms of $\mathcal{G}^*$ and its derivatives for the metric tensor. He derived two equivalent forms of the stress-energy pseudo-tensor of the gravitational field from Eq. (1.457). Considering condition (1.467), Eq. (1.457) takes the form:

$$-\left(g^{\mu\nu}\frac{\partial \mathcal{G}^*}{\partial g^{\mu\sigma}} + g_\alpha^{\mu\nu}\frac{\partial \mathcal{G}^*}{\partial g_\alpha^{\mu\sigma}}\right) = \frac{1}{2}\left(\delta_\sigma^\nu \mathcal{G}^* - g_\sigma^{\mu\alpha}\frac{\partial \mathcal{G}^*}{\partial g_\nu^{\mu\alpha}}\right). \tag{1.477}$$

The terms on the right-hand side of the Eq. (1.477) lead to the following form of the stress-energy pseudo-tensor equation [E4-16]:

$$t_\sigma^\nu = \frac{1}{2}\left(\delta_\sigma^\nu \mathcal{G}^* - g_\sigma^{\mu\alpha}\frac{\partial \mathcal{G}^*}{\partial g_\nu^{\mu\alpha}}\right). \tag{1.478}$$

The terms on the left-hand side of the Eq. (1.477) are also t_σ^ν, and represent the second form of the stress-energy pseudo-tensor t_σ^ν [E4-16]:

$$t_\sigma^\nu = -\left(g^{\mu\nu}\frac{\partial \mathcal{G}^*}{\partial g^{\mu\sigma}} + g_\alpha^{\mu\nu}\frac{\partial \mathcal{G}^*}{\partial g_\alpha^{\mu\sigma}}\right). \tag{1.479}$$

---

to the pre-factor $\frac{\partial \mathcal{G}^*}{\partial g_\alpha^{\mu\sigma}} g^{\mu\nu}$, we separate the integral into boundary terms and a remaining volume integral with reduced derivative order.

The second term vanishes because it is a total divergence over the entire volume, assuming appropriate boundary conditions (that the variations Δx_σ and their derivatives vanish on the boundary). Therefore, we are left with the boxed term in Eq. (1.474). 2. *Integration by parts (second time):* Next, we integrate by parts again the boxed term in Eq. (1.474) to remove the remaining derivative from $\frac{\partial \Delta x_\sigma}{\partial x_\alpha}$:

$$0 = \int \frac{\partial}{\partial x_\nu}\left(\frac{\partial \mathcal{G}^*}{\partial g_\alpha^{\mu\sigma}} g^{\mu\nu}\right)\frac{\partial \Delta x_\sigma}{\partial x_\alpha} d\tau = \boxed{-\int \frac{\partial^2}{\partial x_\nu \partial x_\alpha}\left(\frac{\partial \mathcal{G}^*}{\partial g_\alpha^{\mu\sigma}} g^{\mu\nu}\right)\Delta x_\sigma d\tau}$$
$$+\int \frac{\partial}{\partial x_\alpha}\left(\frac{\partial}{\partial x_\nu}\left(\frac{\partial \mathcal{G}^*}{\partial g_\alpha^{\mu\sigma}} g^{\mu\nu}\right)\Delta x_\sigma\right) d\tau. \tag{1.475}$$

Again, the second term vanishes under the assumption that the variations Δx_σ and their derivatives vanish on the boundary. The boxed term in Eq. (1.475) vanishes. For the integral (1.470) to be zero for arbitrary variations Δx_σ, the integrand of the boxed term must vanish. If the integrand did not vanish, we could choose a specific Δx_σ such that the integral would not be zero, which contradicts the assumption that the left-hand side is zero. The integrand of the boxed term in Eq. (1.475) $= 0$ is the contracted Bianchi identity Eq. (1.476).

Einstein interconnected three equations to yield new field equations:
1. The definitions for the stress-energy pseudo tensor t_σ^ν [Eq. (1.478) and the alternate form, expressing a canonical stress-energy tensor due to $\mathcal{G}^*$ (1.479)].
2. The expression involving $\mathcal{M}$, which represents the interaction of matter fields with the metric:

$$-\frac{\partial \mathcal{M}}{\partial g^{\mu\sigma}} g^{\mu\sigma} = \mathcal{I}_\sigma^\nu, \tag{1.480}$$

He substituted the forms of t_σ^ν and $\mathcal{I}_\sigma^\nu$ into his field equations Eq. (1.441). Combining the forms and their derivatives and equating terms with similar tensorial structures led him to the general field equation for a metric-dependent Lagrangian $\mathcal{G}^*$ [E4-16]:

THE NEW FIELD EQUATIONS:

$$\boxed{\frac{\partial}{\partial x_\alpha}\left(g^{\mu\nu}\frac{\partial \mathcal{G}^*}{\partial g_\alpha^{\mu\sigma}}\right) = \mathcal{I}_\sigma^\nu + t_\sigma^\nu.} \tag{1.481}$$

Einstein then applied the contracted Bianchi identities (1.476). The divergence of the left-hand side of the field Eq. (1.481) vanishes. Since the left-hand side of the equation has a vanishing divergence due to the Bianchi identities, the divergence of the right-hand side must also vanish. This implies the conservation of the combined quantities $\mathcal{I}_\sigma^\nu$ and t_σ^ν, the conservation of the total energy-momentum tensor, Eq. (1.258), which includes contributions from both the matter fields and the gravitational field. By employing the contracted Bianchi identities, Einstein could anchor the conservation of energy and momentum not just as a separate, standalone law of physics but as a natural consequence of his field equations.

Next, Einstein contracted the field Eq. (1.441) by multiplying both sides by $g^{\mu\nu}$ and summing over μ and ν.[196]

$$g^{\mu\nu}\left(\frac{\partial}{\partial x_\alpha}\left(\frac{\partial \mathcal{G}^*}{\partial g_\alpha^{\mu\nu}}\right) - \frac{\partial \mathcal{G}^*}{\partial g_{\mu\nu}}\right) = g^{\mu\nu}\frac{\partial \mathcal{M}}{\partial g_{\mu\nu}}, \tag{1.484}$$

The stress-energy pseudo tensor (1.478), and the energy momentum conservation law (1.258), yield [E4-16]:

[196] The first term becomes:

$$g^{\mu\nu}\frac{\partial}{\partial x_\alpha}\left(\frac{\partial \mathcal{G}^*}{\partial g_\alpha^{\mu\nu}}\right), \tag{1.482}$$

and the second term is the trace of the derivative of $\mathcal{G}^*$ with respect to the metric tensor (this derivative is obtained by contracting it with the metric tensor):

$$g^{\mu\nu}\frac{\partial \mathcal{G}^*}{\partial g_{\mu\nu}}. \tag{1.483}$$

The right-hand side of Eq. (1.441) becomes Eq. (1.480), and can be replaced with $\mathcal{I}_\sigma^\nu$. Combining both terms [(1.482)) and (1.483)] leads to the left-hand side of Eq. (1.484).

THE ENERGY- MOMENTUM BALANCE (GENERALLY COVARIANT):

$$\boxed{\frac{\partial \mathcal{I}_\sigma^\nu}{\partial x_\nu} - \frac{1}{2} g_\sigma^{\mu\nu} \mathcal{I}_{\mu\nu} = 0.} \tag{1.485}$$

At first glance, the equations of the 1914 *Entwurf* theory [E2-14] and Einstein's 1916 theory [E4-16] may seem very similar, especially when comparing the stress-energy pseudo-tensors for gravitational field from both the *Entwurf* theory and the 1916 equations and the equations representing the energy-momentum balance in a gravitational field [Eqs. (1.203) and (1.485)] from both theories. Einstein's 1916 Eq. (1.457) bears a resemblance to Eq. (1.221) from his 1914 *Entwurf* paper [E2-14]. It is even tempting to say that $\mathcal{G}^*$ is the venerable 1914 *Entwurf* gravitational tensor (1.217) rescued from obscurity, and now it has been adorned into an invariant quantity. In his 1914 *Entwurf* theory, Einstein had already employed a variational approach that yielded Euler-Lagrange equations—field equations constrained to adapted coordinates, thereby lacking general covariance. These *Entwurf* field equations included an extra term. To rectify this limitation, Einstein imposed the restriction $B_\sigma = 0$ to eliminate this superfluous term and imposed the condition $S_\sigma^\nu = 0$ to simplify the equations. Scholars have noticed that the contracted Bianchi identity (1.476) and the 1914 *Entwurf* $B_\sigma = 0$ restriction (1.222) have a similar mathematical form. They both involve second derivatives of the metric and some functional of the metric for the coordinates [Jan-07].

In his 1916 paper, "Hamilton's Principle and the General Theory of Relativity," Einstein integrated significant theoretical elements from both his and Hilbert's prior work [Hil15-2] to develop the framework of general relativity further. This integration encompassed Hilbert's AXIOM I and his variational methodology, alongside Einstein's own approaches outlined in his 1914 *Entwurf* theory, "The Formal Foundation of the General Theory of Relativity" [E2-14] and his 1915 general relativity. Einstein's primary objective was to eliminate reliance on the unimodular coordinate condition, simplifying his earlier derivations in 1915. Adopting Hilbert's methodological advances allowed Einstein the necessary tools to achieve this aim.

Einstein's realization in 1915 that the general Lagrangian he posited in 1914 [Eq. (1.212) (1914), and Eq. (1.249) (1915)] could yield the field equations of general relativity marked a turning point. This insight meant that the variational principle from 1914 could indeed underpin generally covariant field equations. The synthesis of his 1914 methodology [E2-14] with Hilbert's variational approach [Hil15-2] culminated in Einstein's elegant exposition over six pages in his 1916 paper [E4-16], achieving a significant milestone in the development of general relativity.

1.8 Neurath's Boat and Einstein the Sailor

Einstein's use of the machinery from the *Zurich Notebook* and the *Entwurf* theory in developing general relativity illustrates a critical scientific strategy: leveraging previous frameworks as stepping stones.

The *Zurich Notebook* encapsulated foundational equations pivotal to formulating the 1915 general theory of relativity. It included the line element, the contravariant stress-energy tensor, and the energy-momentum balance in a gravitational field. The harmonic restriction, the flat (Minkowski) spacetime metric coupled with a perturbation and the conservation law for the gravitational field were also documented. Furthermore, it contained the Riemann curvature tensor, the Ricci tensor, the Christoffel symbols, the November tensor, and a Laplacian-like term. Early formulations of the geodesic equation, linearized forms resembling the final 1915 field equations, and even the *Entwurf* field equations, along with a version of the stress-energy pseudo tensor, were meticulously detailed in the notebook. These initial mathematical formulations underpinned the development of Einstein's subsequent scholarly works, serving as critical components in the framework of his theoretical edifice.

Einstein concluded the *Zurich Notebook* with the *Entwurf* theory. While the *Entwurf* theory had limitations, the underlying principles and mathematical structures of this theory provided a foundation for Einstein to build corrected ideas. The shift from the *Entwurf* theory to general relativity was not just a simple correction of errors. Instead, Einstein re-conceptualized his approach based on the shift to generally covariant field equations. While the derivation of the *Entwurf* field equations relied on specific coordinate restrictions, the "Hamilton's Principle and the General Theory of Relativity" paper turned them into identities. In the *Entwurf* theory, the equations were more about coordinate conditions and restrictions. In contrast, in general relativity, they represented a deep connection to energy-momentum conservation embodied in the variational procedure. This reflects a deeper conceptual shift rather than a mere mathematical adjustment. The *Entwurf* theory served as an inspiration and a conceptual springboard for Einstein. It is a testament to his scientific creativity and ability to transcend beyond the limitations of his previous work. By re-deriving and reinterpreting these equations under the new principles of general relativity, Einstein demonstrated a remarkable capacity to adapt and evolve his theoretical understanding.

Otto Neurath compares the development of our thinking and concepts to the shifting structure of a boat. As we integrate new concepts and connections, the structure of the boat changes, including the relationships between concepts and their central importance. This process is dynamic and continuous, affecting each concept to varying degrees. He suggests that with every advancement in thought, our previous understandings and expressions should ideally be revised to reflect these changes. However, this is challenging because our language and expression are linear, while conceptual changes are more complex.

Neurath uses the analogy of a miner with a lamp in a mine to illustrate how our understanding illuminates only parts of a broader, mostly unknown field. As we shift our focus, what was once illuminated can become obscure, and new areas emerge. This represents the selective and changing nature of our understanding and cognition. Neurath emphasizes that our cognitive process involves a network of interrelated concepts, not isolated ones. He argues that conceptual relations are not just tools for understanding (as posited by Ernst Mach) but are cognition in themselves. This interconnectedness of concepts burdens thinkers, who must always consider the entire mass of concepts, even though they cannot fully comprehend them all at once. New ideas must grow out of old ones.

Finally, the metaphor of sailors who must rebuild their ship at sea conveys Neurath's view. Sailors cannot start from scratch but must replace parts while using the rest of the ship for support. This process symbolizes how our knowledge and understanding are constantly rebuilt using what we already know in a gradual, ongoing process. Neurath's boat metaphor illustrates the dynamic, complex, and interconnected nature of knowledge and conceptual change. It emphasizes the continuous evolution of our understanding, the challenges in fully grasping our conceptual frameworks, and the necessity of building on existing knowledge. Neurath's boat metaphor suggests that just like repairing a boat at sea, where we cannot rebuild it from the bottom but must fix it piece by piece while staying afloat, knowledge and theories are also constructed and adjusted while in use. There is no opportunity to start from a solid foundation; we must continuously adapt and adjust our knowledge based on what works in practice [Neur21, Car-99].[197]

[197] In his paper, "Protocol Sentences," Neurath emphasizes the necessity of a *universal jargon* in unified science. He acknowledges that while efforts are made to increase precision, no term in unified science is entirely free of imprecision because all terms are ultimately linked to imprecise *protocol sentences.Protocol sentences* are basic statements that describe immediate observations and experiences. They are intended to serve as foundational elements in scientific language, recording raw empirical data. Neurath critiques the idea of an ideal language built from precise, foundational elements—similar to what he calls the *Laplacian spirit. Laplace's spirit* represents a deterministic view where perfect knowledge of initial conditions allows for precise future predictions. He states that scientific language cannot be seen as approximating such an ideal language even as it becomes more systematically symbolized. Everyday language, filled with imprecise and unanalyzed terms, which he refers to as *Ballungen* (clusters), serves as the starting point. He notes that imprecision in scientific language is inevitable and persistent. Even as precision increases in one area, imprecision might reappear more strongly elsewhere. Neurath discusses *protocol sentences* concerning *Ballungen*. He emphasizes that even fundamental *protocol sentences* are not free from imprecision. The terms used in these sentences are derived from everyday language, which is inherently imprecise and filled with *Ballungen*. Neurath invokes the boat metaphor, highlighting that imprecise *Ballungen* are always part of scientific knowledge. Just as the sailors cannot fully dismantle their ship, scientists cannot eliminate imprecision from their language and concepts. Scientific language is inherently imprecise and interconnected, like a ship that must be continuously rebuilt at sea without being fully dismantled. There is no perfect starting point for science, just as there is no *tabula rasa*. Scientific language evolves from and is intertwined with the imprecise *Ballungen*. This metaphor illustrates that scientists must work with the existing imprecise language and concepts, continuously refining them. Just as sailors rebuild their ships at sea without completely dismantling them, scientists improve their language and theories without eliminating the foundational imprecision. Efforts to increase precision in one area may lead to increased imprecision in another, reflecting scientific knowledge's dynamic and holistic nature [Neur32].

Einstein's approach to developing general relativity was recursive and involved revising and refining his ideas. His journey can be viewed through Neurath's boat metaphor. Neurath's boat metaphor suggests a process of continuous reconstruction where parts are replaced and adjusted while the whole structure remains afloat. Einstein's pathway mirrors this approach. He did not discard his entire theoretical framework when encountering problems but replaced and refined its parts. Between 1913 and 1915, Einstein engaged in a dynamic construction of his papers, meticulously adapting and adjusting his derivations. He built these papers upon the foundational equations recorded in his *Zurich notebook*, illustrating a process of continuous refinement and evolution in his theoretical development.

In the elucidation provided in Sect. 1.3, the truncated variant of the November tensor, particularly its second constituent, manifests within the ambit of the 1913 *Entwurf* field equations. This recurrence suggests that Einstein may not have categorically abandoned the November tensor approach. During the early tenure of the *Entwurf* period, it appears that Einstein sought a harmonization between the methodologies espoused in the *Entwurf* and those underpinning the Ricci tensor. This assertion implies that Einstein's detachment from the November tensor was not absolute, even in the face of conviction alongside Grossmann that the Newtonian limit was irreconcilable with the November tensor framework. Such a stance underscores continuity in Einstein's theoretical pursuits, reflecting an ongoing endeavor to integrate disparate methodological elements within his evolving gravitational theory.

Just as a sailor on Neurath's boat cannot start from scratch but must adapt and repair using what is already in place, Einstein adapted his theories (the *Entwurf* theory and the *Zurich Notebook*) based on new insights and corrections. This reflects the process of scientific progress, where old ideas are not entirely discarded but modified and integrated into new frameworks. Einstein's return to his 1912 covariant tensor in 1915, after realizing the limitations of the non-covariant *Entwurf* field equations, exemplifies the idea of using "old beams" (previous theories) to shape the "ship" (theoretical framework) anew.

Moreover, the evolutionary trajectory of Einstein's theoretical formulations, from his early work, depicted within the *Zurich Notebook* to the culminating achievement of general relativity in 1915, bears a striking resemblance to the artistic process of *pentimento* observed in the works of Leonardo da Vinci. The term *pentimento*,

---

Neurath rejected the foundationalist view that there are certain basic, indubitable propositions upon which all scientific knowledge is built. Instead, he believed that scientific knowledge is more like a web of statements that support each other, which aligns with his holistic view of science. He advocated for an encyclopedic approach to knowledge, where different fields of science are not seen as separate but as parts of a larger, interconnected whole. His work on the *International Encyclopedia of Unified Science* reflects this approach, which aimed to present a coherent picture of scientific knowledge as a unified whole. The *Ballungen* are not confined to a single scientific discipline. They involve concepts and theories from various fields that address complex problems and phenomena. The clusters are not static. As scientific knowledge progresses, the relationships within and between these clusters change. New connections are formed, and old ones may become less relevant. The relevance and importance of certain clusters can vary depending on the context in which they are considered. Different scientific problems and research questions can highlight different clusters of concepts and theories.

derived from the Italian for *repentance*, refers to the alterations an artist makes during the painting process, which are often obscured by subsequent layers of paint. Similar to conceptual drafts, these underlying sketches and changes offer a window into the artist's evolving creative thought process. Leonardo da Vinci employed this technique extensively, laying initial sketches in *sinopia* on the gesso before overlaying them with the final painting layers. Though concealed beneath the surface, these preliminary designs, or *pentimenti*, play a pivotal role in understanding the artist's developmental process. The primary method for uncovering these hidden layers is through infrared reflectography. This technique has revealed how motifs sketched in the initial stages of one artwork recurrently appear across Leonardo's oeuvre. This practice of reusing and adapting motifs across different compositions illustrates Leonardo's iterative approach to creativity, wherein he repurposed and recontextualized existing ideas within new artistic endeavors.

Analogously, Einstein's journey from the preliminary formulations found in the *Zurich Notebook* through to the *Entwurf* theory, and ultimately to the comprehensive framework of general relativity, exemplifies a scientific form of *pentimento*. The *Zurich Notebook*, a conceptual underpainting, contains the embryonic forms of equations and operators that Einstein would refine and reconfigure in his subsequent theoretical explorations. This recursive process, characterized by the adaptation and transformation of initial ideas, mirrors the artistic method of employing *pentimenti*. Just as Leonardo da Vinci revisited and reshaped motifs across his paintings, Einstein reexamined and modified his initial mathematical constructs, adapting them to the evolving landscapes of his theories.

This parallel between the artistic process of *pentimento* and Einstein's theoretical development underscores the fluidity of creativity, whether manifested in the visual arts or theoretical physics. It highlights the importance of initial, exploratory manuscripts like the *Zurich Notebook* in the gestation of revolutionary ideas. It demonstrates that the path to mastery, whether in art or science, is continual revision, reevaluation, and reinvention.

Chapter 2
Schwarzschild, Droste, and Nordström Solve Einstein's Field Equations

2.1 Introduction

This chapter examines the contributions of Karl Schwarzschild, Johannes Droste, Hans Reissner, and Gunnar Nordström.

I start with Schwarzschild. Schwarzschild communicated with Einstein, providing two exact solutions to Einstein's field equations: an exterior solution for the vacuum field equations and an interior solution for field equations with sources. Einstein presented these solutions to the Prussian Academy. The exterior solution featured a physical singularity at $r = 0$ and a coordinate singularity at $r = 2GM$. On the other hand, the interior solution did not have any singularity.

Schwarzschild took several ingenious steps to solve Einstein's vacuum field equations under spherical symmetry and static fields. His approach culminated in deriving the now-famous Schwarzschild solution (the exterior solution). Schwarzschild started with the most general line element satisfying Einstein's conditions of static fields, spherical symmetry, and the determinant condition (the metric determinant equals -1). He cleverly introduced a coordinate transformation to simplify calculations. This involved changing from rectangular to polar coordinates and then defining new variables that maintained the determinant condition while retaining the advantages of polar coordinates. Schwarzschild derived the field equations for these new variables and used the determinant condition to relate four functions in the metric tensor. He simplified the field equations by focusing on the equator and solving them under the determinant condition. This led to expressions for the metric functions in terms of the new radial coordinate. Schwarzschild introduced a redefined radial coordinate (now known as the Schwarzschild radius) to express the solution in a more familiar form, returning it to standard spherical coordinates. The final form of the Schwarzschild line element describes spacetime around a spherical, non-rotating mass in a vacuum.

In his doctoral dissertation at the University of Leiden under the mentorship of Hendrik Lorentz, Johannes Droste independently derived a solution to Einstein's field equations concurrent with Schwarzschild. In May 1917, Einstein expressed to Paul

G. Weinstein, *Einstein's Legacy*, History of Physics,
https://doi.org/10.1007/978-3-031-73572-1_2

Ehrenfest that he found Droste's dissertation to be remarkably impressive [CPAE5], Doc. 344. While both Droste and Schwarzschild converged on analogous solutions, their methodologies diverged significantly, reflecting their interpretative approaches to addressing the problem of solving Einstein's equations under conditions of spherical symmetry in a vacuum. Droste commenced his thesis by declaring his departure from the unimodular coordinate system (1.299) frequently utilized by Einstein in his seminal works on general relativity [Dro-16]. The unimodular coordinate condition predominantly characterized this system, $\sqrt{-g} = 1$. Droste articulated a deliberate choice to eschew this framework, which was emblematic of Einstein's general theory of relativity in 1915–1916.[1] By abstaining from using the unimodular condition, Droste's methodology followed a distinct path through the complexity of Einstein's equations. Droste relied on a variational principle, which allowed him to bypass the constraints imposed by the unimodular condition. This resulted in fewer mathematical manipulations than Schwarzschild's approach. In deriving the field equations, Droste adopted David Hilbert's methodology (see Sect. 1.7.2), particularly Hilbert's use of a variational principle in "Foundations of Physics" [Hil15-2] (see Sect. 1.7.2). Droste equated α in his line element to $\frac{2Gm}{c^2}$, resulting in the well-known expression for the Schwarzschild radius and the Schwarzschild line element. By 1935, Einstein and Nathan Rosen had adopted this form in their Einstein-Rosen bridge paper, which had become the standard form of the Schwarzschild line element by then [ER].

Conversely, Schwarzschild's solution was constrained by the conditions outlined in Einstein's earlier paper on the perihelion precession of Mercury [E3-15], which demanded adherence to Einstein's determinant condition. These constraints necessitated that Schwarzschild employ a mathematical stratagem to address the inherent limitations of spherical coordinates, which do not naturally satisfy the determinant condition of unity.

This chapter delineates how Droste's derivation, through a separate line of reasoning and distinct from the one employed by Schwarzschild, ultimately converged on the same solution, underscoring the robustness and potential diversity of approaches in solving Einstein's field equations. Through this comparative analysis, the chapter highlights the innovative aspects of Schwarzschild's and Droste's works.

Schwarzschild used his solution to find Mercury's orbit around the Sun. In the context of Schwarzschild's paper, Mercury is treated as a point-like object[2] moving within the gravitational field created by the Sun [S-1]. Schwarzschild's approach to deriving the orbit of Mercury involved exploiting the symmetries of his spacetime solution, as described in his 1916 paper [S-1]. Schwarzschild assumed that space is isotropic and that the motion of a point (like Mercury) is confined to the equatorial plane. This assumption simplifies the mathematical treatment significantly. Due to the symmetries in time and the azimuthal angle ϕ, Schwarzschild derived three constants of motion corresponding to conservation laws. These are: 1. From the independence of the Lagrangian on the azimuthal angle, Schwarzschild deduced

[1] References [E1-15, E2-15, E3-15, E4-15, E1-16].

[2] A point is a test particle in the gravitational field. It is a point-like object whose gravitational influence on the field is negligible compared to the central mass (the Sun).

the conservation of angular momentum. 2. The lack of explicit time dependence in the line element implies energy conservation. 3. Combining the first two conservations with the normalization condition of the four-velocity, Schwarzschild obtained a relationship involving the energy, angular momentum, and the radial component of the motion. Droste utilized the Euler-Lagrange equations to derive conservation laws from the Schwarzschild metric. Although not explicitly invoking the concepts later formalized in Emmy Noether's theorem, this method aligns with the underlying principles that Noether would eventually establish. Noether's theorem, developed later, systematically connects symmetries (such as time independence and rotational symmetry) with conservation laws. Schwarzschild's and Droste's works intuitively understand these relationships; they predate and are distinct from Noether's formal theorem.

Using the three integrals he obtained from the line element, Schwarzschild manipulated them to derive the orbit equation of Mercury in the Sun's gravitational field. He transformed the integrals to express everything in terms of $\frac{dR}{d\phi}$ and used chain rule and substitution methods to arrive at an orbit equation similar to Einstein's (in his 1915 paper on the precession of Mercury's perihelion). This equation accurately predicted the perihelion precession of Mercury's orbit.

Unlike Einstein's earlier approximation methods, Schwarzschild showed that his solution was unique and exact. He addressed the continuity condition, warranting that the solution was physically valid and that any potential discontinuity was appropriately located at the origin. Schwarzschild demonstrated that approximation methods (like those used by Einstein) involve simplifications that can obscure certain solution features, such as singularity's precise nature and location. Schwarzschild's exact solution, by satisfying the continuity condition, provides a clearer and more accurate description of the gravitational field, particularly near the singularity. Schwarzschild explained that the uniqueness of his solution is not an imposed condition but rather emerges naturally from the mathematics of the solution. He demonstrated that the continuity condition ensures this uniqueness by defining how the two constants emerging from his derivation α and ρ are related, preventing many possible but physically irrelevant solutions. In other words, Schwarzschild aimed to show that one can find multiple approximate solutions, but the problem becomes physically indeterminate.

After discussing the exterior line element, I concentrate on Schwarzschild's derivation of the interior solution for a spherically symmetric, incompressible fluid sphere. This derivation involves several clever steps. Schwarzschild used Einstein's field equations with the sources for unimodular coordinates. For the interior of the fluid sphere, he considered the energy-momentum tensor of an incompressible fluid with constant density and pressure as described by Einstein's model of a perfect fluid in his 1914 *Entwurf* paper [E2-14]. Schwarzschild wrote the line element in transformed coordinates to simplify calculations, leading to the components of the metric tensor. This line element was adapted to fit Einstein's four conditions for a static and spherically symmetric field and satisfied the determinant condition. Schwarzschild applied Einstein's field equations to the interior of the fluid sphere, using the derived metric components and considering the properties of the perfect fluid. He employed the equilibrium condition relating the divergence of the stress-energy tensor to the Christoffel symbols and the stress-energy tensor. This condition was necessary to link the sphere's pressure and density to the spacetime geometry. After solving the field equations in transformed coordinates, Schwarzschild translated the solution back to standard spherical coordinates, showing that the solution was free of singularities within the sphere.

Schwarzschild's solution for the sphere's interior was designed to match smoothly with the exterior Schwarzschild's solution at the sphere's boundary, demonstrating a consistent overall description of spacetime around and inside the sphere. Through his derivations, Schwarzschild determined specific forms for the functions describing the spacetime geometry inside the fluid sphere in terms of a spherical variable and the constant parameters. Schwarzschild explored the implications of his solution, such as the behavior of light velocity within the sphere, the redshift of radiation emitted from the sphere, and the conditions for the sphere's collapse under increasing density.

Einstein, however, believed that discussing the region beyond the singularity ($r = 2GM$) was futile, as the field equations did not provide physical insights for that area. He considered the *Schwarzschild singularity* ($r = 2GM$) as the critical boundary, not $r = 0$. Interestingly, in 1916, following Schwarzschild's proposal, Einstein was not deeply concerned about singularities, a stark contrast to his later years, especially during his work on unified field theory.[3] The Einstein of 1916 preferred approximate methods over Schwarzschild's exact solution, primarily due to his focus on unimodular coordinates, as Schwarzschild's solution did not have a determinant of 1. This and the next chapter also delve into other reasons for Einstein's preference. Moreover, they highlight that despite being an exact and unique solution, the Schwarzschild solution was not universally applicable to solving all problems, particularly when many mathematical tools were still developing.

Schwarzschild's and Droste's research was further developed to include charged spherically symmetric solutions. In 1918, Gunnar Nordström published two papers aiming to extend the spherically symmetric solution in Einstein's theory of gravitation [Nor-18-1, Nor-18-2]. Nordström's work expanded Droste's solution to include electrically charged, spherically symmetric bodies. His analysis detailed the gravitational field's energy density impacted by electrical charges, extending the scope of gravitational field analysis beyond non-charged entities.

In parallel, Hans Reissner, another contemporary physicist, investigated the implications of electric charges on gravitational fields in 1916 [Reis]. Reissner's exploration centered on how an electric charge's gravitational field influences its electric field. Reissner and Nordström independently derived a solution to Einstein's field equations known as the *Reissner-Nordström solution*, which describes the gravitational field around a charged, non-rotating, spherically symmetric body. Reissner's

[3] As I show in the second part of this book, this change is evident in his 1935 collaboration with Nathan Rosen on the Einstein-Rosen bridge theory, where they successfully circumvented singularities and avoided representing material particles as field singularities. This achievement served a dual purpose: it eliminated the *Schwarzschild singularity* ($r = 2m$) from general relativity's field equations. It dismissed the notion of particles as singularities within the gravitational field [ER].

approach utilized the general covariance of Einstein's field equations and the Maxwell stress-energy tensor.

Reissner showed that the validity of his solution could be verified because it reduces to Schwarzschild's solution (2.49) when transformed into the specific Schwarzschild coordinate system and when the charge is set to zero. While Reissner directly utilized Einstein's 1915–1916 field equations (1.290) without the restriction to unimodular coordinates—Nordström's approach was distinctly characterized by the implementation of a variational principle to derive the field equations, which incorporated effects from both matter and electromagnetism. Reissner's contributions were precocious, with his paper "On the Self-Gravitation of the Electric Field according to Einstein's Theory" submitted on March 20, 1916 [Reis], antedating Einstein's publication "Hamilton's Principles and the General Theory of Relativity" [E4-16]. In this work, recall that Einstein similarly derived his field equations via a variational methodology (see the discussion in Sect. 1.7.3).

Nordström and Reissner presented their initial findings through direct communications with Einstein before publicly releasing their theories, highlighting the foundational exchanges that influenced their theoretical developments. Recall that Schwarzschild communicated his results to Einstein in two letters. Specifically, in June 1915, Reissner communicated with Einstein, sharing his theory's foundational equations and soliciting his insights [CPAE8], Doc. 90. This act of exchange was paralleled by Nordström [CPAE8], Doc. 382, albeit two years later, who similarly highlighted to Einstein the fundamental equations of his derivation and the condition $t_4^4 = 0$ in equilibrium scenarios within his theoretical construct. Unfortunately, the archival record preserved in the *Collected Papers of Albert Einstein* [CPAE8] reveals no documented response to these communications.[4]

Einstein's position at the nexus of general relativity facilitated a unique scholarly environment. During this period, the field was synonymous with Einstein, supplemented by a few close collaborators and the broader community of physicists who engaged with him on these pioneering topics. This dynamic indicates the nascent stage of general relativity as a scientific discipline, where Einstein's contributions and his theoretical framework provided the foundation upon which others built and refined their ideas. Moreover, the fact that prominent theorists like Schwarzschild, Reissner, and Nordström sought feedback from Einstein before going public with their research indicates a scientific culture that valued his critique and approval, reflecting his influence over the direction and acceptance of new ideas within the field. This consultative process enhanced the theoretical robustness of their work through Einstein's scrutiny. Their innovations aligned with the prevailing theoretical landscape, heavily shaped by Einstein's thinking and ideas. Such interactions between Einstein and his contemporaries illuminated scientific work's highly collaborative aspects during this era before Einstein died in 1955. While highlighting Einstein's central role, this collaboration demonstrates the collective effort in advancing general

---

[4] Einstein addressed Nordström's concerns in his paper on gravitational waves. This will be discussed further in Sect. 3.2.3.

relativity, where Einstein's contributions are critically examined and integrated into a larger theoretical corpus.

It is important to note that Reissner engaged in discussions with Einstein. In his paper, "On the Self-Gravitation of the Electric Field according to Einstein's Theory," Reissner explained that, with support from Einstein through various clarifications and critiques, the task's computational execution had no restrictions regarding the choice of a coordinate system for setting up the field equations, particularly concerning the self-gravitation of the electric field. Reissner noted that the standard spherical polar coordinate system could be used, contrasting this with Schwarzschild's cleverly chosen coordinate system [see Eq. (2.9) and the ensuing derivations in Sect. 2.2.2] that maintains the value 1 of the determinant $\sqrt{-g}$ for the line element [Eq. (2.13)] [Reis].

Thus, Reissner's solution to Einstein's field equations for a charged, static, spherically symmetric body is not just a standalone accomplishment; it is also interwoven with Einstein's evolving perspective on coordinate conditions. When Reissner developed his work, Einstein reconsidered his earlier commitment to the unimodular coordinate condition (1.299). Recall that in his review paper, "The Foundation of the General Theory of Relativity," Einstein explicitly apologized for this constraint, noting it as an unnecessary limitation in developing covariant field equations. This apology, which appears in the 1916 manuscript following the equation of the electromagnetic stress-energy tensor, suggests a pivotal shift in Einstein's approach to general relativity (see the discussion in Sect. 1.7.3 and footnote 180). Reissner's work, which benefited from Einstein's guidance, reflects this broader, more flexible approach to solving Einstein's field equations. Einstein's note on page 40*a* of the review paper manuscript, coupled with Reissner's acknowledgment of Einstein's assistance, suggests that Einstein's evolving thoughts on coordinate systems could have played a pivotal role in shaping Reissner's approach.

Moreover, Reissner's paper presents an interesting situation regarding coordinate systems and determinant conditions. In this chapter, I compare his approach with Schwarzschild's to explain this. Reissner asserts that his method does not impose restrictions on the choice of the coordinate system for setting up the field equations, especially concerning the electric field's self-gravitation. This implies a more general approach than Schwarzschild's specific coordinate choice. Reissner mentions that the standard spherical polar coordinate system could be used, contrasting it with Schwarzschild's coordinate system, which was chosen to maintain the value 1 for the determinant $\sqrt{-g}$ of the line element. We will observe that, in practice, Reissner ultimately imposes constraints on the coordinate system, albeit not the unimodular constraint.

Nordström, in 1918, further scrutinized the energy density of the gravitational field in a charged sphere, concluding that in unimodular coordinates or in a coordinate system where $\sqrt{-g} = c = uw$ (u and w are Droste's two metric functions), this energy density is zero. As said above, his findings were first communicated to Einstein in 1917. They were subsequently elaborated upon in his 1918 paper, which referenced Einstein's earlier work on gravitational waves, further tying his analysis into the framework of Einstein's linearized theory. In this chapter, and continuing next, I will discuss Reissner's and Nordström's derivation of the Reissner-Nordström metric in detail and Nordström's interactions with Einstein concerning the vanishing energy density components. I will then examine the implications of these interactions for Einstein's gravitational waves paper, providing a deeper insight into Nordström's contributions to general relativity.

Nordström was an integral member of the Leiden physics community, including prominent figures such as Paul Ehrenfest, Hendrik Antoon Lorentz, Willem de Sitter, and the younger physicists Johannes Droste and Adriaan Fokker. This group maintained a robust epistolary exchange with Einstein. Meanwhile, Reissner was associated with the center of physicists in Berlin. Beyond Berlin, the center in Leiden was one of the foremost physics hubs during this period. In 1916, Einstein spent several weeks in the Netherlands, collaborating with Lorentz, Ehrenfest, Nordström, Droste, and Fokker. During his stay, Einstein entrusted Ehrenfest with the galley proofs of his paper, "Hamilton's Principles and the General Theory of Relativity," which outlined the derivation of the field equations from a Lagrangian framework without the imposition of the unimodular coordinate condition (1.299) [E4-16], asking that his Leiden colleagues view them [Gio]. On October 24, 1916, Einstein requested Ehrenfest to circulate his paper [E4-16] among the Leiden Brotherhood [CPAE8], Doc 269. In this chapter, I demonstrate the influence of Einstein's paper [E4-16] on Nordström's formulation of his solution to Einstein's field equations when considering an electrically charged source, the Reissner-Nordström metric (communicated in a meeting on January 26, 1918). I also show that Nordström's interaction with his Leiden colleague, Droste, is reflected by distinctively addressing his work over Schwarzschild's.

2.2 Schwarzschild's Exterior and Interior Solutions

2.2.1 The Schwarzschild-Einstein Letters

Karl Schwarzschild, who was leading the Astrophysical Observatory in Potsdam, was the first to present an exact solution to Einstein's vacuum field equations, which Einstein had introduced in his November 18, 1915, paper, "Explanation of the Perihelion Motion of Mercury from the General Theory of Relativity" [E3-15].

Schwarzschild, while stationed at the Russian front, wrote to Einstein on December 22, 1915, indicating his intention to refine Einstein's earlier calculations on Mercury's perihelion movement, which Einstein had presented on November 18, 1915, as evidence of the accuracy of his general theory of relativity. In response to Einstein's solution (1.300) (see Sect. 1.6.3), Schwarzschild discovered another similar solution [CPAE8], Doc. 169:

$$g_{\rho\sigma} = \delta_{\rho\sigma}\left[\frac{\beta}{3r^3}\right] - \frac{\alpha(x_\rho x_\sigma)}{r^3} - \frac{\beta(x_\rho x_\sigma)}{r^5}, \quad g_{44} = 1 \tag{2.1}$$

Following this, Schwarzschild excluded the first term (setting $\alpha = 0$) in Eq. (2.1) and presented his approximate solution that closely resembled the form of Einstein's solution (1.300):

$$g_{\rho\sigma} = \delta_{\rho\sigma}\left[\frac{\beta}{3r^3}\right] - \frac{\beta(x_\rho x_\sigma)}{r^5}, \quad g_{44} = 1. \tag{2.2}$$

Schwarzschild aimed to show that if there are multiple approximate solutions [Eqs. (2.1), (2.2)], the problem becomes physically indeterminate.

Subsequently, Schwarzschild transitioned to addressing the complete solution. He realized that, unlike the first-order approximate solutions, only one specific line element (2.3) met all of Einstein's criteria for the Sun's gravitational field-static, spherically symmetric, satisfying certain metric conditions, and reverting to a flat metric at great distances (defined in Sect. 2.2.2)-as well as the vacuum field Eq. (1.288) from Einstein's November 11, 1915 paper. Schwarzschild conveyed the following line element to Einstein in his correspondence [CPAE8], Doc. 169:

THE EXTERIOR SCHWARZSCHILD SOLUTION:

$$ds^2 = \left(1 - \frac{\gamma}{R}\right)dt^2 - \frac{dR^2}{1 - \frac{\gamma}{R}} - R^2\left(d\theta^2 + \sin^2\theta\, d\phi^2\right). \tag{2.3}$$

The solution's symmetry and static nature meant that the only independent variable was the radial distance from the center. Near the origin ($R = 0$), Schwarzschild's solution exhibited a mathematical singularity that could not be eliminated through coordinate transformations, marking a point where spacetime dynamics cease. Schwarzschild clarified to Einstein that his solution had just one true singularity at $R = 0$. He further pointed out that his spherical coordinates are not admissible, i.e., do not conform to the requirements of Einstein's field Eq. (1.289). Indeed, these coordinates do not align with Einstein's condition for unimodular coordinates, where the square root of the negative determinant is equal to 1 [Eq. (1.299)].

Schwarzschild stressed that the spherical coordinates provide the most suitable representation for spherical symmetry. This was evident as Schwarzschild's solution precisely matched Einstein's first-order approximation for Mercury's orbit (1.353) (as explored in detail in Sect. 2.2.4) [CPAE8], Doc. 169.

On December 29, 1915, Einstein responded by recognizing Schwarzschild's solution as uniquely significant [CPAE8], Doc. 176. Schwarzschild further communicated this solution to Einstein in a manuscript, deriving it from Einstein's November 18, 1915, field equations for a single mass. At the start of January 1916, Einstein keenly received and reviewed Schwarzschild's manuscript. He was pleasantly surprised by Schwarzschild's ability to formulate a precise solution to the problem, something he had not anticipated to be so straightforward. In mid-January 1916, Einstein presented Schwarzschild's paper at the Prussian Academy, offering a brief explanation to accompany it (as mentioned in a letter from Einstein to Schwarzschild on January 9, 1916) [CPAE8], Doc. 181. Schwarzschild's paper, "On the Gravitational Field of a Point-Mass according to Einstein's Theory," was published a month later [S-1].

2.2.2 *Schwarzschild's Trick in Solving Einstein's Equations*

Schwarzschild sought an exact solution to Einstein's field equations that satisfied the four criteria outlined in Einstein's paper on Mercury's perihelion [E3-15] [see Sect. 1.6.3, Chap. 1]. According to Schwarzschild, his challenge lay in identifying a line element with coefficients that simultaneously satisfy Einstein's vacuum field Eq. (1.289), the determinant condition (1.298), and the four requirements [S-1].

Schwarzschild first expressed *the most general line element satisfying the four conditions* in terms of rectangular coordinates (x, y, z) and time (t)[5]:

THE GENERAL LINE ELEMENT IN CARTESIAN COORDINATES:

$$ds^2 = Fdt^2 - G(dx^2 + dy^2 + dz^2) - H(xdx + ydy + zdz)^2. \tag{2.4}$$

Subsequently, he changed coordinates from rectangular to polar coordinates (r, θ, ϕ) according to:

$$\begin{aligned} x &= r\sin\theta\cos\phi, \\ y &= r\sin\theta\sin\phi, \\ z &= r\cos\theta. \end{aligned} \tag{2.5}$$

Schwarzschild's line element (2.4) in polar coordinates reads [S-1][6]:

THE LINE ELEMENT IN POLAR COORDINATES:

$$ds^2 = Fdt^2 - (G + Hr^2)dr^2 - Gr^2(d\theta^2 + \sin^2\theta d\phi^2). \tag{2.7}$$

---

[5] The line element ds^2 is described using functions F, G, and H, which depend on the radial distance r defined by Eq. (1.315). At $r \to \infty$, the conditions $F = G = 1$ and $H = 0$ are imposed. This aligns with condition 4), i.e., with the expectation that the spacetime should resemble flat Minkowski spacetime at great distances.

[6] The line element:

THE LINE ELEMENT IN POLAR COORDINATES:

$$ds^2 = Fdt^2 - G\left(dr^2 + r^2d\theta^2 + r^2\sin^2\theta d\phi^2\right) - Hr^2dr^2, \tag{2.6}$$

is equivalent to (2.7). ds^2 represents the square of the differential element of spacetime distance, dt^2, dr^2, $d\theta^2$, and $d\phi^2$ are the differential elements of time, radial distance, polar angle, and azimuthal angle, respectively. If we assume spherical symmetry, which simplifies the problem significantly, F, G, H would only be functions of r, as the symmetry implies no dependence on θ and ϕ.

Schwarzschild faced a problem transforming the line element (2.4) to polar coordinates. The transformed volume element in polar coordinates:

THE VOLUME ELEMENT IN SPHERICAL COORDINATES:

$$\boxed{r^2 \sin\theta \, dr \, d\theta \, d\phi,} \tag{2.8}$$

does not maintain the determinant condition of 1 [Eq. (1.298)].

While at the Russian front, Schwarzschild ingeniously resolved this issue through a clever approach, a trick.[7] This solution is a testament to his mathematical prowess and ability to navigate complex theoretical issues. Specifically, Schwarzschild introduced a coordinate transformation, i.e., new variables x_1, x_2, x_3:

SCHWARZSCHILD'S COORDINATE TRANSFORMATION:

$$\boxed{x_1 = \frac{r^3}{3}, x_2 = -\cos\theta, x_3 = \phi, x_4 = t,} \tag{2.9}$$

which he called: *polar coordinates with the determinant* 1 [S-1], and he obtained the new volume element in Cartesian coordinates[8]:

THE VOLUME ELEMENT IN CARTESIAN COORDINATES:

$$\boxed{dV = dx_1 \, dx_2 \, dx_3.} \tag{2.10}$$

Schwarzschild emphasized that the new variables retain the advantages of polar coordinates for solving the problem (since they are modified polar coordinates). He stressed that the altered radial and angular relationships in Eq. (2.9) simplify calculations but do not alter the form of the field Eq. (1.289) or the determinant condition (1.298). By including time as $t = x_4$, both the field equations and the determinant equation of the metric tensor remained unaltered in form.

Schwarzschild then rewrote the line element (2.7) using the new variables (2.9) and with four functions f_1, f_2, f_3, f_4 [S-1][9]:

---

[7] A trick: a clever and non-obvious method used to simplify and solve the complex problem of Einstein's determinant condition (1.298) in the form of coordinate transformations.

[8] The differentials of the new coordinates (2.9) in terms of the old ones are:

$$dx_1 = r^2 \, dr, \quad dx_2 = \sin\theta \, d\theta, \quad dx_3 = d\phi. \tag{2.11}$$

Substituting these into the volume element in spherical coordinates (2.8), we get Eq. (2.10).

[9] Rewriting the line element (2.7) using Eq. (2.9) gives:

$$ds^2 = F dx_4^2 - \left(\frac{G}{r^4} + \frac{H}{r^2}\right) dx_1^2 - Gr^2 \left[\frac{dx_2^2}{1 - dx_2^2} + dx_3^2(1 - dx_2^2)\right]. \tag{2.12}$$

.

THE LINE ELEMENT IN CARTESIAN COORDINATES:

$$\boxed{ds^2 = f_4\,dx_4^2 - f_1\,dx_1^2 - f_2\,\frac{dx_2^2}{1-x_2^2} - f_3\,dx_3^2\,(1-x_2^2),} \quad \text{where:} \tag{2.13}$$

THE METRIC COMPONENTS:

$$g_{11} = -f_1, \quad g_{22} = -\frac{f_2}{1-x_2^2}, \quad g_{33} = -f_2(1-x_2^2), \quad g_{44} = f_4. \tag{2.14}$$

He derived three (4, due to one redundancy) functions f_1, f_2, f_3, f_4 that satisfied Einstein's conditions and the vacuum field Eq. (1.289). These equations are simplified and *constrained by the determinant condition*, which affects the metric components.

Schwarzschild first wrote the *differential equations of the geodesic line* [Eq. (2.15)] for the line element (2.13) [S-1][10]: He then compared Eq. (2.15) with the general geodesic Eq. (1.210), and identified *the components of the gravitational field* Γ (represented by the Christoffel symbols) [S-1][11]:

[10] These equations are obtained using the variational principle, leading to expressions in terms of the second derivative of the coordinates with respect to the proper time s and partial derivatives of the metric functions f_1, f_2, and f_4 with respect to x_1:

SCHWARZSCHILD'S EQUATIONS OF THE GEODESIC LINE:

$$\begin{aligned}
0 &= f_1\frac{d^2x_1}{ds^2} + \frac{1}{2}\frac{\partial f_4}{\partial x_1}\left(\frac{dx_4}{ds}\right)^2 + \frac{1}{2}\frac{\partial f_1}{\partial x_1}\left(\frac{dx_1}{ds}\right)^2 \\
&\quad - \frac{1}{2}\left[\frac{\partial f_2}{\partial x_1}\frac{1}{1-x_2^2}\left(\frac{dx_2}{ds}\right)^2 + (1-x_2^2)\left(\frac{dx_3}{ds}\right)^2\right], \\
0 &= \frac{f_2}{1-x_2^2}\frac{d^2x_2}{ds^2} + \frac{\partial f_2}{\partial x_1}\frac{1}{1-x_2^2}\frac{dx_1}{ds}\frac{dx_2}{ds} + \frac{f_2x_2}{(1-x_2^2)^2}\left(\frac{dx_2}{ds}\right)^2 + f_2x_2\left(\frac{dx_3}{ds}\right)^2, \\
0 &= f_2(1-x_2^2)\frac{d^2x_3}{ds^2} + \frac{\partial f_2}{\partial x_1}(1-x_2^2)\frac{dx_1}{ds}\frac{dx_3}{ds} - 2f_2x_2\frac{dx_2}{ds}\frac{dx_3}{ds}, \\
0 &= f_4\frac{d^2x_4}{ds^2} + \frac{\partial f_4}{\partial x_1}\frac{dx_1}{ds}\frac{dx_4}{ds}.
\end{aligned} \tag{2.15}$$

[11] The gravitational field components are derived by *matching the geodesic equations* (2.15) [in footnote 10] *with the general geodesic equation* (1.210). Each term in the specific geodesic equations corresponds to a particular Christoffel symbol.

For instance: $\frac{1}{2}\frac{\partial f_4}{\partial x_1}\left(\frac{dx_4}{ds}\right)^2$ in the first equation of (2.15) contributes to Γ^1_{44}.

The components of the gravitational field involve only derivatives of the functions f_1, f_2, f_3, f_4 with respect to x_1, simplifying the computation of the field equations:

Schwarzschild refers to *simplifying Einstein's vacuum field equations* (1.289) due to the rotational symmetry around the origin, specifically *for the equator* where $x_2 = 0$ [S-1][12]:

THE SIMPLIFIED FIELD EQUATIONS:

$$\frac{\partial}{\partial x_1}\left(\frac{1}{f_1}\frac{\partial f_1}{\partial x_1}\right) = \frac{1}{2}\left(\frac{1}{f_1}\frac{\partial f_1}{\partial x_1}\right)^2 + \left(\frac{1}{f_2}\frac{\partial f_2}{\partial x_1}\right)^2 + \frac{1}{2}\left(\frac{1}{f_4}\frac{\partial f_4}{\partial x_1}\right)^2,$$
$$\boxed{\frac{\partial}{\partial x_1}\left(\frac{1}{f_1}\frac{\partial f_2}{\partial x_1}\right) = 2 + \frac{1}{f_1 f_2}\left(\frac{\partial f_2}{\partial x_1}\right)^2,} \tag{2.17}$$
$$\frac{\partial}{\partial x_1}\left(\frac{1}{f_1}\frac{\partial f_4}{\partial x_1}\right) = \frac{1}{f_1 f_4}\left(\frac{\partial f_4}{\partial x_1}\right)^2.$$

Given the line element (2.13), the determinant of the metric tensor depends on the functions f_1, f_2, f_4.

The unimodular coordinate condition (1.298) hence translates into a specific relationship among these functions. The equation, as per the condition, can be written as [S-1][13]:

THE COMPONENTS OF THE GRAVITATIONAL FIELD:

$$\begin{aligned}
&\Gamma^1_{11} = -\frac{1}{2}\frac{1}{f_1}\frac{\partial f_1}{\partial x_1},\ \Gamma^1_{22} = \frac{1}{2}\frac{1}{f_1}\frac{\partial f_2}{\partial x_1}\frac{1}{1-x_2^2},\\
&\Gamma^1_{33} = \frac{1}{2}\frac{1}{f_1}\frac{\partial f_2}{\partial x_1}(1-x_2^2),\ \Gamma^1_{44} = -\frac{1}{2}\frac{1}{f_1}\frac{\partial f_4}{\partial x_1}.\\
&\Gamma^2_{21} = -\frac{1}{2}\frac{1}{f_2}\frac{\partial f_2}{\partial x_1},\ \Gamma^2_{22} = -\frac{x_2}{1-x_2^2},\ \Gamma^2_{33} = -x_2(1-x_2^2),\\
&\Gamma^3_{31} = -\frac{1}{2}\frac{1}{f_2}\frac{\partial f_2}{\partial x_1},\ \Gamma^3_{32} = \frac{x_2}{1-x_2^2}.\ \ \Gamma^4_{41} = -\frac{1}{2}\frac{1}{f_4}\frac{\partial f_4}{\partial x_1}.
\end{aligned} \tag{2.16}$$

[12] Terms in Eq. (2.16) involving $1 - x_2^2$ can be simplified since $1 - x_2^2 = 1$. The term equator corresponds to the traditional equator in a spherically symmetric space. In a spherically symmetric system, the equator is typically the plane that divides the sphere into two halves, which in classical spherical coordinates corresponds to a polar angle $\theta = \frac{\pi}{2}$. In Schwarzschild's transformed coordinate system, he defines the equator in terms of the new coordinates (x_1, x_2, x_3 and x_4). Schwarzschild's statement that the equator is at x_2 in his coordinate system means that this value of x_2 represents points lying on the equatorial plane of the system he describes. So, given the rotational symmetry and focusing on the equator, all terms involving x_2 disappear.

[13] We calculate the determinant of $g_{\mu\nu}$ using the Eq. (2.14):

$$g_{\mu\nu} = \begin{pmatrix} -f_1 & 0 & 0 & 0 \\ 0 & -\frac{f_2}{1-x_2^2} & 0 & 0 \\ 0 & 0 & -f_2(1-x_2^2) & 0 \\ 0 & 0 & 0 & f_4 \end{pmatrix}. \tag{2.18}$$

The determinant of the metric tensor is the product of its diagonal elements:

THE DETERMINANT OF THE METRIC CONDITION:

$$\boxed{f_1 f_2^2 f_4 = 1.} \tag{2.21}$$

Equation (2.21) implies [S-1][14]:

$$\frac{1}{f_1}\frac{\partial f_1}{\partial x_1} + \frac{2}{f_2}\frac{\partial f_2}{\partial x_1} + \frac{1}{f_4}\frac{\partial f_4}{\partial x_1} = 0. \tag{2.22}$$

For now, Schwarzschild neglects the second field equation involving f_2 (the boxed equation) in Eq. (2.17) and rewrites the third one involving f_4 [S-1][15]:

$$\frac{\partial}{\partial x_1}\left(\frac{1}{f_4}\frac{\partial f_4}{\partial x_1}\right) = \frac{1}{f_1 f_4}\frac{\partial f_1}{\partial x_1}\frac{\partial f_4}{\partial x_1}. \tag{2.23}$$

He integrates Eq. (2.23) to yield a relationship between the derivatives of f_1 and f_4. This results in:

$$\frac{1}{f_4}\frac{\partial f_4}{\partial x_1} = \alpha f_1, \quad \text{where } \alpha \text{ is an integration constant.} \tag{2.24}$$

By combining the field equation for f_1 in Eq. (2.17) with the transposed Eq. (2.23) for f_4, Schwarzschild got another equation relating the derivatives of f_1, f_2, and f_4:

$$\frac{\partial}{\partial x_1}\left(\frac{1}{f_1}\frac{\partial f_1}{\partial x_1} + \frac{1}{f_4}\frac{\partial f_4}{\partial x_1}\right) = \left(\frac{1}{f_2}\frac{\partial f_2}{\partial x_1}\right)^2 + \frac{1}{2}\left(\frac{1}{f_1}\frac{\partial f_1}{\partial x_1} + \frac{1}{f_4}\frac{\partial f_4}{\partial x_1}\right)^2. \tag{2.25}$$

This combination leads to an equation involving only the derivative of f_2. In other words, *using the determinant condition* (2.22), *Schwarzschild eliminated terms*

$$|g_{\mu\nu}| = (-f_1) \times \left(-\frac{f_2}{1-x_2^2}\right) \times \left(-f_2(1-x_2^2)\right) \times f_4, \quad \text{from which we get:} \tag{2.19}$$

$$|g_{\mu\nu}| = -f_1 f_2^2 f_4 = -1. \tag{2.20}$$

Notice that $x_2 = 0$ and the actual determinant, considering the signature of the metric, is negative. In Schwarzschild's expression (2.21), the absolute value of the determinant of the metric tensor is 1. However, the actual determinant could be either $+1$ or -1, depending on the signature of the metric.

[14] Equation (2.22) balances the derivatives of the metric functions f_1, f_2, f_4 to maintain the metric tensor's determinant as a constant (i.e., unity). It is a constraint that these functions must satisfy in addition to the field Eq. (2.17). The Eq. (2.22) guarantees that the rate of change of f_1, f_2, f_4 in the coordinates of line element (2.13) is such that their combined effect keeps the determinant of the metric tensor equal to one, Eq. (2.21).

[15] The thrid equation of Eq. (2.17) can be rewritten in the form of Eq. (2.23) by setting $f_1 = f_4$.

involving the derivatives of f_1 and f_4 to get the equation that involves only the derivative of f_2[16]:

$$-2\frac{\partial}{\partial x_1}\left(\frac{1}{f_2}\frac{\partial f_2}{\partial x_1}\right)=3\left(\frac{1}{f_2}\frac{\partial f_2}{\partial x_1}\right)^2. \tag{2.27}$$

EQUATION FOR f_2:
Next, Schwarzschild integrated the Eq. (2.27) for f_2:

$$\frac{1}{\frac{1}{f_2}\frac{\partial f_2}{\partial x_1}}=\frac{3}{2}x_1+\frac{\rho}{2}.\ \text{Alternate form of the equation:}\ \frac{1}{f_2}\frac{\partial f_2}{\partial x_1}=\frac{2}{3x_1+\rho}, \tag{2.28}$$

where ρ is a constant of integration.

Einstein's condition 4 *that the solution vanishes at infinity*, the *condition at infinity*[17] is used to set integration constants, guaranteeing that the function f_2 approaches expected values as x_1 goes to infinity. Hence, Schwarzschild first integrated Eq. (2.28) and got:

$$f_2=\lambda(3x_1+\rho)^{2/3}\quad \text{where } \lambda \text{ is an integration constant.} \tag{2.29}$$

Then the *condition at infinity* demands: $\lambda=1$, and Eq. (2.29) gives f_2:

$$\boxed{f_2=(3x_1+\rho)^{\frac{2}{3}}.} \tag{2.30}$$

EQUATION FOR f_4:
Using the Eq. (2.24) for f_4 and *the determinant condition* (2.22), Schwarzschild found an expression for $\frac{\partial f_4}{\partial x_1}$ in terms of f_1, f_2, f_4:

$$\frac{\partial f_4}{\partial x_1}=\alpha f_1 f_4=\alpha\left(\frac{1}{f_2}\right)^2=\frac{\alpha}{(3x_1+\rho)^{4/3}}. \tag{2.31}$$

Integrating this with respect to x_1 and applying the *condition at infinity* gives the final form of f_4:

$$\boxed{f_4=1-\alpha(3x_1+\rho)^{-\frac{1}{3}}.} \tag{2.32}$$

[16] Schwarzschild rewrote Eq. (2.22) as follows:

$$\frac{1}{f_1}\frac{\partial f_1}{\partial x_1}+\frac{1}{f_4}\frac{\partial f_4}{\partial x_1}=-\frac{2}{f_2}\frac{\partial f_2}{\partial x_1}, \tag{2.26}$$

and plugged Eq. (2.26) into Eq. (2.25).

[17] Refer to Eqs. (2.35), (2.36), and (2.37) further below.

EQUATION FOR f_1:

Schwarzschild then substituted the expressions (2.30) and (2.32) into *the determinant condition* (2.22). The terms involving f_2 and f_4 are now known functions of x_1. The remaining term, involving f_1, must be such that the entire expression sums to zero.[18] The result is:

$$\boxed{f_1 = \frac{(3x_1+\rho)^{-\frac{4}{3}}}{1-\alpha(3x_1+\rho)^{-\frac{1}{3}}}.} \tag{2.33}$$

Finally, recall that Schwarzschild neglected the second field equation involving f_2, the boxed equation among Eq. (2.17). He noted that the two Eqs. (2.30) and (2.33) inherently fulfill this equation.[19]

Schwarzschild found from Eqs. (2.29) and (2.31) expressions for: f_1, f_2, and f_4. Combining Eqs. (2.30), (2.32), and (2.33) gives [S-1]:

$$\boxed{\begin{aligned} f_2 &= f_3 = (3x_1+\rho)^{\frac{2}{3}},\\ f_4 &= 1-\alpha(3x_1+\rho)^{-\frac{1}{3}},\\ f_1 &= \frac{(3x_1+\rho)^{-\frac{4}{3}}}{1-\alpha(3x_1+\rho)^{-\frac{1}{3}}}. \end{aligned}} \tag{2.34}$$

The following criteria guarantee the validity of Eq. (2.34) [S-1]:

1. *The condition at infinity*: The metric functions f_1, f_2, f_3, and f_4 are functions of x_1. Schwarzschild assumes a scaling of the spacetime metric components with respect to x_1 in his model.

For $x_1 = \infty$:

$$f_1 = \frac{1}{r^4} = (3x_1)^{-\frac{4}{3}}. \tag{2.35}$$

Hence, the influence of the metric component f_1 diminishes at large distances from the gravitational source, supporting the condition of asymptotic flatness.[20] This

[18] The condition (2.22) forms a differential equation for f_1 in terms of x_1. By integrating this differential equation, we get an expression for f_1. Schwarzschild's main scheme involves expressing $\frac{\partial f_2}{\partial x_1}$ and $\frac{\partial f_4}{\partial x_1}$ in terms of x_1. We rearrange the determinant condition to isolate the term involving $\frac{\partial f_1}{\partial x_1}$ and integrate this term to find f_1.

[19] From Eq. (2.30), we compute the derivative of f_2 with respect to x_1. We plug this derivative into the second [boxed] field equation involving f_2 in Eq. (2.17). With f_1 defined in Eq. (2.33), we substitute it into the second field equation involving f_2 in Eq. (2.17) and finally show that the left-hand side of this field equation simplifies to the right-hand side.

[20] The condition Eq. (2.35) sets f_1 as inversely proportional to the fourth power of the radius r defined by Eq. (1.315). In the numerator of the Eq. (2.33), $(3x_1+\rho)^{-\frac{4}{3}} \to 0$ as $x_1 \to \infty$, because its base term $(3x_1+\rho)$ becomes infinitely large, making the expression approach zero. The denominator, $1-\alpha(3x_1+\rho)^{-\frac{1}{3}}$, as $x_1 \to \infty$ approaches $1-0=1$. Thus, the denominator (*equal to* f_4,

behavior is consistent with Einstein's condition 4 of asymptotic flatness, where spacetime becomes increasingly flat as one moves away from the gravitational source.

Next, the metric functions $f_2 = f_3$, for $x_1 = \infty$:

$$f_2 = f_3 = r^2 = (3x_1)^{\frac{2}{3}}. \tag{2.36}$$

The asymptotic behavior of these metric components is consistent with a spacetime that becomes flat at large distances from a spherically symmetric mass.[21]

Finally, for f_4:

$$f_4 = 1. \tag{2.37}$$

This condition sets f_4 as 1 when $x_1 = \infty$. In other words, as one moves away from the gravitational source to infinity, the spacetime behaves more like flat spacetime.

2. *The determinant condition*: After transforming to the new variables (2.9) and defining the functions:

$$f_1, f_2, f_3, f_4,$$

Schwarzschild needed the product of these functions:

$$\sqrt{-g} = \sqrt{f_1 \times f_2 \times f_3 \times f_4}, \tag{2.38}$$

to be equal to 1, thereby satisfying the determinant condition (1.298), necessary for solving Einstein's vacuum field Eq. (1.289). To fulfill this requirement, Schwarzschild specifically chose and determined the functions such that their product, when considered with respect to the metric tensor for the line element (2.13) [Eq. (2.7)] and its determinant, resulted in -1. This was a key step in Schwarzschild's derivation, as it allowed him to successfully apply Einstein's field Eq. (1.289) to find his exact solution.

3. *The continuity condition*: the functions f_1, f_2, f_3, f_4 are continuous, except at $x_1 = 0$.

---

see Eq. (2.37)) converges to 1 as $x_1 \to \infty$, and f_1 approaches $\frac{0}{1} = 0$. Therefore, as x_1 goes to infinity, f_1 converges to 0. which is consistent with the statement (2.35) when considering x_1 around a gravitational source.

[21] While f_1 decreases to zero at infinity, both f_2 and f_3 will grow without bounds. This suggests that the spatial components of the metric associated with f_2 and f_3 expand as one moves further away from the central region. The specific power of $\frac{2}{3}$ indicates how the spatial dimensions expand as a function of x_1. f_2 and f_3 approach infinity as x_1 goes to infinity: as x_1 the term $3x_1 + \rho$ also approaches infinity, regardless of the value of ρ (since ρ is a constant and its impact becomes negligible compared to the infinitely large x_1). Therefore, $(3x_1 + \rho)^{\frac{2}{3}} = \infty^{\frac{2}{3}} = \infty$. This means that f_2 and f_3 approach infinity as x_1 goes to infinity. This aligns with the statement about expanding spatial dimensions (2.36), contextualized by Einstein's condition of asymptotic flatness.

Schwarzschild realized that the functions f_1, f_2, f_3, f_4 violate the continuity condition. He identified a mathematical singularity in the expression for f_1[22]:

$$\text{If: } \alpha(3x_1 + \rho)^{-\frac{1}{3}} = 1, \text{ and} \tag{2.41}$$

$$3x_1 = \alpha^3 - \rho, \text{ then } f_1 \text{ is singular.} \tag{2.42}$$

Let us rewrite Eq. (2.42) as follows:

$$3x_1 + \rho = \alpha^3. \tag{2.43}$$

To explore the discontinuity coinciding with the origin, Schwarzschild set $x_1 = 0$ in Eq. (2.43) and obtained:

THE DISCONTINUITY CONDITION:

$$\boxed{\rho = \alpha^3.} \tag{2.44}$$

Schwarzschild redefined the function f_1 by relating the integration constants α and ρ [S-1]. In other words, Schwarzschild found that *the continuity condition at the origin relates the two integration constants* in the specific way of (2.44). This condition warrants that any potential discontinuity in the solution is located at the origin, a requirement for physical validity.[23] This discontinuity occurs regardless of the specific value of ρ, as long as it follows the relationship Eq. (2.44).

After obtaining the metric functions,[24] Schwarzschild rewrote them. With the relationship (2.44) established, he reformulated the functions f_1, f_2, f_3 and f_4 in terms of a new transformed *radial coordinate* R [S-1].

---

[22] Plugging the second condition Eq. (2.43) into the first one Eq. (2.41). We get:

$$\alpha \cdot \alpha^{-1} = 1. \tag{2.39}$$

Thus, the two conditions (2.41) and (2.42) are consistent, confirming that the denominator of f_1 Eq. (2.34) is zero under these conditions, making the function singular. Let us confirm this by substituting Eq. (2.43) into the expression for f_1 Eq. (2.34). The simplified expression for f_1 under the conditions (2.41) and (2.42) becomes:

$$f_1 = \frac{(\alpha^3)^{-\frac{4}{3}}}{1 - \alpha \cdot (\alpha^3)^{-\frac{1}{3}}} = \frac{\alpha^{-4}}{1 - \alpha \cdot \alpha^{-1}} = \frac{0}{\alpha^4}. \tag{2.40}$$

The function becomes undefined and singular.

[23] Thus, evaluating at the origin (where $x_1 = 0$) reveals that the discontinuity coincides with the origin:

$$f_1(0) = \frac{\rho^{-\frac{4}{3}}}{1 - \alpha\rho^{-\frac{1}{3}}} = \frac{\alpha^{-4}}{1 - \alpha \cdot \alpha^{-1}} = \frac{0}{\alpha^4}. \tag{2.45}$$

[24] Equations (2.30), (2.32), and (2.33) or Eq. (2.34), satisfying the determinant condition (2.21).

First, Schwarzschild used the transformation Eq. (2.9) to rewrite the condition Eq. (2.43). This led to the formulation of the Schwarzschild radius R equation[25]:

$$\boxed{R = (3x_1 + \rho)^{\frac{1}{3}} = \left(r^3 + \alpha^3\right)^{\frac{1}{3}},} \tag{2.46}$$

Substituting the transformation Eq. (2.9) (for x_1) into Eq. (2.34), we can express the metric functions f_1, $f_2 = f_3$, and f_4 in terms of r[26]:

$$\begin{aligned} f_1 &= \frac{(r^3 + \rho)^{-\frac{4}{3}}}{1 - \alpha(r^3 + \rho)^{-\frac{1}{3}}}, \\ f_4 &= 1 - \alpha(r^3 + \rho)^{-\frac{1}{3}}, \\ f_2 &= f_3 = (r^3 + \rho)^{\frac{2}{3}}. \end{aligned} \tag{2.47}$$

Second, Schwarzschild *substituted the expression for R* [Eq. (2.46)] into the functions f_1, $f_2 = f_3$ and f_4, Eq. (2.34), and obtained:

$$\begin{aligned} f_1 &= \frac{1}{R^4} \cdot \frac{1}{1 - \frac{\alpha}{R}}, \\ f_2 &= f_3 = R^2, \\ f_4 &= 1 - \frac{\alpha}{R}. \end{aligned} \tag{2.48}$$

Plugging these metric functions back into the line element (2.13) and *reverting to the usual polar coordinates (the standard spherical coordinates)*, Schwarzschild derived the final form of the line element (metric), the now-famous Schwarzschild line element [S-1][27]:

THE EXTERIOR SCHWARZSCHILD LINE ELEMENT:

$$\boxed{ds^2 = (1 - \frac{\alpha}{R})dt^2 - \frac{1}{1 - \frac{\alpha}{R}}dR^2 - R^2(d\theta^2 + \sin^2\theta d\phi^2).} \tag{2.49}$$

---

[25] Recall that Schwarzschild introduced the transformation Eq. (2.9) as a mathematical maneuver to simplify the solution of Einstein's field equations under the conditions of spherical symmetry. In his derivation of Eq. (2.46), Schwarzschild redefined Eq. (2.43). He substituted the transformation Eq. (2.9) into the condition Eq. (2.43) and expressed the condition in terms of the original radial coordinate $x_1 = \frac{r^3}{3}$ or $3x_1 = r^3$, thereby linking R with the original radial coordinate r. Since the f_1, $f_2 = f_3$, and f_4 are only functions of x_1, $x_2 = x_3 = 0$.

[26] r represents the standard radial coordinate in spherical coordinates, derived from Cartesian coordinates, defined by Eq. (1.315).

[27] The Schwarzschild line element describes the spacetime interval ds^2 in terms of the coordinates t, R, θ and ϕ. This new form of the line element is a unique and exact solution to the Einstein field Eq. (1.288) for a spherically symmetric and static mass in a vacuum.

Schwarzschild discussed the uniqueness of his solution to Einstein's field Eq. (1.288) and the difficulties that would arise in ascertaining this uniqueness through Einstein's approximation method [S-1].

In Schwarzschild's exact solution, the uniqueness emerged naturally during the derivation of Eq. (2.47). The uniqueness of the solution emerges from the analysis of Eq. (2.44). Schwarzschild did not have to impose additional conditions to ensure uniqueness; it was a direct result of his approach and the mathematical structure of the problem. Schwarzschild points out that using an approximation method (like Einstein's approach) to solve the field equations would not easily reveal the uniqueness of the solution. In approximations, one typically expands the functions (2.47) in a series and keeps terms up to a certain order, assuming the parameters α and ρ are small.

In Schwarzschild's exact solution, the continuity condition (2.44) played a pivotal role. It linked the constants α and ρ so that the potential discontinuity in the solution would occur at the origin (the location of the mass point) (see footnote 23). Without this condition, the solution would appear to have arbitrary constants, leading to physical indeterminacy. Recognizing the necessary link between α and ρ would be challenging in an approximation using series expansion. The general form of the series expansion (up to second order) of f_1 [in Eq. (2.47)] in terms of α and ρ, which would look like this:

$$f_1 \approx \frac{1}{r^4}\left[1+\frac{\alpha}{r}-\frac{4}{3}\frac{\rho}{r^3}+\cdots\right], \tag{2.50}$$

would satisfy all the conditions specified by Einstein only up to a certain accuracy. Up to the second order in α and ρ we first expand each term in the equation as a power series in α and ρ. Given that this is a complex expansion, Eq. (2.50) involves a Taylor series expansion of each component of the function around $\alpha = 0$ and $\rho = 0$.

Hence, Schwarzschild demonstrates that in the case of approximate solutions, the continuity condition would not seem to add new information since discontinuities naturally occur only at the origin. However, he emphasizes that the exact solution shows that the discontinuity does not occur at the origin; it happens at $r = (\alpha^3 - \rho)^{\frac{1}{3}}$, necessitating the setting of the Eq. (2.44) for the discontinuity to be correctly positioned at the origin. The exact solution thus provides a deeper insight that might be missed in an approximation approach. It shows how certain parameters must be linked to guarantee physical accuracy and the correct behavior of the solution, particularly regarding the location of discontinuities and the behavior near the origin. It follows from the exact solution (2.49) that the constants α and ρ are not arbitrary but are related in a specific way. Without the continuity condition, the solution would appear physically undetermined [S-1].

The determinant of the Schwarzschild metric (2.49) is not -1; it varies with the radial coordinate. Hence, the Schwarzschild solution is valid in the broader context of Einstein's field equations, not limited to unimodular coordinates (1.288). For the resulting Schwarzschild metric (2.49), the determinant g is negative, but $\sqrt{-g}$ simplifies to $R^2 \sin\theta$ rather than 1 [S-1].[28]

According to the four conditions imposed by Einstein, as $\lim_{r\to\infty}$:

THE MINKOWSKI LINE ELEMENT IN SPHERICAL COORDINATES:

$$\boxed{ds^2 = dt^2 - dR^2 - R^2\left(d\theta^2 + \sin^2\theta\, d\phi^2\right),} \tag{2.53}$$

and we obtain the Minkowski line element in the spherical symmetric form. A mathematical singularity is seen to occur at the origin of the Schwarzschild line element (2.49) when $R = 0$. Coordinate transformations cannot remove this singularity.

2.2.3 Droste Independently Derives the Same Solution

Johannes Droste, a student of Hendrik Lorentz at the University of Leiden, independently derived a solution to Einstein's field equations around the same time as Schwarzschild. Both Droste and Schwarzschild arrived at the same destination—what is now known as *the Schwarzschild metric*, Eq. (2.49)—but they took different routes, reflecting their approaches to the problem of solving Einstein's field equations under the assumption of spherical symmetry in a vacuum. While Droste solved the field Eq. (1.290), Schwarzschild solved the field Eq. (1.289).

At the beginning of his dissertation, Droste explicitly stated that he did not adopt the coordinate system commonly used by Einstein in his work on general relativity, characterized by the unimodular coordinate condition (1.299), $\sqrt{-g} = 1$ [Dro-16]. As I show here, Droste's derivation is not dependent on Schwarzschild's derivation. His approach differs from Schwarzschild's explicit handling of the determinant condition (1.299); this difference in methodology demonstrates that Droste's derivation was developed through a separate line of reasoning, independently arriving at a

[28] Calculating the determinant for the Schwarzschild metric (2.49) involves the product of the diagonal elements of the metric tensor. The diagonal elements are:

$\left(1 - \frac{\alpha}{R}\right), -\frac{1}{1-\frac{\alpha}{R}}, -R^2, -R^2\sin^2\theta$. The determinant g is the product of these diagonal elements, giving:

$$g = \left(1 - \frac{\alpha}{R}\right)\cdot\left(-\frac{1}{1-\frac{\alpha}{R}}\right)\cdot(-R^2)\cdot(-R^2\sin^2\theta) = -R^4\sin^2\theta. \tag{2.51}$$

The square root of the negative determinant $\sqrt{-g}$ simplifies to: $\sqrt{-g} = \sqrt{-\left(-R^4\sin^2\theta\right)} = R^2\sin\theta$.

$$\text{Thus: } \sqrt{|-g|} = R^2\sin\theta. \tag{2.52}$$

similar solution. Specifically, this is evident, as mentioned above, by Schwarzschild's method, which directly addressed the determinant [Eq. (1.298)] issue, introducing new variables (2.9) to transform the volume element and which does not rely on a variational principle.

In 1916, Droste started with the action integral that spans a region from r_1 to r_2 where he integrated the Lagrangian density $\mathcal{L} = \sqrt{-g}R$ over this interval [Dro-16][29]:

THE ACTION INTEGRAL:

$$\delta \int_{r_1}^{r_2} \mathcal{L}r^2 dr = 0, \;\; \mathcal{L} = -\frac{d}{dr}\left(\frac{v^2 w'}{u}\right) - \frac{d}{dr}\left(\frac{2vwv'}{u}\right) + \frac{wv'^2 + 2vv'w'}{u} + uw. \tag{2.54}$$

The variation of the action integral (2.54), assuming $\mathcal{L}$ involves metric functions u, v, and w, results in the following equation [Dro-16]:

$$\delta \int_{r_1}^{r_2} \left[-\frac{d}{dr}\left(\frac{v^2 w'}{u}\right) - \frac{d}{dr}\left(\frac{2vwv'}{u}\right) + \frac{wv'^2 + 2vv'w'}{u} + uw \right] dr = 0. \tag{2.55}$$

We can write the variation of the integral (2.55) in the simplified form, focusing on the most essential terms after integration by parts. This represents the dynamics without explicitly considering the total derivative terms, as they contribute only to boundary terms that vanish under appropriate boundary conditions.[30] Therefore, the variation of the action integral can be expressed as:

$$\delta \int_{r_1}^{r_2} \left[\frac{wv'^2 + 2vv'w'}{u} + uw \right] dr = \delta \int_{r_1}^{r_2} \mathcal{L}_{\text{eff}}\, dr = 0. \tag{2.56}$$

Equation (2.56) simplifies our problem, as we are now dealing with the key terms that contribute to the Euler-Lagrange equations. This expression states that the variation of the integral of the effective Lagrangian density, which includes the necessary terms after integration by parts, is set to zero.[31]

---

[29] Droste's notation is more verbose than necessary by modern standards. His way to explicitly indicate integration over the four-dimensional spacetime is by four integrals $\int\int\int\int$, and dx_1, dx_2, dx_3, dx_4 rather than the modern d^4x to represent differential elements along each of the four dimensions (three spatial dimensions and one temporal dimension) [Dro-16], [Dro-17]. Einstein wrote the integral compactly using a single integral sign and a four-dimensional volume element $d\tau$ [El-16]. I adopt a succinct notation here, which does not feature four separate integrals.

[30] The first two terms in the Lagrangian (2.54) are total derivatives and contribute only boundary terms when integrated. If the variations of the metric functions u, v and w are zero at the boundaries (i.e., $\delta u = \delta v = \delta w = 0$ at r_1 and r_2), these boundary terms do not contribute to the field equations.

[31] Consider the general form of the second-order integration by parts formula, similar to the Eq. (1.432) in footnote 181. This form is used to handle higher-order derivatives in the Lagrangian (2.54):

Equation (2.56) is used to derive the Euler-Lagrange equations for the metric functions u, v, and w. By setting the variations to zero for each metric function u, v, and w, we obtain the vacuum field equations from the simplified form of the Lagrangian density (2.56) [Dro-16][32]:

SECOND-ORDER INTEGRATION BY PARTS FORMULA:

$$\boxed{\int \frac{\partial^2 f}{\partial x_\alpha \partial x_\beta} g d\tau = \left[\frac{\partial f}{\partial x_\alpha} g\right]_{\text{boundary}} - \int \frac{\partial f}{\partial x_\beta} \frac{\partial g}{\partial x_\alpha} d\tau.} \tag{2.57}$$

This formula states that integrating a second partial derivative of a function f multiplied by another function g over a volume τ can be transformed into a boundary term and a volume integral involving the first derivatives of both f and g (I adhere to the notation in footnote 181 to maintain consistency with this footnote).

We now apply Eq. (2.57) to $\mathcal{L}$ in Eq. (2.54).

Consider the first term on the right-hand side of the Lagrangian density in Eq. (2.54), $-\frac{d}{dr}\left(\frac{v^2 w'}{u}\right)$:

$$\int_{r_1}^{r_2} -\frac{d}{dr}\left(\frac{v^2 w'}{u}\right) dr = -\left[\frac{v^2 w'}{u}\right]_{r_1}^{r_2} + \boxed{\int_{r_1}^{r_2} \frac{d}{dr}\left(\frac{v^2 w'}{u}\right) \delta r dr.} \tag{2.58}$$

The boundary term $-\left[\frac{v^2 w'}{u}\right]_{r_1}^{r_2}$ vanishes if the variations δv, δw, δu are zero at the boundaries, and we are left with the integral of the term involving variations.

Similarly, consider the second term on the right-hand side of the Lagrangian density in Eq. (2.54), $-\frac{d}{dr}\left(\frac{2vwv'}{u}\right)$:

$$\int_{r_1}^{r_2} -\frac{d}{dr}\left(\frac{2vwv'}{u}\right) dr = -\left[\frac{2vwv'}{u}\right]_{r_1}^{r_2} + \boxed{\int_{r_1}^{r_2} \frac{d}{dr}\left(\frac{2vwv'}{u}\right) \delta r dr.} \tag{2.59}$$

Again, the boundary term vanishes under appropriate boundary conditions.

The third term of the Lagrangian (2.54) remains after performing integration by parts on the first and second terms:

$$\int_{r_1}^{r_2} \left(\frac{wv'^2 + 2vv'w'}{u} + uw\right) \delta r dr. \tag{2.60}$$

In other words, we can focus on the effective Lagrangian $\mathcal{L}_{\text{eff}}$ to write the Euler-Lagrange equations, ignoring the total derivative terms that contribute only to boundary terms (see footnote 30). Thus, we can derive the same field equations by focusing on the third term of Droste's specific Lagrangian (2.54).

[32] Consider $\mathcal{L}_{\text{eff}}$ in Eq. (2.56). The Euler-Lagrange equation for w is:

$$\frac{\partial \mathcal{L}}{\partial w} - \frac{d}{dr}\left(\frac{\partial \mathcal{L}}{\partial w'}\right) = 0. \text{ Calculating these derivatives, we get:} \tag{2.61}$$

$$\frac{\partial \mathcal{L}}{\partial w} = \boxed{\frac{v'^2}{u} + u.} \quad \frac{d}{dr}\left(\frac{\partial \mathcal{L}}{\partial w'}\right) = \frac{d}{dr}\left(\frac{2vv'}{u}\right) = \boxed{\frac{2v'v''}{u} - \frac{2vv'u'}{u^2}.} \tag{2.62}$$

$$\text{Setting the variation to zero, we get: } \boxed{\frac{v'^2}{u} + u - \left(\frac{2v'v''}{u} - \frac{2vv'u'}{u^2}\right) = 0.} \tag{2.63}$$

DROSTE'S VACUUM FIELD EQUATIONS FOR A SPHERICALLY SYMMETRIC MASS:

$$\frac{wv'^2 + 2vv'w'}{u^2} - w = 0,$$
$$\frac{wv'' + v'w' + vw''}{u} - (vw' + wv')\frac{u'}{u^2} = 0,$$
$$\frac{2vv'' + v'^2}{u} - u - 2vv'\frac{u'}{u^2} = 0. \qquad (2.70)$$

Droste wrote the general line element [Dro-16, Dro-17]:

DROSTE'S GENERAL LINE ELEMENT:

$$\boxed{ds^2 = w^2dt^2 - u^2dr^2 - v^2(d\vartheta^2 + \sin^2\vartheta d\varphi^2).} \qquad (2.71)$$

Setting $uw = 0$ [Eq. (2.56)], the Euler-Lagrange equation for v is:

$$\frac{\partial\mathcal{L}}{\partial v} - \frac{d}{dr}\left(\frac{\partial\mathcal{L}}{\partial v'}\right) = 0. \text{ Calculating these derivatives, we get:} \qquad (2.64)$$

$$\frac{\partial\mathcal{L}}{\partial v} = \boxed{\frac{2wv'w'}{u}.} \quad \frac{\partial\mathcal{L}}{\partial v'} = \frac{2wv'}{u} + \frac{2ww'}{u}. \qquad (2.65)$$

$$\frac{d}{dr}\left(\frac{\partial\mathcal{L}}{\partial v'}\right) = \frac{d}{dr}\left(\frac{2wv'}{u} + \frac{2ww'}{u}\right) = \boxed{\frac{2wv''}{u} + \frac{2w'v'}{u} - \frac{2wv'u'}{u^2}.} \qquad (2.66)$$

Setting the variation to zero, we get: $$\boxed{\frac{2wv'w'}{u} - \left(\frac{2wv''}{u} + \frac{2w'v'}{u} - \frac{2wv'u'}{u^2}\right) = 0.} \qquad (2.67)$$

Consider again the term in brackets in Eq. (2.56). The Euler-Lagrange equation for u is:

$$\frac{\partial\mathcal{L}}{\partial u} - \frac{d}{dr}\left(\frac{\partial\mathcal{L}}{\partial u'}\right) = 0. \text{ Calculating these derivatives, we get: } \frac{\partial\mathcal{L}}{\partial u} = \boxed{-\frac{wv'^2 + 2vv'w'}{u^2} + w.} \qquad (2.68)$$

Since $\mathcal{L}$ does not explicitly depend on u', $\frac{\partial\mathcal{L}}{\partial u'} = 0$. Therefore, the boxed equation is equal to zero.

$$-\frac{wv'^2 + 2vv'w'}{u^2} + w = 0. \qquad (2.69)$$

Simplifying Eqs. (2.63) and (2.67) leads to the second and third field equations in Eq. (2.70). Equation (2.69) gives us the first field equation in Eq. (2.70).

This form is adapted to accommodate spherical symmetry in a gravitational field. The metric (2.71) is spherically symmetric if the functions w, u, and v depend only on r and not on ϑ and φ. Thus, setting:

$$v = r, \tag{2.72}$$

Equation (2.71) becomes [Dro-16]:

DROSTE'S SPHERICAL SYMMETRIC LINE ELEMENT:

$$\boxed{ds^2 = w^2 dt^2 - u^2 dr^2 - r^2(d\vartheta^2 + \sin^2\vartheta d\varphi^2).} \tag{2.73}$$

Droste substituted $v' = 1$ and $v'' = 0$ into the first and third original field equations of (2.70) and considered Eq. (2.72). This allowed the first and the third original field equations to simplify to [Dro-16]:

$$\frac{w + 2rw'}{u^2} - w = 0, \quad \frac{1}{u} - u - 2r\frac{u'}{u^2} = 0. \tag{2.74}$$

Differentiating the second equation of (2.74) with respect to r leads to[33]:

$$\frac{d}{dr}\left(\frac{r}{u^2}\right) = 1. \tag{2.76}$$

Integrating Eq. (2.76) leads to:

$$\frac{r}{u^2} = r - \alpha, \tag{2.77}$$

[33] We differentiate $\frac{r}{u^2}$ using the quotient rule, where $f(r) = r$ and $g(r) = u^2$. The quotient rule states:

THE QUOTIENT RULE: $$\boxed{\frac{d}{dr}\left(\frac{f}{g}\right) = \frac{f'g - fg'}{g^2}, \quad \frac{\text{lo de hi - hi de lo}}{\text{lolo}}.} \tag{2.75}$$

We substitute $f = r$, $f' = 1$, $g = u^2$, and $g' = 2uu'$ into the quotient rule formula:

$$\frac{d}{dr}\left(\frac{r}{u^2}\right) = \frac{1 \cdot u^2 - r \cdot 2uu'}{(u^2)^2}.$$

We simplify the expression: $\frac{d}{dr}\left(\frac{r}{u^2}\right) = \frac{u^2 - 2ruu'}{u^4}.$

$$\frac{d}{dr}\left(\frac{r}{u^2}\right) = \frac{1}{u^2} - \frac{2ruu'}{u^4}. \quad \text{Thus:} \quad \frac{d}{dr}\left(\frac{r}{u^2}\right) = \frac{1}{u^2} - \frac{2ru'}{u^3}.$$

Setting $u^2 = 1$ and $u' = 0$, this simplifies to Eq. (2.76). u' represents the derivative of u with respect to r. The derivative of a constant is zero. Since 1 is a constant, its derivative with respect to r is zero.

where α is the constant of integration. Dividing (2.77) by r, Droste obtained the first metric function [Dro-16]:

DROSTE'S FIRST METRIC FUNCTION:

$$\boxed{\frac{1}{u^2} = 1 - \frac{\alpha}{r}.} \tag{2.78}$$

Droste plugged this metric function into the first field equation of (2.74), the w field equation[34]:

[34] First recall that Droste substituted $v' = 1$ and $v'' = 0$ into the first field equation of (2.70). This led to the two field equations (2.74). Now we isolate w' from Eq. (2.88), this leads to:

$$w' = \frac{\alpha}{2r(r-\alpha)}w. \tag{2.79}$$

The derivative of w^2 with respect to r can be expressed as:

$$\frac{d(w^2)}{dr} = ww'. \quad [\frac{d(w^2)}{dr} = 2w \cdot \frac{dw}{dr}.] \tag{2.80}$$

Substituting w' from equation (2.79) into equation (2.80), we obtain:

$$\frac{d(w^2)}{dr} = 2w\left(\frac{\alpha}{2r(r-\alpha)}w\right) = \frac{\alpha}{r(r-\alpha)}w^2. \tag{2.81}$$

Dividing both sides of equation (2.81) by w^2 and moving dr to the other side, we get:

$$\boxed{\frac{1}{w^2}dw^2 = \frac{\alpha}{r(r-\alpha)}dr.} \tag{2.82}$$

Integrating both sides of equation (2.82) gives:

$$\int \frac{1}{w^2}dw^2 = \int \frac{\alpha}{r(r-\alpha)}dr. \tag{2.83}$$

However, integrating $\frac{1}{w^2}dw^2$ directly, which is more aligned with our variable substitution, yields:

$$\int \frac{1}{w^2}d(w^2) = \boxed{\ln|w^2| + C_1.} \tag{2.84}$$

For the right-hand side, the partial fraction decomposition gives:

$$\frac{\alpha}{r(r-\alpha)} = \frac{1}{r-\alpha} - \frac{1}{r}. \quad \text{Integrating this leads to:} \tag{2.85}$$

$$\int\left(\frac{1}{r-\alpha} - \frac{1}{r}\right)dr = \ln|r-\alpha| - \ln|r| = \boxed{\ln\left|\frac{r-\alpha}{r}\right| + C_2.} \tag{2.86}$$

Combining the boxed integrated results we get:

$$\ln|w^2| + C_1 = \ln\left|\frac{r-\alpha}{r}\right| + C_2. \tag{2.87}$$

$$2(r-\alpha)w' = \frac{\alpha}{r}w. \tag{2.88}$$

This suggested an exponential form solution. Integrating both sides of Eq. 2.82 gives an equation for $w^2 = f(r)$:

$$|w^2| = e^C \left|\frac{r-\alpha}{r}\right|. \tag{2.89}$$

Assuming $c = e^C$ where c is adjusted to normalize the expression at infinity:

$$w^2 = c\left(\frac{r-\alpha}{r}\right). \tag{2.90}$$

At infinity, the metric approaches the Minkowski flat metric. Thus, setting $c = 1$, and assuming the constant of integration is adjusted accordingly based on the boundary conditions, we obtain:

DROSTE'S SECOND METRIC FUNCTION:

$$\boxed{w^2 = 1 - \frac{\alpha}{r}.} \tag{2.91}$$

Plugging the metric functions (2.78) and (2.91) into the metric (2.73), Droste derived the line element (2.49). In his dissertation, Droste noted that Schwarzschild had derived the same metric [Dro-16].

Droste discussed what happens when general relativity reduces to Newtonian gravity in the weak-field limit, where the gravitational field is not too strong, and α is given by Eq. (1.164) [see Sect. 1.4.1, Chap. 1]. In modern terminology, we define α in Eq. (1.164) as the Schwarzschild radius r_s:

$$\text{THE SCHWARZSCHILD RADIUS:} \quad \boxed{r_s = \frac{2GM}{c^2},} \tag{2.92}$$

and the parameter r_g is used to denote the gravitational radius, which is half of the Schwarzschild radius:

$$\text{THE GRAVITATIONAL RADIUS:} \quad \boxed{r_g = \frac{GM}{c^2}.} \tag{2.93}$$

It is a measure of the strength of the gravitational field of the mass.

Assuming $C = C_2 - C_1$, we obtain equation (2.89). Assuming $c = e^C$, we obtain equation (2.90) and setting $c = 1$, leads to equation (2.91).

Substituting $\alpha = 2Gm$ into the line element (2.49), Droste showed that it takes the known form [Dro-16]:

THE DROSTE LINE ELEMENT:

$$ds^2 = \left(1 - \frac{2Gm}{r}\right) dt^2 - \frac{dr^2}{1 - \frac{2Gm}{r}} - r^2 \left(d\theta^2 + \sin^2\theta \, d\phi^2\right). \tag{2.94}$$

Using Eq. (2.92), the Droste line element (2.94) takes the modern form:

$$ds^2 = \left(1 - \frac{r_s}{r}\right) dt^2 - \frac{dr^2}{1 - \frac{r_s}{r}} - r^2 \left(d\theta^2 + \sin^2\theta \, d\phi^2\right). \tag{2.95}$$

The substitution of $\alpha = 2Gm$ guarantees that the metric reduces to the Newtonian form under these conditions. Specifically, in the limit as the radial coordinate r becomes much larger than α the metric should approximate the metric of weak gravitational fields. The Droste line element (2.94) quickly became the standard formulation for expressing what we call today the *Schwarzschild line element* [Kox].

2.2.4 Conservation Laws and Einstein's Orbit Equation

Droste derived conservation laws from the Lagrangian using Euler-Lagrange equations. He started with a Lagrangian $\mathcal{L}$ [Eq. (2.97)] that describes the system. For a particle in the gravitational field of a spherically symmetric body, the Lagrangian is constructed from the Schwarzschild metric (2.49). The Euler-Lagrange equations are derived from the principle of least action. Applying these equations to the Lagrangian (2.97) derived from the metric yields the geodesic equations. If the Lagrangian does not explicitly depend on a particular coordinate (implying a symmetry), the corresponding momentum (conjugate to that coordinate) is conserved. This method is powerful for identifying conserved quantities associated with symmetries of the spacetime (like energy and angular momentum in Schwarzschild spacetime).

Schwarzschild followed a different path. He considered the normalization condition for the four-velocity of a particle, combined with his metric (2.49)[35]:

$$g_{\mu\nu} \frac{dx^\mu}{ds} \frac{dx^\nu}{ds} = -1, \tag{2.96}$$

[35] According to this condition, the magnitude of the four-velocity vector remains constant along the particle's trajectory, reflecting the invariant speed of light in the particle's rest frame.

for timelike geodesics, representing the paths of particles with non-zero rest mass. This directly leads to conservation laws due to the symmetries of the metric. The Schwarzschild metric, being independent of the time coordinate t and azimuthal angle ϕ implies conservation of energy and angular momentum, respectively. The independence of the Schwarzschild metric (2.49) from the time coordinate t means that the metric is stationary, leading to the conservation of a quantity related to the particle's energy. The independence from the azimuthal angle ϕ indicates axial symmetry, resulting in the conservation of angular momentum around the symmetry axis. These conservation laws can be derived by examining how the metric components enter the equations of motion (e.g., the geodesic equation) and identifying the conserved quantities.

Both methods ultimately rely on the fundamental symmetries of the spacetime under consideration. Using the Lagrangian and Euler-Lagrange equations, Droste's approach provides a systematic way to derive equations of motion and identify conserved quantities. Schwarzschild's approach, leveraging the normalization condition and specific properties of the spacetime metric, offers a direct route to identifying conserved quantities based on the symmetries of the spacetime.

DROSTE'S METHOD:

The Lagrangian $\mathcal{L}$ is symmetric with respect to continuous transformation (time translation and rotation)[36]:

$$\mathcal{L} = -\left(1 - \frac{\alpha}{R}\right)\left(\frac{dt}{ds}\right)^2 + \left(1 - \frac{\alpha}{R}\right)^{-1}\left(\frac{dR}{ds}\right)^2 - R^2\left(\frac{d\phi}{ds}\right)^2. \tag{2.97}$$

If the Lagrangian does not explicitly depend on time, Noether's theorem implies energy conservation Eq. (2.104). Invariance under rotations results in the conservation of angular momentum Eq. (2.103).

[36] The Lagrangian for a particle moving in the equatorial plane around a spherically symmetric mass described by the Schwarzschild metric (2.49), focusing on radial $\dot{r} = \frac{dr}{ds}$ [how fast the particle moves radially] and azimuthal $\dot{\phi} = \frac{d\phi}{ds}$ [how fast the particle moves around the central mass] motions, along with the time component $\dot{t} = \frac{dt}{ds}$ [how the coordinate time elapses relative to the particle's proper time s]. The rate of change of the radial coordinate, the azimuthal angle, and coordinate time with respect to the proper time s.

Using the Euler-Lagrange equations, Droste derived from the line element (2.49) conservation laws [Dro-17].[37,38]

The system's symmetries (i.e., time independence and rotational symmetry) lead to conserved quantities through these equations, per Emmy Noether's theorem (for every differentiable symmetry of the action of a physical system, there is a corresponding conservation law).

Droste published his solution in 1916–1917, a year before Emmy Noether published her groundbreaking paper in 1918. Though Droste's work predated Noether's theorem, symmetries and conservation laws were already central to physics. Noether's work provided a more formal and general mathematical framework for understanding these principles [Noe].

SCHWARZSCHILD'S METHOD:

Schwarzschild assumed that space is isotropic and that the motion of a point (test particle) in the gravitational field of a spherically symmetric body is confined to the equatorial plane ($\theta = d\theta = 0$ or $x_2 = 0$). He then obtained three integrals from these assumptions and his line element (2.49), necessary for describing the motion of the point in this gravitational field. Further, Schwarzschild assumed that the coefficients of the line element (2.49) do not depend on the time coordinate t (which implies time translation symmetry) and the azimuthal angle ϕ (which implies

---

[37] The following generally outlines Droste's derivation of Einstein's area law [Dro-17]. For ϕ, the Euler-Lagrange equation is expressed as:

$$\frac{d}{ds}\left(\frac{\partial \mathcal{L}}{\partial \dot{\phi}}\right) - \frac{\partial \mathcal{L}}{\partial \phi} = 0, \tag{2.98}$$

In the Schwarzschild metric, the Lagrangian $\mathcal{L}$ (2.97) does not explicitly depend on the azimuthal angle ϕ: This Lagrangian does not include ϕ explicitly (it appears only as $\dot{\phi}^2$ term). This means that:

$$\frac{\partial \mathcal{L}}{\partial \phi} = 0. \tag{2.99}$$

Given Eq. (2.99), the Euler-Lagrange equation (2.98) simplifies to:

$$\frac{d}{ds}\left(\frac{\partial \mathcal{L}}{\partial \dot{\phi}}\right) = 0. \tag{2.100}$$

This implies that $\frac{\partial \mathcal{L}}{\partial \dot{\phi}}$ is a constant of motion. Given the form of the Schwarzschild Lagrangian:

$$\frac{\partial \mathcal{L}}{\partial \dot{\phi}} \text{ is proportional to: } \boxed{R^2\dot{\phi} = R^2\frac{d\phi}{ds}.} \tag{2.101}$$

$R^2\frac{d\phi}{ds}$ is a constant [Dro-17]. We obtain conservation of angular momentum (Einstein's area law) (2.103).
From the Euler-Lagrange equation for t:

$$\frac{\partial \mathcal{L}}{\partial t} - \frac{d}{ds}\left(\frac{\partial \mathcal{L}}{\partial \dot{t}}\right) = 0, \tag{2.102}$$

we get the energy conservation (2.104).

[38] $\dot{\phi}$ is the derivative of ϕ with respect to the proper time s and $\mathcal{L}$ is the Lagrangian of the system.

rotational symmetry). Schwarzschild pointed out that due to the homogeneity of the line element in the differentials and the symmetries in time and azimuthal angle, there are three constants of motion in the system [S-1]:

The first integral is *Einstein's area law*, stating that the product $R^2\frac{d\phi}{ds}$ is constant:

$$\boxed{R^2\frac{d\phi}{ds} = \text{const.} = c.} \tag{2.103}$$

The quantity $R^2\frac{d\phi}{ds}$ (where ϕ is the azimuthal angle) remains constant along the geodesic, representing *conservation of angular momentum* for a particle moving in this spacetime (the conservation of angular momentum in a spherically symmetric gravitational field).[39]

Similarly, the lack of explicit time dependence in the line element implies *the conservation of energy*[40]:

$$(1 - \frac{\alpha}{R})\frac{dt}{ds} = \text{const.} = 1. \tag{2.104}$$

The third integral provides a relationship between the energy, angular momentum, and the radial component of the motion[41]:

[39] For a planet (e.g., Mercury) in orbital motion around the Sun, its path is governed by the geodesic equation derived from the Schwarzschild metric (2.49). This planet has a stable orbit, but the predictions for its orbit include relativistic corrections (e.g., the precession of the perihelion, which Newtonian mechanics does not predict).

[40] This gives rise to the integral involving $\frac{dt}{ds}$, which remains constant along the particle's path.

[41] This integral comes from the normalization condition of the four-velocity (the derivative of the coordinates with respect to proper time ds) for timelike geodesics with h being a constant of integration. For motion in the equatorial plane ($\theta = \frac{\pi}{2}$), the Schwarzschild metric (2.49) simplifies to:

$$ds^2 = -\left(1 - \frac{\alpha}{R}\right)dt^2 + \left(1 - \frac{\alpha}{R}\right)^{-1} dR^2 + R^2 d\phi^2. \tag{2.105}$$

The four-velocity $\frac{dx^\mu}{ds}$ must satisfy the normalization condition (2.96). Substituting the Schwarzschild metric components from Eq. (2.105) into the normalization condition (2.96) and solving for the motion of a particle in this geometry leads to the equation:

$$(1 - \frac{\alpha}{R})\left(\frac{dt}{ds}\right)^2 - \frac{1}{1-\frac{\alpha}{R}}\left(\frac{dR}{ds}\right)^2 - R^2\left(\frac{d\phi}{ds}\right)^2 = -1. \tag{2.106}$$

Using the conserved quantities (2.103) and (2.104), we can substitute them back into Eq. (2.118):

$$-\left(1 - \frac{\alpha}{R}\right)\left(\frac{k}{1-\frac{\alpha}{R}}\right)^2 + \left(1 - \frac{\alpha}{R}\right)^{-1}\left(\frac{dR}{ds}\right)^2 + R^2\left(\frac{c}{R^2}\right)^2 = -1, \tag{2.107}$$

$$\text{where: } \frac{dt}{ds} = \frac{k}{1-\frac{\alpha}{R}}, \text{ and } \frac{d\phi}{ds} = \frac{c}{R^2}. \tag{2.108}$$

We simplify Eq. (2.107):

$$(1-\frac{\alpha}{R})\left(\frac{dt}{ds}\right)^2-\frac{1}{1-\frac{\alpha}{R}}\left(\frac{dR}{ds}\right)^2-R^2\left(\frac{d\phi}{ds}\right)^2=\text{const.}=h. \tag{2.118}$$

Schwarzschild showed that the integrals (2.103), (2.104), (2.118) lead to the correct prediction of Mercury's perihelion advance, confirming the validity of his solution (2.49). Schwarzschild manipulated the integrals (2.103), (2.104), and (2.118) and obtained Mercury's orbit equation in the Sun's gravitational field. After substi-

---

$$-\frac{k^2}{1-\frac{\alpha}{R}}+\frac{1}{1-\frac{\alpha}{R}}\left(\frac{dR}{ds}\right)^2+\frac{c^2}{R^2}=-1. \tag{2.109}$$

We multiply this through by $(1-\frac{\alpha}{R})$:

$$-k^2+\left(\frac{dR}{ds}\right)^2+\left(1-\frac{\alpha}{R}\right)\frac{c^2}{R^2}=-\left(1-\frac{\alpha}{R}\right), \tag{2.110}$$

and rearrange this equation to get:

$$\left(\frac{dR}{ds}\right)^2=\boxed{k^2}-\frac{c^2}{R^2}\left(1-\frac{\alpha}{R}\right)+\boxed{1}-\frac{\alpha}{R}. \tag{2.111}$$

We introduce the constant h as a constant of integration:

$$h=k^2-1. \tag{2.112}$$

We substitute back Eqs. (2.112) and (2.108) into Eq. (2.111). Thus, the equation becomes:

$$\boxed{\left(\frac{dR}{ds}\right)^2=h+\frac{\alpha}{R}-\frac{c^2}{R^2}\left(1-\frac{\alpha}{R}\right).} \tag{2.113}$$

Equation (2.113) transforms into the form (2.118). We can show this by reversing the derivation: we substitute (2.108) back into the normalization condition (2.107) and simplify each term:

$$(1-\frac{\alpha}{R})\left(\frac{k^2}{(1-\frac{\alpha}{R})^2}\right)=\frac{k^2}{1-\frac{\alpha}{R}},\quad (1-\frac{\alpha}{R})^{-1}\left(\frac{dR}{ds}\right)^2,\quad \frac{c^2}{R^2}. \tag{2.114}$$

Now, we substitute these simplified terms back into Eq. (2.107):

$$\frac{k^2}{1-\frac{\alpha}{R}}-(1-\frac{\alpha}{R})^{-1}\left(\frac{dR}{ds}\right)^2-\frac{c^2}{R^2}=1, \tag{2.115}$$

and multiply through by $(1-\frac{1}{\alpha})$:

$$k^2-\left(\frac{dR}{ds}\right)^2-\frac{c^2}{R^2}(1-\frac{\alpha}{R})=1-\frac{\alpha}{R}. \tag{2.116}$$

Substituting Eq. (2.112) into Eq. (2.116) and simplifying further, we get Eq. (2.113):

$$\boxed{k^2-\left(\frac{dR}{ds}\right)^2-\frac{c^2}{R^2}(1-\frac{\alpha}{R})+\frac{\alpha}{R}=1.} \tag{2.117}$$

.

tution and rearrangement, Schwarzschild arrived at an orbit equation similar to that of Einstein for the perihelion precession of Mercury [S-1][42]:

$$\left(\frac{dR}{d\phi}\right)^2 + R^2\left(1-\frac{\alpha}{R}\right) = \frac{R^4}{c^2}\left[1-h\left(1-\frac{\alpha}{R}\right)\right]. \tag{2.125}$$

Schwarzschild then wrote the orbit equation in terms of a new variable $x = \frac{1}{r} \equiv \frac{1}{R}$ expressing his equation per Einstein's notation [S-1]:

SCHWARZSCHILD'S ORBIT EQUATION:

$$\boxed{\left(\frac{dx}{d\phi}\right)^2 = \frac{1-h}{c^2} + \frac{h\alpha}{c^2}x - x^2 + \alpha x^3.} \tag{2.126}$$

This equation incorporates the key features of Einstein's orbit equation, including the precession of Mercury's perihelion.[43]

---

[42] First, we isolate $\frac{dt}{ds}$ from Eq. (2.104):

$$\frac{dt}{ds} = \frac{1}{1-\frac{\alpha}{R}}. \tag{2.119}$$

Then, we isolate $\frac{d\phi}{ds}$ from the area law, Eq. (2.103):

$$\frac{d\phi}{ds} = \frac{c}{R^2} \tag{2.120}$$

This expression involves squaring the fraction $\frac{c}{R^2}$: $\frac{c^2}{R^4}$. We then plug Eqs. (2.119) and (2.120) into Eq. (2.118). Schwarzschild wanted to express everything in terms of $\frac{dR}{d\phi}$. We use the chain rule:

$$\frac{dR}{ds} = \frac{dR}{d\phi}\frac{d\phi}{ds}. \tag{2.121}$$

Substituting Eq. (2.120) into the expression (2.121) we get:

$$\frac{dR}{ds} = \frac{dR}{d\phi}\frac{c}{R^2}. \tag{2.122}$$

Substituting Eqs. (2.119) and (2.122) back into the Eq. (2.118), we can solve for $\frac{dR}{d\phi}$. We get:

$$\frac{1}{(1-\frac{\alpha}{R})} - \frac{1}{1-\frac{\alpha}{R}}\left(\frac{dR}{d\phi}\right)^2\frac{c^2}{R^4} - R^2\left(\frac{d\phi}{ds}\right)^2 = h \tag{2.123}$$

Since $\frac{1}{(1-\frac{\alpha}{R})}$ is the reciprocal of $(1-\frac{\alpha}{R})$, multiplying these terms together simplifies to 1. Thus multiplying by $(1-\frac{\alpha}{R})$, we get the following:

$$-\left(\frac{dR}{d\phi}\right)^2\frac{c^2}{R^4} - R^2\left(1-\frac{\alpha}{R}\right)\frac{c^2}{R^4} = h\left(1-\frac{\alpha}{R}\right) - 1. \tag{2.124}$$

[43] Let us derive Eq. (2.126) from Eq. (2.125). We start with the left-hand side of the orbit Eq. (2.125) and substitute $R = \frac{1}{x}$:

Equation (2.126) matches Einstein's formulation from his Mercury perihelion paper [E3-15], Eq. (1.353):

$$\text{only if:}\quad \frac{2E}{B^2} = \frac{1-h}{c^2} \text{ and } B^2 = \frac{c^2}{h},$$

verifying the anomaly in Mercury's orbit that had puzzled astronomers for decades [S-1].

Schwarzschild notes that Einstein's approximation for the orbit of a planet (like Mercury) aligns with the exact solution (2.49) when substituting for r in the Eq. (1.353) the relationship Eq. (2.46) [S-1]:

$$\boxed{R = \left(r^3 + \alpha^3\right)^{\frac{1}{3}} \approx r\left(1 + \frac{\alpha^3}{r^3}\right),} \tag{2.133}$$

valid when $\frac{\alpha^3}{r^3}$ is much less than 1. Schwarzschild points out that in the case of an approximate solution like Einstein's, the term $\frac{\alpha^3}{r^3}$ is extremely small. Recall that the term α is defined by Eq. (1.164). $\frac{\alpha}{r}$ is nearly equal to twice the square of the velocity of the planet (when the speed of light = 1), which is a small fraction. Due to the smallness of $\frac{\alpha^3}{r^3}$, the expression $\left(1 + \frac{\alpha^3}{r^3}\right)$ on the right-hand side of Eq. (2.133) is very close to 1. Thus, for practical purposes, R is almost identical to r as the difference

---

$$\boxed{R^2\left(1 - \frac{\alpha}{R}\right) = \frac{1}{x^2}\left(1 - \alpha x\right).} \tag{2.127}$$

The transformation of the derivative part of the orbit equation is:

$$dR = -\frac{dx}{x^2}\frac{dR}{d\phi} = -\frac{1}{x^2}\frac{dx}{d\phi}. \quad \text{Thus:} \tag{2.128}$$

$$\boxed{\left(\frac{dR}{d\phi}\right)^2 = \left(-\frac{1}{x^2}\frac{dx}{d\phi}\right)^2 = \frac{1}{x^4}\left(\frac{dx}{d\phi}\right)^2.} \tag{2.129}$$

For the right-hand side, we again substitute $R = \frac{1}{x}$ and simplify:

$$\boxed{\frac{R^4}{c^2}\left[1 - h\left(1 - \frac{\alpha}{R}\right)\right] = \frac{1}{c^2 x^4}\left[1 - h\left(1 - \alpha x\right)\right].} \tag{2.130}$$

Now, we substitute Eqs. (2.127), (2.129) and (2.130) into the original orbit Eq. (2.125) and obtain:

$$\frac{1}{x^4}\left(\frac{dx}{d\phi}\right)^2 + \frac{1}{x^2}\left(1 - \alpha x\right) = \frac{1}{c^2 x^4}\left[1 - h\left(1 - \alpha x\right)\right]. \tag{2.131}$$

Next, we multiply through by x^4 to eliminate the denominators:

$$\left(\frac{dx}{d\phi}\right)^2 + x^2 - \alpha x^3 = \frac{1}{c^2} - \frac{h}{c^2} + \frac{h\alpha x}{c^2} = \frac{1-h}{c^2} + \frac{h\alpha}{c^2}x - x^2 + \alpha x^3. \tag{2.132}$$

between r and R is extremely small. In such cases, $R \approx r$ is extremely accurate. Schwarzschild affirms that Einstein's approximation, which treats R as equivalent to r in the approximate methodology, is more than sufficient for the accuracy required in practical applications, like calculating the orbit of Mercury. The difference between $\left(1+\frac{\alpha^3}{r^3}\right)$ and 1 is so minute (of the order 10^{-12}) that it does not significantly impact the calculations of Mercury's orbit to a high degree of precision [S-1].

Einstein's original calculations for the perihelion shift of Mercury were performed using an approximation method and with the assumption of vacuum field equations under the condition that the determinant of the metric tensor was -1 [Eq. (1.298)], see Sect. 1.6.3. This specific choice simplifies calculations but raises an important question: does this choice limit the generality or correctness of the result? The choice of coordinates (including the unimodular condition) does not affect the physical predictions of Einstein's general relativity. Thus, while Einstein's calculations were simplified by his choice of unimodular coordinates (1.299), the physical predictions do not depend on this choice.

The exact Schwarzschild solution (2.49) describes the gravitational field outside a spherically symmetric mass. This solution does not assume a specific determinant for the metric tensor and is expressed in spherical coordinates. Importantly, when Schwarzschild calculated the precession of Mercury's perihelion using solution (2.49), the result agreed with Einstein's calculation, which presupposed the determinant to be equal to one (1.299). Specifically, Schwarzschild's analysis elucidated that the discrepancy between the two computational approaches is negligible to such an extent that Einstein's approximation yields a precision practically indistinguishable from the exact solution. This revelation underscores the remarkable accuracy of Einstein's 1915 preliminary calculations and affirms their validity in general relativity. This agreement between Einstein's (1.353) and Schwarzschild's (2.126) equations and conclusions is because of the following: the physical situation described by both approaches is the same, and according to general relativity's general covariance, the choice of coordinate conditions (like unimodular coordinates) does not affect physical predictions. The fact that the determinant of the metric tensor g is not set to -1 in the Schwarzschild solution but still yields the same result for Mercury's perihelion precession in Einstein's original calculation underlines a fundamental aspect of general relativity: physical results are independent of the coordinate system used. The choice of the determinant 1 condition (1.298) or any other condition is a mathematical convenience, not a fundamental theory restriction. The use of unimodular coordinates and the determinant 1 assumption (1.298) by Einstein in his initial calculations of the precession of the perihelion of Mercury was a methodological choice that simplified the mathematics but did not limit the generality or validity of the result.

2.2.5 *The Interior Metric: Schwarzschild Employs the Same Trick*

After reading Einstein's November 25, 1915 paper, which extended his November 18, 1915 field equations, Schwarzschild approached the singularity problem in his line element. Acquainting himself with Einstein's stress-energy tensor, Schwarzschild extended his calculation from the gravitational field of a mass point to the gravitational field of an incompressible homogeneous and isotropic fluid sphere. Utilizing Einstein's revised field equations from his November 25 paper, Schwarzschild derived an exact solution for the interior of an incompressible fluid sphere. Schwarzschild's further research demonstrated that his line element (2.49) applied to the region outside an incompressible fluid sphere. Schwarzschild communicated his new findings to Einstein in a letter dated February 6, 1916, attaching his paper titled "On the Gravitational Field of a Sphere of Incompressible Fluid, According to Einstein's Theory" [S-2, CPAE8], Doc. 188. Einstein replied to Schwarzschild's letter on February 19 and presented Schwarzschild's paper to the Prussian Academy on February 24, 1916.

In his further research, Schwarzschild began with Einstein's field Eq. (1.290) *valid only in unimodular coordinates* (1.299) [E4-15]. Recall that Eq. (1.290) valid in unimodular coordinates are Eq. (1.292). Equations (1.292) simplify in regions without matter (outside the fluid sphere), i.e., the components of $R_{\mu\nu}$ vanish for areas without matter (see Eq. (1.289)). Inside the fluid sphere, the components of $R_{\mu\nu}$ are defined using the *mixed stress-energy tensor of an incompressible fluid*, with the pressure p and constant density ρ_0 as:

$$\boxed{T_1^1 = T_2^2 = T_3^3 = -p, \quad T_4^4 = \rho_0,} \tag{2.134}$$

and all other components of T_μ^ν are zero.

In his discussion about Eq. (2.134), Schwarzschild mentions Einstein's mixed energy tensor while referencing Einstein's 1914 *Entwurf* review paper, "The Formal Foundation of the General Theory of Relativity" [S-1]. Einstein devised his stress-energy tensor of a perfect fluid, defined in terms of pressure p and constant density ρ_0 in the following manner [E2-14][44]:

---

[44] In this equation, p represents the pressure in the stress-energy tensor, P is a derived quantity from p, and $\rho_0(1 + P)$ represents the energy density. Schwarzschild describes the structure of the tensor T_μ^ν, where one index is covariant (subscript) and the other is contravariant (superscript). This specific arrangement of indices in the tensor illustrates the mixed nature of the energy tensor per Einstein's formulation in his 1914 paper. However, in his 1914 *Entwurf* review paper, Einstein referred to his stress-energy tensor T_σ^ν as the stress-energy tensor density. A tensor density incorporates the factor $\sqrt{-g}$. The $\sqrt{-g}$ allows the tensor density to transform differently under coordinate transformations than a regular tensor. However, in the case of the tensor Eq. (2.135), we would not include the $\sqrt{-g}$. Hence, in the specific case of the matrix representing a perfect fluid (2.135), the distinction between the mixed stress-energy tensor and the tensor density is not immediately relevant.

$$\begin{pmatrix} -p & 0 & 0 & 0 \\ 0 & -p & 0 & 0 \\ 0 & 0 & -p & 0 \\ 0 & 0 & 0 & \rho_0(1+P) \end{pmatrix}. \tag{2.135}$$

To achieve equilibrium, the fluid sphere must satisfy Einstein's equilibrium condition (1.246) [E1-15] (see Sect. 1.6.3, Chap. 1).

In his second paper, Schwarzschild used the line element (2.13) and the associated metric tensor (2.14) to apply Einstein's field Eq. (1.292) specifically to the interior of the incompressible fluid sphere.[45] This involves solving the field equations using the derived metric components, considering the properties of the perfect fluid (i.e., considering p and ρ_0) and Eq. (2.134) [S-2][46]:

$$T_{11} = T_1^2(-pg_{11}),\ T_{22} = T_2^2(-pg_{22}),\ T_{33} = T_3^2(-pg_{33}).\ T_{44} = T_4^2(\rho_0 g_{44}), \tag{2.136}$$

$$G_{11} = \frac{\kappa f_1}{2}(p-\rho_0),\ G_{22} = \frac{\kappa f_2}{2}\frac{1}{1-x_2^2}(p-\rho_0). \tag{2.137}$$

$$G_{33} = \frac{\kappa f_2}{2}(1-x_2^2)(p-\rho_0),\ G_{44} = -\frac{\kappa f_4}{2}(\rho_0+3p).$$

Schwarzschild's derivation aims to find *a solution for the sphere's interior that matches smoothly with the exterior* Schwarzschild's solution (2.49) at the sphere's boundary. This matching guarantees a consistent overall description of spacetime around and inside the sphere. Similar to the derivation of the external line element (2.49), *the determinant condition* (2.21) plays a significant role in Schwarzschild's derivation of the line element for the interior of an incompressible fluid sphere; it is useful for simplifications and substitutions.[47]

Schwarzschild considered *the Christoffel symbols* in terms of *the functions f from his previous paper* [S-1] [Eq. (2.16)], which involve only derivatives of the

---

[45] Recall that Schwarzschild transformed the usual spherical coordinates into new coordinates to simplify the mathematical treatment (refer to Eq. (2.9) in Sect. 2.2.2). The transformation (2.9) is chosen to facilitate the application of Einstein's field equations in a spherically symmetric scenario. With these transformed coordinates, Schwarzschild writes the line element (2.13)(ds^2) in a form adapted to Einstein's four conditions and the determinant condition Eq. (1.298), where f_1, f_2, and f_4 are functions of x_1. This line element leads to the components of the metric tensor $g_{\mu\nu}$ (2.14). These components and Eqs. (2.30), (2.32), and (2.21) reflect the curvature of spacetime outside the fluid sphere.

[46] g_{11}, g_{22}, g_{33} and g_{44} are represented by Eq. (2.14).

[47] Recall that Schwarzschild derived an additional condition (2.22) (also called the determinant condition), see Sect. 2.2.2, which relates the derivatives of the functions f_1, f_2, and f_4 to each other, providing a direct way to connect changes in these functions.

functions f_1, f_2, f_3, f_4 with respect to x_1. In his second paper, again at the equator, Schwarzschild set $x_2 = 0$, thus[48]:

$$-pg_{22} = -pf_2 \text{ (at } x_2 = 0), \tag{2.138}$$

and the simplified field equations [S-2][49]:

$$-\frac{1}{2}\frac{\partial}{\partial x_1}\left(\frac{1}{f_1}\frac{\partial f_1}{\partial x_1}\right) + \frac{1}{4}\frac{1}{f_1^2}\left(\frac{\partial f_1}{\partial x_1}\right)^2 + \frac{1}{2}\frac{1}{f_2^2}\left(\frac{\partial f_2}{\partial x_1}\right)^2 + \frac{1}{4}\frac{1}{f_4^2}\left(\frac{\partial f_4}{\partial x_1}\right)^2 = -\frac{\kappa}{2}f_1(\rho_0 - p). \tag{2.139}$$

$$\frac{1}{2}\frac{\partial}{\partial x_1}\left(\frac{1}{f_1}\frac{\partial f_2}{\partial x_1}\right) - 1 - \frac{1}{2}\frac{1}{f_1 f_2}\left(\frac{\partial f_2}{\partial x_1}\right)^2 = -\frac{\kappa}{2}f_2(\rho_0 - p). \tag{2.140}$$

$$-\frac{1}{2}\frac{\partial}{\partial x_1}\left(\frac{1}{f_1}\frac{\partial f_4}{\partial x_1}\right) + \frac{1}{2}\frac{1}{f_1 f_4}\left(\frac{\partial f_4}{\partial x_1}\right)^2 = -\frac{\kappa}{2}f_4(\rho_0 + 3p). \tag{2.141}$$

Schwarzschild rewrote *the equilibrium condition* (1.246) in terms of the properties of the perfect fluid sphere p and ρ_0 [S-2][50]:

$$-\frac{\partial p}{\partial x_1} = -\frac{p}{2}\left[\frac{1}{f_1}\frac{\partial f_1}{\partial x_1} + \frac{2}{f_2}\frac{\partial f_2}{\partial x_1}\right] + \frac{\rho_0}{2}\frac{1}{f_4}\frac{\partial f_4}{x_1}. \tag{2.142}$$

---

[48] Recall that in [S-1], Schwarzschild set $x_2 = 0$ at the equator as a simplifying assumption streamlining the field equations because the equations involve rotational symmetry around the origin.

[49] The field equations relate the partial derivatives of the functions f_1, f_2, and f_4 to the pressure p and the constant density ρ_0.

[50] For this purpose, Schwarzschild followed the following steps:
(1) He used the equilibrium condition (1.246) to connect the pressure p and constant density ρ_0 with the geometry of spacetime embodied in the Christoffel symbols. Schwarzschild started with the equilibrium condition (1.246) and expressed the divergence of the stress-energy tensor $\frac{\partial T^{\alpha}_{\sigma}}{\partial x_{\alpha}}$ in terms of the metric functions f_1, f_2, and f_4 and their derivatives. These metric functions correspond to the components of the metric tensor: $-pg_{11}$, $-pg_{22}$, $-pg_{33}$ and $\rho_0 g_{44}$ (see Eq. (2.14)). Remember that the derivatives are with respect to the coordinate x_1. This process involves the Christoffel symbols Eq. (2.16).
(2) He considered a simplified scenario where $x_2 = 0$ and *applied the simplified field Eqs.* (2.139), (2.140), (2.141) to replace expressions involving p and ρ_0 with expressions involving the derivatives of the metric functions f_1, f_2, and f_4.
(3) Finally, he *employed the determinant condition* (2.21) to simplify expressions and substitute one metric function in terms of others. He used the additional (determinant) condition Eq. (2.22) to simplify and eliminate redundancies. Recall that this equation relates the derivatives of the functions f_1, f_2, and f_4 to each other and connects changes in these functions, allowing for further simplifications in the derivation.
Combining the components (1), (2), and (3), we arrive at a new form of the equilibrium condition (2.142). This equilibrium condition expresses p and ρ_0 in terms of the derivatives of the f functions.

With *the determinant condition* (2.21), *the equilibrium condition* (2.142) becomes[51]:

THE EQUILIBRIUM CONDITION FOR THE PERFECT FLUID SPHERE:

$$\boxed{-\frac{\partial p}{\partial x_1} = \frac{\rho_0 + p}{2}\frac{1}{f_4}\frac{\partial f_4}{\partial x_1}.} \tag{2.143}$$

Now, integrating both sides with respect to x_1 gives:

$$\boxed{(\rho_0 + p)\sqrt{f_4} = const. = \gamma,} \text{ where } \gamma \text{ is a constat.} \tag{2.144}$$

Schwarzschild then multiplied each of *the field Eqs.* (2.139), (2.140), (2.141) by specific factors and simplified them.[52] The resulting transformed equations are (where for simplicity $x_1 \equiv x$) [S-2]:

$$\begin{aligned}
&\frac{\partial}{\partial x}\left(\frac{1}{f_1}\frac{\partial f_1}{\partial x}\right) = \frac{1}{2f_1^2}\left(\frac{\partial f_1}{\partial x}\right)^2 + \frac{1}{f_2^2}\left(\frac{\partial f_2}{\partial x}\right)^2 + \frac{1}{2f_4^2}\left(\frac{\partial f_4}{\partial x}\right)^2 + \kappa f_1(\rho_0 - p). \text{ (EQ. 1.)}\\
&\frac{\partial}{\partial x}\left(\frac{1}{f_2}\frac{\partial f_2}{\partial x}\right) = 2\frac{f_1}{f_2} + \frac{1}{f_1 f_2}\frac{\partial f_1}{\partial x}\frac{\partial f_2}{\partial x} - \kappa f_1(\rho_0 - p). \quad \text{(EQ. 2.)}\\
&\frac{\partial}{\partial x}\left(\frac{1}{f_4}\frac{\partial f_4}{\partial x}\right) = \frac{1}{f_1 f_4}\frac{\partial f_1}{\partial x}\frac{\partial f_4}{\partial x} + \kappa f_1(\rho_0 + 3p). \quad \text{(EQ. 3.)}
\end{aligned} \tag{2.145}$$

The next step was combining the transformed field Eq. (2.145) in such a way that they simplify further, using *the determinant condition* (2.21) to facilitate this process.[53] After combining and simplifying, Schwarzschild arrived at two equations:

$$\begin{aligned}
0 &= 4\frac{f_1}{f_2} - \frac{1}{f_2^2}\left(\frac{\partial f_2}{\partial x}\right)^2 - 2\frac{1}{f_2 f_4}\frac{\partial f_2}{\partial x}\frac{\partial f_4}{\partial x} + 4\kappa f_1 p,\\
0 &= 2\frac{\partial}{\partial x}\left(\frac{1}{f_2}\frac{\partial f_2}{\partial x}\right) + 3\frac{1}{f_2^2}\left(\frac{\partial f_2}{\partial x}\right)^2 + 2\kappa f_1(\rho_0 + p).
\end{aligned} \tag{2.146}$$

---

[51] This equation relates the pressure gradient to the constant density ρ_0, the pressure p, and the derivative of the metric function f_4.

[52] Schwarzschild multiplied Eq. (2.139) by -2 and this multiplication cancels the negative sign and the factor of $\frac{1}{2}$ in the first term, simplifying the equation. He then multiplied Eq. (2.140) by $2\frac{f_1}{f_2}$. This multiplication is intended to simplify the terms involving f_2, particularly the first term on the left-hand side. He multiplied the third Eq. (2.141) by $-2\frac{f_1}{f_4}$. Similar to the second field equation, this multiplication simplifies terms involving f_4.

[53] First, Schwarzschild combined Eq. (2.145) (EQ 1), twice Eq. (2.145) (EQ 2), and Eq. (2.145) (EQ 3). *The determinant condition* (2.21) *plays a role in eliminating certain terms and simplifying the equations* further. Subsequently, again, Schwarzschild combined Eq. (2.145) (EQ 1) and Eq. (2.145) (EQ 3) similarly. The process was similar to the first combination, where he aligned terms involving the derivatives and used the determinant condition (2.21) to simplify.

So, the *field Eq.* (2.146), and the *equilibrium condition* (2.144) finally lead to the following values for the functions [S-2][54]:

$$\boxed{f_2 = \frac{3}{\kappa\rho_0}\sin^2\sigma,\ f_4 = \left(\frac{3\cos\sigma_{ext} - \cos\sigma}{2}\right)^2,\ \text{valid with the condition (2.21).}} \tag{2.147}$$

Schwarzschild combined f_4 [Eq. (2.147)], and the equilibrium condition, Eq. (2.144), and obtained an equation that relates the pressure and density inside the static, spherically symmetric body (i.e., a non-rotating star).

From the equilibrium equation, we have: $\rho_0 + p = \frac{\gamma}{\sqrt{f_4}}$. Substituting f_4 from Eq. (2.147) we get:

$$\rho_0 + p = \frac{2\gamma}{3\cos\sigma_{ext} - \cos\sigma}. \tag{2.148}$$

Let us consider the boundary condition where $\sigma = \sigma_{ext}$. At this boundary, the equilibrium condition, Eq. (2.144), must hold. Plugging $\sigma = \sigma_{ext}$ into f_4 [Eq. (2.147)] we get:

$$\boxed{f_4 = \left(\frac{3\cos\sigma_{ext} - \cos\sigma_{ext}}{2}\right)^2 = \cos\sigma_{ext}^2.} \tag{2.149}$$

Therefore, at the boundary, the equilibrium condition, Eq. (2.144) becomes:

$$(\rho_0 + p)\cos\sigma_{ext} = \gamma. \tag{2.150}$$

Assuming $p = 0$ at the boundary, we get from Eq. (2.150):

$$\rho_0\cos\sigma_{ext} = \gamma. \tag{2.151}$$

Substituting Eq. (2.151) back into Eq. (2.148), we get [S-2][55]:

$$\boxed{\rho_0 + p = \rho_0\frac{2\cos\sigma_{ext}}{3\cos\sigma_{ext} - \cos\sigma}.} \tag{2.152}$$

The specific forms of f_2 and f_4 in Eq. (2.147) are designed to *describe the space-time geometry inside the fluid sphere*. Schwarzschild defines the functions f_2 and f_4 in terms of the *spherical variable* σ and the constant parameters κ (Einstein's gravitational constant) and ρ_0. $\sigma = 3x - \rho$ is valid outside the sphere, and as said

[54] I will streamline the derivation because I want to highlight only the key steps relevant to the subject matter of this book.

[55] ρ_0, the constant density, and p, the pressure, are measured within the spherically symmetric body. The equation describes how the pressure p changes with the radial distance inside a star.

above, $\sigma = \sigma_{\text{ext}}$ at the sphere's surface. The functions f_2 and f_4 are metric tensor components in the interior line element (2.153).

After establishing the solution in the coordinates (2.9), Schwarzschild then *transformed back to standard spherical coordinates* σ, θ, ϕ. While spherical coordinates do not satisfy the determinant condition (2.21), the derived functions f_2 and f_4 [(2.147)] *are still applicable in spherical coordinates*. Schwarzschild's skillful application of the determinant condition (2.21) and Cartesian coordinates and subsequent transformation back to spherical coordinates demonstrate his ability to navigate the complexities of Einstein's 1915 formulation.

Using spherical variables σ, θ, ϕ instead of x_1, x_2, x_3, Schwarzschild wrote the line element for the region inside the sphere, showing it to be *free of singularities* [S-2][56]:

SCHWARZSCHILD'S INTERIOR LINE ELEMENT:

$$\boxed{ds^2 = \left(\frac{3\cos\sigma_{\text{ext}} - \cos\sigma}{2}\right)^2 dt^2 - \frac{3}{\kappa\rho_0} d\sigma^2 + \frac{3}{\kappa\rho_0}\sin^2\sigma\,(d\theta^2 + \sin^2\theta\, d\phi^2).} \tag{2.153}$$

The Schwarzschild interior solution (2.153) is regular everywhere inside the incompressible sphere (for $0 \leq \sigma \leq \sigma_{\text{ext}}$), including at the center (where $\sigma = 0$). The metric behaves regularly at the center, with no physical singularity.

Schwarzschild analyzed the pressures within and outside a collapsing sphere. He defined two parameters α and ρ [S-2][57]:

$$\text{THE FIRST PARAMETER } \alpha : \ \alpha = \sqrt{\frac{3}{\kappa\rho_0}} \sin^3\sigma_{\text{ext}}, \tag{2.155}$$

[56] The terms σ, θ, ϕ involving these coordinates do not introduce singularities as long as the density ρ_0, and the fluid characteristics are well-defined and non-zero.

[57] The second parameter:

$$\text{THE SECOND PARAMETER } \rho : \ \rho = A^{-3/2}\left[\frac{3}{2}\sin^3\sigma_{\text{ext}} - \frac{9}{4}\cos\sigma_{\text{ext}} B\right], \tag{2.154}$$

relating it to the sphere's density and radius, where:

$$A = \frac{\kappa\rho_0}{3} \quad \text{and: } B = \sigma_{\text{ext}} - \frac{1}{2}\sin 2\sigma_{\text{ext}}.$$

This is a key parameter defined in terms of the fluid's density ρ_0 and the spherical variable σ_{ext} in Eq. (2.153).[58]

When describing the internal properties of the sphere, Schwarzschild stressed that the characteristics of spherical geometry dominate. The curvature radius of this spherical space is established as [S-2]:

$$R_{int} = \sqrt{\frac{3}{\kappa\rho_0}}\sigma_{ext}. \tag{2.161}$$

R_{int} denotes the sphere's radius as measured inside from the sphere's center to the boundary.

Schwarzschild considered an object that can fall freely from rest at an infinite distance to the surface of the spherical sphere. The object starts from rest at an infinite distance. As the object falls towards the sphere, it follows a geodesic path, and its velocity v_{ext} increases due to the gravitational attraction of the sphere. Upon reaching the sphere's surface, the object attains its maximum velocity. This final velocity is calculated using the metric [Eq. (2.153)].[59]

Schwarzschild observed that as the sphere collapses at a constant mass but increasing density, it transitions to a smaller radius than before and emits radiant

---

[58] Recall that at the boundary $\sigma = \sigma_{ext}$. Schwarzschild therefore defined [S-2]:

$$R_{ext} = \sqrt{\frac{3}{\kappa\rho_0}}\sin\sigma_{ext}, \quad \text{thus:} \tag{2.156}$$

$$\frac{3}{\kappa\rho_0} = \frac{R_{ext}^2}{\sin^2\sigma_{ext}}, \quad \text{and:} \tag{2.157}$$

$$\frac{\alpha}{R_{ext}} = \sin^2\sigma_{ext}. \tag{2.158}$$

We can substitute $\sin^2\sigma_{ext}$ from Eq. (2.158) into Eq. (2.157). We get:

$$\frac{3}{\kappa\rho_0} = \frac{{R_{ext}}^3}{\alpha}. \tag{2.159}$$

Now if we combine Eqs. (2.156) and (2.155), we get at the boundary, Eq. (2.159):

$$\alpha = \frac{\kappa\rho_0}{3}{R_{ext}}^3. \tag{2.160}$$

[59] Schwarzschild calls this velocity the *naturally measured* fall velocity [S-2]:

$$v_{ext} = \sqrt{\frac{\alpha}{R_{ext}}}. \tag{2.162}$$

Plugging Eq. (2.158) into this equation, we get:

$$v_{ext} = \sin\sigma_{ext}. \tag{2.163}$$

energy. Radiation that is emitted from the surface of the massive sphere is redshifted. Schwarzschild calculated that the pressure becomes infinite at the center of the sphere and the velocity of fall v_{ext} [Eq. (2.162)] approaches approximately $0.94c$, where c is the speed of light, signifying a density limit beyond which the sphere collapses.[60] Schwarzschild thought that he had thus set a boundary beyond which an incompressible fluid sphere ceases to exist. For $\cos\sigma_{ext} < \frac{1}{3}$, the sphere becomes a point mass with infinite density at the center, and Eq. (2.44) is valid. At this stage, the exterior Schwarzschild solution (2.49) becomes relevant.

It follows from Eq. (2.162) that for an external observer, a sphere of gravitational mass cannot have a radius measured from outside whose numerical value is smaller than $R_{ext} = \alpha$ or less than $R = 2Gm$ (recall that $\alpha = 2Gm$). If it does have a radius $R_{ext} < \alpha$, or $R < 2Gm$, then the equations fail, indicating a sphere collapse [S-2]. $R = 2Gm$ limits the size of the incompressible fluid sphere to the size $R > 2Gm$.

Schwarzschild demonstrated that the two solutions (2.153) and (2.49) coincide at the sphere's boundary. He related the radius and density of the sphere to the spherical variables, both inside and at the sphere's surface. He showed that the physical parameters are continuous across the boundary. He demonstrated that his specific line element (2.49) accurately described the external space of an incompressible fluid sphere, the vacuum space outside the sphere [Eis]. This description hinged on the radius vector of the sphere (r), which was linked to the sphere's density (ρ), and the constant α was associated with the uniform density of the fluid sphere. Accordingly, the external cube of the radius R^3_{ext} equated to the internal cube of the radius $r^3_{int} + \rho$, and also α equaled ρ for a point mass, indicate that the exterior properties of the sphere depend on its density.

[60] Radiation is redshifted according to the gravitational redshift formula:

$$z = \frac{1}{\cos\sigma} - 1. \tag{2.164}$$

The velocity of light within the sphere is given by [S-2]:

$$v_{int} = \frac{2}{3\cos\sigma_{ext} - \cos\sigma}. \tag{2.165}$$

This equation indicates that the velocity of light varies within the sphere, depending on the position (as represented by σ_{ext} and $\cos\sigma$). $\frac{1}{\cos}\sigma_{ext}$ represents the speed of light at the sphere's surface. As we move toward the center of the sphere, the speed of light increases, reaching a maximum value of $\frac{2}{3\cos\sigma_{ext}-1}$ at the center. At the sphere's center, where $\cos\sigma_{ext} = \frac{1}{3}$, and $\sigma = 0$, $v_{int} \to \infty$. Using Eqs. (2.152) and (2.144), at the sphere's center $\rho_0 + p \to \infty$. Equation (2.152) gives:

$$\rho_0 + p = \frac{2}{3}\rho_0\frac{1}{1-1} = 0. \tag{2.166}$$

The pressure at the center is infinite, and beyond this critical density point, an incompressible fluid sphere cannot sustain its structure and collapses.

The interior Schwarzschild solution (2.153) should match smoothly with the exterior Schwarzschild solution (2.49) at the sphere's boundary.[61] I adhere to the subsequent steps to demonstrate the alignment of the two line elements.

From the Pythagorean identity, we know that $\cos^2\sigma + \sin^2\sigma = 1$. Hence, Eq. (2.163) becomes:

$$v_{\text{ext}}^2 = \frac{1}{\cos^2\sigma_{\text{ext}}}. \tag{2.167}$$

For an object falling radially (i.e., moving directly towards the center of the sphere without any angular motion), the relevant part of the metric is the radial component. Considering only the radial component of the motion and given Eq. (2.167) for $\sigma = \sigma_{\text{ext}}$, the line element (2.153) becomes:

$$\boxed{ds^2 = \left(\frac{3\cos\sigma_{\text{ext}} - \cos\sigma}{2}\right)^2 dt^2 - \frac{1}{\cos^2\sigma}d\sigma^2 + \frac{3}{\kappa\rho_0}\sin^2\sigma\,(d\theta^2 + \sin^2\theta\, d\phi^2).} \tag{2.168}$$

At the boundary, the interior solution (describing the spacetime inside the sphere) (2.176) match smoothly with the exterior Schwarzschild solution (describing the spacetime outside the sphere)[62]:

[61] The term smooth refers to mathematical smoothness. This means there should be no discontinuities or abrupt changes in the values of physical quantities across the boundary. It does not mean that the equations themselves must look the same. Instead, it means that the solutions provided by these equations must align and match at the boundary to ensure a smooth transition. The interior and exterior solutions come from different equations that describe different physical situations. The key requirement is the continuity of the solutions at the boundary. The boundary is where the interior and exterior solutions must meet boundary conditions. These conditions calculated by both solutions should be the same.

[62] Consider the coordinate transformation (See "Interior Schwarzschild metric" for more information):

$$\sigma = \arcsin\left(\frac{r}{\mathcal{R}}\right), \text{ and:} \tag{2.169}$$

$$\sigma_{\text{ext}} = \arcsin\left(\frac{\alpha}{2\mathcal{R}}\right), \quad \text{which simplifies the line element 2.168.} \tag{2.170}$$

$\mathcal{R} = \frac{r_g^3}{\alpha} \equiv \frac{r_g^3}{r_s}$, where r_s is the Schwarzschild radius. The gravitational radius r_g is given by Eq. (2.93) (from Sect. 2.2.2):

$$r_g = \frac{\alpha}{2} \equiv \frac{r_s}{2}. \tag{2.171}$$

Transformation (2.169) and (2.170) helps express the interior metric in a form that can easily match the exterior Schwarzschild metric at the boundary. Let us do the derivation. We will apply the coordinate transformation (2.169) and (2.170) to the Schwarzschild interior line element (2.168), and replace σ and σ_{ext} with their respective expressions in terms of r, $\mathcal{R}$, and Schwarzschild's notation α or the modern r_s.

Firstly, let us express $\cos\sigma$ and $\cos\sigma_{\text{ext}}$ in terms of r and α. Using the identity $\cos(\arcsin(x)) = \sqrt{1-x^2}$, we have:

$$\boxed{ds^2 = \left(1 - \frac{r_s}{r}\right) dt^2 - \frac{1}{1 - \frac{r_s}{r}} dr^2 - r^2 \left(d\theta^2 + \sin^2\theta d\phi^2\right), \text{ if } r_s \leq r.} \tag{2.177}$$

Birkhoff's theorem warrants that the Schwarzschild exterior solution (2.49) will apply at the boundary and the exterior of any spherically symmetric mass distribution, including an incompressible fluid sphere characterized by the Schwarzschild interior solution (2.168). John Synge acknowledges three contributors in his textbook on general relativity, *Relativity: The General Theory*, to the discovery and formalization of the theorem commonly known as *Birkhoff's theorem* [Syn]: Jørg Tofte Jebsen, W. Alexandrow, and George David Birkhoff. Jebsen, a Norwegian physicist, published a paper in 1921 that contained the key results of what would later be known as *Birkhoff's theorem*. His work was among the earliest to articulate the idea underlying the Birkhoff theorem [Jeb]. In 1923, Alexandrow independently formulated the theorem [Ale]. Birkhoff independently discovered and formulated the theorem in his

---

$$\cos\sigma = \cos\left(\arcsin\left(\frac{r}{\mathcal{R}}\right)\right) = \sqrt{1 - \left(\frac{r}{\mathcal{R}}\right)^2}, \tag{2.172}$$

$$\cos\sigma_{\text{ext}} = \cos\left(\arcsin\left(\frac{\alpha}{2\mathcal{R}}\right)\right) = \sqrt{1 - \left(\frac{\alpha}{2\mathcal{R}}\right)^2}, \text{ and:} \tag{2.173}$$

$$\cos^2\sigma = 1 - \left(\frac{r}{\mathcal{R}}\right)^2. \tag{2.174}$$

To transform the expression $\cos^2\sigma$ using the transformation Eq. (2.169), we need to express $\cos\sigma$ in terms of r and $\mathcal{R}$. Again using the Pythagorean identity, $\cos^2\sigma + \sin^2\sigma = 1$, since $\sin\sigma = \frac{r}{\mathcal{R}}$ from the transformation (2.169), we write:

$$\cos^2\sigma = 1 - \sin^2\sigma = 1 - \left(\frac{r}{\mathcal{R}}\right)^2. \quad \text{This is Eq. (2.174).}$$

Substituting Eqs. (2.172), (2.173), and (2.174) back into the line element (2.168), and considering Eq. (2.157), we get the interior line element:

$$\begin{aligned} ds^2 =& \frac{1}{4}\left(3\sqrt{1 - \left(\frac{\alpha}{2\mathcal{R}}\right)^2} - \sqrt{1 - \left(\frac{r}{\mathcal{R}}\right)^2}\right)^2 dt^2 \\ & - \frac{1}{1 - \left(\frac{r}{\mathcal{R}}\right)^2} dr^2 + r^2\left(d\theta^2 + \sin^2\theta d\phi^2\right), \quad \text{if } 0 \leq r \leq r_g. \end{aligned} \tag{2.175}$$

This equation expresses the metric in terms of the coordinates r, θ, and ϕ, with the parameters $\mathcal{R}$ and α representing a specific form of the Schwarzschild interior solution, particularly focused on the boundary condition where $\sigma = \sigma_{\text{ext}}$. Equation (2.175) is the resulting line element after applying the coordinate transformation (2.169) and (2.170). Substituting Eq. (2.171) into Eq. (2.175), we get:

THE INTERIOR LINE ELEMENT IN TERMS OF r_s AND r_g :

$$\begin{aligned} ds^2 =& \frac{1}{4}\left(3\sqrt{1 - \left(\frac{r_g}{\mathcal{R}}\right)^2} - \sqrt{1 - \left(\frac{r}{\mathcal{R}}\right)^2}\right)^2 dt^2 \\ & - \frac{1}{1 - \left(\frac{r}{\mathcal{R}}\right)^2} dr^2 + r^2\left(d\theta^2 + \sin^2\theta d\phi^2\right), \quad \text{if } 0 \leq r \leq r_g. \end{aligned} \tag{2.176}$$

1923 textbook *Relativity and Modern Physics*. His theorem formulation provided a more general proof and was more accessible to a broad audience, contributing to its integration into the canon of general relativity [Bir]. The theorem, attributed to Birkhoff, posits that any spherically symmetric solution of Einstein's vacuum field equations (1.288) is static. In other words, if we have a spherically symmetric mass distribution, the solution to how the gravitational field behaves outside of this mass, where there is a vacuum, is uniquely given by the Schwarzschild solution. This is true regardless of its internal structure or dynamics as long as it maintains spherical symmetry and a vacuum is outside. It follows from Birkhoff's theorem that no gravitational waves are emitted during a perfectly spherically symmetric collapse, such as that of an idealized incompressible fluid sphere.[63]

2.3 The Reissner-Nordström Metric

2.3.1 Reissner's Contribution to the Reissner-Nordström Metric

In 1916, Hans Reissner examined the influence of the gravitational field generated by the mass-energy of an electric field on the field itself, focusing on a spherically symmetric example. He sought a specific solution for the field equations, paying particular attention to how the gravitational field of an electrically charged object influences its electric field. Reissner explored the mathematical implications of these effects. By applying Einstein's field Eq. (1.290) to a spherically symmetric electric field and deriving the energy tensor for the electrostatic field, Reissner formulated a solution to Einstein's field equations that describes the gravitational field outside a charged, non-rotating, spherically symmetric body [Reis]. He employed the generally covariant Maxwell stress-energy tensor and Einstein's gravitational field equations without restrictions on the determinant $\sqrt{-g}$ value.

Reissner starts with the metric for a static, spherically symmetric spacetime:

THE METRIC FOR A STATIC, SPHERICALLY SYMMETRIC SPACETIME:

$$ds^2 = -g_1(r)dr^2 + g_2(r)d\vartheta^2 + g_3(r)\sin^2\vartheta d\varphi^2 + g_4(r)dt^2. \tag{2.178}$$

[63] In the 1940s, Birkhoff presented a new gravitational framework that challenged Einstein's general theory of relativity. A defender of Birkhoff's theory, Carlos Graef Fernandez, recounts a personal visit to Einstein's home on Mercer Street in Princeton, New Jersey, specifically intending to discuss and defend Birkhoff's theory. Birkhoff's theory included a new approach to explaining gravitational phenomena, building upon Einstein's special relativity to offer a unified explanation of a wide array of physical phenomena. As described by Fernandez, Einstein's reaction to Birkhoff's theory was skepticism. Despite Fernandez's efforts to defend Birkhoff's theory by emphasizing its simplicity, logical consistency, and potential to unify various phenomena under a more straightforward mathematical framework, Einstein maintained his reservations [Fer].

First, he writes the form of Einstein's field equations that take into account the overall scalar curvature [Reis]:

THE GENERAL TRACE FORM OF EINSTEIN'S FIELD EQUATIONS:

$$G_{\mu\mu} = -\kappa T_{\mu\mu}. \tag{2.179}$$

Reissner then writes the general Einstein field equations in the presence of an electrostatic field [Reis]:

THE GENERAL FIELD EQUATIONS IN AN ELECTROSTATIC FIELD:

$$\boxed{G_{11}} = \frac{d^2}{dr^2}\left(\ln(g_2\sqrt{g_4})\right) - \frac{d}{dr}\ln\sqrt{g_1}\frac{d}{dr}\left(\ln(g_2\sqrt{g_4})\right) + \frac{1}{2}\left(\frac{d\ln g_2}{dr}\right)^2 + \left(\frac{d\ln\sqrt{g_4}}{dr}\right)^2$$
$$= \boxed{-\kappa T_{11}.}$$
$$\boxed{G_{22}} = -1 + \frac{d}{dr}\left(\frac{1}{2g_1}\frac{dg_2}{dr}\right) + \frac{1}{4g_1^2}\frac{dg_1}{dr}\frac{dg_2}{dr} + \frac{1}{4g_1g_4}\frac{dg_2}{dr}\frac{dg_4}{dr} = \boxed{-\kappa T_{22},}$$
$$\boxed{G_{33}} = G_{22}\sin^2\vartheta = \boxed{-\kappa T_{33}} = -\kappa T_{22}\sin^2\vartheta,$$
$$\boxed{G_{44}} = \frac{d}{dr}\left(\frac{1}{2g_1}\frac{dg_4}{dr}\right) - \frac{1}{4g_1g_4}\left(\frac{dg_4}{dr}\right)^2 + \frac{1}{4g_1^2}\frac{dg_4}{dr}\frac{dg_1}{dr} + \frac{1}{2g_1g_2}\frac{dg_2}{dr}\frac{dg_4}{dr} = \boxed{-\kappa T_{44}.} \tag{2.180}$$

The second set of Eq. (2.180) provided by Reissner is a concrete realization of Eq. (2.179); it is a more detailed and specific expression of the general form given in the field Eq. (2.179), particularly in the presence of an electromagnetic field.[64]

The terms on the right-hand side of Eq. (2.180) include the stress-energy tensor components T_{11}, T_{22}, T_{33}, T_{44} as derived by Reissner [Reis][65]:

[64] Equations (2.179) relate the Einstein tensor's trace to the stress-energy tensor's trace. They are more general than (2.180) because they apply to more scenarios. The expressions (2.180) specifically describe the components of the Einstein tensor $G_{\mu\mu}$ for a system with a static, spherically symmetric metric in an electrostatic field. They are more specific to this case and detailed expressions showing how the Einstein tensor components relate to the stress-energy tensor components $T_{\mu\mu}$ in this context. They provide explicit component-by-component relationships for $G_{\mu\mu}$ and $T_{\mu\mu}$.

[65] Reissner derives the components of the stress-energy tensor $T_{\mu\nu}$ (2.184) from the electromagnetic field using Einstein's equations as formulated in his paper, "A New Formal Interpretation of Maxwell's Field Equations of Electrodynamics" [E2-16]. He references this paper in his own derivation [Reis]:

$$T_{11} = -\frac{\mathcal{G}^2}{2g_1g_4}, \quad T_2^2 = T_3^3 = \frac{\mathcal{G}^2}{2g_1g_4}, \quad T_4^4 = -\frac{\mathcal{G}^2}{2g_1g_4}, \quad T_{\sigma\sigma} = T_\sigma^\sigma g_{\sigma\sigma},$$
$$T_{11} = -\mathcal{G}^2\frac{g_1}{2g_1g_4}, \quad T_{22} = \mathcal{G}^2\frac{g_2}{2g_1g_4}, \quad T_{33} = \mathcal{G}^2\frac{g_3}{2g_1g_4}, \quad T_{44} = -\mathcal{G}^2\frac{g_4}{2g_1g_4}. \tag{2.181}$$

He provides the expressions for the electromagnetic field components used in his Eq. (2.192) [Reis]:

ELECTROMAGNETIC TENSOR COMPONENTS:

$$\mathcal{I}^{14} = \mathcal{G}\frac{\sqrt{-g}}{g_1g_4}, \quad \mathcal{I}^4 = \rho\frac{\sqrt{-g}}{g_4} = \frac{\partial F^{14}}{\partial x_1} = \frac{d}{dr}\left(\mathcal{G}\frac{\sqrt{-g}}{g_1g_4}\right). \tag{2.182}$$

THE STRESS – ENERGY TENSOR COMPONENTS FOR AN ELECTROSTATIC FIELD:

$$T_{11} = \frac{1}{2}\left(\frac{e}{4\pi c}\right)^2 \frac{g_1}{g_2^2}, \quad T_{22} = -\frac{1}{2}\left(\frac{e}{4\pi c}\right)^2 \frac{g_2}{g_2^2}, \quad T_{33} = -\frac{1}{2}\left(\frac{e}{4\pi c}\right)^2 \frac{g_3}{g_2^2},$$
$$T_{44} = \frac{1}{2}\left(\frac{e}{4\pi c}\right)^2 \frac{g_4}{g_2^2}. \tag{2.184}$$

The metric tensor components are:

$$g_1 = -1, \; g_2 = -r^2, \; g_3 = -r^2 \sin^2 \vartheta, \; \text{and} \; g_4 = c^2. \tag{2.185}$$

The components of the stress-energy tensor for the electrostatic field of a point charge are:

THE COMPONENTS OF THE STRESS – ENERGY TENSOR FOR THE ELECTROSTATIC FIELD:

$$\boxed{T_{44} = \frac{e^2}{8\pi r^4}, \quad T_{11} = -\frac{e^2}{8\pi r^4},} \quad T_{22} = \frac{e^2}{8\pi r^4}, \quad T_{33} = T_{22} \sin^2 \vartheta. \tag{2.186}$$

Equation (2.186) is derived utilizing Einstein's equation for the electromagnetic stress-energy tensor (1.386) (see Sect. 1.7.1).[66]

The components $\mathcal{I}^{14}$ and $\mathcal{I}^4$ represent the components of the electromagnetic field tensor, which are related to the electric field and its potential. ρ is the charge density, and $\mathcal{G}$ is the field strength derived from the components of the metric tensor $g_{\mu\nu}$ and the charge e:

ELECTROMAGNETIC FIELD STRENGTH:

$$\mathcal{G} = \text{const.} \cdot \frac{\sqrt{-g_1 g_4}}{-g_2}, \quad \mathcal{G} = F_{14} = -\frac{e}{4\pi c}\frac{\sqrt{-g_1 g_4}}{-g_2}, \quad \mathcal{F}^{14} = F_{14}\frac{\sqrt{-g}}{g_1 g_4} = -\frac{e}{4\pi c}. \tag{2.183}$$

For an electrostatic field, the non-zero component of the electromagnetic field tensor is F_{14}, representing the electric field. $\mathcal{G} = F_{14}$ [Eq. (2.183)] represents the electrostatic component of the electromagnetic field. This component of the electromagnetic field tensor corresponds to the electric field in the radial direction in the electrostatic case. We substitute this into Einstein's equation (1.386) for the electromagnetic tensor (see Sect. 1.7.1), and we can derive the components of $T_{\mu\nu}$ [Eq. (2.186)], see the derivation in footnote 66. As I show in footnote 66, Reissner's expressions for the stress-energy tensor components, derived using the specific metric and field strength $\mathcal{G}$ forms, align with the general form provided by Einstein (1.386). The integration constant in Eq. (2.183) is determined from the condition at infinity, where e is the total charge at the center and the metric tensor components (2.185). Thus, const. $= -\frac{e}{4\pi c}$.

[66] Consider Einstein's equation for the electromagnetic stress-energy tensor (1.386). For an electrostatic field, the non-zero component of the electromagnetic field tensor is $F_{14} = E = \frac{e}{r^2}$ [see Eq. (2.183)]. First, we calculate the scalar invariant $F_{\alpha\beta}F^{\alpha\beta}$ in Einstein's equation (1.386):

$$F_{\alpha\beta}F^{\alpha\beta} = 2(F_{14}F^{14})^2 g^{11}g^{44} = 2(F_{14})^2 g^{11}g^{44} = 2\left(\frac{e}{r^2}\right)^2 g^{11}g^{44}. \tag{2.187}$$

Using the inverse metric tensor components [the inverse of Eq. (2.185)] $g^{11} = 1$ and $g^{44} = \frac{1}{c^2}$, we get:

Reissner simplifies the field Eq. (2.180) to three field equations [Reis]:

$$\boxed{G_{11}} = \frac{d^2}{dr^2}\ln(g_2\sqrt{g_4}) + \frac{1}{2g_1^2}\left(\frac{dg_2}{dr}\right)^2 + \frac{1}{4g_4^2}\left(\frac{dg_4}{dr}\right)^2 = \boxed{+\frac{\lambda^2}{g_2^2}},$$
$$\boxed{G_{22}} = -1 + \frac{1}{2}\frac{dg_2^2}{dr^2} + \frac{1}{4g_4}\frac{dg_4}{dr}\frac{dg_2}{dr} = \boxed{-\frac{\lambda^2}{g_2}}, \tag{2.192}$$
$$\boxed{G_{44}} = \frac{1}{2}\frac{d^2g_4}{dr^2} - \frac{1}{4g_4}\left(\frac{dg_4}{dr}\right)^2 + \frac{1}{2g_1}\frac{dg_1}{dr}\frac{dg_4}{dr} = \boxed{+\frac{\lambda^2 g_4}{g_1^2}},$$

$$\text{where (see footnote 65): } \lambda = \frac{e}{4\pi c}. \tag{2.193}$$

Recall that Schwarzschild transformed the general line element from polar [(2.7)] to rectangular coordinates [(2.13)] and introduced new variables to maintain the determinant condition (see Sect. 2.2.2). Reissner explicitly stated that, unlike Schwarzschild, he initially calculated without imposing the determinant condition (1.299). Reissner mentions that by choosing:

$$g_{11} = -1, \tag{2.194}$$

$$\boxed{F_{\alpha\beta}F^{\alpha\beta}} = 2\left(\frac{e}{r^2}\right)^2 \cdot 1 \cdot \frac{1}{c^2} = \boxed{\frac{2e^2}{c^2r^4}}. \tag{2.188}$$

Now, we will substitute $F_{\alpha\beta}F^{\alpha\beta}$ and $F_{\mu\nu}$ into Einstein's equation (1.386) to get the components of the stress-energy tensor T_{11}, T_{22}, T_{33} and T_{44}. For instance, let us consider the component T_{44}.

$$T_{44} = \frac{1}{4}g_{44}F_{\alpha\beta}F^{\alpha\beta} - F_{4\alpha}F_4^{\ \alpha}. \tag{2.189}$$

We substitute $g_{44} = c^2$, Eq. (2.188) and:

$$\boxed{F_{4\alpha}F_4^{\ \alpha}} = (F_{14})^2 g^{11} = \left(\frac{e}{r^2}\right)^2 \cdot 1 = \boxed{\frac{e^2}{r^4}}, \tag{2.190}$$

into Eq. (2.189):

$$\boxed{T_{44}} = \frac{1}{4} \cdot c^2 \frac{2e^2}{c^2r^4} - \frac{e^2}{r^4} = \frac{e^2}{2r^4} - \frac{e^2}{r^4} = \boxed{\frac{e^2}{2r^4}}. \tag{2.191}$$

We multiply $\frac{e^2}{2r^4}$ by 4π [see Eq. (2.183). In the same manner, we calculate T_{11}, T_{22}, and T_{33}, and write the simplified components of the stress-energy tensor for the electrostatic field of a point charge (2.186).

he obtains the above three Eq. (2.192) with only two unknowns g_2 and g_4. His statement about choosing condition (2.194) refers to a specific choice of a coordinate system, which is different from the unimodular condition[67]:

$$g_1 g_2^2 g_4 = 1. \tag{2.196}$$

Reissner's condition simplifies the metric. By imposing condition (2.194), he is fixing the radial component of the metric tensor to be constant, simplifying the field Eq. (2.180). This choice affects the form of the metric and the resulting field equations. It reduces the complexity of the field equations by eliminating the variation in the radial component. Reissner simplifies the field Eq. (2.180) without imposing additional constraints on the metric determinants and other components. On the other hand, condition (2.196) indirectly affects all metric components by imposing a relationship between them; the product of certain metric components is 1, simplifying Einstein's field equations. Both conditions (2.194) and (2.196) simplify Einstein's field equations. However, they do so in different ways. Reissner's condition (2.194) is a direct choice for one specific metric component, simplifying the field Eq. (2.180) by making the radial component constant. It does not impose direct constraints on the metric tensor's determinant. On the other hand, the unimodular condition (2.196) is a global constraint on the metric components. It affects the relationship between multiple metric components rather than fixing a single component.

Reissner highlights that, despite having three Eq. (2.192) with only two unknowns, solving any two of these equations will automatically satisfy the third [Reis]. This property is quite significant as it simplifies the solution process. These constraints are akin to those used by Schwarzschild [see Eqs. (2.17), (2.23), (2.25), (2.27), (2.30) and (2.33) that automatically satisfy the second equation of the three simplified field equations in Eq. (2.17)]. In Reissner's case, solving for two metric functions automatically satisfies the third equation. This could be any of the equations not necessarily tied to a specific coordinate, like the equator in the unimodular condition.

---

[67] For a spherically symmetric metric in polar coordinates $(t, r, \vartheta, \varphi)$, the metric determinant g in spherical coordinates looks like this [see Eq. (2.52) in Sect. 2.2.2]:

$$|-g| = -g_1 g_2^2 g_4 \sin^2 \vartheta. \tag{2.195}$$

For the condition (1.299) $\sqrt{-g} = 1$, we need:

$$-\left(-g_1 g_2^2 g_4 \sin^2 \vartheta\right) = 1, \quad \text{which simplifies to:}$$

$$g_1 g_2^2 g_4 \sin^2 \vartheta = 1.$$

For the determinant condition (2.196) to be satisfied uniformly, the following condition must hold [see Schwarzschild's equation (2.21) in Sect. 2.2.2]: If Eq. (2.196) is valid, then $\sin^2 \vartheta$ must equal 1. This occurs at specific values of ϑ, namely, $\vartheta = \frac{\pi}{2} + n\pi$, where n is an integer. When $\vartheta = \frac{\pi}{2}$, it corresponds to the equator in spherical coordinates. At the equator, $\sin(\frac{\pi}{2}) = 1$. This means the determinant condition simplifies to Eq. (2.196). Since $\sin^2 \vartheta = 1$ at the equator, we do not have to consider variations in ϑ or $\sin \vartheta$. This simplification allowed Schwarzschild to ignore the coordinate x_2 at the equator (see footnotes 12 and 13 in Sect. 2.2.2).

In Schwarzschild's calculations, Eqs. (2.30) and (2.33) that automatically satisfy the second field equation do so at the equator where $\vartheta = \frac{\pi}{2}$ because $\sin(\frac{\pi}{2}) = 1$ (refer to footnote 67). In this case, the satisfaction of the third equation happens specifically due to the symmetry and the imposed condition in spherical coordinates.

Equation (2.192) reduce to the simplified forms of Einstein's field equations in the presence of an electromagnetic field. Using Eq. (2.186), we get[68]:

EINSTEIN'S FIELD EQUATIONS IN THE PRESENCE OF AN ELECTROMAGNETIC FIELD:

$$G_{11} = \frac{1}{r^2}\left(1 - g_{11} - r\frac{dg_{44}}{dr}\frac{1}{g_{44}}\right) = \kappa T_{11} = -\frac{e^2}{8\pi r^4},$$
$$G_{44} = \frac{1}{r^2}\left(1 - \frac{d}{dr}\left(r(1 - g_{11}^{-1})\right)\right) = \kappa T_{44} = \frac{e^2}{8\pi r^4}. \tag{2.197}$$

These help solve the metric functions by considering specific contributions from the electromagnetic field. Integrating Eq. (2.197) to find g_{11} and g_{44}, gives:

$$\frac{1}{g_{11}} = \left(1 - \frac{\alpha}{r} + \frac{e^2}{r^2}\right)^{-1}, \quad \text{and:} \quad g_{44} = 1 - \frac{\alpha}{r} + \frac{e^2}{r^2}. \tag{2.198}$$

In Reissner's paper, the explicit final metric functions (2.198) are not directly presented in this simplified form. Reissner's formulation can be seen as a more complex solution that reduces to the Schwarzschild metric when the parameter λ (2.193) (related to the charge of the object) is set to zero. This generalization includes the effects of charge, leading to what is now known as the Reissner-Nordström metric.

Reissner's derivation integrates the electromagnetic field into general relativity by assuming spherical symmetry and static conditions and generalizing the classical electrostatic laws within the framework of curved spacetime. The derived field equations reflect the interactions between the electromagnetic and gravitational fields, showing how Reissner's work extends classical electromagnetism into general relativity. According to Reissner, Einstein informed him during their discussions of the derivation of the Reissner-Nordström metric that he was now able to show generally that the momentum-energy conservation law was contained within his final field equations and, therefore, did not need to be specially applied to any problem [Reis].[69] This result, which Einstein presented in his paper "Hamilton's Principles and the General Theory of Relativity" [E4-16], showed that the field equations were derived from considerations of general covariance.

[68] Equation (2.197) are derived by simplifying and isolating specific components from the set of Eq. (2.192). The simplified Eq. (2.197) combine the set of field Eq. (2.192), focusing on first-order terms and primary components without explicitly including the second derivatives of Eq. (2.192).

[69] See Einstein's derivation of the Bianchi identity (1.476) (refer to Sect. 1.7.3).

2.3.2 *Nordström's First Paper: A Sphere with No Charge*

In 1918, in two papers, "On the mass of a material system according to the gravitation theory of Einstein" and "On the Energy of the Gravitational Field in Einstein's Theory," Gunnar Nordström sought to extend Johannes Droste's solution [Dro-17]. He extended the analysis of the gravitational field around a spherically symmetric body to consider electrically charged centers. Nordström also aimed to quantify the energy density of the gravitational field as influenced by electrical charges. Nordström's work is delineated within Einstein's framework in his paper "Hamilton's Principles and the General Theory of Relativity" [E4-16] (see Sect. 1.7.3), which (like Reissner's work) expounds upon Einstein's field Eq. (1.290) without confining these to unimodular coordinates. Einstein's work in his paper [E4-16] serves as the basis upon which Nordström analyzes the gravitational field of an electrically charged, spherically symmetric sphere.

In his first paper, "On the mass of a material system according to the gravitation theory of Einstein," Nordström writes Droste's metric (2.71) [Nor-18-1] [Dro-16]. Using transformations similar to Eq. (2.5):

$$\begin{aligned} x_1 &= r\cos\vartheta\cos\varphi, \\ x_2 &= r\cos\vartheta\sin\varphi, \\ x_3 &= r\sin\vartheta. \end{aligned} \tag{2.199}$$

Nordström converts the spherical (polar) coordinates (r, ϑ, φ) into Cartesian coordinates (x_1, x_2, x_3), and redefines the radial coordinate v as:

$$v = rp. \tag{2.200}$$

He then writes the metric tensor components $g_{\mu\nu}$ in Cartesian coordinates [Nor-18-1, CPAE8], Doc. 382[70]:

THE METRIC IN A CARTESIAN FORM:

$$\begin{aligned} & g_{12} = -\frac{x_1 x_2}{r^2}(u^2 - p^2), \quad g_{11} = -p^2 - \frac{x_1^2}{r^2}(u^2 - p^2), \\ & g_{14} = g_{24} = g_{34} = 0, \quad g_{44} = w^2. \end{aligned} \tag{2.201}$$

---

[70] The spatial components are affected by the transformation (2.199) and the functions u, v, w, and p. The matrices (2.201) and (2.202) do not have a form of spherical symmetry. Because the metric components of Eq. (2.201) and (2.202) explicitly depend on the Cartesian coordinates x_1, x_2 and x_μ, rather than just the radial distance r, these metrics do not exhibit spherical symmetry.

And the inverse metric tensor components $g^{\mu\nu}$:

THE INVERSE METRIC IN A CARTESIAN FORM:

$$g^{12} = -\frac{x_1 x_2}{r^2}\left(\frac{1}{u^2} - \frac{1}{p^2}\right), \quad g^{11} = -\frac{1}{p^2} - \frac{x_\mu^2}{r^2}\left(\frac{1}{u^2} - \frac{1}{p^2}\right), \tag{2.202}$$

$$g^{14} = g^{24} = g^{34} = 0, \quad g^{44} = \frac{1}{w^2}.$$

The determinant of these metrics is not 1 [Nor-18-1].

Setting $p = 1$ [refer to Eq. (2.200)] means the metric has spherical symmetry in its form and the determinant:

THE DETERMINANT OF THE INVERSE METRIC TENSOR IS NOT EQUAL TO 1:

$$\sqrt{-g} = uwp^2. \tag{2.203}$$

is $\sqrt{-g} = c$ (this is explained in Sect. 2.3.4).

Nordström wrote to Einstein that to derive the Schwarzschild line element (2.49), it is necessary to set $p = 1$ in Eqs. (2.201) and (2.202). Subsequently, the Droste metric functions

THE SCHWARZSCHILD-DROSTE METRIC FUNCTIONS:

$$\boxed{u^2 = \frac{1}{1 - \frac{\alpha}{r}}, \quad w^2 = 1 - \frac{\alpha}{r}.} \tag{2.204}$$

must be substituted into these equations, thereby facilitating the derivation of the Schwarzschild solution [CPAE8], Doc. 382.[71]

---

[71] Setting $p = 1$, both p^2 and $\frac{1}{p^2}$ in Eqs. (2.201) and (2.202) simplify to 1:

$$g_{12} = -\frac{x_1 x_2}{r^2}(u^2 - 1), \quad g_{11} = -1 - \frac{x_1^2}{r^2}u^2 + \frac{x_1^2}{r^2}, \tag{2.205}$$

$$g^{12} = -\frac{x_1 x_2}{r^2}\left(\frac{1}{u^2} - 1\right), \quad g^{11} = -1 - \frac{x_1^2}{r^2}\frac{1}{u^2} + \frac{x_1^2}{r^2}. \tag{2.206}$$

The final substitution $w^2 = 1 - \frac{\alpha}{r}$ into g_{44} gives:

$$\boxed{g_{44} = 1 - \frac{\alpha}{r}, \quad g^{44} = \frac{1}{1 - \frac{\alpha}{r}}.} \tag{2.207}$$

We consider the relationship between Cartesian and spherical coordinates:

$$r = \sqrt{x_1^2 + x_2^2 + x_3^2} \tag{2.208}$$

We consider x_1 along the radial direction, where ϑ and φ define the position on a sphere. When $\vartheta = 0$, then $\cos\vartheta = 1$ and $\sin\vartheta = 0$. Plugging this in Eq. (2.199), we get: $x_1 = r$, $x_2 = x_3 = 0$.

Subsequently, Nordström incorporates the stress-energy tensor components to describe the distribution of energy, momentum, and stress within the matter comprising the spherically symmetric body, as well as in its surrounding field [Nor-18-1][72]:

THE STRESS- ENERGY TENSOR OUT AND INSIDE THE SPHERE IN CARTESIN FORM:

$$\begin{aligned} &\mathcal{I}'_p = \frac{x_\mu x_\nu}{r^2}(\mathcal{I}'_r - \mathcal{I}^p_p) \quad \text{for} \quad \mu \neq \nu \\ &\mathcal{I}^\mu_\mu = \mathcal{I}^p_p + \frac{x^2_\mu}{r^2}(\mathcal{I}'_r - \mathcal{I}^p_p), \\ &\mathcal{I}^4_\mu = \mathcal{I}^\mu_4 = 0, \quad \text{for: } \mu = \nu = 1, 2, 3. \end{aligned} \tag{2.211}$$

Spherical symmetry implies that a system's physical properties are invariant under any rotation about a central point. This symmetry simplifies the metric tensor's form and Einstein's field equations. Nordström leveraged spherical symmetry to simplify the Eqs. (2.202) and (2.211), combining them for $x_1 = r, x_2 = x_3 = 0$ [Nor-18-1][73]:

Thus: $x_1^2 = r^2$. This radial component removes the angular dependence in spherical coordinates represented by ϑ.

We substitute this into the metric components g_{11} and g^{11} in Eqs. (2.205) and (2.206), and get:

$$g_{11} = -1 - u^2 + 1 = -u^2, \quad g^{11} = -\frac{1}{u^2} \quad \text{where: } u^2 = \frac{1}{1 - \frac{\alpha}{r}}. \tag{2.209}$$

$$\begin{aligned} &\text{Thus: } g_{11} = -u^2 = -\frac{1}{1-\frac{\alpha}{r}}. \quad \text{Hence: } \boxed{g_{11} = u^2 = \frac{1}{1-\frac{\alpha}{r}}.} \\ &g^{11} = -\frac{1}{u^2} = -\frac{1}{1-\frac{\alpha}{r}} \quad \text{Hence: } \boxed{g^{11} = \frac{1}{u^2} = 1 - \frac{\alpha}{r}.} \end{aligned} \tag{2.210}$$

[72] The terms involving $\mathcal{I}'_r - \mathcal{I}^p_p$ scaled by the position-related factors reflect the internal structure or the field within the sphere. The radial dependency $\frac{x^2_\mu}{r^2}$ represents change with distance from the center, i.e., this term is related to the outside of the sphere. $\mathcal{I}^4_\mu = \mathcal{I}^\mu_4 = 0$ reflects the system's stationary nature, indicating no energy or momentum flux in the spatial directions, consistent with a static gravitational field. Such components in a tensor represent the absence of energy-momentum transfer in those directions. Surrounding a spherically symmetric mass, the static nature indicated by these zero components would apply outside the sphere. Outside the sphere, particularly in a vacuum, the gravitational field is entirely determined by the mass distribution within the sphere. Since no momentum or energy flux exists in these spatial directions, the external gravitational field is purely radial and not dependent on time or angular coordinates.

[73] Nordström selects a specific point or X-axis as he calls it in a coordinate system, which is part of examining and verifying spherical symmetry. He reduces his problem to one dimension by focusing on the X-axis, a specific radial direction from the origin. The diagonal components of the stress-energy tensor in Eq. (2.213) are equal, where $\mathcal{I}^2_2 = \mathcal{I}^3_3$, indicates that the system has isotropic properties. The $\mathcal{I}^1_1 = \mathcal{I}^r_r$ equation reflects that the radial component of the stress-energy tensor is aligned with the radial direction. Setting $p = 1$, we obtain (see derivation in footnote 71):

$$g^{11} = -\frac{1}{u^2}, \quad g^{22} = g^{33} = -1, \quad g^{44} = \frac{1}{w^2}. \tag{2.212}$$

THE SPHERICALLY SYMMETRIC INVERSE METRIC AND STRESS-ENERGY TENSOR:

$$g^{11} = -\frac{1}{u^2}, \quad g^{22} = g^{33} = -\frac{1}{p^2} = -\frac{r^2}{v^2}, \quad g^{44} = \frac{1}{w^2},$$
$$I_1^1 = I_r^r, \quad I_2^2 = I_3^3 = I_p^p. \tag{2.213}$$

The Lagrangian density function $\mathcal{L}$ is expressed by Eq. (1.440) (refer to Sect. 1.7.3). It is used in the action integral (1.431) to derive the field equations and is divided into two distinct parts: the gravitational part $\mathcal{G}$ and the matter part $\mathcal{M}$.

Nordström then applies the variation principle to both parts [Nor-18-1][74]:

$$\delta \int_{r_1}^{r_2} \mathcal{G} r^2 dr = \kappa \delta \int_{r_1}^{r_2} \mathcal{M}. \tag{2.214}$$

Given Eq. (2.214), the variations of the terms inside the integral give two integrals [Nor-18-1]. The first integral is similar to Droste's equation (2.56)[75]:

$$\boxed{\delta \int_{r_1}^{r_2} \mathcal{G} r^2 dr = 2\delta \int_{r_1}^{r_2} \left(\frac{wv'^2 + 2vv'w'}{u} + uw \right) dr,} \quad \text{and:} \tag{2.215}$$

$$\boxed{\kappa\delta \int_{r_1}^{r_2} \mathcal{M}(m) r^2 dr = 2\kappa \int_{r_1}^{r_2} \left(I_r^r \frac{du}{u} + 2I_p^p \frac{dv}{v} + I_4^4 \frac{dw}{w} + \right) r^2 dr.} \tag{2.216}$$

Finally, Nordström performs the variation of (2.215) and (2.216) and derives the field equations for a spherically symmetric uncharged mass [Nor-18-1]:

---

[74] Similar to Droste's notation (see footnote 29), Nordström's notation is considered more verbose than contemporary conventions. Specifically, he denoted integration over the four-dimensional spacetime using four separate integrals $\int\int\int\int$ and expressed the differential elements as dx_1, dx_2, dx_3, dx_4 rather than employing the modern compact form d^4x. This older notation explicitly represents integration over each of the four dimensions (three spatial and one temporal). Nordström integrated the gravitational Lagrangian density $\mathcal{G} = \sqrt{-g}R$ (R is the Ricci scalar) and the matter Lagrangian density $\mathcal{M}$ over a region from r_1 to r_2 and applied the variation to find how changes in the metric functions u, v, w and the charge e affect this integral. This led to deriving the field Eq. (2.217) from the Lagrangian.

[75] Recall that the first integral represents how the dynamic properties of the metric functions u, v, and w contribute to the total action. The second integral connects these contributions to specific components of the stress-energy tensor, showing how variations in metric field functions influence the overall body. See metric (2.211) for the stress-energy tensor components.

THE FIELD EQUATIONS FOR AN UNCHARGED SPHERICALLY SYMMETRIC MASS:

$$
\begin{aligned}
&-\frac{wv'^2+2vv'w'}{u^2}+w=\frac{r^2}{u}\kappa \mathcal{T}^r_r,\\
&-\frac{wv''+v'w'+vw''}{u}+(vw'+wv')\frac{u''}{u^2}=\frac{r^2}{v}\kappa \mathcal{T}^p_p,\\
&-\frac{2vv''+v'^2}{u}+u+2vv'\frac{u'}{u^2}=\frac{r^2}{w}\kappa \mathcal{T}^4_4.
\end{aligned}
\tag{2.217}
$$

The derivation of Nordström's field Eq. (2.217) heavily draws from Droste's 1916 work on the Droste (Schwarzschild) exterior metric, as presented in his dissertation [Dro-16]. Like Nordström, Droste began with the variational principle (2.215). However, Droste derived vacuum field Eq. (2.70) and diverged in his approach by not separating the Lagrangian into distinct gravitational and matter components. Before Einstein published his paper, "Hamilton's Principles and the General Theory of Relativity," Droste applied the variation principle to the gravitational Lagrangian. Although influenced by David Hilbert's work [Hil15-2], he did not embrace the conceptual division between gravitational and matter-electromagnetic fields.

2.3.3 Nordström's Second Paper: A Charged Sphere

In his second paper, "On the Energy of the Gravitational Field in Einstein's Theory," Nordström extended his equations from the first paper to include the effects of an electromagnetic field.

He begins his paper by expressing the matter Lagrangian in terms of the matter and the electromagnetic fields [Nor-18-2]:

THE MATTER LAGRANGIAN = THE MATTER AND THE ELECTROMAGNETIC FIELDS:

$$
\mathcal{M}=\mathcal{M}(e)+\mathcal{M}(m).
\tag{2.218}
$$

He then isolates the effects of electromagnetic fields. $\mathcal{M}(e)$ represents the electromagnetic Lagrangian [Nor-18-2][76]:

[76] This Lagrangian incorporates the metric tensor $g^{\alpha\mu}$, and the electrostatic potential φ, as well as the current w^{σ}, into its formulation. The first term of Eq. (2.219) is related to Eq. (1.387), see Sect. 1.7.1.

THE ELECTROMAGNETIC LAGRANGIAN:

$$\mathcal{M}(e) = \frac{1}{8\pi}\sqrt{-g}\sum_{\alpha\beta\mu\nu} g^{\alpha\mu}g^{\beta\nu}\left(\frac{\partial\varphi_\beta}{\partial x_\alpha} - \frac{\partial\varphi_\alpha}{\partial x_\beta}\right)\left(\frac{\partial\varphi_\nu}{\partial x_\mu} - \frac{\partial\varphi_\mu}{\partial x_\nu}\right) + 2\sum_\sigma \varphi_\sigma w^\sigma. \tag{2.219}$$

Nordström further simplifies Eq. (2.219) by assuming a static field and spherical symmetry.[77] The above simplifications and plugging Eq. (2.203) into Eq. (2.219) lead to the equation [Nor-18-2]:

THE ELECTROMAGNETIC LAGRANGIAN UNDER THE SPHERICAL SYMMETRY AND STATIONARY ASSUMPTIONS:

$$\mathcal{M}(e) = -\frac{p^2}{4\pi uw}\left(\frac{d\varphi}{dr}\right)^2 + 2\varphi\rho. \tag{2.220}$$

Integrating $\mathcal{M}(e)$ (2.220) over the four-dimensional spacetime gives [Nor-18-2][78]:

$$\int \mathcal{M}(e)d^4x = 4\pi(t_2 - t_1)\int_{r_1}^{r_2} \mathcal{M}(e)r^2dr, \quad \text{where:} \tag{2.221}$$

$$4\pi\int_{r_1}^{r_2} \mathcal{M}(e)r^2dr = \int_{r_1}^{r_2}\left[\frac{-r^2p^2}{uw}\left(\frac{d\varphi}{dr}\right)^2 + 8\pi\varphi\rho r^2\right]dr. \tag{2.222}$$

By Eq. (2.262) (see the next Sect. 2.3.4), Einstein's $\mathcal{G}^*$ (the gravitational part, independent of second order derivatives, see Sect. 1.7.3) and the matter part $\mathcal{M}(m)$ are

---

[77] When simplifying under a stationary and static field assumption, $\varphi_1 = \varphi_2 = \varphi_3 = 0$ and $w^1 = w^2 = w^3 = 0$. Nordström substitutes $\varphi_4 = \varphi$, representing the electrostatic potential, and $w^4 = \rho$, representing the charge density. He further assumes that the field possesses spherical symmetry and chooses the time coordinate so that $g_{r4} = 0$. In other words, Nordström chooses a coordinate system with no cross-terms between radial and temporal components. In this coordinate system, condition (2.203) is valid. The simplification (2.220) to a scenario where the field is stationary and all-electric charges are at rest reduces the problem to one dependent only on the radial coordinate r. This reduction simplifies the mathematical treatment significantly and aligns with the physical intuition that the radial direction is the primary concern in a spherically symmetric situation.

[78] Nordström takes the integral of the electromagnetic Lagrangian over 4-dimensional space, considering the variations of the metric functions [u, w, and Eq. (2.200)] while keeping the electrostatic potential φ and charge density ρ constant. Equation (2.222) breaks down $\mathcal{M}(e)$ into components within the region between r_1 and r_2 that represent contributions from two terms. The first term on the left-hand side is the electrostatic field's energy density contribution. Here, $\left(\frac{d\varphi}{dr}\right)^2$ quantifies the square of the electric field, where $E = -\frac{d\varphi}{dr}$. The second term $8\pi\varphi\rho r^2$ represents the interaction energy density between the electrostatic potential and a charge density distribution ρ. This term aligns with the classic electrostatic energy term, integrating the product of potential and charge density over the volume of a spherically symmetric sphere.

independent of the potential φ. From Eq. (2.222), Nordström obtains by variation [Nor-18-2][79]:

$$2\frac{d}{dr}\left(\frac{r^2p^2}{uw}\frac{d\varphi}{dr}\right)+8\pi\rho r^2=0. \tag{2.231}$$

[79] Applying the calculus of variations to Eq. (2.222), we use the Euler-Lagrange equation:

$$\frac{\partial f}{\partial\varphi}-\frac{d}{dr}\left(\frac{\partial f}{\partial(\frac{d\varphi}{dr})}\right)=0. \tag{2.223}$$

Calculating $\frac{\partial f}{\partial\varphi}$, the term:

$$\frac{-r^2p^2}{uw}\left(\frac{d\varphi}{dr}\right)^2, \tag{2.224}$$

in Eq. (2.222) does not depend on φ, so its partial derivative with respect to φ is zero. The only term that depends on φ is $8\pi\varphi\rho r^2$. Therefore:

$$\boxed{\frac{\partial f}{\partial\varphi}}=\frac{\partial}{\partial\varphi}\left(8\pi\varphi\rho r^2\right)=\boxed{8\pi\rho r^2.} \tag{2.225}$$

Now we calculate: $\frac{\partial f}{\partial\left(\frac{d\varphi}{dr}\right)}$. Consider the term that involves $\frac{d\varphi}{dr}$, Eq. (2.224).

In Eq. (2.224), $\frac{-r^2p^2}{uw}$ is a constant with respect to $\frac{d\varphi}{dr}$. To differentiate $\left(\frac{d\varphi}{dr}\right)^2$ with respect to $\frac{d\varphi}{dr}$, we use the power rule:

THE POWER RULE:

$$\frac{d}{dx}x^n=nx^{n-1}. \tag{2.226}$$

Applying this rule, we get: $$\frac{d}{d\left(\frac{d\varphi}{dr}\right)}\left(\left(\frac{d\varphi}{dr}\right)^2\right)=2\left(\frac{d\varphi}{dr}\right). \tag{2.227}$$

Now, including the constant $\frac{-r^2p^2}{uw}$, we have:

$$\frac{\partial}{\partial\left(\frac{d\varphi}{dr}\right)}\left(\frac{-r^2p^2}{uw}\left(\frac{d\varphi}{dr}\right)^2\right)=2\left(\frac{-r^2p^2}{uw}\right)\left(\frac{d\varphi}{dr}\right). \tag{2.228}$$

Thus, the derivative is:

$$\boxed{\frac{\partial f}{\partial\left(\frac{d\varphi}{dr}\right)}=-2\frac{r^2p^2}{uw}\frac{d\varphi}{dr}.} \tag{2.229}$$

Substituting Eqs. (2.225) and (2.229) into the Euler-Lagrange equation (2.223), we get:

$$8\pi\rho r^2-\frac{d}{dr}\left(-2\frac{r^2p^2}{uw}\frac{d\varphi}{dr}\right)=0. \tag{2.230}$$

The constant factor 2 in the derivative term can be factored out because it is a multiplicative constant, which does not affect the differentiation operation. This leads to Eq. (2.231).

Nordström integrated the Eq. (2.231) and arrived at [Nor-18-2][80]:

$$-\frac{r^2 p^2}{uw}\frac{d\varphi}{dr} = e(r), \qquad -\frac{d\varphi}{dr} = \frac{uw}{p^2}\frac{e(r)}{r^2}. \tag{2.234}$$

Recall that $\mathcal{M}$ is defined by Eq. (2.218). The variational result for $\mathcal{M}(e)$ is [Nor-18-2][81]:

$$\boxed{\delta \int_{r_1}^{r_2} \mathcal{M}(e) r^2 dr = \int_{r_1}^{r_2} \frac{uwe^2}{4\pi v^2}\left(\frac{\delta u}{u} - 2\frac{\delta v}{v} - \frac{\delta w}{w}\right) dr.} \tag{2.235}$$

Nordström combined Eqs. (2.235) and (2.216), and this led to an expression for $\mathcal{M}$ [Eq. (2.218)].

He then took the variation of (2.214) and derived the field equations of a spherically symmetric charged mass [Nor-18-2][82]:

[80] We begin by simplifying and rearranging terms in Eq. (2.231):

$$\frac{d}{dr}\left(\frac{r^2 p^2}{uw}\frac{d\varphi}{dr}\right) = -4\pi\rho r^2. \tag{2.232}$$

Integrating both sides with respect to r, we obtain:

$$\frac{r^2 p^2}{uw}\frac{d\varphi}{dr} = -\int 4\pi\rho r^2\, dr + \text{const}. \tag{2.233}$$

The integral $\int 4\pi\rho r^2\, dr$ represents $e(r)$, the total charge enclosed within a sphere of radius r (the integral of the charge density over the volume of a sphere). Outside the sphere, $e(r)$ is constant and equal to the sphere's total charge. Therefore, we can write the first equation of (2.234). This is a statement of how the gradient of the potential scales with r, modified by p, u, w and $e(r)$. From the first equation of (2.234), solving for $\frac{d\varphi}{dr}$ gives the second equation of (2.234). This relationship describes how the potential φ changes are affected by the spatial distribution of ρ, adjusted by p, u, and w.

[81] Given Eq. (2.222), which represents the energy contributions from the electromagnetic field characterized by the potential, Nordström obtained the Eq. (2.235). This equation describes how the integral of the electromagnetic energy contribution varies with changes in the metric functions u, v, and w. The term $\frac{uwe^2}{4\pi v^2}$ indicates how the energy density depends on u, v, and w. $\frac{\delta u}{u}$, $\frac{\delta v}{v}$, and $\frac{\delta w}{w}$ show the relative changes in the metric functions and their impact on the integral. Equation (2.235) expresses how changes in u, v, w, and e affect the total electromagnetic energy calculated over the specific region $r_1 - r_2$.

[82] This approach allowed Nordström to derive field equations that account for the gravitational effects of both mass-energy distributions and electromagnetic fields as described by $\mathcal{M}$ (expressed in terms of $\mathcal{I}_r^r$, $\mathcal{I}_p^p$, $\mathcal{I}_4^4$, the charge of the mass e, and u, w, v, p, v', w', v'', w'', where recall that v', w', v'', w'' are derivatives of v and w.

THE FIELD EQUATIONS FOR A CHARGED SPHERICALLY SYMMETRIC MASS:

$$
\begin{aligned}
&-\frac{wv'^2+2vv'w'}{u^2}+w=\frac{\kappa}{u}\left(\frac{uw}{8\pi}\frac{e^2}{v^2}+r^2\mathcal{I}_r^r\right),\\
&-\frac{wv''+v'w'+vw''}{u}+(vw'+wv')\frac{u''}{u^2}=\frac{\kappa}{v}\left(\frac{uw}{8\pi}\frac{e^2}{v^2}+r^2\mathcal{I}_p^p\right),\\
&-\frac{2vv''+v'^2}{u}+u+2vv'\frac{u'}{u^2}=\frac{\kappa}{w}\left(\frac{uw}{8\pi}\frac{e^2}{v^2}+r^2\mathcal{I}_4^4\right).
\end{aligned}
\tag{2.236}
$$

Compare these field equations to the field equations for an uncharged spherically symmetric mass (2.217) (see Sect. 2.3.2).

By the definition:

$$\boxed{\frac{\kappa e^2}{8\pi}=\varepsilon^2.} \tag{2.237}$$

Nordström then derived the metric functions from the field equations, applying Droste's methodology.

He first derived the radial metric function $\frac{1}{u^2}$ [Nor-18-2].[83]

---

[83] Given the third field equation of Eq. (2.236), we set $v=r$ and apply the derivatives $v'=1$ and $v''=0$. Substituting these:

$$-\frac{1}{u}+u+2r\frac{u'}{u^2}=\frac{\kappa}{8\pi}\frac{ue^2}{r^2}+\frac{\kappa r^2\mathcal{I}_4^4}{w}. \tag{2.238}$$

In the vacuum region outside the electric mass distribution [Nor-18-2]:

$$\mathcal{I}_r^r=\mathcal{I}_p^p=\mathcal{I}_4^4=0,\quad e=\text{const}. \tag{2.239}$$

Thus, focusing only on the electromagnetic component, we get:

$$-\frac{1}{u}+u+2r\frac{u'}{u^2}=\frac{\kappa}{8\pi}\frac{ue^2}{r^2}. \tag{2.240}$$

Dividing the entire Eq. (2.240) by $-u$ and rearranging terms, we arrive at:

$$\frac{1}{u^2}-2r\frac{u'}{u^3}=1-\frac{\kappa}{8\pi}\frac{e^2}{r^2}. \tag{2.241}$$

Differentiating Eq. (2.241) with respect to r gives:

$$\frac{d}{dr}\left(\frac{r}{u^2}\right)=1-\frac{\kappa}{8\pi}\frac{e^2}{r^2}. \tag{2.242}$$

Integrating Eq. (2.242) leads to:

$$\frac{r}{u^2}=r-\alpha+\frac{\kappa}{8\pi}\frac{e^2}{r}. \tag{2.243}$$

$$\boxed{\frac{1}{u^2} = 1 - \frac{\alpha}{r} + \frac{\kappa}{8\pi}\frac{e^2}{r^2}.} \tag{2.244}$$

And then he derived the temporal metric function w^2 [Nor-18-2].[84] Combining the two metric functions leads to the Reissner-Nordström metric functions [Nor-18-2]:

Solving for $\frac{1}{u^2}$, leads to Eq. (2.244).

[84] Setting $v = r$, and applying the derivatives $v' = 1$ and $v'' = 0$ to the first field equation in Eq. (2.236), and reversing the sign, we get:

$$\frac{w + 2rw'}{u^2} = w - \frac{\kappa w}{8\pi}\frac{e^2}{r^2}. \tag{2.245}$$

Connecting (2.244) to (2.245) [using Eq. (2.237)] leads to:

$$(w + 2rw')(1 - \frac{\alpha}{r} + \frac{\varepsilon^2}{r^2}) = w - w\frac{\varepsilon^2}{r^2}. \tag{2.246}$$

From this, we get:

$$2\frac{w'}{w} = \frac{\frac{\alpha}{r^2} - 2\frac{\varepsilon^2}{r^2}}{1 - \frac{\alpha}{r} + \frac{\varepsilon^2}{r^2}}. \tag{2.247}$$

In Eq. (2.247), the right-hand side's numerator:

$$f'(r) = \frac{\alpha}{r^2} - 2\frac{\varepsilon^2}{r^2}, \quad \text{is the derivative of the denominator:} \tag{2.248}$$

$$\boxed{f(r) = 1 - \frac{\alpha}{r} + \frac{\varepsilon^2}{r^2}.} \tag{2.249}$$

Given that the numerator is the derivative of the denominator, the Eq. (2.247) simplifies to:

$$2\frac{w'}{w} = \frac{f'(r)}{f(r)}. \quad \text{Integrating both sides gives:} \tag{2.250}$$

$$\int 2\frac{w'}{w}\,dr = \int \frac{f'(r)}{f(r)}dr. \quad \text{The left-hand side integrates to:} \tag{2.251}$$

$$2\ln|w| + A = \ln|f(r)| + B. \tag{2.252}$$

where A and B are constants of integration. Simplifying this, we get:

$$\ln w^2 = \ln f(r) + C. \quad \text{Solving for } w^2 \text{ gives:} \tag{2.253}$$

$$\boxed{w^2 = e^C f(r),} \quad \text{were, } C \text{ is the integration constant.} \tag{2.254}$$

Assuming $w^2 = c^2$ at infinity, we need the denominator $f(r)$ to behave such that $f(\infty) = 1$, so:

$$\boxed{e^C = c^2.} \tag{2.255}$$

Plugging Eqs. (2.255) and (2.249) into (2.254) give the metric function w^2 in Eq. (2.256).

THE METRIC FUNCTIONS OF THE REISSNER-NORDSTRÖM METRIC:

$$\boxed{\frac{1}{u^2} = 1 - \frac{\alpha}{r} + \frac{\varepsilon^2}{r^2}, \quad w^2 = c^2\left(1 - \frac{\alpha}{r} + \frac{\varepsilon^2}{r^2}\right).} \tag{2.256}$$

These expressions for the metric functions show how mass and electric charge influence the spacetime geometry outside the spherical body. The resulting equation for u as a function of r incorporates both the Droste-Schwarzschild term ($\frac{\alpha}{r}$) and a charge-term ($\frac{\varepsilon^2}{r^2}$) fully representing the gravitational and electromagnetic contributions to the field. $c = uw$ [see Eq. (2.203)] and, therefore, $\sqrt{-g} = c$. If $uw = 1$ or $c = 1$, then $\sqrt{-g} = 1$. Setting $g_{44} = w^2 = c$ at an infinite distance far away from the massive charged body, the metric [Eqs. (2.201) or (2.213) and the metric functions representing a charged, spherically symmetric body (2.256)] approaches the value it would have in flat Minkowski space (as $c = 1$), indicating no effect of mass or charge—spacetime is asymptotically flat.

Nordström presented the expression for the mass of the sphere [see Eq. (1.164)] [Nor-18-2]:

$$m = \frac{4\pi\alpha}{\kappa}. \tag{2.257}$$

Subsequently, he analyzed Eq. (2.256), writing the condition $w^2 = 0$.[85] This condition gives us the radii at which the metric component g_{44} (the temporal component in the static, spherically symmetric metric) becomes zero. This occurs at [Nor-18-2][86]:

$$r = \frac{\alpha}{2} \pm \sqrt{\frac{\alpha^2}{4} - \varepsilon^2}. \tag{2.260}$$

$\frac{\alpha^2}{4} - \varepsilon^2$ must be greater than or equal to zero for real solutions. This conditions the relationship between the object's mass (through α) and the charge (ε). For $\frac{\alpha^2}{4} > \varepsilon^2$,

[85] In modern terms, we would express Nordström's condition as indicating the locations of horizons of a charged, spherically symmetric (Reissner-Nordström) black hole.

[86] Setting $w^2 = 0$ in (2.256) and solving for r yields:

$$c^2\left(1 - \frac{\alpha}{r} + \frac{\varepsilon^2}{r^2}\right) = 0. \tag{2.258}$$

We multiply through by r^2 (assuming $r \neq 0$) to clear the fractions, resulting in a quadratic equation in terms of r:

$$r^2 - \alpha r + \varepsilon^2 = 0. \tag{2.259}$$

Solving Eq. (2.259) in r we find Eq. (2.260).

this scenario indicates two specific r, where w^2 becomes zero.[87] For $\frac{\alpha^2}{4} < \varepsilon^2$, there are no real roots, meaning w^2, and hence g_{44} remain positive for all r.[88] Nordström described this scenario as a mass and charge concentrated in a mathematical point [Nor-18-2].

2.3.4 *The Energy Density in Unimodular Coordinates*

Nordström ends his second paper by computing the energy density component t_4^4 of the stress-energy pseudo tensor in the gravitational field surrounding the spherically symmetric sphere, which may also be electrically charged. He grounded his calculations on the equations in Einstein's paper, "Hamilton's Principles and the General Theory of Relativity" [E4-16].

§ *The first stage*: First, Nordström derives an equation for static fields that relates the t_4^4 component of the pseudo-tensor, which can be interpreted as the energy density of the gravitational field, to the trace of the stress-energy tensor of matter $\mathcal{T}_\alpha^\alpha$ and the potential $\mathcal{U}_\tau$ [Nor-18-2][89]:

---

[87] In modern terminology, the equation for $w^2 = 0$ has two real and distinct roots, implying the existence of two horizons: the outer event horizon (given by $r = \frac{\alpha}{2} + \sqrt{\frac{\alpha^2}{4} - \varepsilon^2}$) is analogous to the event horizon in a Schwarzschild black hole; the inner Cauchy event horizon (given by $r = \frac{\alpha}{2} - \sqrt{\frac{\alpha^2}{4} - \varepsilon^2}$) is unique to the charged Reissner-Nordström black hole. Between these two r, w^2 becomes negative. The two roots become equal for $\frac{\alpha^2}{4} = \varepsilon^2$, implying a degenerate horizon where the outer and inner horizons coincide. This specific case marks the transition point between a non-extreme black hole and a naked singularity, forbidden by cosmic censorship conjectures. A Reissner-Nordström black hole with degenerate horizons is an extremal black hole. The black hole's mass and charge are perfectly balanced in this case.

[88] This indicates that no horizons exist. This scenario can describe a naked singularity, where a horizon does not enclose the singularity at $r = 0$.

[89] Nordström outlines Einstein's equation (1.478) (see Sect. 1.7.3): This equation represents a relationship involving a pseudo-tensor t_σ^ν related to $\mathcal{G}^*$ [Nor-18-1]:

$$\kappa t_\sigma^\nu = \frac{1}{2}\left(\delta_\sigma^\nu \mathcal{G}^* - \sum_{\mu\tau} g_\sigma^{\mu\nu} \frac{\partial \mathcal{G}^*}{\partial g_\nu^{\mu\tau}}\right). \tag{2.261}$$

He then writes Einstein's variation principle (1.434) (see Sect. 1.7.3):

$$\delta \int (\mathcal{G}^* + \kappa \mathcal{M})\, d^4x = 0. \tag{2.262}$$

He rewrites the terms on the right-hand side of Eq. (2.262) [Nor-18-1]:

$$-\mathcal{G}^* = \mathcal{G} - \sum_\tau \frac{\partial \mathcal{U}_\tau}{\partial x_\tau}. \tag{2.263}$$

$\mathcal{G}^*$ is defined by Eq. (1.443) (see Sect. 1.7.3), and the divergence term of $\mathcal{U}_\tau$ represents how these additional fields or potential contribute to the overall dynamics of the gravitational field. He

$$\boxed{\kappa t_4^4 = -\frac{1}{2}\kappa \sum_\alpha \mathcal{I}_\alpha^\alpha + \frac{1}{2}\sum_\tau \frac{\partial \mathcal{U}_\tau}{\partial x_\tau}.} \tag{2.269}$$

§ *The second stage*: Nordström calculates $\mathcal{U}_\tau$. He derives an expression for the gravitational potential by setting $x_1 = r, x_2 = x_3 = 0$ [Nor-18-2][90]:

---

substitutes Eq. (2.263) into Eq. (2.262):

$$\delta \int (-\mathcal{G} + \kappa \mathcal{M})\, d^4x = 0. \tag{2.264}$$

Nordström plugged the expression $\sum_{\mu\tau} \frac{\partial \mathcal{G}^*}{\partial g_\nu^{\mu\tau}} = 2\mathcal{G}^*$ from Einstein's paper [E4-16] into Eq. (2.261), and obtained the trace of the pseudo-tensor components t_μ^μ [Nor-18-1]:

$$\kappa \sum t_\nu^\nu = \mathcal{G}^*. \tag{2.265}$$

Subsequently, he observed that Eqs. (2.261) and (2.265) in a static field, $g_4^{\mu\tau} = 0$ (indicating no time variation in the components of the metric tensor), lead to the energy density component t_4^4:

$$\kappa t_4^4 = \frac{1}{2}\mathcal{G}^* = \frac{1}{2}\kappa \sum_\alpha t_\alpha^\alpha. \tag{2.266}$$

Plugging Eq. (2.265) and [Nor-18-1]:

$$\kappa \sum \mathcal{I}_\nu^\nu = \mathcal{G}, \tag{2.267}$$

into Eq. (2.263), Nordström obtains:

$$\sum (I_\nu^\nu + t_\nu^\nu) = \frac{\partial \mathcal{U}_\tau}{\partial x_\tau}. \tag{2.268}$$

[90] Recall that Nordström reduced his problem to one dimension by placing the focus on the X-axis, a specific radial direction from the origin, i.e., he used the condition, $x_1 = r, x_2 = x_3 = 0$ (see footnotes 71 and 73). Under spherical symmetry, the summation $\kappa \sum_\alpha \mathcal{I}_\alpha^\alpha$ would only consider radial components since the angular components do not contribute. The divergence $\sum_\tau \frac{\partial \mathcal{U}_\tau}{\partial x_\tau}$ simplifies to the radial derivative of a radially dependent potential $\mathcal{U}_\tau$. Nordström again selected the coordinate $g_{r4} = 0$, see footnote 77, Sect. 2.3.3 for explanation. Equations (2.201), (2.213) lead to [Nor-18-2, CPAE8], Doc. 382:

$$\begin{aligned} g^{11} &= -\frac{1}{u^2}, \quad g^{22} = g^{33} = -\frac{1}{p^2}, \quad g^{12} = g^{13} = g^{23} = 0, \\ \frac{\partial g_{12}}{\partial x_2} &= \frac{\partial g_{13}}{\partial x_3} = -\frac{1}{r}(u^2 - p^2), \\ \frac{\partial g_{22}}{\partial x_1} &= \frac{\partial g_{33}}{\partial x_1} = -2pp', \quad \frac{\partial g_{44}}{\partial x_1} = 2ww'. \end{aligned} \tag{2.270}$$

$p' = \frac{dp}{dr}$. The specific forms of g^{11}, g^{22}, and g^{33} reflect the spherical symmetry, where the spatial part of the metric depends only on the radial distance r. The derivatives of the metric components are with respect to spatial coordinates (x_1, x_2, x_3). Starting from the definition for the gravitational potential $\mathcal{U}_r$:

$$\mathcal{U}_r = -4\frac{wpp'}{u} + \frac{2}{r}uw\left(1 - \frac{p^2}{u^2}\right) - 2\frac{p^2 w'}{u}. \tag{2.273}$$

This equation holds for all coordinate systems where the metric has spherical symmetry if $g_{r4} = 0$.[91] Nordström focuses on t_4^4, outside a spherically symmetric body ($p = 1$ when the coordinate system is chosen such that either $\sqrt{-g} = 1$ or $\sqrt{-g} = c$). Inserting $p = 1$ and the metric functions (2.256) into Eq. (2.273), it reduces to the τth component of the potential $\mathcal{U}_\tau$ [Nor-18-2]:

$$\mathcal{U}_r = \frac{2c}{r}\left(\frac{\alpha}{r} - \frac{\varepsilon^2}{r^2}\right) - c\left(\frac{\alpha}{r^2} - 2\frac{\varepsilon^2}{r^3}\right) = \frac{c\alpha}{r^2}, \quad \text{leading to: } \mathcal{U}_\tau : \tag{2.274}$$

$$\mathcal{U}_1 \equiv \mathcal{U}_\tau = \frac{x_\tau}{r}\frac{\alpha}{r^2} \qquad \tau = 1, 2, 3. \tag{2.275}$$

This equation shows the directional dependence of the gravitational potential in space, modulated by the distance squared (r^2) from the source. $\mathcal{U}_\tau$ is the gravitational potential that approximates the effects of gravity in a way similar to the Newtonian gravitational potential or the 44 component of the potential. $c = 1$ because $\sqrt{-g} = 1$.

From Eq. (2.275), it follows that the divergence of $\mathcal{U}_\tau$ [Nor-18-2]:

$$\sum_\tau \frac{\partial \mathcal{U}_\tau}{\partial x_\tau} = 0. \tag{2.276}$$

In a vacuum outside the spherically symmetric body ($r > R$ with R being the radius of the body), where there is no matter, the stress-energy tensor components for matter $\mathcal{T}_\alpha^\alpha$ vanish. Thus, $\mathcal{G} = 0$, and according to Eq. (2.263) $\mathcal{G}^* = 0$, simplifying Eq. (2.269) to [Nor-18-2]:

$$\mathcal{U}_r = \frac{1}{2}\sqrt{-g}\sum_{\mu\nu\sigma}\left(g^{\mu 1}g^{\nu\sigma} - g^{\sigma 1}g^{\mu\nu}\right)\left(\frac{\partial g_{\mu\sigma}}{\partial x_\nu} + \frac{\partial g_{\nu\sigma}}{\partial x_\mu} - \frac{\partial g_{\mu\nu}}{\partial x_\sigma}\right). \tag{2.271}$$

Nordström wrote for $\mathcal{U}_\tau$ ($\tau = 1$) two contributions:

$$\begin{aligned} &\frac{1}{2}\sqrt{-g}g^{11}\left(g^{22}\frac{\partial g_{22}}{\partial x_1} + g^{33}\frac{\partial g_{33}}{\partial x_1} - g^{44}\frac{\partial g_{44}}{\partial x_1}\right) - \frac{1}{2}\sqrt{-g}g^{11} \\ &\left[g^{22}\left(2\frac{\partial g_{12}}{\partial x_2} - \frac{\partial g_{22}}{\partial x_1}\right) + g^{33}\left(2\frac{\partial g_{13}}{\partial x_3} - \frac{\partial g_{33}}{\partial x_1}\right) - g^{44}\frac{\partial g_{44}}{\partial x_1}\right]. \end{aligned} \tag{2.272}$$

Since $\mathcal{U}_2 = \mathcal{U}_3 = 0$, then $\mathcal{U}_1 = \mathcal{U}_r$. Combining Eq. (2.270), and the second term of Eq. (2.272), led Nordström to derive Eq. (2.273).

[91] See footnote 77, Sect. 2.3.3. Nordström used $g_{r4} = 0$ for deriving Eq. (2.274) (see footnote 90).

$$t_4^4 = -\frac{1}{2}\sum_\alpha T_\alpha^\alpha. \tag{2.277}$$

The right-hand side of Eq. (2.277) vanishes due to Eq. (2.267), simplifying Eq. (2.277) to:

$$t_4^4 = 0. \tag{2.278}$$

This indicates that for $p = 1$, the energy density attributed to the gravitational field vanishes in the vacuum, outside a spherically symmetric (charged) mass [Nor-18-2, CPAE8], Doc. 382.

From Eq. (2.266), and the realization that $t_4^4 = 0$ in a vacuum, Nordström was led to the broader conclusion that all components of the stress-energy pseudo tensor of the gravitational field t_μ^ν vanish outside the mass/charge distribution for which, as stated in [Nor-18-2]:

$$\sum_\alpha t_\alpha^\alpha = 0. \tag{2.279}$$

Recall from Sect. 1.7.1 that in Einstein's 1916 review paper on general relativity [E1-16], he presented the foundational equations of the theory, including the field equations, by restricting them to the specific unimodular coordinate system [Eq. (1.299)]. Nordström pointed towards this specific formulation that Einstein used. He highlighted that we can easily prove that all t_μ^ν are zero by choosing a reference system in which $p = 1$ (or a coordinate system such that either $\sqrt{-g} = 1$ or $\sqrt{-g} = c$) and with the aid of equations from Einstein's 1916 review paper [E1-16]. Nordström references the paper published by Erwin Schrödinger in 1918, "The Energy Components of the Gravitational Field," in which Schrödinger demonstrated that the expression (2.279) holds for a non-electric sphere [Sch].

2.4 The Tower and the Unimodular Blueprint

Long ago, humanity spoke a single language and worked together harmoniously. They built a tower to reach the heavens, symbolizing their unity and ambition. However, the plan for the tower, devised by the great architect Einstein, was based on a blueprint known as the Unimodular Blueprint, which imposed significant limitations on the design and construction process.

Four eminent builders stepped forward despite these constraints to take on the challenge.

Schwarzschild was the first to tackle the problem. Knowing the limitations of the Unimodular Blueprint, he ingeniously devised a series of transformations. By converting the restrictive grid into a more flexible system, he laid a solid and stable foundation for the tower, overcoming the blueprint's inherent limitations through sheer brilliance.

Working independently, Droste approached the problem differently. He implemented advanced architectural principles that allowed him to bypass the constraints of the Unimodular Blueprint altogether. His plans mirrored Schwarzschild's foundation but were grounded in a methodology that freed him from the initial restrictions.

As the tower began to rise, Reissner joined the effort. He wove intricate iron reinforcements into the structure and infused it with an electric spark. Through correspondence with Einstein and leveraging new insights, he managed to sidestep the Unimodular Blueprint restrictions and extend Schwarzschild's work by including additional structural elements that enhanced the tower's stability.

Finally, Nordström arrived with visionary ideas. Learning from Droste's methods and building upon Einstein's new work, he integrated advanced technologies and electrical systems into the tower. His contributions added a new dimension to the project.

Over time, Einstein observed the builders' struggles and triumphs. Recognizing the limitations imposed by the Unimodular Blueprint, he revised his original plans to allow for more flexibility and innovation. This endorsement paved the way for even more creative and effective designs, celebrating the brilliance and tenacity of those who had dared to build under such challenging conditions.

In the end, the tower stood as a testament to human ingenuity, collaboration, and the ability to overcome limitations. What began as a daunting challenge became a beacon of innovation, symbolizing the power of unity and the relentless pursuit of progress.[92]

[92] This story is a reimagined version of the Tower of Babel, set to highlight human ingenuity and collaboration. It uses historical figures—Einstein, Schwarzschild, Droste, Reissner, and Nordström—to showcase overcoming significant constraints with creativity and progress. While respecting the original symbolism of human ambition, the narrative introduces new elements that emphasize the power of innovation under challenging conditions. The positive twist celebrates unity and adaptability, turning a tale of disunity into triumph and inspiration. The engaging language and vivid descriptions aim to make the story memorable and impactful.

Chapter 3
Ripples, Rotations, and the Kerr Black Hole

3.1 Introduction

I start this chapter by examining Einstein's work on gravitational waves. Initially, Einstein favored unimodular transformations and coordinates, as they were mathematically convenient and familiar. However, as seen at the end of Chap. 1, Einstein realized the need to generalize his theory beyond unimodular coordinates. In his 1916 review paper, "The Foundation of the General Theory of Relativity," he acknowledged the importance of formulating the theory in a way that was not restricted to any specific coordinate system [E1-16]. This marked a significant advancement in deriving his field equations and affirming energy-momentum conservation without resorting to simplifications or constraints in the derivation. Schwarzschild suggested abandoning the determinant condition in Einstein's field equations, prompting Einstein to revise his approach.

Willem de Sitter suggested that Einstein use harmonic coordinates to study gravitational fields. Harmonic coordinates simplify the mathematical treatment of wave equations and are particularly suited for analyzing wave-like solutions in general relativity [Ken-07]. This advice was pivotal in Einstein's development of a consistent theory of gravitational waves. In his pursuit of understanding gravitational waves, Einstein developed a linearized version of his field equations. This linearization approximated weak gravitational fields as small perturbations on the flat spacetime described by the Minkowski metric. The full, nonlinear field equations of general relativity were simplified to a linear form that resembled classical wave equations. Einstein derived a solution for gravitational waves using these linearized field equations. In formulating the solution for gravitational waves, Einstein used retarded potentials borrowed from electromagnetism.[1] By applying the harmonic coordinate

[1] This is an equation used to express the gravitational field at a given point and time in terms of the distribution and motion of mass energy at earlier times, accounting for the finite speed at which gravitational influences propagate (the speed of light).

G. Weinstein, *Einstein's Legacy*, History of Physics,
https://doi.org/10.1007/978-3-031-73572-1_3

condition to the full non-linear field equations and incorporating retarded potentials, Einstein obtained solutions for plane gravitational waves.

Initially, in 1916, Einstein made an error in his analysis of gravitational waves, believing that certain types of these waves did not transport energy. This misunderstanding led him back to considering unimodular coordinates. However, comments from Gunnar Nordström, Erwin Schrödinger, and Tullio Levi Civita and re-examining his work in 1918 made Einstein redefine and correct the results from his first gravitational waves paper. In this chapter, I will illustrate how, in 1918, Einstein manipulated his initial error, ultimately transforming it into a correct result and uncovering gravitational waves capable of carrying energy. I demonstrate that Einstein rebranded the equations, presenting old wine in a new bottle and giving it a fresh twist.

The Schwarzschild solution, which describes the spacetime geometry outside a spherical, static, uncharged mass, does not apply to rotation scenarios. Exploring problems involving rotation, particularly those not adequately addressed by the Schwarzschild solution, required alternative approaches to convey the complexities of these scenarios. The Kerr solution, more appropriate for rotating massive bodies, addresses spacetime geometry around rotating masses. It predicts phenomena like frame dragging, which the Schwarzschild solution does not cover. I show that Einstein's collaborators overcame the absence of mathematical tools for discussing rotation using Einstein's approximate methods rather than the exact Schwarzschild solution. This approach allowed them to analyze relativistic effects in non-inertial frames.

Hans Thirring and Josef Lense's work in 1918 was a significant step in understanding the effects of rotation in general relativity, preceding the discovery of the Kerr solution for several decades. Thirring and Lense had to rely on Einstein's approximation methods from his gravitational waves paper, namely, the linearized treatment of the field equations of general relativity. This treatment allowed them to derive results like the approximate frame-dragging effect without needing a full solution for a rotating body.[2] Thirring's research focused on the gravitational field inside a rotating hollow sphere, examining how rotation could influence centrifugal and Coriolis forces. This exploration was related to frame-dragging, where the rotation of a mass, such as a hollow sphere, drags the spacetime around it. Thirring sent a letter to Einstein, outlining two paradoxes related to the hollow sphere, and sought Einstein's response. Einstein's response to Thirring's first paradox corrected Thirring's approach to the motion equation in a rotating frame, suggesting the correct condition for force-free circular motion. Thirring and Lense expanded upon these ideas in their joint paper and shifted their focus from a hollow sphere to a rotating solid sphere. This change allowed them to more broadly explore the rotational effects on gravitational fields and their implications for celestial bodies. They used Einstein's

[2] When using linearized general relativity, an approximation that neglects higher-order terms, the predicted frame-dragging effect is also approximate. In this approximation, the effects of rotation are simplified. They may not fully account for all the complexities the full Kerr solution describes in systems with strong gravitational fields, such as around rapidly rotating black holes.

linearized equations from his paper on gravitational waves as a foundation for their analysis and discovered frame dragging.

Recall from Chap. 2 that Schwarzschild, Reissner, and Nordström communicated the main equations of their derivations to Einstein in letters before publishing their papers, seemingly seeking his approval and perhaps guidance. The limited evidence, including a few response letters from Einstein, suggests that he was receptive to assisting them, providing patient guidance, and correcting their derivations. Moreover, as demonstrated in Chap. 2, this pattern of consulting Einstein before publication reflected the collaborative structure of the scientific community in the field of general relativity before Einstein's death. This community comprised Einstein, his assistants, and several colleagues who developed his theory and sought his consultation.

After Einstein's death, significant developments in black hole physics were made, including Roy Kerr's discovery of the Kerr metric in 1963. I show that Kerr's formulation, predicated on the sophisticated mathematical methodologies developed during the 1950s and early 1960s, underscores a rich theoretical exploration focused on the physics of gravitational waves (e.g., Pirani [Pir], Newman-Penrose [NP1], and Robinson and Truatman [RT]). The historical arc of this narrative intriguingly intersects with the contributions of Lense and Thirring, who, leveraging analytical tools derived from Einstein's papers on gravitational waves, elucidated the frame-dragging effect within the confines of a linearized approximation of general relativity. The juxtaposition of Kerr's reliance on advanced mathematical frameworks against the backdrop of Lense and Thirring's application of more rudimentary tools from the same conceptual lineage highlights not only a methodological continuum but also an ironic twist in the evolution of classical general relativity. This confluence of theoretical endeavors, spanning the investigation of gravitational waves to the nuanced understanding of the Kerr metric and frame-dragging, emblematically reflects the pivotal role that gravitational wave research has played within the broader discourse of general relativity. Far from being a narrowly focused area of inquiry, research into gravitational waves has catalyzed significant breakthroughs across a spectrum of topics ostensibly tangential to directly detecting these waves. Such developments poignantly illustrate the interconnectedness of areas in general relativity, where innovations in mathematical and conceptual tools in one area can precipitate foundational advances in another.

Accordingly, the historical and intellectual trajectories of Kerr's metric derivation and the Lense-Thirring frame-dragging effect underscore a broader thematic narrative within general relativity: the indispensable role of gravitational wave research not merely as a quest for empirical verification but as a crucible for theoretical innovation that transcends its immediate application. The sophisticated tools devised for gravitational wave research have served as the linchpin in unraveling the complexities of the Kerr metric. The grand finale of Kerr's tireless endeavors culminated in a reward. LIGO's detection of GW150914 provided strong empirical support for the Kerr metric: The match between the observed gravitational wave signal and the theoretical models based on the Kerr metric suggests that the black holes involved in GW150914 were Kerr black holes.

3.2 Einstein's Odyssey with Gravitational Waves

3.2.1 Schwarzschild, de Sitter, and the Departure from Unimodularity

In February 1916, Schwarzschild suggested to Einstein a condition different from the determinant condition (1.298) that could further simplify the field equations of general relativity [CPAE8], Doc 188. Einstein's response to Schwarzschild's letter about two and a half weeks later contained the following explanations [CPAE8], Doc. 194[3]: He imposed the harmonic condition (1.274) and derived the Eq. (1.275) from his November 4. 1915 field equations. These equations retained their form under unimodular transformations (see Sect. 6, Chap. 1). However, in his letter addressed to Schwarzschild, Einstein elucidated that the strategy of selecting a coordinate system based on the coordinate condition (1.274) [E1-15] is incompatible with postulating and setting $\sqrt{-g} = 1$. Following this insight, Einstein told Schwarzschild that his approach to integrating Newtonian principles within the comprehensive framework of his theory has evolved. After the November 4, 1915 paper, in his subsequent papers [E3-15, E4-15] and review paper on general relativity [E1-16], he shifted to an alternative approach when dealing with the Newtonian limit, consistent with his field equations (1.292) valid only in unimodular coordinates (refer to Sects. 1.6.2 and 1.6.3).

After imposing the unimodular coordinate condition (1.299), Einstein focused on the Newtonian limit using an approach more suitable for drawing parallels with Newtonian mechanics, especially for weak gravitational fields and the Newtonian approximation. In contrast, the harmonic condition (1.274)) aligns more with harmonic coordinates, which are helpful in wave solutions and special relativistic analogies. Eventually, Einstein recognized that a departure from unimodular coordinates was necessary to explore gravitational waves. The harmonic condition (1.274)) was more appropriate for analyzing wave-like solutions in the theory, as it facilitated the comparison with Maxwell's equations of electromagnetism [Ken-07].[4]

---

[3] In the Addendum to his first paper on general relativity (i.e., the November 4, 1915 paper [E1-15]) published on November 11, 1915, Einstein revised the restriction to unimodular transformations to the condition for unimodular coordinates (1.299) [E2-15]. He justified the change by stating that his comments in the November 4, 1915 paper [E1-15] were outdated. In the first paper on general relativity [E1-15], Einstein wrestled with the weak field approximation. He postulated that $\sqrt{-g} = 1$, so this condition was a starting point for his theory. It was a foundational assumption on which the rest of the theoretical framework was built [CPAE8], Doc. 194.

[4] However, Einstein's journey in developing and refining his theory was marked by revisions, reconsiderations, and occasional missteps. After initially exploring and setting aside the unimodular coordinate condition in his work on gravitational waves, Einstein revisited this condition. This revisit was partly due to an error in his calculations (see discussion in Sect. 3.2). Einstein's involvement and subsequent complications with the unimodular coordinate condition culminated in him informing Schwarzschild that this could imply the non-existence of gravitational waves, similar to light waves [CPAE8], Doc. 194.

Willem de Sitter found that Einstein's 1916 gravitational field equations could be solved to first-order approximation by (for the fields emanating from the body) [E3-16][5]:

$$\boxed{ds^2 = \left(1 - \frac{\kappa M}{4\pi r}\right) dl^2 - \left(1 + \frac{\kappa M}{4\pi r}\right) dt^2,} \text{ where: } \frac{\kappa M}{4\pi} = \alpha \text{ [Eq. (1.164)].} \tag{3.2}$$

Like Schwarzschild, de Sitter advised Einstein through a letter to avoid using the unimodular coordinate condition (1.299). As Einstein noted, de Sitter suggested using a different reference system (harmonic coordinate system) from the unimodular coordinates Einstein initially used, as this would simplify the representation of a stationary mass point's gravitational field. De Sitter discovered that by not using the unimodular coordinate condition (1.299) Einstein's field equations could be resolved for the fields produced by a mass. This led Einstein to realize that if he did not use the unimodular coordinate condition and used instead harmonic coordinates, the field equations could be precisely solved using retarded potentials in a first-order approximation [see Eq. (3.40) in Sect. 3.2.2]. Consequently, Einstein based his 1916 gravitational waves paper [E3-16] on the generally covariant field equations. In his correspondence with de Sitter, Einstein acknowledged that de Sitter's first-order solution [Eq. (3.2)] differed from his own earlier solution [Eq. (1.300)] only in the choice of the coordinate system, not in any fundamental aspect [CPAE8], Doc. 227. Indeed, Eq. (3.2) is equivalent to the first two terms of Einstein's first-order solution (1.300) when it is expressed in the format of (1.322) (refer to Sect. 1.6.3):

$$\boxed{ds^2 = (1 - \frac{\alpha}{r})dt^2 - (1 + \frac{\alpha}{r})dr^2.} \tag{3.3}$$

3.2.2 *Linearized General Relativity*

In June 1916, Einstein presented a paper on gravitational waves to the Prussian Academy of Sciences, which was later published under the title "Approximate Integration of the Field Equations of Gravitation." In this paper, assuming that the gravitational field is infinitely weak, he solved a linearized approximation version of his field equations. Einstein started his 1916 paper on gravitational waves with the first

---

[5] De Sitter's line element can be expressed as follows [E-22]:

$$\begin{cases} \gamma_{11} = \gamma_{22} = \gamma_{33} = -\frac{\kappa}{4\pi} \int \frac{\rho \, dV_0}{r}, \\ \gamma_{44} = \frac{\kappa}{4\pi} \int \frac{\rho \, dV_0}{r}, \end{cases} \quad ds^2 = \left(1 - \frac{\kappa}{4\pi} \int \frac{\rho \, dV_0}{r}\right) dl^2 - \left(1 + \frac{\kappa}{4\pi} \int \frac{\rho \, dV_0}{r}\right) dt^2. \tag{3.1}$$

where $dl^2 = dx_1^2 + dx_2^2 + dx_3^2$ if the coordinate condition (1.299) was abandoned. According to Einstein, de Sitter's metric led him to the approximate solution given in his gravitational waves 1916 paper (3.2) [E3-16].

approximation of the total metric tensor $g_{\mu\nu}$ when dealing with weak gravitational fields [E3-16]:

$$\text{THE FLAT (MINKOWSKI) SPACETIME METRIC + A PERTURBATION:}$$
$$\boxed{g_{\mu\nu} = -\delta_{\mu\nu} + \gamma_{\mu\nu}.} \tag{3.4}$$

$\gamma_{\mu\nu}$ is small compared to 1 (in the linear approximation regime), and $\delta_{\mu\nu}$ represents the Minkowski flat metric (1.43) (see Sect. 1.2.3, footnote 24). Einstein represented the flat spacetime metric with $\delta_{\mu\nu}$ instead of the modern $\eta_{\mu\nu}$. When Einstein worked on his new theory, the choice of notation came down to his preference. Over time, as the theoretical framework of relativity and gravitational physics has evolved, so too have the notational conventions, leading to the widespread adoption of $\eta_{\mu\nu}$.

Einstein clarified that unimodular coordinates with the condition (1.299) were not favorable for calculating fields in the first approximation. On the other hand, Eq. (3.4) is particularly useful for calculating gravitational fields when the deviation from Minkowski spacetime is not large [E3-16]. I will elaborate on this topic in Sect. 3.2.2.

Subsequently, Einstein linearizes his field equations (1.290) (without the restriction $\sqrt{-g} = 1$) for a weak gravitational field by expanding $\gamma_{\mu\nu}$. This expansion leads to a linearized form of the field equations (1.290) [E3-16]:

$$\text{A LINEARIZED FORM OF EINSTEIN'S FIELD EQUATIONS:}$$
$$\sum_\alpha \left(\frac{\partial^2 \gamma_{\mu\alpha}}{\partial x_\nu \partial x_\alpha} + \frac{\partial^2 \gamma_{\nu\alpha}}{\partial x_\mu \partial x_\alpha} - \frac{\partial^2 \gamma_{\mu\nu}}{\partial x_\alpha^2} \right) - \boxed{\frac{\partial^2}{\partial x_\mu \partial x_\nu} \left(\sum_\alpha \gamma_{\alpha\alpha} \right)} = - 2\kappa \left(T_{\mu\nu} - \frac{1}{2}\delta_{\mu\nu} \sum_\alpha T_{\alpha\alpha} \right). \tag{3.5}$$

The boxed term in this equation accounts for the trace of $\gamma_{\mu\nu}$. Solving these equations in the weak field limit involves finding $\gamma_{\mu\nu}$ that satisfies the expanded field equations for a given distribution of mass and energy, represented by the stress-energy tensor $T_{\mu\nu}$. The linearized approximate solution $\gamma_{\mu\nu}$ is further decomposed[6]:

$$\gamma_{\mu\nu} = \text{SCALAR PERTURBATIONS + OTHER PERTURBATIONS:}$$
$$\gamma_{\mu\nu} = \psi\delta_{\mu\nu} + \gamma'_{\mu\nu}. \tag{3.6}$$

$\gamma'_{\mu\nu}$ (other perturbations) obeys the additional condition[7]:

---

[6] This equation introduces a scalar field ψ and a traceless tensor $\gamma'_{\mu\nu}$ as the perturbation components. $\delta_{\mu\nu}$ represents a form of metric scaling distinct from the Minkowski metric.

[7] This is the condition (1.274)). Equation (3.7) states that the divergence of $\gamma'_{\mu\nu}$ vanishes.

THE HARMONIC CONDITION:

$$\sum_{\nu} \frac{\partial \gamma'_{\mu\nu}}{\partial x_{\nu}} = 0. \tag{3.7}$$

When the expression (3.6) is substituted into the left-hand side of the linearized field equations (3.5), each term involving $\gamma_{\mu\nu}$ is replaced with $\psi\delta_{\mu\nu} + \gamma'_{\mu\nu}$.

I will substitute $\gamma_{\mu\nu}$ from Eq. (3.6) into each term of the left-hand side of equation (3.5). Recall that $\delta_{\mu\nu}$ is given by Eq. (1.43).

The result is: the first two terms lead to[8]:

$$\boxed{2\frac{\partial^2 \psi}{\partial x_{\mu} \partial x_{\nu}}} \quad \text{for: } \mu = \nu = \alpha. \tag{3.14}$$

[8] The first term on the left-hand side of equation (3.5):

$$\sum_{\alpha} \frac{\partial^2 \gamma_{\mu\alpha}}{\partial x_{\nu} \partial x_{\alpha}} = \sum_{\alpha} \frac{\partial^2 (\psi\delta_{\mu\alpha} + \gamma'_{\mu\alpha})}{\partial x_{\nu} \partial x_{\alpha}}. \quad \text{Simplification leads to:} \tag{3.8}$$

$$\sum_{\alpha} \frac{\partial^2 \gamma'_{\mu\alpha}}{\partial x_{\nu} \partial x_{\alpha}} + \sum_{\alpha} \frac{\partial^2 \psi}{\partial x_{\nu} \partial x_{\alpha}} \delta_{\mu\alpha}. \tag{3.9}$$

When $\mu = \alpha$, $\delta_{\mu\nu} = 1$ and the second term simplifies to:

$$\frac{\partial^2 \psi}{\partial x_{\nu} \partial x_{\mu}}. \tag{3.10}$$

The second term on the left-hand side of equation (3.5):

$$\sum_{\alpha} \frac{\partial^2 \gamma_{\nu\alpha}}{\partial x_{\mu} \partial x_{\alpha}} = \sum_{\alpha} \frac{\partial^2 (\psi\delta_{\nu\alpha} + \gamma'_{\nu\alpha})}{\partial x_{\mu} \partial x_{\alpha}}. \quad \text{Simplification leads to:} \tag{3.11}$$

$$\sum_{\alpha} \frac{\partial^2 \gamma'_{\nu\alpha}}{\partial x_{\mu} \partial x_{\alpha}} + \sum_{\alpha} \frac{\partial^2 \psi}{\partial x_{\mu} \partial x_{\alpha}} \delta_{\nu\alpha}. \tag{3.12}$$

When $\nu = \alpha$, the second term simplifies to:

$$\frac{\partial^2 \psi}{\partial x_{\mu} \partial x_{\nu}}. \tag{3.13}$$

The contributions from the first (3.10) and second (3.13) terms of equation (3.5) add together, resulting in Eq. (3.14). Since the second derivatives are symmetric with respect to their indices:

$$\frac{\partial^2}{\partial x_{\nu} \partial x_{\mu}} = \frac{\partial^2}{\partial x_{\mu} \partial x_{\nu}},$$

(The first terms of equations (3.9) and (3.12) cancel out).

The third term[9]:

$$\boxed{\sum_\alpha \frac{\partial^2 \gamma'_{\mu\nu}}{\partial x_\alpha^2}} + \boxed{\delta_{\mu\nu} \sum_\alpha \frac{\partial^2 \psi}{\partial x_\alpha^2}}. \tag{3.16}$$

The fourth term[10]:

$$\boxed{\frac{\partial^2}{\partial x_\mu \partial x_\nu}\left(\sum_\alpha \gamma'_{\alpha\alpha}\right)} + \frac{\partial^2 \psi}{\partial x_\mu \partial x_\nu} \sum_\alpha \delta_{\alpha\alpha}. \tag{3.18}$$

The second term of equation (3.18) is rewritten as[11]:

$$\frac{\partial^2 \psi}{\partial x_\mu \partial x_\nu} \sum_\alpha \delta_{\alpha\alpha} = \boxed{4 \frac{\partial^2 \psi}{\partial x_\mu \partial x_\nu}}. \tag{3.19}$$

Now, we combine all the boxed terms to form the left-hand side of equation (3.5) resulting in [E3-16]:

THE LINEARIZED FIELD EQUATION FOR SCALAR + OTHER PERTURBATIONS:

$$\boxed{-\sum_\alpha \frac{\partial^2 \gamma'_{\mu\nu}}{\partial x_\alpha^2}} - \frac{\partial^2}{\partial x_\mu \partial x_\nu}\left(\sum_\alpha \gamma'_{\alpha\alpha}\right) + 2\frac{\partial^2 \psi}{\partial x_\mu \partial x_\nu} - \delta_{\mu\nu} \sum_\alpha \frac{\partial^2 \psi}{\partial x_\alpha^2} - 4\frac{\partial^2 \psi}{\partial x_\mu \partial x_\nu}. \tag{3.20}$$

Einstein then demonstrates that certain terms in Eq. (3.20) cancel out when applying the condition:

$$-\sum_\alpha \gamma'_{\alpha\alpha} = 2\psi, \tag{3.21}$$

[9] The third term on the left-hand side of equation (3.5):

$$\sum_\alpha \frac{\partial^2 \gamma_{\mu\nu}}{\partial x_\alpha^2} = \sum_\alpha \frac{\partial^2 (\psi \delta_{\mu\nu} + \gamma'_{\mu\nu})}{\partial x_\alpha^2}. \tag{3.15}$$

Simplification leads to Eq. (3.16).

[10] Fourth term of Eq. (3.5):

$$\frac{\partial^2}{\partial x_\mu \partial x_\nu}\left(\sum_\alpha \gamma_{\alpha\alpha}\right) = \frac{\partial^2}{\partial x_\mu \partial x_\nu}\left(\sum_\alpha (\psi \delta_{\alpha\alpha} + \gamma'_{\alpha\alpha})\right). \tag{3.17}$$

Simplification leads to Eq. (3.18).

[11] The term $\sum_\alpha \delta_{\alpha\alpha}$ in the fourth term's contribution evaluates to 4, as it sums over the spacetime indices α, which run from 0 to 3. This leads to the contribution in Eq. (3.20) expressed in Eq. (3.19).

simplifying Eq. (3.20). Specifically, the second, third, fourth, and fifth terms in Eq. (3.20) involve derivatives of ψ and $\gamma'_{\alpha\alpha}$ that, when summed over α and considering the condition (3.21), cancel each other out. This leads to the linearized field equations[12]:

THE LINEARIZED FIELD (WAVE) EQUATION FOR $\gamma'_{\mu\nu}$:

$$\boxed{\sum_\alpha \frac{\partial^2}{\partial x_\alpha^2}\gamma'_{\mu\nu} = 2\kappa T_{\mu\nu}.} \tag{3.27}$$

This equation represents *the linearized version of Einstein's field equations under the weak-field approximation*. It is a wave equation for $\gamma'_{\mu\nu}$ with a source term $2\kappa T_{\mu\nu}$.[13]

Einstein demonstrates the consistency of equation (3.27) with the harmonic condition (3.7) and the conservation of energy-momentum [E3-16]:

---

[12] After applying the condition (3.21) to Eq. (3.20) and canceling out terms, we are left with:

$$\boxed{-\sum_\alpha \frac{\partial^2\gamma'_{\mu\nu}}{\partial x_\alpha^2}} - \delta_{\mu\nu}\sum_\alpha \frac{\partial^2\psi}{\partial x_\alpha^2}. \tag{3.22}$$

Applying Eq. (3.21), this condition directly impacts terms involving ψ and $\gamma'_{\alpha\alpha}$ in Eq. (3.20):

$$\begin{aligned} &-\frac{\partial^2}{\partial x_\mu \partial x_\nu}(\sum_\alpha \gamma'_{\alpha\alpha}) + 2\frac{\partial^2\psi}{\partial x_\mu \partial x_\nu} - 4\frac{\partial^2\psi}{\partial x_\mu \partial x_\nu} = \\ &-\frac{\partial^2}{\partial x_\mu \partial x_\nu}(\sum_\alpha \gamma'_{\alpha\alpha}) - \frac{\partial^2}{\partial x_\mu \partial x_\nu}(\sum_\alpha \gamma'_{\alpha\alpha}) + 2\frac{\partial^2}{\partial x_\mu \partial x_\nu}(\sum_\alpha \gamma'_{\alpha\alpha}) = 0. \end{aligned} \tag{3.23}$$

Hence, these terms cancel each other out because they are equal and opposite. However, we can further simplify the second term in Eq. (3.22) using Eq. (3.21):

$$-\delta_{\mu\nu}\sum_\alpha \frac{\partial^2\psi}{\partial x_\alpha^2} = \boxed{\frac{1}{2}\delta_{\mu\nu}\sum_\alpha \frac{\partial^2}{\partial x_\alpha^2}(\sum_\alpha \gamma'_{\alpha\alpha}).} \tag{3.24}$$

Adding the boxed term in Eq. (3.22) to the boxed term in Eq. (3.24), we get:

$$-\sum_\alpha \frac{\partial^2\gamma'_{\mu\nu}}{\partial x_\alpha^2} - \frac{1}{2}\delta_{\mu\nu}\sum_\alpha \frac{\partial^2}{\partial x_\alpha^2}(\sum_\alpha \gamma'_{\alpha\alpha}). \tag{3.25}$$

Substituting Eq. (3.25) into the left-hand side of equation (3.5) we get:

$$\sum_\alpha \frac{\partial^2}{\partial x_\alpha^2}\left(\gamma'_{\mu\nu} - \frac{1}{2}\delta_{\mu\nu}\sum_\alpha \gamma'_{\alpha\alpha}\right) = 2\kappa(T_{\mu\nu} - \frac{1}{2}\delta_{\mu\nu}\sum T_{\alpha\alpha}). \tag{3.26}$$

Taking into consideration Eq. (1.43) we get Eq. (3.27) or (1.275).

[13] The expression:

$$\frac{\partial^2\gamma'_{\mu\nu}}{\partial x_1^2} + \frac{\partial^2\gamma'_{\mu\nu}}{\partial x_2^2} + \frac{\partial^2\gamma'_{\mu\nu}}{\partial x_3^2} - \frac{1}{c^2}\frac{\partial^2\gamma'_{\mu\nu}}{\partial x_4^2} = 2\kappa T, \tag{3.28}$$

represents a wave equation in Eq. (3.27).

THE CONSERVATION OF ENERGY-MOMENTUM FOR MATTER:

$$\boxed{\sum_{\nu} \frac{\partial T_{\mu\nu}}{\partial x_{\nu}} = 0.} \tag{3.29}$$

Equation (3.29) implies that the stress-energy tensor $T_{\mu\nu}$ has zero divergence. Einstein demonstrates that the linearized field equation (3.27) correctly incorporates this fundamental principle.[14] Einstein's derivation of equation (3.27) and the demonstration that this simplified, linearized equation is consistent reflects the underlying general covariance of his field equations.

Einstein introduced a pair of transformation rules defining $\gamma'_{\mu\nu}$ and $\gamma_{\mu\nu}$ in terms of each other and sums involving $\gamma'_{\alpha\alpha}$ and its trace, $\gamma_{\alpha\alpha}$[15]:

THE PAIR OF TRANSFORMATIONS:

$$\boxed{\gamma_{\mu\nu} = \gamma'_{\mu\nu} - \frac{1}{2}\delta_{\mu\nu}\sum_{\alpha}\gamma'_{\alpha\alpha}.} \tag{3.32}$$

$$\boxed{\gamma'_{\mu\nu} = \gamma_{\mu\nu} - \frac{1}{2}\delta_{\mu\nu}\sum_{\alpha}\gamma_{\alpha\alpha}.} \tag{3.33}$$

[14] Einstein first applies the divergence $\sum_{\nu}\frac{\partial}{\partial x_{\nu}}$ to both sides of equation (3.27), the left-hand side becomes:

$$\sum_{\alpha\nu} \frac{\partial^3}{\partial x_{\nu}\partial x_{\alpha}^2}\gamma'_{\mu\nu}. \tag{3.30}$$

The left-hand side involves third derivatives of $\gamma'_{\mu\nu}$. By the harmonic condition [Eq. (3.7)], the sum of the derivatives of $\gamma'_{\mu\nu}$ with respect to x_{ν} is zero. This implies that the sum of the third derivatives will also be zero. Hence, the left-hand side of equation (3.27) vanishes. The right-hand side becomes:

$$\sum_{\alpha\nu} \frac{\partial}{\partial x_{\nu}}\left(2\kappa T_{\mu\nu}\right). \tag{3.31}$$

The right-hand side involves the derivative of $T_{\mu\nu}$ with respect to x_{ν}. The conservation law Eq. (3.29) states that Eq. (3.31) is zero. Hence, the right-hand side of equation (3.27) also vanishes.

Since both the left-hand side and the right-hand side of equation (3.27) vanish, we can say that applying $\sum_{\nu}\frac{\partial}{\partial x_{\nu}}$ to Eq. (3.27) yields a consistent result, confirming that Eq. (3.27) is in line with both the harmonic condition Eq. (3.7) and the conservation of energy-momentum Eq. (3.29) [E3-16].

[15] Recall that Eq. (3.6) defines $\gamma_{\mu\nu}$ as a sum of $\psi\delta_{\mu\nu}$ and $\gamma'_{\mu\nu}$, and Eq. (3.21) relates the sum over $\gamma'_{\alpha\alpha}$ to 2ψ.

This re-definition is a mathematical trick Einstein uses in his linearized theory to simplify the equations by removing the trace part of $\gamma_{\mu\nu}$. According to Einstein, Eqs. (3.6) and (3.21) support Eqs. (3.32) and (3.33) [E3-16].[16]

Let us briefly summarize the derivation that Einstein carried out. Suppose $\gamma_{\mu\nu}$ are very small compared to 1 everywhere. In that case, we obtain a first-order approximation solution of the field equations by neglecting in the full non-linear field equations (1.290) [not restricted by the condition (1.299)] the higher powers of $\gamma_{\mu\nu}$ and their derivatives, compared with the lower ones. By imposing on the full non-linear field equations the coordinate condition [Eq. (3.7)], one is then led to the linearized gravitational field equations (3.27).

In the new harmonic coordinate system, Einstein calculated $\gamma'_{\mu\nu}$ in a manner analogous to that of retarded potentials in electrodynamics. Einstein solved Eq. (3.27), which, as said above, is a wave equation using Green's functions.[17]

---

[16] Let us see how the two equations (3.6) and (3.21) support the pair of transformations (3.32) and (3.33). We start with Eq. (3.6). We rearrange this equation to express $\gamma'_{\mu\nu}$ in terms of $\gamma_{\mu\nu}$ and ψ:

$$\gamma'_{\mu\nu} = \gamma_{\mu\nu} - \psi\delta_{\mu\nu}. \tag{3.34}$$

We substitute ψ from Eq. (3.21) into this Eq. (3.34) and obtain Eq. (3.33). Now we substitute:

$$\psi = -\frac{1}{2}\sum_{\alpha}\gamma'_{\alpha\alpha},$$

from Eq. (3.21) into Eq. (3.6) and get Eq. (3.32).

[17] Green's function $G(x, x')$ satisfies the following wave equation:

$$\sum_{\alpha}\frac{\partial^2}{\partial x_\alpha^2}G(x, x') = \delta(x - x'). \tag{3.35}$$

$\delta(x - x')$ is the delta function and Green's function $G(x, x')$ represents the influence at point x due to a point source located at x'. The solution $\gamma'_{\mu\nu}$ to the original Eq. (3.27) is expressed as an integral involving the Green's function and the source term $T_{\mu\nu}$:

$$\gamma'_{\mu\nu}(x) = 2\kappa\int G(x, x')T_{\mu\nu}(x'), d\tau. \tag{3.36}$$

In this expression, the integration is over the spacetime coordinates x'. $T_{\mu\nu}(x')$ is evaluated at these coordinates. For the wave equation in a 3-dimensional space, Green's function is written as:

$$G(x, x') = -\frac{1}{4\pi}\frac{\delta(t - t' - \frac{|x-x'|}{c})}{|x - x'|} \tag{3.37}$$

This is Green's function for the retarded potential, where the δ function guarantees causality. Substituting Green's function into the integral (3.36), we get:

$$\gamma'_{\mu\nu}(x) = -\frac{\kappa}{2\pi}\int\frac{\delta(t - t' - \frac{|x-x'|}{c})}{|x - x'|}T_{\mu\nu}(x')\,d\tau \tag{3.38}$$

The integral (3.38) evaluates the influence of $T_{\mu\nu}$ at all points x' at the retarded time:

RETARDED POTENTIALS FROM THE STRESS- ENERGY TENSOR:

$$\gamma'_{\mu\nu} = -\frac{\kappa}{2\pi} \int \frac{T_{\mu\nu}(x_0, y_0, z_0, t - \frac{r}{c})}{r} dV_0. \tag{3.40}$$

The expression on the right-hand side has to be taken at the time of emission, at the retarded time $t - \frac{r}{c}$, reaching a point at the time t; and [E3-16]:

$$r = \sqrt{(x - x_0)^2 + (y - y_0)^2 + (z - z_0)^2}, \quad \text{is the distance to the source.}$$

In his 1916 paper on gravitational waves, Einstein showed that in the linearized approximation, choosing harmonic coordinates simplifies his field equations, making them resemble wave equations. This was useful for studying gravitational waves (i.e., wave-like solutions to the field equations).

Einstein multiplied both sides of the linearized field equation (3.27) by $\frac{\partial \gamma'_{\mu\nu}}{\partial x_\sigma}$ and summed over μ and ν:

$$\sum_{\mu\nu} \left(\sum_\alpha \frac{\partial^2}{\partial x_\alpha^2} \gamma'_{\mu\nu} \right) \frac{\partial \gamma'_{\mu\nu}}{\partial x_\sigma} = 2\kappa \sum_{\mu\nu} T_{\mu\nu} \frac{\partial \gamma'_{\mu\nu}}{\partial x_\sigma}, \tag{3.41}$$

and demonstrated consistency with the conservation law for the total energy-momentum for matter and gravitational field (1.258) (see Sect. 1.6.1, Chap. 1) by transforming both sides of the equation (3.41) [E3-16]. To show that his linearized field equations respect energy-momentum conservation, he introduced the Eq. (1.203) (see Sect. 1.5.2, Chap. 1) for energy-momentum balance in a gravitational field from his first 1915 general relativity paper [E3-16]:

$$\sum_\sigma \frac{\partial \sqrt{-g} T_\mu^\sigma}{\partial x_\sigma} + \frac{1}{2} \sum_{\rho\sigma} \frac{\partial g_{\rho\sigma}}{\partial x_\sigma} \sqrt{-g} T_{\rho\sigma} = 0. \tag{3.42}$$

$$t' = t - \frac{|x - x'|}{c}.$$

The integral (3.38) becomes an integration over the spatial volume $dV_0 = d\tau$ and involves the stress-energy tensor $T_{\mu\nu}$ evaluated at the retarded time, leading to:

$$\gamma'_{\mu\nu} = -\frac{\kappa}{2\pi} \int \frac{T_{\mu\nu}(t - \frac{|x-x'|}{c})}{|x - x'|} dV_0, \quad \text{i.e., Eq. (3.40).} \tag{3.39}$$

Einstein noted that in the linear approximation, $\sqrt{-g}$ can equal 1. So the above equation reduces to a form akin to Eq. (1.257) (see Sect. 1.6.1, Chap. 1)[18]:

THE ENERGY-MOMENTUM BALANCE IN A GRAVITATIONAL FIELD:

$$\boxed{\sum_\sigma \frac{\partial T_{\mu\sigma}}{\partial x_\sigma} + \frac{1}{2}\sum_{\rho\sigma} \frac{\partial g_{\rho\sigma}}{\partial x_\sigma} T_{\rho\sigma} = 0.} \tag{3.43}$$

THE RIGHT-HAND SIDE OF EQUATION (3.41):

Einstein rearranged Eq. (3.43)[19]:

$$2\kappa \sum_{\mu\nu} T_{\mu\nu} \frac{\partial \gamma'_{\mu\nu}}{\partial x_\sigma} = \boxed{-4\kappa \sum_\sigma \frac{\partial T_{\mu\sigma}}{\partial x_\sigma}.} \tag{3.46}$$

He then wrote the energy-momentum balance in a gravitational field as follows[20]:

---

[18] Under the assumption of weak gravitational fields, spacetime is nearly flat, and the metric $\gamma'_{\mu\nu}$ is close to the Minkowski metric with small deviations. Since the determinant of a matrix is the product of its diagonal elements and the diagonal elements of $\gamma'_{\mu\nu}$ are close to those of the flat metric, which are ± 1 $[(-1) \times 1 \times 1 \times 1 = -1]$, the determinant of g is close to -1 (the determinant of the Minkowski metric). Therefore, $\sqrt{-g}$ is approximately 1 in Eq. (1.203). Here, the $\sqrt{-g} = 1$ is used in the context of a linear approximation, and it is distinct from the unimodular coordinate condition (1.299). Unlike the case for linear approximation, the unimodular condition refers to the constraint on the form of the metric tensor, where the determinant of the metric tensor is set to be -1.

[19] We move the factor of 2 from one side of equation (3.43) to the other:

$$-2\sum_\sigma \frac{\partial T_{\mu\sigma}}{\partial x_\sigma} = \sum_{\rho\sigma} \frac{\partial g_{\rho\sigma}}{\partial x_\sigma} T_{\rho\sigma}. \tag{3.44}$$

Now, we substitute the right-hand side of the field equation (3.41) with the left-hand side of Eq. (3.44):

$$2\kappa \sum_{\mu\nu} T_{\mu\nu} \frac{\partial \gamma'_{\mu\nu}}{\partial x_\sigma} = 2\kappa\left(-2\sum_\sigma \frac{\partial T_{\mu\sigma}}{\partial x_\sigma}\right) - 4\kappa \sum_\sigma \frac{\partial T_{\mu\sigma}}{\partial x_\sigma}. \tag{3.45}$$

With this substitution, the right-hand side of equation (3.41) transforms into Eq. (3.46).

[20] We can rearrange Eq. (3.48) to express $\frac{\partial t_{\mu\sigma}}{\partial x_\sigma}$ in terms of $\frac{\partial g_{\rho\sigma}}{\partial x_\sigma} T_{\rho\sigma}$:

$$2\sum_\sigma \frac{\partial t_{\mu\sigma}}{\partial x_\sigma} = \sum_{\rho\sigma} \frac{\partial g_{\rho\sigma}}{\partial x_\sigma} T_{\rho\sigma}. \tag{3.47}$$

Using Eq. (3.47), we can replace:

$$\sum_{\mu\nu} T_{\mu\nu} \frac{\partial \gamma'_{\mu\nu}}{\partial x_\sigma} \text{ from Eq. (3.41) with: } 2\sum_\sigma \frac{\partial t_{\mu\sigma}}{\partial x_\sigma} \text{ from Eq. (3.47).}$$

$$\boxed{\sum_{\sigma} \frac{\partial t_{\mu\sigma}}{\partial x_{\sigma}} = \frac{1}{2} \sum_{\rho\sigma} \frac{\partial g_{\rho\sigma}}{\partial x_{\sigma}} T_{\rho\sigma}.} \tag{3.48}$$

Einstein transformed the stress-energy tensor on the right-hand side of equation (3.41) to the energy components of the gravitational field[21]:

$$2\kappa \sum_{\mu\nu} T_{\mu\nu} \frac{\partial \gamma'_{\mu\nu}}{\partial x_{\sigma}} = \boxed{4\kappa \sum_{\sigma} \frac{\partial t_{\mu\sigma}}{\partial x_{\sigma}}.} \tag{3.49}$$

The right-hand side of equations (3.46) and (3.49) contribute to the energy-momentum conservation in a linearized theory (1.258).

THE LEFT-HAND SIDE OF EQUATION (3.41).

The left-hand side of equation (3.41) involves the second derivative of $\gamma'_{\mu\nu}$ and its first derivative. We can express this as a divergence of a tensor, which has the form: $\sum_{\alpha} \frac{\partial}{\partial x_{\alpha}}$(something), leading to an expression that resembles a stress-energy pseudo tensor for the gravitational field. Using Eq. (3.49) we get [E3-16]:

THE 1916 STRESS-ENERGY PSEUDO TENSOR IN LINEARIZED GRAVITY:

$$\boxed{t_{\mu\nu} = \frac{1}{4\kappa} \left(\frac{\partial \gamma'_{\alpha\beta}}{\partial x_{\mu}} \frac{\partial \gamma'_{\alpha\beta}}{\partial x_{\nu}} - \frac{1}{2} \delta_{\mu\nu} \left(\frac{\partial \gamma'_{\alpha\beta}}{\partial x_{\tau}} \right)^2 \right).} \tag{3.50}$$

In the derivation of equation (3.50), Einstein cleverly applied the product rule:

THE PRODUCT RULE WHEN APPLIED TO SECOND DERIVATIVES AND TENSOR FIELDS:

$$\boxed{\frac{\partial}{\partial x_{\alpha}} \left(\frac{\partial \gamma'_{\mu\nu}}{\partial x_{\sigma}} \frac{\partial \gamma'_{\mu\nu}}{\partial x_{\alpha}} \right) = \frac{\partial \gamma'_{\mu\nu}}{\partial x_{\sigma}} \frac{\partial^2 \gamma'_{\mu\nu}}{\partial x_{\alpha}^2} + \frac{\partial \gamma'_{\mu\nu}}{\partial x_{\alpha}} \frac{\partial^2 \gamma'_{\mu\nu}}{\partial x_{\alpha} \partial x_{\sigma}}.} \tag{3.51}$$

and the ingenious transformation (3.33). By applying the product rule (3.51) to the second derivative term of the field equation (3.41) and recognizing how it can be expressed as a divergence, he leveraged a fundamental mathematical tool to simplify the expression and reveal the underlying structure of the equations. Einstein's transformation (3.33) isolates the trace part of the metric perturbation, simplifying the equations and making the physical interpretation more straightforward. By separating the trace part of the metric perturbation $\gamma'_{\mu\nu}$, the transformation allows the field equation (3.41) to be expressed to reveal the divergence of a tensor quan-

[21] With the substitution described in footnote 20 the stress-energy tensor on the right-hand side of equation (3.41) is transformed to the energy components of the gravitational field.

tity. Applying this transformation and the product rule (3.51) shows how the field equation can be reinterpreted as a conservation law for the energy-momentum tensor of the gravitational field and the stress-energy pseudo-tensor. The term involving the divergence:

$$\sum_{\alpha} \frac{\partial t_{\mu\nu}}{\partial x_{\alpha}}. \tag{3.52}$$

is recognized as representing the energy components of the gravitational field in the linearized theory. Thus, Einstein's use of the transformation (3.33) as an ingenious mathematical trick simplified the problem and revealed deeper physical insights. His approach exemplifies how clever mathematical techniques can illuminate complex physical theories.[22] In 1918, Einstein corrected Eq. (3.50). However, Einstein's recognition and subsequent correction of a mistake in the equation for the stress-energy pseudo-tensor later do not diminish the ingenuity of his path to derive it.

3.2.3 *Einstein Discovers Plane Gravitational Waves*

Einstein calculated the gravitational field of a mass point of mass m, resting at the origin of the coordinate system. In first-order approximation, the contravariant stress-energy tensor reduces to:

$$T^{\mu\nu} = \rho \frac{dx^{\mu}}{ds}\frac{dx^{\nu}}{ds}, \tag{3.56}$$

with the density of a material point.

---

[22] We rearrange Eq. (3.33):

$$\gamma_{\mu\nu} = \gamma'_{\mu\nu} - \frac{1}{2}\delta_{\mu\nu}\sum_{\alpha}\gamma_{\alpha\alpha} = \gamma'_{\mu\nu} - \frac{1}{2}\delta_{\mu\nu}\gamma'. \tag{3.53}$$

The general formula for the product rule, when applied to a divergence of partial derivatives, is given by Eq. (1.31) (see Sect. 1.2.2). Applying the product rule (3.51), we can express the second derivative term in Eq. (3.41) as a divergence. Applying the product rule (3.51) to the second derivative term in Eq. (3.41) simplifies the field equation:

$$\sum_{\mu\nu}\sum_{\alpha}\left(\frac{\partial}{\partial x_{\alpha}}\left(\frac{\partial \gamma'_{\mu\nu}}{\partial x_{\sigma}}\frac{\partial \gamma'_{\mu\nu}}{\partial x_{\alpha}}\right) - \frac{\partial \gamma'_{\mu\nu}}{\partial x_{\sigma}}\frac{\partial^2 \gamma'_{\mu\nu}}{\partial x_{\alpha}^2}\right), \tag{3.54}$$

because each term involving a second derivative can be split into a product of the first and their respective partial derivatives. Now considering Eq. (3.53), the expression (3.54) becomes:

$$\sum_{\alpha}\frac{\partial}{\partial x_{\alpha}}\left[\left(\sum_{\mu\nu}\frac{\partial \gamma'_{\mu\nu}}{\partial x_{\sigma}}\frac{\partial \gamma'_{\mu\nu}}{\partial x_{\alpha}} - \frac{1}{2}\delta_{\sigma\alpha}\sum_{\mu\nu\beta}\left(\frac{\partial \gamma'_{\mu\nu}}{\partial x_{\beta}}\right)^2\right)\right]. \tag{3.55}$$

The term in the brackets of equation (3.55) can be interpreted as the energy components of the gravitational field $t_{\mu\nu}$ in the weak-field limit (3.52).

Consider the gravitational field of a static material point of mass M, at rest at the origin of the coordinates system, and another material point at a great distance r from M (from the origin of the coordinates). It follows from $\gamma'_{\mu\nu}$ calculated in the manner of retarded potentials [Eqs. (3.40)] and (3.56) that in the first-order approximation, all $\gamma'_{\mu\nu}$ except γ_{44} vanish. The latter component is [E3-16]:

$$\boxed{\gamma'_{44} = \frac{\kappa M}{2\pi r},} \tag{3.57}$$

the time-time component of the de Sitter metric, Eq. (3.1) (see footnote 5 in Sect. 3.2.1). Einstein concluded that it also follows from the linearized field Eqs. (3.27) and (3.40) that gravitational fields always propagate with the velocity of light (velocity 1) [E3-16].

In the context of the linearized version of his field equations (3.27), Einstein was investigating plane gravitational waves, a simplified model of gravitational waves.[23] He wrote a plane wave-like solution to the linearized field equations (3.27) [E3-16][24]:

$$\boxed{\gamma'_{\mu\nu} = \alpha f(x_1 + ix_4) = \alpha f(x - ct).} \tag{3.58}$$

Applying the harmonic coordinate condition (3.7) to $\gamma'_{\mu\nu}$ we get[25]:

$$\frac{\partial \gamma'_{\mu 1}}{\partial x_1} + \frac{\partial \gamma'_{\mu 4}}{\partial x_4} = 0. \tag{3.59}$$

This equation implies that:

$$\alpha_{\mu 1}\frac{\partial f_{\mu 1}}{\partial x_1} + i\alpha_{\mu 4}\frac{\partial f_{\mu 4}}{\partial x_4} = 0. \tag{3.60}$$

---

[23] Plane waves are waves whose wavefronts (surfaces of constant phase) are infinite, parallel planes of constant amplitude perpendicular to the direction of wave propagation. Plane gravitational waves, especially when observed far from their source, are perturbations of the Minkowski flat metric [Eq. (3.4)]. Thus, far from the source of the gravitational waves, where the waves are weak, the spacetime can be approximated as flat with small deviations. These perturbations represent the gravitational waves. This approximation falls under Einstein's linearized field equation (3.27). These gravitational waves are analogous to plane waves in electromagnetism, representing waves propagating in a single direction (for instance, the x direction) with uniform wavefronts.

[24] α is a constant, $ix_4 = t$, the terms $(x_1 + ix_4)$ or $(x - ct)$ indicate the wave's propagation in the positive x-direction at the speed of light c, and f is a function of $(x_1 + ix_4)$.

[25] Einstein substituted Eq. (3.58) into the vacuum linearized version of the field equations (3.27) and imposed the harmonic condition [Eq. (3.7)], which yielded relations between the $\alpha_{\mu\nu}$. We substitute $\gamma'_{\mu\nu}$ from Eq. (3.58) into Eq. (3.7).

This leads to the following conditions [E3-16][26]:

1. $\alpha_{11} + i\alpha_{14} = 0$	has the symmetry properties of a longitudinal wave.
2. $\alpha_{12} + i\alpha_{24} = 0$	has the symmetry properties of a transverse wave.
3. $\alpha_{13} + i\alpha_{34} = 0$	has the symmetry properties of a longitudinal wave.
4. $\alpha_{14} + i\alpha_{44} = 0$	has the symmetry properties of a longitudinal wave.

(3.61)

Einstein found solutions that describe plane gravitational waves under the constraints of his linearized field equations.

He *predicted* three different types of plane gravitational waves:

1. LONGITUDINAL WAVES: In these waves, the oscillations of spacetime are in the direction of propagation.[27]

2. TRANSVERSE WAVES: In these waves, spacetime oscillations are perpendicular to the direction of wave propagation. These are similar to electromagnetic waves.[28]

3. A NEW TYPE OF WAVE: This refers to a wave with a different symmetry property; unlike longitudinal or transverse waves, these waves indicate a significant component, i.e., effect in the second and third spatial dimensions:

$$\boxed{\alpha_{22} \neq 0, \alpha_{23} \neq 0, \alpha_{33} \neq 0 \text{ correspond to a new type of symmetry.}} \tag{3.62}$$

When gravitational waves propagate, they carry energy with them. This energy is part of the gravitational field itself. Einstein was, therefore, keen on understanding how the above three types of waves could carry energy across spacetime (carry energy away from their source). Energy transport in gravitational waves is closely related to the energy components of the gravitational field (3.50).

---

[26] The enumerated conditions (3.61) imply certain relationships between the real and imaginary parts of the α components. There are initially 10 independent constants $\frac{4\times(4+1)}{2} = 10$. Only 6 of the 10 constants $\alpha_{\mu\nu}$ can be chosen freely because $\alpha_{\mu\nu}$ is symmetric ($\alpha_{12} = \alpha_{21}, \alpha_{13} = \alpha_{31}, \alpha_{14} = \alpha_{41}, \alpha_{23} = \alpha_{32}$). The harmonic coordinate condition (3.7) imposes additional restrictions on these components. The condition (3.7) relates components of $\alpha_{\mu\nu}$ to each other, as seen in Eqs. (3.59) and (3.61). These relations reduce the number of independent components that can be freely chosen. This corresponds to choosing the harmonic coordinate system that simplifies the problem.

[27] Equations 1 in (3.61) involves the 11 (spatial-spatial) and 14 components [1 (spatial), 4 (temporal)]. This reflects a coupling between spatial and temporal changes, which is characteristic of longitudinal waves where the disturbance is in the direction of propagation. Equation 4 in (3.61) involves spatial-temporal 14 and temporal-temporal 44 components. Including a temporal-temporal component alongside a spatial-temporal component suggests a change along the direction of wave propagation, hence longitudinal.

[28] Equations 2 and 3 in (3.61) involve a spatial-spatial component 12 and 13 and a spatial-temporal component 24 and 34. A purely spatial component 12 perpendicular to the wave's direction (assuming the wave travels in the 1-direction) indicates a transverse nature.

Einstein used these to compute the energy transported by the longitudinal and transverse waves. He considered the component $\frac{t_{41}}{i}$ from the equation for $t_{\mu\nu}$ (3.50) and obtained the energy pseudo-tensor [E3-16][29]:

$$\boxed{4\kappa t_{41} = \left(\frac{\partial \gamma'_{\alpha\beta}}{\partial x_4}\frac{\partial \gamma'_{\alpha\beta}}{\partial x_1}\right).} \tag{3.66}$$

Therefore:

$$\boxed{\frac{t_{41}}{i} = 0.} \tag{3.67}$$

This equation is the density of the energy current in the x direction.

Since the components of $\alpha_{\mu\nu}$ in the relations (3.62) $\neq 0$, Einstein found that only the third type of gravitational waves transport energy. In contrast, longitudinal and transverse gravitational waves do not transport any energy (see explanation in footnote 29).

[29] To express the component $\frac{t_{41}}{i}$ from the equation for $t_{\mu\nu}$, we first evaluate t_{41} using the Eq. (3.50). We substitute $\mu = 4$ and $\nu = 1$:

$$t_{41} = \frac{1}{4\kappa}\left(\frac{\partial \gamma'_{\alpha\beta}}{\partial x_4}\frac{\partial \gamma'_{\alpha\beta}}{\partial x_1} - \frac{1}{2}\delta_{41}\left(\frac{\partial \gamma'_{\alpha\beta}}{\partial x_\tau}\right)^2\right). \tag{3.63}$$

The second term disappears since $\delta_{41} = 0$. Thus, we have:

$$t_{41} = \frac{1}{4\kappa}\left(\frac{\partial \gamma'_{\alpha\beta}}{\partial x_4}\frac{\partial \gamma'_{\alpha\beta}}{\partial x_1}\right). \tag{3.64}$$

Now, dividing t_{41} [Eq. (3.63)] by i, we get:

$$\frac{t_{41}}{i} = \frac{1}{4\kappa i}\left(\frac{\partial \gamma'_{\alpha\beta}}{\partial x_4}\frac{\partial \gamma'_{\alpha\beta}}{\partial x_1}\right). \tag{3.65}$$

Simplifying and rearranging this equation gives Eq. (3.66). If one or more of the $\alpha_{\mu\nu}$ components are such that they lead to zero when substituted into the expression for $\frac{t_{41}}{i}$ in Eq. (3.65), then we obtain Eq. (3.67). Equation (29) is derived based on the behavior of the $\gamma_{\alpha\beta}$ components in the gravitational field, which are represented in terms of $\alpha_{\mu\nu}$ components in the plane wave solution (3.58). When the α components are used in the expression for $\frac{t_{41}}{i}$ [Eq. (29)] and if these components lead to the terms inside the parentheses summing to zero, then the whole expression for $\frac{t_{41}}{i}$ becomes zero [Eq. (3.67)].

Given the constraints (3.62), and $(\alpha_{\alpha\beta})^2 \neq 0$, we obtain from the stress-energy pseudo-tensor (3.50) and the component (3.67), the 1916 Eq. [E3-16][30]:

$$\frac{t_{41}}{i} = \frac{1}{4\kappa}(\alpha_{\alpha\beta})^2 f'^2 = \frac{1}{4\kappa}\left(\frac{\partial \gamma'_{\alpha\beta}}{\partial x_4}\right)^2, \quad \text{for } (\alpha, \beta) \in (2, 2), (2, 3), (3, 3), \ \ x_4 \equiv t. \tag{3.77}$$

---

[30] First, we determine $\gamma'_{\alpha\beta}$. Given Eq. (3.83), we can write:

$$\gamma'_{\alpha\beta} = \alpha_{\alpha\beta} f(x_1 + ix_4). \tag{3.68}$$

Now, we calculate the partial derivatives. Since f depends on $x_1 + ix_4$, then $\frac{\partial f}{\partial x_1} = f'$ and $\frac{\partial f}{\partial x_4} = if'$, where: $f' = \frac{df}{d(x_1+ix_4)}$. Thus:

$$\frac{\partial \gamma'_{\alpha\beta}}{\partial x_1} = \boxed{\alpha_{\alpha\beta} f'}, \quad \text{and} \quad \frac{\partial \gamma'_{\alpha\beta}}{\partial x_4} = \boxed{i\alpha_{\alpha\beta} f'}. \tag{3.69}$$

Substituting the boxed terms into the first term on the right-hand side of (3.50), we get:

$$\frac{\partial \gamma'_{\alpha\beta}}{\partial x_4}\frac{\partial \gamma'_{\alpha\beta}}{\partial x_1} = (i\alpha_{\alpha\beta} f')(\alpha_{\alpha\beta} f') = \boxed{i(\alpha_{\alpha\beta})^2 (f')^2.} \tag{3.70}$$

For the second term on the right-hand side of equation (3.50):

$$\boxed{\left(\sum_{\tau}\frac{\partial \gamma'_{\alpha\beta}}{\partial x_\tau}\right)^2} = \left(\frac{\partial \gamma'_{\alpha\beta}}{\partial x_1}\right)^2 + \left(\frac{\partial \gamma'_{\alpha\beta}}{\partial x_4}\right)^2. \quad \text{Thus:} \tag{3.71}$$

$$\left(\frac{\partial \gamma'_{\alpha\beta}}{\partial x_1}\right)^2 = \boxed{(\alpha_{\alpha\beta} f')^2,} \quad \left(\frac{\partial \gamma'_{\alpha\beta}}{\partial x_4}\right)^2 = (i\alpha_{\alpha\beta} f')^2 = \boxed{-(\alpha_{\alpha\beta} f')^2,} \quad \text{where } i^2 = -1. \tag{3.72}$$

Plugging the boxed terms of equation (3.72) into the boxed term of equation (3.71), we get:

$$\left(\sum_{\tau}\frac{\partial \gamma'_{\alpha\beta}}{\partial x_\tau}\right)^2 = (\alpha_{\alpha\beta} f')^2 - (\alpha_{\alpha\beta} f')^2 = 0. \tag{3.73}$$

Substituting Eq. (3.73) and the boxed term in Eq. (3.70) into Eq. (3.50) for t_{41}, we get:

$$t_{41} = \frac{1}{4\kappa}\left(i(\alpha_{\alpha\beta})^2 (f')^2 - 0\right) = \boxed{\frac{i}{4\kappa}(\alpha_{\alpha\beta})^2 (f')^2.} \tag{3.74}$$

Now we evaluate Eq. (3.67):

$$\frac{t_{41}}{i} = \frac{1}{4\kappa}(\alpha_{\alpha\beta})^2 (f')^2. \tag{3.75}$$

Under the constraints (3.61), $-i\alpha_{\alpha 4}$ for $\beta = 1, 2, 3, 4$ and $\alpha \in 1, 2, 3$. Thus $(\alpha_{\alpha\beta})^2 = 0$, leading to Eq. (3.79). Given the constraints Eq. (3.62), $\alpha_{\alpha\beta}$ for $(\alpha, \beta) \in (2, 2), (2, 3), (3, 3)$. Thus $(\alpha_{\alpha\beta})^2 \neq 0$, and:

$$\left(\sum_{\tau}\frac{\partial \gamma'_{\alpha\beta}}{\partial x_\tau}\right)^2 \neq 0. \tag{3.76}$$

Therefore plugging the left-hand side of equation (3.76) and the boxed term in Eq. (3.70) into Eq. (3.50) for t_{41}, we get Eq. (3.77).

Einstein concluded that the energy flux I_z in the z direction of the third type of gravitational waves is represented by [E3-16]:

$$I_z = \frac{1}{4x}\left[\left(\frac{\partial\gamma'_{22}}{\partial t}\right)^2 + 2\left(\frac{\partial\gamma'_{23}}{\partial t}\right)^2 + \left(\frac{\partial\gamma'_{33}}{\partial t}\right)^2\right]. \tag{3.78}$$

However, the right-hand side of equation (3.77) indicates that the equation is inconsistent because the difference between two identical quantities is zero.[31] This implies that this is the correct form [E3-16]:

$$\boxed{\frac{t_{41}}{i} = \frac{1}{4\kappa}(\alpha_{\alpha\beta})^2 f'^2 = 0.} \quad \text{for } \beta = 1, 2, 3, 4 \text{ and } \alpha \in 1, 2, 3. \tag{3.79}$$

Einstein was troubled by de Sitter's suggestion and metric, which seemed to have led to waves transporting no energy! [E3-16]. He decided to return to unimodular coordinates, the first-order approximation of $g_{\rho\sigma}$ valid for unimodular coordinate system (1.299) [E3-16]. Einstein even found a very interesting physical justification for this coordinate condition. He explained to de Sitter why he returned to the coordinate condition (1.299). Einstein introduced two coordinate systems: a coordinate system K, with respect to which (1.299) holds everywhere, and de Sitter's system denoted by K'. Einstein was now searching for plane gravitational waves. In de Sitter's system K', he found three types of gravitational waves (longitudinal, transverse, and an additional type), of which only the third type was connected to energy transportation. In Einstein's system K, by contrast, only this energy-carrying type was present. Einstein explained to de Sitter that this meant that the first two types of waves, the longitudinal and transverse waves, obtained for system K' did not exist in reality but were simulated by the coordinate system's wavelike motions with respect to an inertial system. Einstein concluded that unimodular coordinates exclude systems where we find waves with no energy and, therefore, waves that do not exist [CPAE8], Doc. 227.

Subsequently, Einstein added an addendum to his 1916 paper on gravitational waves. He explained that there was a simple way to clarify the strange result he obtained: gravitational waves that transport no energy. These were not real gravitational waves but rather apparent waves. They were fictitious waves reflecting vibrations of the reference system, i.e., de Sitter's system K' presented in Einstein's paper on gravitational waves, from which they had been derived. If, from the start, one selects a coordinate system in which (1.299) holds, then with this choice, only the third type of waves that transport energy exists. The coordinate condition (1.299) eliminates gravitational waves that do not transport energy, so this choice of coordinates eliminates these types of waves; in this sense, they are not real waves.

[31] See explanation in footnote 30. The expression on the right-hand side of equation (3.77) should logically evaluate to zero. In other words, the equation as written (3.76) in footnote 30 is contradictory because Eq. (3.73) is violated.

The unimodular coordinate condition (1.299) in the linearized theory is equivalent to:

THE UNIMODULAR COORDINATE CONDITION:

$$\boxed{\gamma = \sum_\alpha \gamma_{\alpha\alpha} = 0.} \tag{3.80}$$

The field equations (3.5) in a vacuum are written as

$$\sum_\alpha \frac{\partial^2 \gamma_{\mu\alpha}}{\partial x_\nu \partial x_\alpha} + \sum_\alpha \frac{\partial^2 \gamma_{\nu\alpha}}{\partial x_\mu \partial x_\alpha} - \sum_\alpha \frac{\partial^2 \gamma_{\mu\nu}}{\partial x_\alpha^2} - \frac{\partial^2}{\partial x_\mu \partial x_\nu}\left(\sum_\alpha \gamma_{\alpha\alpha}\right) = 0. \tag{3.81}$$

Upon applying the condition [Eq. (3.80)] to Eq. (3.81) we get:

$$\sum_\alpha \frac{\partial^2 \gamma_{\mu\alpha}}{\partial x_\nu \partial x_\alpha} + \sum_\alpha \frac{\partial^2 \gamma_{\nu\alpha}}{\partial x_\mu \partial x_\alpha} - \sum_\alpha \frac{\partial^2 \gamma_{\mu\nu}}{\partial x_\alpha^2} = 0. \tag{3.82}$$

Einstein substituted the solution representing a plane gravitational wave traveling along the x-axis:

$$\gamma_{\mu\nu} = \alpha f(x_1 + i x_4), \tag{3.83}$$

into Eq. (3.82). After plugging the solution (3.83) into Eq. (3.82), Einstein concluded that only the third type of wave [Eq. (3.62)] existed and carried energy (and it was thus real). In contrast, the other two types [Eq. (3.61)] were eliminated (and were thus fictitious).[32]

[32] First consider the following ansatz: the plane wave solution of the form (3.83) suggests that the perturbation propagates in a complex direction given by u:

$$u = x_1 + i x_4. \tag{3.84}$$

So Eq. (3.83) becomes: $\gamma_{\mu\nu} = \alpha f(u)$. Since $\gamma_{\mu\nu}$ depends on u, we first compute the partial derivatives in the field equation (3.82):

$$\frac{\partial^2 \gamma_{\mu\alpha}}{\partial x_\nu \partial x_\alpha} = \alpha \frac{\partial^2 f(u)}{\partial x_\nu \partial x_\alpha} = \alpha f''(u) \frac{\partial u}{\partial x_\nu} \frac{\partial u}{\partial x_\alpha}. \tag{3.85}$$

$$\frac{\partial^2 \gamma_{\mu\nu}}{\partial x_\alpha^2} = \alpha f''(u) \sum_\alpha \left(\frac{\partial u}{\partial x_\alpha}\right)^2. \tag{3.86}$$

Next, we substitute these derivatives back into the field equation (3.82). For the specific case of equation (3.84), we compute the first partial derivatives of u with respect to x_1, x_2, and x_j:

$$\frac{\partial u}{\partial x_1} = 1, \quad \frac{\partial u}{\partial x_4} = i, \quad \frac{\partial u}{\partial x_j} = 0 \text{ for } j \neq 1, 4. \tag{3.87}$$

Thus, in blissful ignorance, Einstein interpreted Eq. (3.77) as portraying the zestful energy of novel waves, unaware of the lurking misstep. Meanwhile, Eq. (3.79), revealing that longitudinal and transverse waves bore no energy, caused him consid-

---

Thus, u changes linearly with x_1 and i times linearly with x_4, and it does not change with respect to other x_j. This simplifies our expressions for the second partial derivatives (3.85) in the field equation (3.82). Substituting the partial derivatives of u into the second derivative:

$$\frac{\partial u}{\partial x_\nu} = (\delta_{\nu 1} + i\delta_{\nu 4}), \qquad \frac{\partial u}{\partial x_\alpha} = (\delta_{\alpha 1} + i\delta_{\alpha 4}). \tag{3.88}$$

Combining these into the second partial derivative expression (3.85):

$$\frac{\partial^2 \gamma_{\mu\alpha}}{\partial x_\nu \partial x_\alpha} = \alpha f''(u)(\delta_{\nu 1} + i\delta_{\nu 4})(\delta_{\alpha 1} + i\delta_{\alpha 4}). \tag{3.89}$$

$$\text{Now, we need to evaluate the sum: } \sum_\alpha (\delta_{\alpha 1} + i\delta_{\alpha 4}). \tag{3.90}$$

$$\begin{aligned} \text{For } \alpha = 1: &\quad \delta_{11} + i\delta_{14} = 1 + i\cdot 0 = 1, \\ \text{For } \alpha = 4: &\quad \delta_{41} + i\delta_{44} = 0 + i\cdot 1 = i, \\ \text{For } \alpha \neq 1 \text{ and } \alpha \neq 4\ (\alpha = 2, 3): &\quad \delta_{\alpha 1} + i\delta_{\alpha 4} = 0 + i\cdot 0 = 0. \end{aligned} \tag{3.91}$$

$$\text{Summing these contributions: } \sum_\alpha (\delta_{\alpha 1} + i\delta_{\alpha 4}) = 1 + i + 0 = 1 + i. \text{ Thus:} \tag{3.92}$$

$$\boxed{\sum_\alpha \frac{\partial^2 \gamma_{\mu\alpha}}{\partial x_\nu \partial x_\alpha} = \alpha f''(u)(1+i)(\delta_{\nu 1} + i\delta_{\nu 4}).} \quad \text{Similarly:} \tag{3.93}$$

$$\boxed{\sum_\alpha \frac{\partial^2 \gamma_{\nu\alpha}}{\partial x_\mu \partial x_\alpha} = \alpha f''(u)(1+i)(\delta_{\mu 1} + i\delta_{\mu 4}).} \tag{3.94}$$

For the last term (3.86) in the field equation (3.82), we have:

$$\sum_\alpha \delta_{\alpha 1} = 1, \ \sum_\alpha \delta_{\alpha 4} = 1, \ \sum_\alpha \delta_{\alpha 1}\delta_{\alpha 4} = 0. \text{ Thus:} \tag{3.95}$$

$$\sum_\alpha \frac{\partial^2 \gamma_{\mu\nu}}{\partial x_\alpha^2} = \alpha f''(u) \sum_\alpha (\delta_{\alpha 1} + i\delta_{\alpha 4})^2 = \delta_{\alpha 1} + i2i\delta_{\alpha 1}\delta_{\alpha 4} - 1 = 1 - 1 + 2i\delta_{14} = 0. \tag{3.96}$$

$$\boxed{\sum_\alpha \frac{\partial^2 \gamma_{\mu\nu}}{\partial x_\alpha^2} = \alpha f''(u)\cdot 0 = 0.} \tag{3.97}$$

Putting the boxed equations together in the field equation (3.82), we get:

$$\alpha_{\mu\alpha}(\delta_{\nu 1} + i\delta_{\nu 4}) = -\alpha_{\nu\alpha}(\delta_{\mu 1} + i\delta_{\mu 4}). \tag{3.98}$$

Given Eq. (3.83), the constraints (3.61) and (3.62), are given by:

$$\alpha_{11} = -i\alpha_{14}, \ \alpha_{12} = -i\alpha_{24}, \ \alpha_{13} = -i\alpha_{34}, \ \alpha_{14} = -i\alpha_{44}, \ \alpha_{22} \neq 0, \ \alpha_{23} \neq 0, \ \alpha_{33} \neq 0. \tag{3.99}$$

We substitute them into Eq. (3.98) and check if it holds. For $\mu = 1$ and $\nu = 4$, we have: $\alpha_{1\alpha}(\delta_{41} + i\delta_{44}) = -\alpha_{4\alpha}(\delta_{11} + i\delta_{14})$. Substituting the δs in this equation gives: $\alpha_{1\alpha}(0 + i) = -\alpha_{4\alpha}(1 + 0i)$. This simplifies to the condition $i\alpha_{1\alpha} = -\alpha_{4\alpha}$, which is satisfied given Eq. (3.99).

erable consternation. This revelation teased at the existence of plane gravitational waves, quietly carrying their hidden payloads of energy, much to Einstein's bemusement. Einstein found a solution to his dilemma through the trusty old unimodular coordinates. He yielded to the unimodular coordinate condition (1.299) and contended that it possessed a profound underlying physical rationale. Succumbing to this condition was a combination of stubbornness and oversight. It was probably the second time Einstein found himself trapped in the pit (the first time was the hole argument and the *Entwurf* theory). Einstein concluded that even though it was not preferable to restrict the choice of coordinates for the calculation of the first-order approximation, his results (3.77) showed that the choice of coordinates under the unimodular restriction to a coordinate system in which (1.299) was valid, was physically justified [E3-16]. Einstein got entangled with the coordinate condition (1.299) and the third type of gravitational wave he had found in 1916.

In the summer of 1917, Tullio Levi-Civita criticized Einstein's use of the t_σ^α pseudo-tensor to represent the energy components of the gravitational field. His concern was rooted in pseudo-tensors and their behavior under coordinate transformations, which differ from tensors. Levi-Civita also had reservations about Einstein's use of unimodular coordinates, suggesting that Einstein's results, which in 1916 he found unusual, were a consequence of using the t_σ^α pseudo-tensor. In response, Einstein acknowledged that the components T_σ^α of the stress-energy tensor in his energy-momentum conservation Eq. (1.258) form a proper tensor. Still, the t_σ^α representing the energy components of the gravitational field do not. Einstein argued that within the framework of his general theory of relativity, it was impossible to use an expression other than the pseudo-tensor he presented for the energy components of the gravitational field. The combined energy-momentum tensor and the gravitational field pseudo-tensor system satisfy an overall conservation law. If a true tensor represented the gravitational field energy, then the expression for energy conservation would not vanish as it should. Using a pseudo-tensor allows for the divergence of the combined energy-momentum (matter plus gravitational field) (1.258) to vanish, thus preserving the principle of energy conservation within the framework of general relativity. Therefore, Einstein reasoned that it was not possible, within his general theory of relativity, to represent the energy of the gravitational field as a true tensor. Einstein's use of a pseudo-tensor to represent gravitational field energy was unconventional, seemingly contradicting the general covariance principle of relativity. However, Einstein saw the pseudo-tensor as a necessary and appropriate tool within his theory, arguing that it was a valid and useful representation in general relativity despite not being a true tensor [E1-18].

Recall from Sect. 2.3.3, Chap. 2 that in September 1917, Gunnar Nordström calculated the t_4^4 outside a spherically symmetric body. He found that the 44-component of the t_σ^α stress-energy pseudo tensor $t_4^4 = 0$ vanishes (see detailed explanation in Sect. 2.3.4). This meant that in the coordinate system (1.299), the gravitational field and gravitational wave of a mass point carry no energy, and this did not agree with Einstein's conclusion in the Addendum of his 1916 gravitational waves paper.

With unyielding determination, Einstein endeavored to convince Nordström of the validity of equation (3.50) within an unimodular coordinate system. Yet, by October

1917, Nordström had not relinquished his resolve. He used the components of the stress-energy pseudo tensor of the gravitational field (1.232) from Einstein's 1914 *Entwurf* review paper [E2-14] and the unimodular general theory of relativity [E1-15, E1-16]. He again demonstrated to Einstein that not only did the energy density t_4^4 vanish, but all components of Einstein's t_σ^α vanish in an unimodular coordinate system (1.299): $t_1^1 = t_2^2 = t_3^3 = t_4^4 = 0$. Nordström conferred with Adriaan Fokker on the matter, and together, they surmised that the issue might reside within the unimodular coordinate system (1.299) [CPAE8], Doc. 393. Nordström's findings received validation from Erwin Schrödinger, who, in 1918, demonstrated that under the choice of the unimodular coordinate system (1.299), not just a single component, but all κt_σ^ν energy components of the gravitational field vanish [E2-18].

It is worth noting that on June 22, 1915, Hans Reissner had also communicated to Einstein about the same issue (that t_4^4 vanishes in an equilibrium scenario) [CPAE8], Doc. 90, one that Nordström would later elaborate upon and which would prompt a response from Einstein in his paper on gravitational waves. However, Reissner's mention of this problem was brief. It did not explicitly reference a specific issue in Einstein's gravitational waves paper (submitted later on June 29, 1916 [E3-16]) that characterized Nordström's correspondence. The timing of Reissner's correspondence was inopportune, as it reached Einstein during the intense focus on his *Entwurf* theory, preceding his transition to the general theory of relativity in 1915.

However, Einstein eventually accepted Nordström's and Schrödinger's results and, in 1918, presented a cleaned-up version of his gravitational waves paper to the Prussian Academy. In his 1918 paper, "On Gravitational Waves," Einstein dropped the result that the unimodular coordinate system (1.299) was privileged because it eliminated gravitational waves that did not transport energy. He mentioned Nordström and Schrödinger's results and explained that Nordström had already pointed out to him that if we choose the unimodular coordinate system (1.299), then all the energy components of the gravitational field vanish when one calculates them to second-order utilizing t_σ^ν. In his 1918 paper on gravitational waves, Einstein revised the derivation (described in Sect. 3.2.2), leading to a corrected version of the equation describing the energy components of the gravitational field $t_{\mu\sigma}$ [E1-18].[33]

[33] Using Eq. (3.32) and after rearrangement (a similar procedure to the one outlined in the Sect. 3.2.2), one arrives at:

$$\sum_\sigma \frac{\partial t_{\mu\sigma}}{\partial x_\sigma} = \sum_\sigma \frac{\partial}{\partial x_\sigma}\left[\frac{1}{4\kappa}\left(\sum_{\alpha\beta}\left(\frac{\partial \gamma'_{\alpha\beta}}{\partial x_\mu}\frac{\partial \gamma'_{\alpha\beta}}{\partial x_\sigma}\right) - \frac{1}{2}\frac{\partial \gamma'}{\partial x_\mu}\frac{\partial \gamma'}{\partial x_\sigma}\right]\right.$$
$$\left. - \frac{1}{8\kappa}\delta_{\mu\sigma}\left(\sum_{\alpha\beta\lambda}\left(\frac{\partial \gamma'_{\alpha\beta}}{\partial x_\lambda}\right)^2 - \frac{1}{2}\sum_\lambda\left(\frac{\partial \gamma'}{\partial x_\lambda}\right)^2\right)\right]. \tag{3.100}$$

γ' represents the trace of $\gamma'_{\mu\nu}$. Einstein employed the same methodology in his 1916 paper to derive the $t_{\mu\sigma}$ components. To derive $t_{\mu\sigma}$, one focuses on expressing $t_{\mu\sigma}$ in terms of $\gamma'_{\alpha\beta}$ and its derivatives. From Eq. (3.47), $T_{\mu\nu}$ is expressed in terms of $\frac{\partial t_{\mu\sigma}}{\partial x_\sigma}$, and $T_{\mu\nu}$ in Eq. (3.27) is replaced with the expression obtained from Eq. (3.47).

Conservation of energy-momentum (1.258) is satisfied if [E1-18][34]:

$$\boxed{\begin{aligned}4\kappa t_{\mu\sigma} = &\left(\sum_{\alpha\beta}\left(\frac{\partial\gamma'_{\alpha\beta}}{\partial x_\mu}\frac{\partial\gamma'_{\alpha\beta}}{\partial x_\sigma}\right) - \frac{1}{2}\frac{\partial\gamma'}{\partial x_\mu}\frac{\partial\gamma'}{\partial x_\sigma}\right)\\ -\frac{1}{2}\delta_{\mu\sigma}&\left(\sum_{\alpha\beta\lambda}\left(\frac{\partial\gamma'_{\alpha\beta}}{\partial x_\lambda}\right)^2 - \frac{1}{2}\sum_{\lambda}\left(\frac{\partial\gamma'}{\partial x_\lambda}\right)^2\right).\end{aligned}} \tag{3.101}$$

Compare this expression to Eq. (3.50). Einstein then did the same derivation as the one he had conducted in the 1916 paper (see Sect. 3.2.3). He wrote Eq. (3.58), which represents gravitational waves.

TWO TYPES OF WAVES:

First, Einstein set himself the task of showing that the first two types of waves do not carry energy. Einstein derived conditions (3.61), the specific relations between the $\alpha_{\mu\nu}$, from the linearized field equations (3.27) by imposing the harmonic coordinate condition (3.7) [E1-18]. Subsequently, he set himself to find the expression for $\frac{t_{41}}{i}$. He found t_{41} by setting $\mu = 4$ and $\sigma = 1$ in Eq. (3.101) [$\delta_{41} = 0$]:

$$\frac{t_{41}}{i} = i\left(\frac{1}{4\kappa}\sum_{\alpha\beta}\frac{\partial\gamma'_{\alpha\beta}}{\partial x_1}\frac{\partial\gamma'_{\alpha\beta}}{\partial x_4} - \frac{1}{8\kappa}\frac{\partial\gamma'}{\partial x_1}\frac{\partial\gamma'}{\partial x_4}\right). \tag{3.102}$$

We can simplify and rearrange this equation:

$$\boxed{4\kappa t_{41} = \left(\frac{\partial\gamma'_{\alpha\beta}}{\partial x_1}\frac{\partial\gamma'_{\alpha\beta}}{\partial x_4} - \frac{1}{2}\frac{\partial\gamma'}{\partial x_1}\frac{\partial\gamma'}{\partial x_4}\right).} \tag{3.103}$$

Compare this equation to Eq. (3.66) derived (in Sect. 3.2.3) from Einstein's 1916 energy pseudo tensor (3.50). As seen from this comparison between the two equations, the new expression derived from Einstein's 1918 pseudo tensor (3.101) incorporates an extra term absent in Eq. (3.66). Given the conditions (3.61) and the expression for $\gamma'_{\mu\nu}$ (3.58), the result of equation (3.102) simplifies to zero. This outcome indicates that the component t_{41} for the pseudo tensor (3.101) is zero when divided by i for the constants (3.61). This suggests that no energy is transported in the direction corresponding to this component in these wave configurations. However, Eq. (3.66)

[34] Compare the Eq. (3.101) to its equivalent from Einstein's 1916 paper [E3-16], Eq. (3.50).

yielded the same result, indicating that practically no changes have occurred. This conclusion is not unexpected, considering the given Eq. (3.102) and the $\alpha_{\mu\nu}$ defined by (3.61).[35]

Einstein did not stop at this point after proving this. If he had stopped here, he would not have shown that although the first two types of waves do not carry energy, we can redefine the third type of waves as transverse waves that carry energy. Specifically, Einstein's solution to address the 1916 discrepancy was proposing that the components α_{22}, α_{33}, and α_{23} are transverse gravitational waves.

Recall that for $\mu = 4$ and $\nu = 1$, $\delta_{41} = 0$. Thus, Eq. (3.101) reduces to the first two terms on the right-hand side:

$$\boxed{4\kappa t_{41} = if'^2\left(\sum_{\alpha\beta}\alpha_{\alpha\beta}^2 - \frac{1}{2}\left(\sum_{\alpha\beta}\alpha_{\alpha\beta}\right)^2\right).} \tag{3.107}$$

The correct expression for the right-hand side is given by Eq. (3.107). Einstein demonstrated that using Eq. (3.107) for the components of the types of waves (the components α_{22}, α_{33}, and α_{23}), the component $\frac{t_{41}}{i}$ of equation (3.101), which represents the density of the energy current in this direction, is equal to [E1-18][36]:

[35] 1. *Equation* (3.103): Since γ' is the trace term and is zero, the second term on the right-hand side in Eq. (3.103) vanishes. Consider the example:

$$\alpha_{11} = -i\alpha_{14}, \quad \text{which implies} \quad \gamma'_{11} = -i\gamma'_{14}. \text{ Thus:} \tag{3.104}$$

$$4\kappa t_{41} = \frac{\partial\gamma'_{\alpha\beta}}{\partial x_1}\frac{\partial\gamma'_{\alpha\beta}}{\partial x_4} = \left(\frac{\partial\gamma'_{11}}{\partial x_1}\frac{\partial\gamma'_{11}}{\partial x_4}\right) = \left(\frac{\partial(-i\gamma'_{14})}{\partial x_1}\frac{\partial(-i\gamma'_{14})}{\partial x_4}\right) = \boxed{\frac{\partial\gamma'_{14}}{\partial x_1}\frac{\partial\gamma'_{14}}{\partial x_4}}. \tag{3.105}$$

2. *Equation* (3.66) : Recall from footnote 30 that applying the constraints (3.104) similarly, leads to:

$$4\kappa t_{41} = \frac{\partial\gamma'_{\alpha\beta}}{\partial x_4}\frac{\partial\gamma'_{\alpha\beta}}{\partial x_1} = \left(\frac{\partial(-i\gamma'_{14})}{\partial x_4}\frac{\partial(-i\gamma'_{14})}{\partial x_1}\right) = \boxed{\frac{\partial\gamma'_{14}}{\partial x_4}\frac{\partial\gamma'_{14}}{\partial x_1}}. \tag{3.106}$$

Thus, both Eqs. (3.66) and (3.103) give the same result. Equation (3.103): the term involving γ' vanishes because the trace γ' is zero, as it does not depend on α. Equation (3.66): the term involving $\delta_{41} = 0$, simplifying the expression to the same form as Eq. (3.103) under the same constraints.

[36] We substitute the constraints (3.62) into Eq. (3.107), and get:

$$\sum_{\alpha\beta}\alpha_{\alpha\beta}^2 = \alpha_{22}^2 + 2\alpha_{23}^2 + \alpha_{33}^2, \quad \left(\sum_{\alpha\beta}\alpha_{\alpha\beta}\right)^2 = (\alpha_{22}+\alpha_{33})^2. \quad \text{Thus:} \tag{3.108}$$

$$4\kappa t_{41} = if'^2\left(\alpha_{22}^2 + 2\alpha_{23}^2 + \alpha_{33}^2 - \frac{1}{2}(\alpha_{22}+\alpha_{33})^2\right). \tag{3.109}$$

$$4\kappa t_{41} = if'^2\left(\alpha_{22}^2 + 2\alpha_{23}^2 + \alpha_{33}^2 - \frac{1}{2}(\alpha_{22}^2 + 2\alpha_{22}\alpha_{33} + \alpha_{33}^2)\right). \tag{3.110}$$

$$4\kappa t_{41} = if'^2\left(\alpha_{22}^2 + 2\alpha_{23}^2 + \alpha_{33}^2 - \frac{1}{2}\alpha_{22}^2 - \alpha_{22}\alpha_{33} - \frac{1}{2}\alpha_{33}^2\right). \tag{3.111}$$

$$\boxed{\frac{t_{41}}{i}=\frac{1}{2\kappa}f'^2\left[\left(\frac{\alpha_{22}-\alpha_{33}}{2}\right)^2+\alpha_{23}^2)\right].} \tag{3.115}$$

Einstein then related α_{22}, α_{33}, and α_{23} to transverse gravitational waves; he reinterpreted the Eq. (3.62) as showing that two types of gravitational waves carry energy:

$$\left(\frac{\gamma'_{22}-\gamma'_{33}}{2}\right)+\gamma'_{23} \text{ or } \left(\frac{\gamma_{22}-\gamma_{33}}{2}\right)+\gamma_{23}. \tag{3.116}$$

The two types of waves, indicated by these components, differ in their orientation but are similar—they both represent distortions in the plane perpendicular to the direction of propagation. Hence, these components represent waves that are transverse to the direction of propagation (along the x-axis) and involve deformations in the plane perpendicular to the wave's travel direction (the $x_2 - x_3$ plane). The expression (3.115) represents the energy current density associated with these waves.

The flow of energy is in the direction of wave propagation. Einstein discovered that only plane transverse waves can carry energy. This energy would manifest as perturbations in spacetime.

In his 1918 paper on gravitational waves, Einstein reproduced de Sitter's 1916 solution [Eq. (3.2)]. He wrote that the speed of light is [E1-18] defined by Eq. (1.362). Therefore Eq. (3.2) becomes:

$$\boxed{\left(1-\frac{\kappa M}{4\pi r}\right)dl^2-\left(1+\frac{\kappa M}{4\pi r}\right)dt^2=0.} \tag{3.117}$$

We obtain from this equation [or Eq. (1.391), see Sect. 1.7.1], the velocity of light [E1-18]:

$$\boxed{c=\sqrt{\frac{dl^2}{dt^2}}=1-\frac{\kappa M}{4\pi r}.} \tag{3.118}$$

In the 1916 review paper, "The Foundation of the General Theory of Relativity," in his derivation of the deflection of light rays, Einstein utilized the metric (1.300) similarly to how he employed the metric (3.2) or (3.117) in the 1918 paper on

---

$$4\kappa t_{41}=if'^2\left(\frac{1}{2}\alpha_{22}^2+2\alpha_{23}^2+\frac{1}{2}\alpha_{33}^2-\alpha_{22}\alpha_{33}\right)=if'^2\left(\frac{1}{2}(\alpha_{22}^2+\alpha_{33}^2-2\alpha_{22}\alpha_{33})+2\alpha_{23}^2\right) \tag{3.112}$$

$$4\kappa t_{41}=if'^2\left(\frac{1}{2}(\alpha_{22}-\alpha_{33})^2+2\alpha_{23}^2\right). \tag{3.113}$$ Dividing both sides by 4 and rearranging:

$$\kappa t_{41}=\frac{i}{2}f'^2\left(\frac{(\alpha_{22}-\alpha_{33})^2}{4}+\alpha_{23}^2\right), \tag{3.114}$$ leading to Eq. (3.115).

gravitational waves. He calculated the change in the velocity of light in curved spacetime, arriving at an equation similar to (1.389) (see Sect. 1.7.1): [E1-16]:

$$\boxed{\sqrt{\left(\frac{dx^1}{dx^4}\right)^2 + \left(\frac{dx^2}{dx^4}\right)^2 + \left(\frac{dx^3}{dx^4}\right)^2} \equiv \sqrt{\frac{dl^2}{dt^2}},} \tag{3.119}$$

using the metric components from Eq. (1.300).

Thus, we may conclude that during his calculations of gravitational waves, Einstein drew inspiration from his earlier work on bending light rays by gravitational fields. In this context, Einstein took the derivation of the velocity of light from the bending of light rays and applied it to the derivation of the velocity of gravitational waves.

3.3 Rotational Effects in Gravitational Theory

3.3.1 Thirring's Correspondence with Einstein That Solved a Puzzle

Hans Thirring and Josef Lense's work in 1918 predated the discovery of the Kerr solution for several decades (Roy Kerr introduced the Kerr solution in 1963 [Ker-63], see discussion in Sect. 3.3.3). When Thirring and Lense conducted their research, the theoretical tools and solutions available for tackling problems in general relativity were limited compared to what we have today. The Schwarzschild solution could not address the complexities introduced by rotation. As I show in Sect. 3.3.2, for their analysis, Thirring and Lense relied on the approximation method introduced by Einstein [see Eq. (3.3) in Sect. 3.2.1]. This approach was more suited to deal with the effects of rotation in a weak gravitational field. Einstein's approximation methods from his Perihelion of Mercury paper [E3-15] and gravitational waves paper [E3-16] allowed for a linearized treatment of the field equations of general relativity, deriving results like the frame-dragging effect without needing a full solution for a rotating body.

I will begin discussing the frame-dragging effect with the exchange of letters between Einstein and Thirring. In December 1917, Thirring wrote a letter to Einstein in which he raised two paradoxes related to applying general relativity to rotating systems. Let us focus on the first paradox. In the first paradox, Thirring discusses how the rotation of a hollow sphere could influence the centrifugal and Coriolis forces experienced by an observer inside the sphere.[37] This is related to the frame-dragging effect Thirring presented later in the paper with Lense, where the rotation of a mass (like a hollow sphere) drags the spacetime around it.[38]

In his letter to Einstein, Thirring focuses on a point mass rotating in the $X - Z$ plane inside the hollow sphere and writes an expression for its motion [CPAE8], Doc. 401:

$$\frac{d^2x}{ds^2} = (C\omega^2 x + C'\omega^2 x)\frac{\partial x_4}{\partial s} + [...\dot{x}, \dot{y}, \dot{z}], \tag{3.120}$$

where, $\dot{x} = \dot{z} = 0$, $\dot{y} = \omega x$. Suppose we substitute this into the expression (3.120). In that case, the equation becomes[39]:

$$\frac{d^2x}{ds^2} = (C\omega^2 x + C'\omega\dot{y}). \tag{3.121}$$

Simplifying Eq. (3.120), we get the relation:

$$\frac{d^2x}{ds^2} = \omega^2 x(C + C'). \tag{3.122}$$

---

[37] There is no gravitational pull experienced by objects inside the cavity of the sphere. The sensation of gravity inside a rotating hollow sphere is entirely due to the artificial effects of rotation, not a gravitational field in the traditional sense. These effects include the following: 1. The centrifugal force, a fictitious force that appears to act on all mass points in a rotating system. It acts outward from the center of rotation and is felt by an object moving in a circle when viewed from within the rotating reference frame. 2. The Coriolis force is another fictitious force that appears in rotating systems. It acts on objects that are moving relative to the rotating reference frame. Still, in a direction perpendicular to the object's motion and the rotation axis. Inside a rotating sphere, this force would cause moving objects not to travel in straight lines but to curve, depending on their direction.

[38] The mechanical system of a rotating, hollow sphere can be seen as an analog for the gravitational frame-dragging observed in the frame-dragging effect. Both involve rotating frames of reference, where the rotation affects the motion of objects within those frames. In the mechanical case, the effects (centrifugal and Coriolis forces) are due to the non-inertial nature of the rotating reference frame. In the gravitational case, the effects are due to the curvature of space-time caused by a massive, rotating body.

[39] Equation (3.121) is part of the geodesic equation describing the acceleration—second derivative of position with respect to the proper time ds—of a point mass in a rotating frame. C and C' are constants, and ω is the angular velocity. $C\omega^2 x$ represents a term related to the centrifugal force (as it is proportional to the square of the angular velocity ω and the distance x), and $C'\omega\dot{y}$ represents a term that is associated with the Coriolis force (as it is proportional to the angular velocity ω and the velocity component $\dot{y}$).

Thirring notes that to balance the two forces, Centrifugal and Coriolis (so that the co-rotating object feels no force), the factors in these forces must be numerically equal, implying $C = -C'$ in Eq. (3.122), which leads to[40]:

$$\frac{d^2x}{ds^2} = 0. \tag{3.124}$$

Thus, for an object rotating with the same angular velocity ω around the Z-axis, no force should be experienced (since everything appears stationary from its perspective). However, this frame's expressions for centrifugal and Coriolis forces lead to a discrepancy. In conventional mechanics, for the expressions $m\omega^2 r$ (centrifugal force) and $2m\omega v$ (Coriolis force), the Newtonian equivalent of the Eq. (3.121) in a rotating frame is expressed as follows[41]:

$$\frac{d^2x}{dt^2} = (Cm\omega^2 r + C'2m\omega v). \tag{3.125}$$

From Eq. (3.125) we get[42]:

$$2C = -C', \tag{3.126}$$

rather than $C = -C$, leading to a paradox [CPAE8], Doc. 401.

---

[40] To have an object feel no net force, we would need it to be stationary with respect to the rotating frame, meaning it rotates at the same angular velocity ω as the sphere and does not move relative to the sphere's interior surface. To express the balance mathematically, F_c (3.125) must be equal and opposite to F_C (3.126) (see footnote 41) for the object to remain stationary relative to the sphere's surface. If the object does not move relative to the sphere, no net force is acting on it. Thus, Thirring matches the motion of the reference frame, and he equates the two equations (3.122) and (3.124):

$$\frac{d^2x}{ds^2} = \omega^2 x(C + C') = 0. \tag{3.123}$$

This leads to $C = -C'$.

[41] *Centrifugal force:* If we are inside a rotating sphere, we feel pushed against the sphere's wall. The formula for centrifugal force is given by:

$$F_c = m\omega^2 r. \tag{3.125}$$

m is the object's mass, and r is the radius from the center of rotation to the object. This force simulates gravity along the sphere's interior surface, allowing objects and people to "stick" to the inside surface.

Coriolis force: It is responsible for the curving paths of objects moving within the rotating system. The formula for the Coriolis force is:

$$F_C = -2m(\omega \times v). \tag{3.126}$$

v is the velocity of the moving object relative to the rotating system.

[42] Given $Cm\omega^2 r + C'2m\omega v = 0,\ \ Cm\omega^2 r = -C'2m\omega v.\ \Rightarrow C\omega r = -C'2v$. Considering $\omega r = v$, we would get Eq. (3.126).

Einstein replied a few days later. In his reply to Thirring's first paradox, he corrects Thirring's expectation about the motion equation. He indicates that the correct condition for force-free circular motion in a rotating frame is not Eq. (3.124). Instead of equation (3.124) Einstein writes [CPAE8], Doc. 405[43]:

$$\frac{d^2x}{ds^2} = -\omega'^2 r = -\omega'^2 x. \tag{3.127}$$

ω' is the rotational velocity of the point mass; this condition reflects the centrifugal acceleration experienced in the rotating frame.

Einstein suggests a relationship between the point mass's rotational velocity ω' and the system's angular velocity ω with [CPAE8], Doc. 405[44]:

EINSTEIN'S CONDITION:

$$\boxed{\omega' = C\omega^2.} \tag{3.128}$$

Using this relation, Eq. (3.120) yields Eq. (3.126).[45] Hence, Eq. (3.126) naturally arises from the conditions of rotational motion and is consistent with the general covariance principle.

According to Thirring's equation (3.124), when centrifugal and Coriolis forces are balanced, this implies that there is no net acceleration experienced by the object in the rotating frame, which is consistent with the object being in a steady state or equilibrium within that frame. The term in equation (3.122), $\omega^2 x(C + C')$, consolidates the effects of both centrifugal and Coriolis forces into a single expression. If these forces are balanced against each other, the expectation is that this expression yields zero acceleration as the forces cancel out. However, in general relativity, interpreting these forces and motions requires considering the effects of spacetime curvature. Thirring's equations suggest a model where centrifugal-like $C\omega^2 x$ and Coriolis-like $C'\omega\dot{y}$ effects in a rotating frame can be described in a way that mirrors classical physics but is framed within the spacetime curvature. The balance of forces,

---

[43] The term $-\omega'^2 x$ suggests a force analogous to a centrifugal force in a rotating reference frame.

[44] This suggests that the effective angular velocity ω' experienced by a point mass is proportional to the square of the system's angular velocity ω, scaled by the constant C.

[45] We can ascertain this by examining the following derivation. By equating Eq. (3.127) to (3.122) we obtain instead of equation (3.123) the following equation:

$$\frac{d^2x}{ds^2} = \omega^2 x(C + C') = -\omega'^2 x. \tag{3.131}$$

Simplifying the right-hand side of this equation to:

$$\omega^2(C + C') = -\omega'^2, \tag{3.132}$$

and substituting Eq. (3.126) into equation (3.132) leads to Einstein's condition, which is expressed as Eq. (3.128).

leading to zero net acceleration, mirrors the classical notion of an object in equilibrium within a rotating system. This blending of classical and relativistic mechanics principles led to the paradox.

Einstein's condition (3.128) bridges the gap between classical expectations and relativistic realities by introducing a scaling factor (C) that modifies the angular velocity. Thirring (and Lense) showed after exchanging letters with Einstein that Einstein's condition (3.128) accounts for additional relativistic effects, such as frame-dragging, by adjusting the perceived angular velocity ω' of the object. This adjustment ensures the calculations considering both centrifugal and Coriolis forces can remain valid in a relativistic framework, where spacetime curvature must be considered.

3.3.2 Frame Dragging in Linearized Relativity

In December 1917, Thirring immediately submitted a paper, "On the effect of rotating distant masses in Einstein's theory of gravitation," in which he expanded on these ideas [Thi]. In this paper, he describes a thought experiment involving a rotating hollow sphere and its gravitational field. Thirring imagines a large, uniformly rotating, hollow sphere with a radius comparable to the orbit of Neptune. This sphere rotates with an angular velocity ω relative to the fixed stars. Thirring is interested in the gravitational field inside this hollow sphere. He suggests that this field could represent the gravitational field at the center of the Sun if such a rotating sphere replaced the Sun and all of the planets. Thirring then imagines an observer at the center of this rotating hollow sphere. This observer is on a world-body (in modern terms, we would say it is a small space station or satellite) that has a negligible gravitational field of its own and is rotating around the same axis as the hollow sphere but with a different angular velocity ω'. The observer in this rotating frame would experience centrifugal and Coriolis forces. The centrifugal force would be due to the rotation of the observer's platform ω'. In contrast, the Coriolis force arises from the combination of the observer's rotation and the rotation of the hollow sphere. Thirring is interested in how the rotation of the hollow sphere influences the centrifugal force experienced due to the observer's rotation. This setup allows for studying how a massive rotating object (the hollow sphere) can affect the local inertial frames and, thus, the forces experienced within those frames. This thought experiment demonstrates how the rotation of a massive object can drag spacetime around it, an effect that would be observable through the modified behavior of objects within that spacetime.

Thirring introduced the *dragging coefficient* in his paper. It measures how much the hollow sphere's rotation (representing a large rotating mass) influences the local frame of reference, particularly affecting the Coriolis force experienced within that frame. The Coriolis force acts perpendicular to the direction of motion and the axis of rotation. Let us revisit the observer inside the rotating hollow sphere on a world-body rotating with an angular velocity ω'. Thirring found that by setting the observer's angular velocity ω' equal to a specific value defined in terms of α [Eq. (1.164)] and a term related to the rotation of the hollow sphere, the Coriolis force acting on objects in the observer's frame could vanish. Thirring refers to this specific value as the *dragging coefficient*. The dragging coefficient is a factor that quantifies the influence of the hollow sphere's rotation on the local inertial frame:

$$\omega' = \omega \times \left(\text{DRAGGING COEFFICIENT}\right). \tag{3.133}$$

When the observer's rotation ω' matches this coefficient times the hollow sphere's rotation ω, the rotational effects of the hollow sphere are exactly countered, eliminating the Coriolis force in the observer's frame.

Thirring's work on the rotational effects and introducing the *dragging coefficient* indicate that Einstein's resolution to his paradox influenced him: The constant C in Einstein's condition (3.128) is equal to the dragging coefficient. As a result, Einstein's condition (3.128) is equivalent to equation (3.133). Einstein's influence on Thirring extended beyond just this aspect. Thirring extensively relied on Einstein's earlier works for inspiration and guidance. The Eqs. (3.4), (1.43), (3.33), and (3.40) from Einstein's 1916 gravitational waves paper [E3-16] are fundamental to the mathematical framework used in Thirring's derivation of effects of rotating masses on gravitational fields [Thi].

In 1918, Thirring published a joint paper with Lense. The joint paper with Lense expands on Thirring's ideas, and in this paper, the focus shifts from the specific case of a hollow sphere to the more general case of a rotating solid sphere. This change allows for a broader exploration of the rotational effects on gravitational fields and the implications for celestial bodies like planets and their moons. Thirring and Lense explore the perturbation effects on planetary and satellite orbits due to the rotation of the central body. They include a more comprehensive analysis of the rotational effects, including both centrifugal and Coriolis forces.

Thirring and Lense start with Einstein's equations (3.4) and (1.43): the Minkowski metric $\delta_{\mu\nu}$, which describes flat spacetime, and the perturbation term, $\gamma_{\mu\nu}$. Both gravitational waves and the Lense Thirring effect (to the first approximation or in linearized general relativity) use a perturbative approach to Einstein's field equations, where the spacetime metric is described as a small perturbation to the flat Minkowski spacetime.[46] Using Einstein's equation (3.33) from his gravitational waves paper, the perturbation $\gamma_{\mu\nu}$ is transformed to incorporate the effects of rotation. Since Thirring and Lense are considering a linear approximation, they neglect the squares and higher-order terms of the velocity components (which include the rotational velocity ω). This simplification is justified as they deal with astronomical bodies like the Sun and planets, where rotational velocities are small compared to the speed of light. Einstein's equation (3.40) of $\gamma'_{\mu\nu}$ from his gravitational waves paper, the retarded potential, is then used to describe how changes in the gravitational field propagate.

---

[46] One might wonder about the distinction between the interpretations of the equation (3.4) in Einstein's and Thirring and Lense's derivations. In the context of gravitational waves, the metric perturbations represent oscillations that travel at the speed of light. In the Lense Thirring effect in the first approximation, the metric perturbations represent the dragging effect around a rotating massive object. This effect is not a wave-like phenomenon but a continuous distortion due to rotation.

Both gravitational waves and the Lense Thirring effect use the retarded potentials, which account for the finite speed of gravitational influence (the speed of light) in general relativity. The perturbation $\gamma'_{\mu\nu}$ is derived from the retarded potential equation (3.40) and is further modified by "Einstein's trick" from his gravitational waves paper, which involves a trace-reversal operation (3.33). This equation simplifies the linearized Einstein field equations by partially decoupling the equations. Thirring and Lense use the retarded potential equation (3.40) to calculate the current state of the gravitational field based on the past state of the rotating mass. It accounts for the time delay between a change in the source and observing it at a distance. They calculate how the rotational motion of a mass influences its gravitational field at different locations.

Thirring and Lense's goal is to find the perturbational terms in the motion of celestial bodies that arise from the rotation of the central body. They seek to add these rotational effects to the existing Einsteinian description of the perihelion precession of planets like Mercury. They argue that the perturbation terms due to the central body's rotation are significant compared to other relativistic effects, such as the perihelion precession. To accurately describe the motion, they must also consider second-order terms involving squares and products of the velocity components. This is a step beyond the first approximation of Einstein's theory.

Including these second-order terms logically leads to considering effects from the second approximation in general relativity. Thirring and Lense argue that while the first-order approximation provides a good description, one must consider these higher-order terms for a complete picture. Despite these complexities, Thirring and Lense suggest a practical approach. They propose starting with Einstein's calculations for the precession of the perihelion [E3-15], the metric equation (3.3), and adding the perturbing terms computed in their analysis. This method would comprehensively describe orbital dynamics, including all significant relativistic effects [Thi-Len].

Thirring and Lense write the components of the velocity field due to the rotation of a massive body (like a sphere) in polar coordinates [Thi-Len][47]:

$$ir'\omega \sin\theta \sin\phi. \tag{3.134}$$

They write Eq. (3.56) (the stress-energy tensor for pressureless dust) from Einstein's gravitational waves paper and construct it for a rotating body. The rotation components (3.134) are part of this construction, representing the rotational motion in the stress-energy tensor. They are substituted into the stress-energy tensor equation (3.56) because they represent the distribution of mass-energy in motion due to rotation. This represents the gravitational effects of a rotating body [Thi-Len].

---

[47] ir' represents the radial distance from the center of the rotating body to a point in the integration space (the field) where the gravitational effects are being calculated. ω is the angular velocity of the rotating body. $\sin\theta \sin\phi$ involve the angular coordinates θ (the polar angle) and ϕ (the azimuthal angle) in polar coordinates.

Thirring and Lense substitute back into Eq. (3.40): 1. The stress-energy tensor with the rotation components (3.134) and: 2. dV_0 formulated in terms of polar coordinates. This finally leads to the final expressions for different components of $\gamma'_{\mu\nu}$ [Thi-Len][48]:

$$\boxed{\gamma'_{44} = \frac{\kappa M}{2\pi r},} \ \gamma'_{14} = 0,$$
$$\boxed{\gamma'_{24} = -i\frac{\kappa M}{2\pi r}\frac{1}{5}\omega l \sin\phi,} \tag{3.135}$$
$$\gamma'_{11} = \gamma'_{22} = \gamma'_{33} = \gamma'_{12} = \gamma'_{13} = \gamma'_{23} = \gamma'_{34} = 0.$$

The first component in Eq. (3.135) is Eq. (3.57) from Einstein's gravitational waves paper[49]:

$$\boxed{\gamma'_{44} = \frac{\kappa M}{2\pi r} = 1 - \frac{\alpha}{r},} \tag{3.136}$$

where α represents the gravitational potential term (1.164).

However, Thirring and Lense also obtained a component of the perturbed metric tensor γ'_{24} with indices $\mu = 2$ and $\nu = 4$ corresponding to angular coordinates and time, respectively. Like Eq. (3.136), γ'_{24} contains a part that represents the gravitational influence of a mass M at a distance r, $\frac{\kappa M}{2\pi r}$. However, the third component in Eq. (3.135), the expression for γ'_{24}, has a term incorporating the rotational effect $\omega l \sin\phi$. The term γ'_{24} would contribute to the frame-dragging effects in the plane perpendicular to the axis of rotation [Thi-Len].

Subsequently, Thirring and Lense used Eqs. (3.4) and (1.48) to rewrite Eq. (3.135) as follows [Thi-Len]:

$$g_{\mu\nu} = \begin{pmatrix} -1 - \frac{2GM}{r} & 0 & 0 & -i\frac{4GM}{5r}\frac{ly}{r^2}\omega l \\ 0 & -1 - \frac{2GM}{r} & 0 & 0 \\ 0 & 0 & -1 - \frac{2GM}{r} & 0 \\ i\frac{4GM}{5r}\frac{ly}{r^2}\omega l - i\frac{4GM}{5r}\frac{lx}{r^2}\omega l & 0 & 0 & -1 + \frac{2GM}{r} \end{pmatrix}. \tag{3.137}$$

This represents the Lense-Thirring metric in the linearized approximation. The diagonal components $-1 - \frac{2GM}{r}$ in the spatial part reflect the influence of mass on the spatial geometry. These components are consistent with Einstein's first order metric (3.3), and the spatial component of Einstein's metric g_{11}, see Eq. (1.319). The

---

[48] Due to the extensive nature of the derivation, which spans two papers ([Thi], and [Thi-Len]) and space constraints within this book, the full derivation has been omitted. However, this does not detract from the overall discussion.

[49] Equation (3.136) and the second component $\gamma'_{14} = 0$ are related to the first two terms of Einstein's first-order solution (1.300) from his perihelion of Mercury paper [E3-15] when it is expressed in a format similar to (3.3) in his gravitational waves paper [E3-16]. Equation (3.136) represents the first term of equation (3.3). Since the metric (3.3) already does not have cross terms involving dl and dr, applying the condition $\gamma'_{14} = 0$ does not change the metric, and it remains the metric (3.3).

components $-1 + \frac{2GM}{r}$ for g_{44} indicate the effect of mass on the time component, again reflecting a weak field approximation [see Einstein's equation (1.320)].

The off-diagonal components $g_{14} = -i\frac{4kM}{5r}\frac{ly}{r^2}\omega l$ and $g_{41} = i\frac{4kM}{5r}\frac{ly}{r^2}\omega l - i\frac{4kM}{5r}\frac{lx}{r^2}\omega l$ incorporate the Lense-Thirring effect, representing the influence of the rotating mass on spacetime, described in the terms involving ω, l, x, and y, describing the frame-dragging effect. The Matrix (3.137) representation is a linearized approximation of the gravitational field around a rotating mass, conveying the essence of the frame-dragging effect within Einstein's general relativity

Next, Thirring and Lense consider how the mass's rotation distorts geodesic paths. They write the geodesic equation (1.367), which represents the equations of motion for a test particle (like a planet) in the gravitational field (3.137) of a rotating massive body. They use the Christoffel symbols and the geodesic equation to calculate a planet's motion in the gravitational field of the central rotating body. Using equation (3.137), they calculate the Christoffel symbols and use them in the geodesic equation. The geodesic equation now includes off-diagonal terms from (3.137) that reflect the rotational motion. The rotation of the central body introduces additional terms in the geodesic equation, which are not present for a non-rotating mass. This addition represents the frame-dragging effect [Thi-Len].

As aforementioned, the derivation is related to Mercury's perihelion precession. Remember that Einstein's theory predicts a small deviation from Newtonian predictions for the orbit of Mercury (refer to the derivation in Sect. 1.6.3). Thirring and Lense's work extends this by considering the additional effect of the Sun's rotation. They mention that the total perturbation $\gamma'_{\mu\nu}$ to the orbit of Mercury can be seen as a sum of two contributions: one due to the direct action of the planet, Einstein's precession of the perihelion of Mercury (its motion and gravitational effects). This effect is unrelated to frame dragging; it results from the static (non-rotating) part of the Sun's gravitational field. The other is due to the perturbing force of the Sun (including effects from its mass and rotation) [Thi-Len]. This is where *frame dragging* comes into play; it refers to the perturbations in the orbit of Mercury due to the rotation of the Sun. The perturbation $\gamma'_{\mu\nu}$ is modified by the trace-reversal transformation [Eq. (3.33)]. It incorporates the deviations from flat spacetime [Eq. (3.4)], including effects like frame dragging caused by rotating masses. In Thirring and Lense's work, rotation leads to distortion, which influences the orbit of objects and the passage of light near the rotating mass.

Thirring and Lense employed in their work Einstein's approximation method from his Mercury perihelion paper rather than the exact Schwarzschild solution (2.49) because the frame-dragging effect is not represented by the static Schwarzschild solution. They explain that, as per the Schwarzschild solution, the external gravitational field of a spherically symmetric, non-rotating object (like a sphere at rest) is equivalent to that of a point mass [Thi-Len]. In other words, outside the sphere, the gravitational effects are indistinguishable from those produced by a point mass at the sphere's center. This equivalence results from the spherical symmetry and the static nature of the mass distribution. This finding is precisely what Schwarzschild demonstrated in his paper on the exterior metric [S-1]. However, when the sphere is rotating, this equivalence no longer holds. The sphere's rotation introduces additional effects that are not present in the case of a point mass or a non-rotating sphere. The rotation of a massive body affects the spacetime geometry in ways that static masses do not. Specifically, it leads to the frame-dragging effect introduced in Thirring and Lense's paper. This effect is a direct consequence of the mass's angular momentum. Thirring and Lense's approach was also grounded in the linearized theory in Einstein's 1916 work on gravitational waves. These methods provided a more manageable framework for deriving and comprehending the physical implications of the frame-dragging effect, particularly when the Kerr exact solution had not yet been developed.

3.3.3 Unveiling The Kerr Metric

In 1963, Roy Kerr discovered a metric for rotating spacetime. He was intrigued by the challenge of finding solutions to Einstein's field equations describing rotating spacetimes. The only exact solution at the time to Einstein's field equations was the Schwarzschild solution. However, this solution could not describe the spacetime around a rotating system. The introduction of rotation necessitated a new solution because rotation alters spacetime in a way that cannot be represented by the Schwarzschild metric, particularly due to the frame-dragging effect. Thirring and Lanse calculated the frame-dragging effect in the approximation where the gravitational fields are assumed to be weak, and the spacetime metric can be considered close to the flat spacetime metric of special relativity (see Sect. 3.3.2). The Kerr metric is an exact solution to Einstein's field equations for the spacetime around a rotating system (a black hole). Using the Kerr metric to calculate the frame-dragging effect provided a comprehensive description that included higher-order gravitational effects.

Kerr's primary challenge in finding a solution for rotating spacetimes was the complexity of Einstein's field equations under these conditions. The equations are nonlinear partial differential equations that are notoriously difficult to solve. The breakthrough came from applying conditions that could reduce these field equations to a more manageable form. Specifically, Kerr assumed the spacetime to be *algebraically special*, allowing this simplification. The classification of space as algebraically special enabled the reduction of Einstein's field equations to more tractable forms.[50] Kerr used *complex null tetrads* to find his solution. The complex null tetrads

[50] *Algebraically special spaces* have certain symmetries that reduce the complexity of the equations governing them. These symmetries can lead to conditions that make it possible to find exact, analytical solutions to the field equations for complex gravitational systems, such as rotating systems (e.g., the Kerr solution). *Algebraically special metrics* refer to a classification that categorizes spacetimes according to the algebraic properties of the *Weyl tensor*. The Weyl tensor describes the tidal force's effect in a gravitational field, representing the part of the curvature of spacetime that is not directly related to the matter content described by Einstein's field equations. This includes the curvature due to gravitational waves and other non-local effects. The curvature of spacetime itself is fully described by the Riemann curvature tensor derived from the metric tensor. An algebraically

were pivotal for simplifying the mathematical framework, exploiting the spacetime's symmetries, and enabling the exact solution of Einstein's field equations for a rotating system [Ker-08].[51]

Kerr discusses the theorem formulated in 1961 by Joshua Goldberg and Ray Sachs as part of his journey toward understanding and utilizing the properties of algebraically special spacetimes. This theorem links the geometric properties of spacetime (*shear-free null geodesic congruences*)[52] to its algebraic structure

---

special space is characterized by having a Weyl tensor that simplifies to have repeated *principal null directions* (directions along which a light ray—*null vector*—experiences no tidal distortion) (for further clarification, see footnote 64). This repetition of null directions indicates *symmetries* in the spacetime's curvature that are not present in more general (algebraically general) spacetimes.

[51] A *tetrad* (*vierbein*) is a set of four vectors that form a basis for the spacetime at each point. These vectors are particularly useful in general relativity, where spacetime geometry is curved. One can locally "flatten" spacetime by using tetrads, simplifying the representation of vectors and tensors. This local flattening allows applying intuition and techniques from flat spacetime (like special relativity) even in curved spacetime. The four vectors serve as a local basis for the tangent space at each point, allowing tensors to be expressed on this local basis rather than on a global coordinate basis. This can simplify calculations, especially in complex geometries. Specifically, these four vectors transform under local Lorentz transformations, which are the transformations preserving the Minkowski metric. Since tetrads provide a locally inertial frame, the components of tensors expressed in this frame follow the rules of special relativity, making tetrads naturally suited to respect local Lorentz invariance. In other words, tetrads allow one to express physical laws in a form that looks identical to special relativity, even in curved spacetime. This local flatness means that the physics locally respects the symmetries of special relativity, i.e., the Lorentz transformations. Using tetrads, one can work in a locally flat frame where the metric is Minkowskian, and the Christoffel symbols vanish, simplifying calculations. The equivalence principle states that locally (in a sufficiently small region of spacetime), the effects of gravity can be transformed away by an appropriate choice of coordinates, making the spacetime locally indistinguishable from flat Minkowski space. This is because, at any point in curved spacetime, it is always possible to choose a locally inertial frame where the metric tensor $g_{\mu\nu}$ approximates the Minkowski metric and the Christoffel symbols (which represent the effects of gravitational forces) vanish. We obtain the complex tetrad formalism by extending the tetrads to include complex vectors. A complex null tetrad consists of four null vectors: two real null vectors and two complex conjugate null vectors. The *complex tetrad formalism* allows a more straightforward expression of Einstein's field equations. By choosing an appropriate complex tetrad, the metric tensor and curvature tensor components can be represented in a way that highlights the problem's symmetries, thus simplifying the equations. *Complex null tetrads* are particularly adept at dealing with *algebraically special spaces*. They make it easier to identify and exploit the *principal null directions* of the *Weyl tensor*, which are key to classifying the spacetime's algebraic properties.

[52] A *null geodesic congruence* is a family of null (light-like) geodesic curves that fill a region of spacetime without intersecting each other. Null geodesics are paths along which massless particles, like photons, travel. At every point along a null geodesic, the tangent vector to the geodesic is null. This means that the tangent vector gives zero when contracted with itself using the spacetime metric. Mathematically, if k^μ is the tangent vector to a null geodesic, then the condition $g_{\mu\nu}k^\mu k^\nu = 0$ holds. The condition that the tangent vector is null (i.e., has zero length according to the spacetime metric) distinguishes null geodesics from timelike or spacelike geodesics. For timelike geodesics, which represent the paths followed by particles with mass, the tangent vector contracted with itself is negative, indicating a nonzero length in the context of the spacetime metric. This contraction is positive for spacelike geodesics, indicating a nonzero length. The zero length of the tangent vector to null geodesics is a direct consequence of the light-like or null nature of these paths, reflecting the principle that the speed of light is constant and represents an invariant speed limit in the universe.

(*algebraically special space*) via the *Petrov classification* [Ker-08] (see footnote 60 for the formulation of the theorem).

Kerr refers to Alexei Zinovievich Petrov's classification.[53] In 1954, Petrov presented a detailed mathematical approach to classify gravitational fields by examining the curvature tensor's algebraic structure in a specially defined *bivector space*.[54] Petrov asserted *three* distinct types of gravitational fields (different configurations of mass-energy that can lead to fundamentally distinct types of spacetime). He classified the three types of gravitational fields based on the algebraic structure of a K matrix and the form of the Ricci tensor $R_{\alpha\beta}$ matrix.[55] This K matrix can exhibit only one of three possible characteristic configurations: [111], [21], or [3].[56] Petrov acknowledged the possibility of further detailed classification within each type, such as distinguishing cases of multiple or real roots, indicating a deep structure to his

---

A *shear-free null geodesic congruence* is a special case of a null geodesic congruence where the shear of the congruence vanishes. Shear, in this context, relates to the distortion that can occur to the cross-sectional area of a bundle of such geodesics as they move through spacetime. A shear-free condition implies that the null geodesic congruence maintains its shape as it evolves without distortion. A shear-free condition helps maintain certain symmetrical properties of spacetime. For example, if the initial cross-section of a congruence is circular, it remains circular as it evolves. This means maintaining spherical symmetry in Schwarzschild's metric.

[53] The Petrov classification involves analyzing the Weyl tensor's principal null directions. By examining the eigenvalues and eigenvectors of the Weyl tensor, spacetimes can be categorized into one of six types (Type I, Type II, Type D, Type III, Type N, and Type 0), with Types I through N being algebraically special to varying degrees, and Type 0 representing conformally flat spacetimes. A conformal transformation changes the metric tensor by a scaling factor varying from point to point in spacetime. A spacetime is conformally flat if its metric can be expressed as a conformal transformation of the flat Minkowski metric. The Weyl tensor is invariant under conformal transformations. Thus, if we start with a flat spacetime (where the Weyl tensor is zero) and apply a conformal transformation, the Weyl tensor remains zero.

[54] Petrov seeks to classify gravitational fields by finding stationary directions in the *bivector space*, which correspond to specific orientations and properties of the gravitational field. *Bivectors* are antisymmetric rank-2 tensors. They can be considered higher-dimensional analogs to vectors, representing directions and magnitudes. A bivector is constructed from two vectors and represents an oriented plane spanned by them. It contains information about the area and the orientation of the parallelogram defined by the two vectors.

[55] K is defined in the bivector space as [Pet]:

$$K = \frac{R_{\alpha\beta} V^{\alpha} V^{\beta}}{g_{\alpha\beta} V^{\alpha} V^{\beta}}, \tag{3.138}$$

where $R_{\alpha\beta} V^{\alpha} V^{\beta}$ represents the projection of the Ricci tensor along the bivector V^{α}, giving a measure of curvature in the direction of V. The denominator, $g_{\alpha\beta} V^{\alpha} V^{\beta}$, normalizes this quantity, providing a curvature measure that considers the metric tensor. The K matrix is defined as follows [Pet]:

$$||R_{\alpha\beta} - K g_{\alpha\beta}||. \tag{3.139}$$

Analyzing this matrix was fundamental for Petrov to classify the geometric structures influenced by mass-energy distributions.

[56] These configurations represent the algebraic multiplicities of the eigenvalues of the matrix, ignoring the cases where elementary divisors share the same basis.

classification scheme.[57] In his paper, Petrov noted that he elaborated on an approach he had previously introduced in 1951 [Pet].

Felix Pirani played a pivotal role in popularizing and extending Petrov's work on classifying spacetimes. Petrov originally articulated a theorem that provided a classification scheme for gravitational fields. According to Petrov's theorem, only *three* distinct types of gravitational fields exist (refer to footnote 57 for details). In his paper, Petrov suggested the possibility of further, more complex classification. Pirani added three more types, relabeling Petrov's types as I, II, D, III, N, and 0 (see explanation in footnote 53). Pirani reinterpreted Petrov's classification for a broader audience, particularly in the English-speaking scientific community, where Petrov's original work (published in Russian) was not widely accessible initially. In his 1957 paper, Pirani shows that the presence of gravitational waves has implications for the motion of test particles and the propagation of disturbances in spacetime. He hints at how these effects differ across the types of spacetime curvature defined by the Petrov classification. Pirani specifies that gravitational radiation is present in empty spacetime if the Riemann tensor is of Type II or III but not Type I. This classification is pivotal for identifying the presence of gravitational waves based on the underlying geometrical structure of spacetime [Pir]. Pirani's efforts helped integrate Petrov's classification into the mainstream of general relativity research, making it more accessible and understood by researchers worldwide. In 1958, Kerr first heard about Petrov's ideas in a relativity seminar at King's College in London, specifically in a lecture by Pirani on his work on gravitational radiation theory (gravitational waves) [Pir], [Ker-08].

In 1962, Ivor Robinson and Andrzej Trautman discovered a specific class of solutions relevant for describing non-rotating objects rather than the rotating cases that Kerr would eventually solve. The absence of twisting simplified their analysis and led to more tractable mathematical models of gravitational waves [RT].[58] Robinson and Trautman explored metrics that can approximate the behavior of actual gravitational waves. Their metrics were simplified models representing how gravitational

[57] 111 indicates a field where the characteristic is of simple type, implying that the Ricci tensor $R_{\alpha\beta}$ has six non-isotropic, pairwise orthogonal principal directions. This suggests a spacetime geometry with distinct and separable curvature properties in different directions. 21 suggests a more complex field with a pair of directions with a more significant curvature characteristic than a singular, less pronounced direction. 3 implies a field where all principal directions of curvature are equivalently significant, suggesting a highly symmetric or uniform gravitational field [Pet].

[58] Robinson and Trautman explored *shear-free*, *hypersurface-orthogonal*, and *diverging null congruences*. Thus, their discussion relates to the absence of twisting; for terminological clarification, see footnote 60. A solution involving *hypersurface orthogonal* geodesics typically assumes a vanishing twist, as orthogonality implies no rotation about the congruence. Since the hypersurface orthogonality condition leads to a vanishing twist, it simplifies the structure of the spacetime. Thus, the vanishing twist condition simplifies the analysis and is consistent with the *Goldberg-Sachs theorem* (see footnote 60), as the hypersurface orthogonal condition implies that if the congruence is also shear-free, then it will correspond to a spacetime with Weyl tensor of Petrov type II or simpler. Unlike Robinson and Trautman's metric, the Schwarzschild metric's shear and divergence are zero, indicating that the cross-sectional area of the congruence of null geodesics is not changing, i.e., no radial expansion or contraction.

waves propagate through spacetime, specifically looking at a type miming spherical radiation. At great distances, these metrics provide good approximations of actual gravitational fields generated by radiating sources.[59]

Robinson and Trautman's solutions demonstrated the practical application and importance of the *Goldberg-Sachs theorem*.[60] The theorem bridges spacetime's geometric properties-specifically, the characteristics of being geodesic and shear-free-with its algebraic structure, manifested in the Weyl curvature tensor. The theorem connects the behavior of *null geodesic congruences* within spacetime to its classification within the Petrov scheme based on the algebraic types defined by the Weyl tensor's properties. It states that a vacuum solution to the Einstein field equations must have a congruence of shear-free null geodesics if and only if the spacetime is of Petrov type D or N.

The above link between geometric and algebraic attributes of spacetime underscored the relevance of null congruences to understanding the nature of algebraically special spacetimes. Types D and N are those algebraically special cases referenced by the Goldberg-Sachs theorem, which relate to rotating (type D, e.g., Kerr black holes) or radiative (type N, e.g., pp-waves) spacetimes.[61] Robinson and Trautman explored and identified classes of solutions to Einstein's field equations that exemplified the above-mentioned geometric properties, particularly focusing on shear-free null geodesic congruences. While much of their work concentrated on non-rotating

---

[59] Robinson and Trautman's work offers a mathematical exploration of how certain simplified metrics can model aspects of spherical gravitational waves. By drawing an analogy to retarded waves (potentials) in their paper, Robinson and Trautman are highlighting a parallel between gravitational waves and electromagnetic waves. Both phenomena are described by waves that carry energy away from a source and are governed by the finite speed at which these waves propagate through the medium (vacuum for electromagnetic waves, spacetime for gravitational waves). This analogy helps understand gravitational waves by comparing them to the more familiar electromagnetic waves, even though the underlying physics and equations governing them are quite distinct. Robinson and Trautman use Maxwell's equations and the *complex null tetrad* and a *complex null bivector* (A null bivector is formed from the exterior product of two null vectors) to derive a specific solution, which helps visualize the characteristics of the radiation fields. They describe exact solutions with properties akin to waves. The solutions exhibit singularities besides the central point singularity representing the source.

[60] The *Goldberg-Sachs theorem*, articulated in 1961, establishes that a vacuum metric is *algebraically special* if and only if it features a *shear-free null geodesic congruence*. Robinson and Trautman discuss *null congruences* that are geodetic, *shear-free*, *hypersurface-orthogonal*, and *diverging*. Recall (footnote 52) that shear-free implies that the cross-sectional shape of a bundle of null geodesics does not distort (except for changes in size) as it propagates. If a congruence of geodesics is hypersurface-orthogonal, it means there exists a family of hypersurfaces to which the geodesics are everywhere orthogonal. This property is related to the potential absence of twisting (i.e., *the absence of a rotation* of the cross-section of the bundle of geodesics around the propagation direction) of the congruence, as twisting would prevent the geodesics from being orthogonal to a family of hypersurfaces. Divergence refers to the spreading out of geodesics from a point or along their paths, an expected property in the context of gravitational waves emanating from a source. A shear-free, hypersurface-orthogonal congruence indicates *no twisting*.

[61] pp-waves spacetimes represent a class of exact solutions to Einstein's field equations that describe plane-fronted gravitational waves (having flat wavefronts) that propagate in a single direction without divergence or convergence, i.e., parallel rays.

solutions, the principles they utilized laid foundational insights for further exploration of spacetime geometries.[62]

In 1963, Ezra T. Newman, T. Unti, and Louis A. Tamburino (NUT) published the NUT solution. They interpreted their results to mean that the NUT metric—an extension of the Schwarzschild solution including a gravitomagnetic monopole moment—was a solution suggesting the presence of angular momentum or rotation in the spacetime structure. The NUT metric introduced a twist in spacetime, which could suggest a form of angular momentum or rotation. However, this twist did not correspond directly to the physical angular momentum of a rotating mass (as in the Kerr solution) [NUT].

NUT employed the *Newman-Penrose formalism* (which simplifies the field equations through the use of *complex null tetrads*). Ezra Newman and Roger Penrose applied their formalism to study the properties of *null geodesic congruences* and the conditions under which the Riemann tensor exhibits specific algebraic properties indicative of the presence of gravitational waves. Specifically, Newman and Penrose proved the Goldberg-Sachs theorem. Recall that this theorem connects the *shear free null geodesic congruence* in vacuum (a characteristic feature of gravitational wave spacetimes) to the algebraic special nature of the Riemann tensor [NP1].[63]

NUT's employment of the Newman-Penrose formalism inadvertently led to a numerical error. Working with Alan Thompson, who lived in the adjoining apartment,

---

[62] Robinson and Trautman explore pp-waves, presenting them as a specialized case within their broader class of shear-free null geodesic congruences and empty-space solutions. They demonstrate the relationship between the more general and pp-wave solutions. The more general solutions encompass various cases: Petrov Type I degenerate, Type II non-null or Type III, and Type II null or flat solutions. Robinson and Trautman derive the conditions under which these metrics satisfy Einstein's field equations for empty space. They explore how the shear-free condition affects the solutions and the resulting curvature tensors, noting the specific cases where the solutions correspond to the above different Petrov types [RT].

[63] The Newman-Penrose spinor formalism uses a *complex null tetrad* approach to break down the complex tensor equations of general relativity into simpler, *scalar equations*. The *complex null tetrad* (see footnote 51 for an explanation of tetrads) consists of two real null vectors and two complex conjugate null vectors, forming a complete basis for spacetime at each point. A *null vector* (or light-like vector) is a vector whose magnitude is zero. In simpler terms, it is a direction in which light could travel. The real null vector l^μ is often chosen as outward pointing and tangent to a *congruence of null geodesics*. It is called the outgoing vector. A null geodesic is a geodesic tangent to a null vector at every point. Another real null vector n^μ, inward pointing, is typically chosen to be tangent to a different congruence of null geodesics, referred to as the ingoing vector. The complex conjugate null vectors [m^μ and $\bar{m}^\mu$] are orthogonal to both real null vectors. They are chosen to lie in the plane perpendicular to the real null vectors, and their complex nature allows for describing spatial directions in a more compact form. A normalization condition [$n_\mu l^\mu = -m^\mu \bar{m}_\mu = 1$] guarantees that the vectors form a *complex null tetrad*. The vectors are null because their norm is zero: $l_\mu l^\mu = m_\mu m^\mu = \bar{m}_\mu \bar{m}^\mu = n_\mu n^\mu = 0$ [NP1]. The utility of a null tetrad comes from its ability to simplify Einstein's field equations. This formalism is particularly useful for analyzing and understanding algebraically special spaces, like those of Type D, as it simplifies the identification of the key properties defining these spaces. Type D spacetimes have two distinct principal null directions, each doubly degenerate (repeated). This means two unique null directions are especially significant in the geometry of such spacetimes. The Newman-Penrose formalism was instrumental in NUT's study of their solution to Einstein's field equations.

Kerr determined the nature of the error. In the work conducted by Newman, Unti, and Tamburino, a constant referred to as the (twisting) *curling term* denoted by $|\rho^0|$ was introduced. This term played a pivotal role in developing the NUT metric, which, under specific conditions, particularly when $|\rho^0|$ was reduced to zero, seamlessly transitioned into the Schwarzschild metric [NUT]. This convergence towards the Schwarzschild metric highlighted an intriguing characteristic of $|\rho^0|$, as its necessity to diminish to zero suggested the presence of a computational discrepancy within the Newman-Penrose formalism equations [Ker-08].[64]

The original manuscript detailing the NUT metric underwent peer review, a process during which Alfred Schild, recognizing the manuscript's significance and complexity, sought the expertise of Kerr. Kerr's involvement proved to be crucial; through meticulous examination, he not only pinpointed the error but also, in collaboration with Newman, traced its origin back to the Newman-Penrose equations. This collaborative effort led Newman and Penrose to identify and rectify a sign error within the equations of the Newman-Penrose formalism [NP2]. According to Newman's account, the correction of this error permitted the curling term to naturally reduce to zero without imposition, a correction that ultimately facilitated the emergence of the Kerr parameter a. This parameter a became a cornerstone for Kerr's development of the correct rotating metric, a formulation distinguished by its reliance on two fundamental parameters: the mass m and the now fundamental, non-zero parameter a [AN]. However, this was not the case [Ker-08]. The NUT metric introduced a twist in spacetime, which could suggest a form of angular momentum or rotation due to its twisting principal null vector. However, the twisting of principal null vectors in the NUT spacetime did not correspond directly to the physical angular momentum of a rotating mass (as in the Kerr solution).[65]

---

[64] A *principal null vector* is a null vector that satisfies the condition of being an eigenvector of the Weyl tensor. When the Weyl tensor acts on principal null vectors, it scales them without changing direction. Principal null vectors define directions, the *principal null directions*. In a given spacetime, the principal null directions are the specific directions in which the Weyl tensor's action on the null vector maintains the vector's direction (up to a scalar multiple). In the Newman-Penrose (NP) formalism (see footnote 63), the principal null vector l^μ is one of the vectors in the *null tetrad* used to express the metric and other physical quantities. To demonstrate angular momentum or twist using the *principal null vector*, we look at the spin coefficients associated with the *congruence of null geodesics* defined by l^μ: *shear* σ (that measures the distortion of the null congruence) (see further clarification in footnote 52) and *twist* ω (that measures the rotation of the null congruence about its axis). For the solution to exhibit angular momentum or twist, the twist component (represented by the imaginary part of $|\rho^0|$, which NUT called the *curling term*) must be non-zero. For a congruence defined by a null vector l^μ, ρ represents the divergence and expansion (real part) and $|\rho^0|$ the twist (imaginary part). NUT focus on the *curling term* of the geodesic rays, with specific references to the conditions under which the solutions are found (e.g., equations involving $|\rho^0|$, σ, and their derivatives). If the NUT solution contains a non-zero twist, it would suggest the presence of angular momentum. In their paper, the NUT metric introduces a kind of twist in spacetime, which could suggest a form of angular momentum or rotation due to its twisting principal null vector. It introduces a gravitomagnetic monopole moment, leading to a twisting spacetime even without mass rotation. However, the twisting of principal null vectors in the NUT spacetime does not correspond directly to the physical angular momentum of a rotating mass (as in the Kerr solution).

[65] Kerr showed that NUT's employment of the Newman-Penrose formalism inadvertently led to a numerical error. Working with Thompson, Kerr determined the nature of the error. When $|\rho^0|$ was

Correcting the error in the NUT paper and building upon and extending the insights of Robinson and Trautman allowed Kerr to derive the field equations and the metric for rotating algebraically special spaces. Kerr used a *complex null tetrad*, which recall is highly useful for analyzing *algebraically special metrics*, particularly those of *Petrov Type D*.[66] This led Kerr to simplify the field equations and present a metric characterized by a *congruence of null geodesics* with *nonvanishing divergence*, which is not necessarily *hypersurface orthogonal*. Kerr's solution was classified as *Petrov Type D* [Ker-63].[67] Kerr discovered the Kerr metric, which describes the

---

reduced to zero, it seamlessly transitioned into the Schwarzschild metric [NUT]. This convergence towards the Schwarzschild metric highlighted an intriguing characteristic of $|\rho^0|$, as its necessity to diminish to zero suggested the presence of a computational discrepancy within the Newman-Penrose formalism equations. This contradicts the following: recall that to demonstrate twist using l^μ, we look at the spin coefficients associated with the *congruence of null geodesics* defined by l^μ: σ and ω. For the solution to exhibit angular momentum or twist, $|\rho^0|$ must be non-zero. If the NUT solution contains a non-zero twist, it would suggest the presence of angular momentum. Working with Thompson, Kerr discovered that the NUT manuscript introduced the $|\rho^0|$ incorrectly. When $|\rho^0|$ was reduced to zero, the NUT metric should seamlessly transition into the Schwarzschild metric, but this was not initially the case due to a numerical error. The error meant that the reduction to the Schwarzschild solution did not occur as expected. For the Schwarzschild metric, the shear $\sigma = 0$, $|\rho^0| = 0$, and the divergence $\rho = 0$. Kerr's meticulous examination, in collaboration with Newman, traced the origin of this error to a sign error in the Newman-Penrose equations. Correcting this error allowed the *curling term* to naturally reduce to zero, facilitating the seamless transition to the Schwarzschild metric. This correction led to the correct formulation of the Kerr metric, distinguished by its reliance on the parameters mass m and the non-zero Kerr parameter a. The NUT metric does not directly correspond to the physical angular momentum of a rotating mass as in the Kerr solution.

[66] The complex null tetrad (see footnote 63) allowed the Newman-Penrose equations, which are significantly simplified in algebraically special spacetimes, to be expressed in terms of scalar quantities.

[67] Kerr's realization that spacetimes could support a rotating *null congruence* marked a significant departure. He corrected a foundational error by adopting a coordinate system that naturally describes these rotating null congruences. This breakthrough underscored the natural alignment of certain coordinates for describing algebraically special spacetimes, building upon and extending the insights of Robinson and Trautman [Ker-08]. In Robinson and Trautman's metric, the congruence of null geodesics has nonvanishing divergence but is hypersurface orthogonal. The congruence is expanding (or contracting) over time, a feature associated with the dynamic nature of the Robinson-Trautman spacetimes. This expansion (or contraction) can be related to gravitational radiation and other dynamic processes within spacetime. Despite the divergence, the congruence remains orthogonal to a family of hypersurfaces. This property indicates that the spacetime can be foliated into hypersurfaces orthogonal to the null geodesics, simplifying the geometric description and analysis of these spacetimes. This characteristic suggests a certain regularity in the spacetime structure. However, there is a primary difference between Robinson and Trautman's spacetime and Kerr's solution: Kerr's metric inherently includes rotation as a fundamental aspect of the spacetime it describes. The null geodesic congruences in Kerr spacetime are generally not hypersurface orthogonal. The lack of hypersurface orthogonality in the Kerr metric is due to the rotation of spacetime. This means that the null geodesics are not perpendicular to any family of hypersurfaces, reflecting the Kerr spacetime's inherent angular momentum and non-static nature. In contrast, Robinson and Trautman's solutions generally do not account for angular momentum similarly, focusing instead on properties like shear and expansion. Moreover, Kerr's work diverged from Robinson and Trautman's in that the Kerr metric is algebraically special and classified as Petrov Type D (see footnote

region outside a rotating mass. This metric inherently includes the effects of angular momentum on the spacetime geometry [Ker-08].

Specifically, in his 1963 paper, Kerr started by defining a line element (ds) in terms of the *complex null tetrad* components [Ker-63][68]:

$$ds^2 = 2tt^* + 2mk. \tag{3.140}$$

He introduced a transformation where the tetrad components were expressed using u and r through new variables that involve P, Ω, D, their derivatives, and the complex conjugate of these terms. Kerr wrote transformation equations with these new variables that involve complex functions and their real parts, differential forms, in a complex coordinate system described by ζ. Kerr's transformation equations involve two coordinates: u, representing a time-like coordinate. r is a radial coordinate representing the distance from a central point in a spherically symmetric or axially symmetric spacetime. The transformation aims to simplify Einstein's field equations and make them more suitable for solving under certain simplifying conditions.

Given the new complex coordinate system, Kerr assumed that $P = 1$, and he further assumed that Ω was complex. He also wrote a simplified version of Einstein's field equations, which I call field equations S. Kerr mentioned a second coordinate system. In this coordinate system, $P \neq 1$ but Ω is purely imaginary. This system gives more complicated field equations but is probably better for analyzing his solution. This suggested to Kerr that while this system may offer a more accurate or insightful representation of the physical situation of spacetime around a rotating, uncharged mass, it could also introduce mathematical complexity. Kerr considered integrability conditions for the parameter m. If m is zero, the field equations S are described as integrable (meaning they can be solved more straightforwardly through analytical methods). These solutions correspond to Petrov Type-III and null spaces with nonzero divergence.[69] However, for $m \neq 0$, certain conditions must be met for the simplified version of Einstein's field equations S to be solvable. These conditions warrant

53 for further explanation). The Kerr metric is a prime example of a Type D spacetime. The NUT metric is also classified as Petrov Type D, similar to the Kerr metric, which means the NUT metric possesses two pairs of repeated principal null directions. Unlike the NUT metric, in the Kerr metric, the *principal null vectors* are related to the direction in which light can orbit the rotating mass or escape to infinity without being pulled in or diverging away. The principal null vectors are related to the angular momentum a in the Kerr metric. The angular momentum a introduces off-diagonal terms in the metric tensor, representing the dragging of spacetime or frame dragging effect around the rotating mass. This effect means that the principal null vectors in the Kerr geometry are not static but instead reflect the dynamic nature of the spacetime caused by rotation. Principal null vectors in the Kerr spacetime correspond to null geodesics (recall that these paths are followed by light or any massless particle near the mass). The angular momentum a influences these geodesics, causing them to spiral around the mass due to frame dragging. The rotation of the mass, as parameterized by a, directly affects the direction and curvature of these principal null geodesics.

[68] With the *complex conjugate null vectors* t and t^* (complex conjugates of one another), alongside m and k, a pair of *real null vectors*.

[69] Petrov Type III spacetimes are more complex than Type D, particularly from the perspectives of their mathematical description, physical interpretation, and the contexts in which they arise. Some Type III solutions represent gravitational waves propagating in curved spacetimes, requiring

that solutions can exist and be found. Kerr suggested expressing m as a function of the variable Ω and its derivatives under specific conditions involving $\dot{\Delta}$ and $\ddot{\Omega}$ (Δ is defined in terms of P, D, and Ω). If the derivatives $\dot{\Delta}$ and $\ddot{\Omega}$ are zero, this leads to a situation where Ω is purely imaginary and $P \neq 1$. This transformation makes field equations S more suitable for analysis of Type D solutions [Ker-63] (the field equations are formulated in a complex representation in the Newman-Penrose formalism):

$$m = cu + A + iB, \tag{3.141}$$

$$P^{-2}\Delta\left[P^{-2}\Delta(\ln P)\right] = 2c, \quad \Delta = \frac{\partial^2}{\partial\zeta\partial\zeta^*}, \tag{3.142}$$

c is a real constant, and A, B and Ω are all independent of u and r. Being independent of u implies that A, B, and Ω do not change over time (at least in the coordinate system used), a property of *stationary spacetimes*. Being independent of r suggests they are the same at any distance from the center in a *radial direction*. Thus, B and A are given by [Ker-63]:

$$iB = \frac{1}{2}P^{-2}\Delta\left(P^{-2}\frac{\partial\Omega}{\partial\zeta}\right) - P^{-4}\left(\frac{\partial\Omega}{\partial\zeta}\right)\Delta(\ln P), \tag{3.143}$$

$$\Delta B = ic\frac{\partial\Omega}{\partial\zeta}, \quad \frac{\partial}{\partial\zeta}(A - iB) = c\Omega. \tag{3.144}$$

Kerr imposed the condition $c = 0$. This implies that there is a time-like killing vector field. The field equations are simplified so that the dependence on the coordinate u drops out of the key expressions defining the metric [Ker-63].[70] Kerr imposed

sophisticated analysis techniques. Unlike the relatively more straightforward Schwarzschild (Type D) and Kerr (also Type D) solutions that describe static and rotating black holes, respectively, Type III solutions often lack simple, exact forms. This is due to their less symmetrical and more dynamic nature, making it challenging to find and classify exact solutions of this type.

[70] The condition $c = 0$ leads to $\frac{\partial}{\partial u}$ being a Killing vector. This condition is rooted in the spacetime symmetries described by Eqs. (3.141)–(3.144). A Killing vector corresponds to a direction along which the metric does not change, indicating an invariance under translations in that direction. This could be a spatial symmetry (like rotation or translation in space) or a temporal symmetry (invariance under shifts in time). The constant c appears in the field equations (3.141), (3.142), and (3.144), including the relation involving A, B, and Ω. When $c = 0$, several consequences emerge from the field equations: The Eq. (3.141) for m reduces to $m = A + iB$, indicating m is independent of u reinforcing the time invariance when $c = 0$. The derivative condition in equation (3.144), $\frac{\partial}{\partial\zeta}(A - iB) = c\Omega$ simplifies to $\frac{\partial}{\partial\zeta}(A - iB) = 0$ when $c = 0$, suggesting that $A - iB$ (and thus A and B individually) is constant with respect to ζ, indicating a form of spatial invariance and symmetry. The first equation in the Eq. (3.144), $\Delta B = ic\frac{\partial\Omega}{\partial\zeta}$, implies that $\Delta B = 0$ and thus, $B = 0$ if $c = 0$, further pointing to spatial symmetry properties. Thus, given that A, B, and Ω are independent of u and r, and with $c = 0$, the structure of the spacetime becomes invariant under translations in u. This invariance under translations in u means the spacetime does not change as u varies, which is what a Killing vector field represents. Thus, $\frac{\partial}{\partial u}$ is a Killing vector, which means the spacetime has a temporal symmetry; it is stationary with respect to u.

another condition on the field equations (3.141)–(3.144): m is a real constant (representing the mass of the rotating system). When $c = 0$ and $B = 0$, and m is a real constant, the field equations simplify significantly:

$$P^{-2}\Delta\left[P^{-2}\Delta(\ln P)\right] = 0, \quad \frac{\partial}{\partial\zeta}A = 0, \quad \Delta = \frac{\partial^2}{\partial\zeta\partial\zeta^*}, \tag{3.145}$$

leading towards a unique stationary and axially symmetric solution.[71]

Finally, Kerr integrated the field equations (3.141)–(3.144) for the functions P and Ω (under the given conditions $c = 0$, $B = 0$, and m being a real constant) into a spacetime metric.[72] For the Kerr metric, this step necessitated transitioning from the complex coordinates ζ and its conjugate ζ^* to a set of real coordinates suitable for describing the Kerr spacetime, i.e., expressing the results in terms of spherical coordinates (r, θ, ϕ) [Ker-63][73]:

---

[71] The reasons behind this include: 1. The condition $c = 0$ directly impacts the dynamics involving u and Ω, eliminating any u-dependent terms from the equation for m. This results in the spacetime geometry becoming independent of the coordinate u, signifying a stationary metric. With $B = 0$, the equation involving B places constraints on Ω in relation to P. Since B is associated with the imaginary part of a complex quantity that could influence Ω and P, setting $B = 0$ simplifies these relationships in a manner that preserves axial symmetry across the spacetime. 2. A real constant m implies a fixed mass parameter, introducing a static mass influence on the spacetime geometry without spatial variations or complex behavior. This significantly restricts the types of solutions to those describing stationary, axially symmetric spacetimes around massive objects. 3. Thus, the combined conditions $c = 0$, $B = 0$, and a constant real m uniquely determine a class of spacetime geometries that are stationary and axially symmetric, leading to a metric devoid of dynamic or complex geometrical behaviors, the Schwarzschild metric and the rotational generalization of it. Kerr's metric (3.153) does not change over time and possesses axial symmetry, matching the conditions from $c = 0$ and $B = 0$.

[72] These conditions simplify the field equations, potentially isolating the effects of mass m and angular momentum a (conceptually linked to Ω on the spacetime structure). The absence of c and B focuses the analysis on the properties of P and Ω, which involve rotational dynamics akin to the role of a.

[73] Consider the transformations:

$$\zeta = e^{i\phi}\cot\left(\frac{\theta}{2}\right), \quad \zeta^* = e^{-i\phi}\cot\left(\frac{\theta}{2}\right). \tag{3.146}$$

To express these in terms of spherical coordinates (r, θ, ϕ), we need to find the relationships between these coordinates. First, we find the relationship between $\zeta\zeta^*$ and θ:

$$\zeta\zeta^* = \left(e^{i\phi}\cot\left(\frac{\theta}{2}\right)\right)\left(e^{-i\phi}\cot\left(\frac{\theta}{2}\right)\right) = \cot^2\left(\frac{\theta}{2}\right). \tag{3.147}$$

Taking the square root on both sides of equation (3.147), we get:

$$\cot\left(\frac{\theta}{2}\right) = \sqrt{\zeta\zeta^*}, \Rightarrow \boxed{\theta = 2\cot^{-1}\left(\sqrt{\zeta\zeta^*}\right).} \tag{3.148}$$

Next, we express ϕ from the definition of ζ:

$$\boxed{\begin{aligned} ds^2 =&(r^2 + a^2\cos^2\theta)(d\theta^2 + \sin^2\theta d\phi^2) \\ &+ 2(du + a\sin^2\theta d\phi)(dr + a\sin^2\theta d\phi) \\ &- \left(1 - \frac{2mr}{r^2 + a^2\cos^2\theta}\right)(du + a\sin^2\theta d\phi)^2. \end{aligned}} \tag{3.153}$$

The Kerr metric emerges as the single solution to the simplified field equations (3.145) under the specific conditions that the spacetime is stationary and axially symmetric.[74] Kerr stressed that the metric (3.153), like the Schwarzschild metric (2.49), which it contains, both belong to the Petrov Type D class [Ker-63].

It is imperative to acknowledge that at the time of Kerr's original derivation of his now-famous solution to Einstein's field equations, the term *black hole* had not been explicitly associated with the resultant spacetime geometry. Kerr's initial discourse did not invoke a black hole. In 1963, physicists did not know that Kerr's solution described a black hole [Pr-Th-Mac]. The profound connection between the Kerr metric and black holes was elucidated subsequently through the seminal work of Subrahmanyan Chandrasekhar and others (I discuss this further below). This historical element is critical for a comprehensive understanding of the development and interpretation of the Kerr metric. In 1965, Ezra Newman and Allen Janis wrote that it had been suggested that the Kerr metric be interpreted as originating from a *spinning particle* (a mass/a celestial body that rotates around its axis) [NJ].

Kerr applied a transformation to bring his metric (3.153) into a form that resembles Cartesian coordinates but in an asymptotically flat spacetime:

$$(r - ia)e^{i\phi}\sin\theta = x + iy, \quad r\cos\theta = z, \quad u = t + r. \tag{3.154}$$

The transformed metric becomes [Ker-63]:

$$e^{i\phi} = \frac{\zeta}{\cot\left(\frac{\theta}{2}\right)}. \tag{3.149}$$

From Eq. (3.148) we get:

$$e^{i\phi} = \frac{\zeta}{\sqrt{\zeta\zeta^*}}. \tag{3.150}$$

Taking the argument on both sides:

$$\phi = \arg\left(\frac{\zeta}{\sqrt{\zeta\zeta^*}}\right), \Rightarrow \boxed{\phi = \arg(\zeta).} \tag{3.151}$$

Finally, we include the radial coordinate r:

$$\boxed{r = \text{radial coordinate in the metric.}} \tag{3.152}$$

[74] Axial symmetry means that the spacetime is symmetric with respect to rotations around the axis of rotation of the rotating mass. In terms of the metric tensor, this symmetry implies that rotating the coordinate system around this axis by any angle does not change the form of the metric. This property distinguishes the Kerr metric from the Schwarzschild metric (2.49).

$$ds^2 = dx^2 + dy^2 + dz^2 - dt^2 + \frac{2mr^3}{r^4 + a^2z^2}(k)^2. \tag{3.155}$$

The first part, $dx^2 + dy^2 + dz^2 - dt^2$, represents the line element in flat (Minkowski) spacetime. The second part, $\frac{2mr^3}{r^4+a^2z^2}k^2$, is a perturbation to the flat spacetime metric due to the mass m and rotation a of the rotating mass. k is a function defined by the expression:

$$(r^2 + a^2)rk = r^2(xdx + ydy) + ar(xdy - ydx) + (r^2 + a^2)(zdz + rdt). \tag{3.156}$$

This function incorporates the rotational effects of the system into the metric. r is defined implicitly by the equation:

$$r^4 - (R^2 - a^2)r^2 - a^2z^2 = 0, \quad \text{where: } R^2 = x^2 + y^2 + z^2. \tag{3.157}$$

This definition of r blends the radial component of spherical coordinates with Cartesian coordinates, adapting to the system's rotation [Ker-63].

The transformation to an asymptotically flat coordinate system makes the Kerr metric more accessible and interpretable. In this form (3.155), the Kerr metric's analytic properties become more apparent, especially regarding how spacetime behaves far away from the mass. The Kerr solution (3.155) is analytic almost everywhere, meaning it can be expressed in terms of well-behaved mathematical functions that are smooth and differentiable at nearly every point. However, there are specific points where the Kerr solution is not analytic, associated with singularities. The denominator of the term $\frac{2mr^3}{r^4+a^2z^2}(k)^2$ in Eq. (3.155) suggests that singularities could arise when $R = 0$ and $z = 0$.

Introducing the rotational parameter a leads to frame-dragging [the $a \sin^2 \theta d\phi$ components in Eq. (3.153)]. The Lense-Thirring effect discussed in Sect. 3.3.2 can be seen as a first-order approximation in the weak-field, slow-rotation limit of the full Kerr solution. Both solutions start with the Minkowski metric and introduce corrections due to mass (in the case of Kerr, angular momentum [Eq. (3.155)]). However, while Thirring and Lense's solution is based on a weak field and slow rotation, the Kerr exact solution [Eq. (3.155)] assumes a fully rotating solution.[75]

Shortly after that, Ezra Newman and his colleagues discovered the Kerr-Newmann solution using a complex null tetrad system, which also formed the starting point of the derivation of the Kerr metric. Newman and colleagues simplified the Weyl tensor and electromagnetic field tensor expressions using a complex null tetrad. They highlight in their paper a *trick* they used, a complex coordinate transformation—to derive the Kerr-Newman metric. In 1964, Newman and Janis discovered the trick. It is a mathematical procedure that generates new solutions to Einstein's field equations

[75] In the case of a Kerr black hole, the black hole's rotation affects the spacetime structure so that the null geodesics exhibit frame-dragging around the black hole. This means that the paths of particles and light rays are twisted around the black hole in the direction of its rotation. This is the frame-dragging effect.

from existing ones, i.e., generates rotating solutions from static ones. First, Newman and Janis started with the known solution to Einstein's field equations in spherical symmetry, the Schwarzschild solution, written using the complex null tetrad system. They then converted it to the Kerr solution.[76] Subsequently, Newman et al.

---

[76] Here is the Newman-Janis algorithm [NJ]:

1. A coordinate transformation is applied to the Schwarzschild metric (2.177) or (2.94):

$$u = t - r - r_0 \ln(r - r_0), \quad r' = r, \quad \theta' = \theta, \quad \phi' = \phi. \tag{3.158}$$

This transformation introduces a new time coordinate u, the retarded time coordinate.

2. After applying the transformation, the Schwarzschild metric (2.94) in terms of the new coordinates (u, r, θ', ϕ') becomes:

$$ds^2 = \left(1 - \frac{r_0}{r}\right) du^2 + 2dudr - r^2 \left(d\theta^2 + \sin^2\theta d\phi^2\right), \quad r_0 = 2m. \tag{3.159}$$

3. The components of $g^{\mu\nu}$ corresponding to the transformed Schwarzschild metric (3.159) are:

$$g^{00} = 0, \quad g^{11} = -\left(1 - \frac{r_0}{r}\right), \quad g^{12} = 1, \quad g^{22} = -\frac{1}{r^2}, \quad g^{33} = -\frac{1}{r^2 \sin^2\theta}. \tag{3.160}$$

4. These components are written in terms of the complex null tetrad, a set of four null vectors, $(l^\mu, n^\mu, m^\mu, \bar{m}^\mu)$, forming the basis of the metric:

$$g^{\mu\nu} = l^\mu n^\nu + l^\nu n^\mu - m^\mu \bar{m}^\nu - m^\nu \bar{m}^\mu. \tag{3.161}$$

5. The detailed form of the null tetrad vectors (3.161) is given by:

$$l^\mu = \delta_1^\mu, \quad n^\mu = \delta_0^\mu - \frac{1}{2}\left(1 - \frac{r_0}{r}\right)\delta_1^\mu. \tag{3.162}$$

The complex conjugate vectors m^μ and $\bar{m}^\mu$ are:

$$m^\mu = \frac{1}{\sqrt{2}r}\left(\delta_2^\mu + \frac{i}{\sin\theta}\delta_3^\mu\right), \quad \bar{m}^\mu = \frac{1}{\sqrt{2}r}\left(\delta_2^\mu - \frac{i}{\sin\theta}\delta_3^\mu\right). \tag{3.163}$$

6. A complex coordinate transformation is performed:

$$r' = r + ia\cos\theta, \quad u' = u - ia\cos\theta, \quad \theta' = \theta, \quad \phi' = \phi. \tag{3.164}$$

The transformation (3.164) involves the parameters r (radial coordinate), u (null coordinate), and a (rotation parameter). By applying transformations (3.164), the Schwarzschild metric [Eqs. (3.162) and (3.163)] is altered to include rotation effects, leading to a new set of tetrad vectors [Eqs. (3.165) and (3.166)]. When expressed in the transformed coordinates, these vectors exhibit the spacetime geometry of a rotating mass.

7. The null tetrad vectors (3.162) and (3.163) are transformed using the new coordinates (3.164) into a new set of tetrad vectors:

$$l'^\mu = \delta_1^\mu, \quad n'^\mu = \delta_0^\mu - \frac{1}{2}\left(1 - \frac{r_0 r'}{r'^2 + a^2\cos^2\theta}\right)\delta_1^\mu, \tag{3.165}$$

$$m'^\mu = \bar{m}'^\mu = \frac{1}{\sqrt{2}(r' + ia\cos\theta)}\left[ia\sin\theta(\delta_0^\mu - \delta_1^\mu) + \delta_2^\mu + \frac{i}{\sin\theta}\delta_3^\mu\right]. \tag{3.166}$$

transformed the Reissner-Nordström metric into a new metric, the Kerr-Newman metric [NJ, NCCEPT].[77] They used the same trick, i.e., the Newman-Janis algorithm[78]:

Newman and Janis represented Einstein's electromagnetic field tensor by the complex tetrad components and applied a complex coordinate transformation. When

---

We apply the transformation given by (3.164). First, transforming l^μ: since $l^\mu = \delta_1^\mu$, it remains unchanged under the given transformation: $l'^\mu = \delta_1^\mu$. This is the first equation in (3.165). Now, transforming n^μ: under the transformation (3.164), r becomes $r' = r + ia\cos\theta$. Therefore, we substitute r with r' in n^μ from Eq. (3.162). Note that the form of n^μ remains the same, but with r replaced by r', we get the second equation in (3.165). Next, we transform m^μ. m^μ involves r, so we need to substitute r with r'. Applying the transformation (3.164), the first equation in (3.163) becomes:

$$m'^\mu = \frac{1}{\sqrt{2}(r + ia\cos\theta)}\left(\delta_2^\mu + \frac{i}{\sin\theta}\delta_3^\mu\right). \tag{3.167}$$

Additionally, we consider the effect of the transformation on δ_0^μ and δ_1^μ, leading to Eq. (3.166). Similarly, for $\bar{m}^\mu$, we substitute r with r'. The second equation in (3.163) becomes:

$$\bar{m}'^\mu = \frac{1}{\sqrt{2}(r' + ia\cos\theta)}\left(\delta_2^\mu - \frac{i}{\sin\theta}\delta_3^\mu\right). \tag{3.168}$$

Additionally, we consider the effect of the transformation on δ_0^μ and δ_1^μ, leading to Eq. (3.166). Thus, the transformed vectors (3.165) and (3.166) are derived by substituting r with $r' = r + ia\cos\theta$ in the expressions (3.162) and (3.163) and appropriately modifying the terms involving δ_0^μ and δ_1^μ.

8. The general transformed metric is given by:

$$g'^{\mu\nu} = l'^\mu n'^\nu + l'^\nu n'^\mu - m'^\mu \bar{m}'^\nu - m'^\nu \bar{m}'^\mu. \tag{3.169}$$

In a private communication, Kerr has shown that the type of complex coordinate transformation (3.164) works for a class of solutions where the metric can be expressed as $g_{\mu\nu} = \eta_{\mu\nu} + \lambda^2 l_\mu l_\nu$. This means the Schwarzschild and Kerr metrics are special cases of this more general form [NJ].

[77] Section 2.3.3 provides a detailed explanation of the Reissner-Nordström metric.

[78] 1. We begin with the static solution to Einstein's field equations, the Reissner-Nordström metric (see Sect. 2.3.2):

$$ds^2 = -\left(1 - \frac{2M}{r} + \frac{e^2}{r^2}\right)dt^2 + \left(1 - \frac{2M}{r} + \frac{e^2}{r^2}\right)^{-1} dr^2 + r^2\left(d\theta^2 + \sin^2\theta\, d\phi^2\right). \tag{3.170}$$

e is the charge of the mass.

2. We transform the metric (3.170) into Eddington-Finkelstein coordinates by introducing the transformation:

$$u = t + \int \left(1 - \frac{2M}{r} + \frac{e^2}{r^2}\right)^{-1} dr. \tag{3.171}$$

This transformation is related to the transformation (3.158). Consider the integral in the general transformation (3.171): For the Schwarzschild case $e = 0$:

$$\int \left(1 - \frac{2M}{r}\right)^{-1} dr. \quad \text{This integral can be computed as:} \tag{3.172}$$

$$\int \frac{r}{r - 2M} dr = r + 2M \ln|r - 2M|. \tag{3.173}$$

3. After applying the transformation (3.171), the Reissner-Nordström metric becomes:

doing so, the electromagnetic field, initially formulated for a non-rotating charged mass, the Reissner-Nordström metric, is adapted to account for the effects of rotation.[79]

$$ds^2 = -\left(1 - \frac{2M}{r} + \frac{e^2}{r^2}\right) du^2 - 2du\,dr + r^2\left(d\theta^2 + \sin^2\theta\,d\phi^2\right). \tag{3.174}$$

4. We write the components of the Reissner-Nordström metric in terms of the complex null tetrad $(l^\mu, n^\mu, m^\mu, \bar{m}^\mu)$, forming the basis of the metric (3.161).

5. The detailed form of the null tetrad vectors (3.161) is given by:

$$l^\mu = \delta^\mu_r, \quad n^\mu = \delta^\mu_u - \frac{1}{2}\left(1 - \frac{2M}{r} + \frac{e^2}{r^2}\right)\delta^\mu_r. \tag{3.175}$$

The complex conjugate vectors m^μ and $\bar{m}^\mu$ are:

$$m^\mu = \frac{1}{\sqrt{2}r}\left(\delta^\mu_\theta + \frac{i}{\sin\theta}\delta^\mu_\phi\right), \quad \bar{m}^\mu = \frac{1}{\sqrt{2}r}\left(\delta^\mu_\theta - \frac{i}{\sin\theta}\delta^\mu_\phi\right). \tag{3.176}$$

6. The null tetrad vectors (3.175) and (3.176) are transformed using the transformation (3.164) into a new set of tetrad vectors (where the general transformed metric is given by (3.169)):

$$l'^\mu = \delta^\mu_r, \tag{3.177}$$

$$n'^\mu = \delta^\mu_u - \frac{1}{2}\left(1 - \frac{2Mr'}{r'^2 + a^2\cos^2\theta} + \frac{e^2}{r'^2 + a^2\cos^2\theta}\right)\delta^\mu_r, \tag{3.178}$$

$$m'^\mu = \bar{m}'^\mu = \frac{1}{\sqrt{2}(r' + ia\cos\theta)}\left[ia\sin\theta(\delta^\mu_u - \delta^\mu_r) + \delta^\mu_\theta + \frac{i}{\sin\theta}\delta^\mu_\phi\right]. \tag{3.179}$$

7. We reconstruct the metric in the new coordinates using the transformed null tetrad vectors. The resulting metric is the Kerr-Newman metric:

$$\begin{aligned} ds^2 = &-\left(1 - \frac{2Mr - e^2}{\rho^2}\right) du^2 - 2dudr + \rho^2 d\theta^2 \\ &+ \left(r^2 + a^2 + \frac{(2Mr - e^2)a^2\sin^2\theta}{\rho^2}\right)\sin^2\theta\,d\phi^2 - 2a\sin^2\theta du d\phi \quad \text{where:} \end{aligned} \tag{3.180}$$

$$\rho^2 = r^2 + a^2\cos^2\theta. \tag{3.181}$$

[79] In developing the Kerr-Newman metric, Newman and Janis [NJ]:

(1) Represented the electromagnetic field tensor $F_{\mu\nu}$ [Eq. (1.387)] by the complex tetrad components ϕ_1, and ϕ_2.

$$(2)\ \phi_1: \quad \phi_1 = \frac{1}{2}F_{\mu\nu}(l^\mu n^\nu + \bar{m}^\mu m^\nu) = -\frac{e}{2r^2}. \tag{3.182}$$

$$(3)\ \phi_2: \quad \phi_2 = F_{\mu\nu}\bar{m}^\mu n^\nu. \tag{3.183}$$

In the Newman-Penrose formalism, the tetrad components, ϕ_0, ϕ_1, and ϕ_2 represent different aspects of the electromagnetic field tensor $F_{\mu\nu}$:

$$(1)\ \phi_0: \quad \phi_0 = F_{\mu\nu}l^\mu m^\nu = 0, \tag{3.184}$$

which indicates a specific polarization condition of the electromagnetic field. The complex tetrad components of the electromagnetic tensor [Eqs. (3.184), (3.182), (3.183)] provide a method for simplifying the incorporation of the electromagnetic field into the Einstein-Maxwell equations,

Newman and his colleagues interpreted the Kerr-Newman solution as representing the fields of a rotating ring of mass and charge. In other words, the Kerr-Newman metric is interpreted as having a ring singularity, similar to the Kerr metric but with the added complexity of an electromagnetic field [NJ] [NCCEPT].

In 1967, Robert H. Boyer and Richard W. Lindquist introduced the *Boyer-Lindquist coordinates*.[80] In 1974, Subrahmanyan Chandrasekhar published a review entitled "The Black Hole in Astrophysics: The Origin of the Concept and its Role" [Cha]. In this review, Chandrasekhar provides an in-depth analysis of the Kerr metric, employing Boyer-Lindquist coordinates to articulate a solution to Einstein's vacuum field equations characterized by two parameters, M and a and exhibiting asymptotic flatness [Cha]. The metric is formally expressed as:

$$ds^2 = -\frac{\Delta}{\rho^2}\left[dt - a\sin^2\theta d\phi\right]^2 + \frac{sin^2\theta}{\rho^2}\left[(r^2+a^2)d\phi - adt\right]^2 + \frac{\rho^2}{\Delta}dr^2 + \rho^2 d\theta^2. \tag{3.187}$$

$$\text{where: } \rho^2 = r^2 + a^2\cos^2\theta \quad \Delta = r^2 + a^2 - 2Mr.$$

with the solution delineated in natural units such that $c = G = 1$.

---

which describe the combined effects of gravity and electromagnetism, Eq. (1.290), where $T_{\mu\nu}$ is the total stress-energy tensor, which includes contributions from matter and the electromagnetic tensor defined by equations (1.386) and (1.387).

(2) Applied the transformation (3.164). The components of the electromagnetic field tensor $F_{\mu\nu}$ in terms of the tetrad vectors ϕ_1 and ϕ_2 are expressed relative to the transformed coordinates (3.164), representing how the rotation modifies the electromagnetic field's structure around the mass. The components (3.182) and (3.183) are influenced by the complex transformation (3.164). We first transform ϕ_1. We substitute $r = r' - ia\cos\theta$, and get:

$$\phi_1' = \frac{1}{2}F_{\mu\nu}(l'^{\mu}n'^{\nu} + \bar{m}'^{\mu}m'^{\nu}) = -\frac{e}{2(r' - ia\cos\theta)^2}. \tag{3.185}$$

We then transform ϕ_2. Under the coordinate transformation (3.164), and given Eqs. (3.165) and (3.166), ϕ_2 becomes:

$$\begin{gathered}\phi_2' = F'_{\mu\nu}\left(\frac{1}{\sqrt{2}(r' + ia\cos\theta)}\left[ia\sin\theta(\delta^{\mu}_{u} - \delta^{\mu}_{r}) + \delta^{\mu}_{\theta} - \frac{i}{\sin\theta}\delta^{\mu}_{\phi}\right]\right) \\ \left(\delta^{\nu}_{u} - \frac{1}{2}\left(1 - \frac{2Mr'}{r'^2 + a^2\cos^2\theta} + \frac{e^2}{r'^2 + a^2\cos^2\theta}\right)\delta^{\nu}_{r}\right).\end{gathered} \tag{3.186}$$

[80] In Boyer-Lindquist coordinates, the Kerr metric takes a form that explicitly showcases the features of the rotating system (black hole's) rotation and its effects on the surrounding spacetime, see Eq. (3.187). The coordinates are (t, r, θ, ϕ). Boyer and Lindquist called their coordinates *Schwarzschild-Like coordinates* because they are reminiscent of Schwarzschild coordinates but are adapted to describe the spacetime around a rotating mass, specifically for the Kerr solution [Bo-Li]. In Schwarzschild coordinates, the symmetry is spherical, meaning the geometry remains unchanged under any rotation about the central mass. The Boyer-Lindquist system uses a time coordinate (t), a radial coordinate (r), and two angular coordinates (θ and ϕ), drawing a direct analogy to spherical coordinates. This structure facilitates the description of radial and angular movements in a central gravitational field. In Boyer-Lindquist coordinates, the symmetry is axial, reflecting the axis of rotation of the mass and its associated effects on the spacetime geometry.

Significantly, setting $a = 0$ simplifies the Kerr metric (3.187) to the Schwarzschild solution (2.94). Furthermore, a test particle tracing a geodesic within the Kerr metric (3.187) from a remote vantage will perceive its motion as akin to the gravitational field exerted by a mass M with angular momentum $J = aM$, a deduction congruent with the Lense-Thirring effect.

For the Kerr-Newman metric, the line element in Boyer-Lindquist coordinates is:

$$\begin{aligned} ds^2 = &-\left(1 - \frac{2mr - e^2}{\rho^2}\right) dt^2 - \frac{(2mr - e^2)2a\sin^2\theta}{\rho^2} dt d\phi + \frac{\rho^2}{\Delta} dr^2 \\ &+ \rho^2 d\theta^2 + \left(r^2 + a^2 + \frac{(2mr - e^2)a^2\sin^2\theta}{\rho^2}\right) \sin^2\theta d\phi^2, \end{aligned} \tag{3.188}$$

$$\text{where: } \rho^2 = r^2 + a^2\cos^2\theta, \quad \Delta = r^2 - 2mr + a^2 + e^2.$$

Chandrasekhar's discourse culminates in the assertion, underpinned by the Carter *uniqueness theorem* that the inevitable conclusion of a massive rotating star's collapse is the formation of a black hole, the external metric of which aligns with Kerr's metric (3.187). The angular momentum of the resultant black hole is denoted as a. Analogous to Schwarzschild's metric, the Kerr metric features an event horizon. For the Kerr metric, the event horizon is defined by:

$$R_K = \left[M + (M^2 - a^2)^{\frac{1}{2}}\right] \frac{G}{c^2}. \tag{3.189}$$

Chandrasekhar underscores that the preceding formulation presumes $M < a$; contrarily, the absence of an event horizon would signify the presence of a naked singularity [Cha].

Kerr demonstrated that his metric (3.153) is algebraically special, belonging to the Petrov Type D class. Type D spacetimes, such as the Kerr metric (describing rotating black holes), have a unique property where the Weyl tensor has two pairs of principal null directions. This special property makes Type D spacetimes highly symmetric and particularly significant in the context of black holes. The principal null directions in the Kerr spacetime are closely related to the null geodesics. One of the principal null directions is associated with the outgoing and ingoing null geodesics that are tangent to the event horizon of the Kerr black hole. The other principal null direction is related to the null geodesics aligned with the black hole's rotation axis. This alignment involves a frame-dragging effect (see footnote 75).

The properties and symmetries associated with these principal null directions, particularly their relationship to the event horizon and the effect of the black hole's rotation, contribute to the uniqueness of the Kerr metric. The alignment of the null geodesics with the black hole's event horizon and rotation axis (and the properties derived from it) allows the Kerr metric to be uniquely identified by its mass and angular momentum. The no-hair theorem (discussed in the next part of the book) asserts that a black hole in (classical) general relativity is fully described by a limited set of parameters: mass, angular momentum, and possibly electric charge (for the Kerr-Newman metric). The properties of the Kerr metric, especially its classification as Petrov Type D and the special alignment of its principal null directions, underpin this theorem.

3.3.4 Neurath's Thousand-Dollar Bill and the LIGO Breakthrough

Otto Neurath suggested that a behaviouristician may predict more accurately in certain situations than a physicist.[81] This is illustrated with a hypothetical scenario: a physicist trying to predict where a thousand-dollar bill, swept away by the wind, will land in Saint Stephen's Square. This task is extremely challenging for a physicist because it involves complex and unpredictable factors like wind patterns, the aerodynamics of the bill, and environmental variables. In contrast, a *behaviouristician*, who studies behavior under controlled conditions, might be able to predict the outcome of a conditioning experiment more accurately [Neur33].

Nancy Cartwright expands on Neurath's example of the thousand-dollar bill swept away by the wind in Saint Stephen's Square to explore broader themes about the nature of scientific laws. She discusses applying scientific laws, particularly in physics, to real-world situations. According to Cartwright, Newton's laws are actually *ceteris paribus* (all else being equal), applicable only in idealized and controlled conditions. In real-world situations where multiple forces and factors interact, these laws alone are insufficient for explanation. Cartwright asserts that universal laws, such as Newton's law of gravitation, are true only under certain conditions and with *ceteris paribus* clauses. Such laws are false without these clauses as they do not apply universally. To illustrate her point, Cartwright uses Neurath's example of the thousand-dollar bill swept away by the wind in Saint Stephen's Square. Cartwright agrees with Neurath's point that, in some cases, a physicist might be less effective at prediction than a behaviorist psychologist due to the complexity and unpredictability of certain physical situations. Cartwright notes that while mechanics provides a partial model for the falling thousand-dollar bill by considering gravitational force, it fails to account for other forces, like the wind. This incompleteness challenges the idea that mechanics can fully predict the bill's behavior. She contrasts the *fundamentalist* belief that mechanics has a model for every physical situation with the practical limitations of such models. She argues that the successes of mechanics in its domain do not justify the belief in its universality.

---

[81] Neurath introduceds the term *behaviouristics*, contrasting it with *behaviorism* because the name *behaviorism* designates a movement in psychology. This reflects the Vienna Circle's rejection of traditional philosophical systems and ideologies, often called "isms," favoring a more empirical and scientific approach to knowledge and philosophy. *Behaviouristics* is a part of unified science and deals with observable, measurable behaviors. This perspective is evident in Neurath's advocacy for using *physicalism* in psychology, which suggests a preference for empirical, observable, and measurable phenomena—characteristics typical of physical sciences.

Cartwright highlights the limitations of applying generalized, simplified laws to complex, specific situations. She illustrates this by stressing that the laws of fluid dynamics might be more appropriate or effective than those of classical mechanics when explaining phenomena like the movement of a bill in the wind. However, modeling the motion of the thousand-dollar bill using fluid dynamics is still problematic because it involves many variables. These include factors like wind speed, wind direction, air density, the bill's shape and material properties, and possibly many others. The complexity and variability inherent in these factors mean that fluid dynamics, while theoretically capable of modeling the bill's motion, may not always offer practical or precise predictions for every real-world situation. This complexity underlines the limitations of using traditional, simplified models in physics to predict complex, real-world phenomena [Car-99].

Cartwright and Neurath's metaphor of the thousand-dollar bill being swept away by the wind is aptly applied to the challenges faced by the LIGO collaboration. The LIGO team directly compared the observed gravitational wave data with numerical relativity simulations of binary black hole coalescence. These simulations provided a detailed and physically accurate representation of the gravitational waveforms expected from such cataclysmic events. However, the LIGO collaboration acknowledges that their computer models, although sophisticated and based on numerical relativity, still do not perfectly match the full predictions of numerical relativity in all aspects. These models involve computational simulations of black hole mergers and gravitational wave generation, which are inherently complex and require substantial computational resources. These models predict and analyze gravitational wave signals detected by LIGO. The goal is to have the most accurate and comprehensive models possible, but given the complexity of the phenomena and computational limitations, some approximation is inevitable. The limitations are computational practicality and the challenges in accurately modeling incredibly complex systems. While these models are based on the full theory of general relativity, they might not represent every complexity due to the discretization of spacetime in simulations, handling boundary conditions, numerical resolution, and other computational constraints [LIG-3].

The metaphor of the bill being swept away by the wind illustrates the unpredictability and complexity of real-world phenomena. In such scenarios, controlled and idealized scientific models struggle to predict outcomes due to many interacting variables and external influences.

In 2016, the LIGO Scientific and Virgo Collaborations announced the first joint detection of gravitational waves (GW150914) using the LIGO detectors.[82] They

[82] The coalescence of the two black holes involves three stages: 1. *Inspiral*: The black holes orbit each other in a quasi-circular orbit, gradually moving closer as they emit gravitational waves with increasing amplitude and frequency. This phase can be described analytically using post-Newtonian approximations. 2. *Merger*: The black holes merge as their orbit shrinks, requiring numerical simulations due to the strong, dynamic gravitational fields. The gravitational wave amplitude peaks during this phase. 3. *Ringdown*: The newly formed black hole settles into a final state, emitting gravitational waves as it oscillates. This phase is described using perturbation theory, with the grav-

reported several quantitative tests on the gravitational wave data from GW150914 to validate the Kerr black hole hypothesis. These tests aimed to confirm that the data were consistent with a rotating black hole solution to Einstein's field equations. The results indicated that the GW150914 inspiral-merger-ringdown (IMR) waveform did not deviate from the predictions of a binary black-hole merger in classical general relativity.[83] LIGO's task of detecting gravitational waves and analyzing their

---

itational waves characterized by specific frequencies and decay rates determined by the black hole's mass and spin.

[83] LIGO performed the following tests on the gravitational-wave signal GW150914 [LIG-2]:

1. The most probable (general relativistic) *GR waveform* (describing the gravitational wave signal expected from two black holes merging, a binary black hole merger) is subtracted from the detected GW150914 data. The remaining signal is the *coherent (GW-like) residual strain* (GW = gravitational wave). The residual strain must be similar in both LIGO detectors and resemble a gravitational wave signal rather than random noise. By constraining this residual strain, LIGO can determine if any leftover signals do not match the general relativistic predictions, which could indicate potential violations of general relativity.

2. Verifying that the mass and spin parameters of the final black hole, as predicted from the binary's inspiral signal, are consistent with the final parameters inferred from the post-inspiral (merger and ringdown) signal. Demonstrating that the data following the peak of GW150914 are consistent with the least-damped quasinormal mode (QNMs) inferred from the final black hole's characteristics. Quasinormal modes (QNMs) are characteristic oscillations of the final black hole formed after the merger. The least-damped QNM loses its energy (or damps) the slowest, making it the most prominent in the signal. The data are compared to this mode because it provides a clear and robust prediction from general relativity about how the final black hole should behave. Consistency with the least-damped QNM means the observed data matches the theoretical predictions for the mass and spin of the black hole, reinforcing that the final black hole behaves as expected according to general relativity.

3. Performing targeted measurements of the post-Newtonian and phenomenological coefficients that parametrize theoretical waveform models. The post-Newtonian coefficients describe the gravitational waveform during the inspiral phase in terms of a series expansion in powers of $\frac{v}{c}$, where v is the relative velocity of the binary components and c is the speed of light. Phenomenological coefficients are parameters derived from numerical relativity simulations and other models to describe the waveform during the merger and ringdown phases.

4. Searching for evidence of dispersion in the propagation of GW150914 towards Earth, which would appear in a theory where the graviton has a finite Compton wavelength (nonzero mass). The graviton Compton wavelength is the wavelength associated with the graviton's mass. If the graviton (the hypothetical particles that mediate gravitational forces) has mass, it implies a finite Compton wavelength. Constraints on this wavelength help to limit or rule out theories with massive gravitons. If gravitons have mass, they travel at different speeds depending on their energy (or frequency), causing dispersion.

5. Investigating the presence of non-general relativistic polarization states in GW150914. General relativity predicts gravitational waves with two polarization states transverse to the direction of wave propagation. Alternative theories of gravity may predict additional polarization states. The BayesWave algorithm (see explanation below) was used to reconstruct the waveform, assuming a pure scalar mode and comparing it with general relativity models. Pure scalar mode is one type of non-general relativistic polarization, where the gravitational wave is a scalar field that causes isotropic expansion and contraction (breathing mode) rather than the transverse stretching predicted by general relativity. The LIGO network configuration could not exclude this possibility.

6. LIGO used LALInference software library—a Bayesian analysis tool for parameter estimation of gravitational wave signals based on models derived from general relativity. It is primarily written in C, but it has Python bindings, which allow it to be used within Python pipelines. The Python

sources (such as black hole mergers) is akin to predicting the bill's path in the wind. Gravitational wave signals arise from astrophysical events of extreme complexity, influenced by numerous factors that must be accounted for in the models used for data analysis. While applicable in controlled conditions, idealized scientific laws and models often fail to convey the full scope of complexity in natural phenomena. This is analogous to using simplified physical models to predict the bill's trajectory, which unpredictable wind patterns and environmental variables can thwart. LIGO uses sophisticated models based on general relativity to interpret the data when detecting gravitational waves. These models are incredibly advanced but still have limitations. They must contend with the intricacies of astronomical events, noise in the data, and the immense computational challenge of simulating black hole interactions. While these models are rooted in the comprehensive framework of general relativity, they, like the physicist with the bill, face the challenge of applying theoretical knowledge to complex, real-world scenarios.

The behavior of the bill in the wind is subject to many unpredictable factors, making accurate prediction challenging. A physicist might make approximations or

---

bindings make it easier to script and automate analyses using the extensive scientific and data manipulation libraries available in Python. It was used to identify the most probable binary black-hole waveform and subtract it from the data to test consistency with general relativity. LIGO used another Bayesian tool, the BayesWave GW-transient analysis algorithm, to identify and characterize gravitational wave transients (short-duration signals). The core computational engine is written in C^{++}. The collaborations used BayesWave to compare models (e.g., Gaussian noise, glitches, gravitational wave signals) to determine which best fits the observed data.

7. The collaborations calculated the signal strength in the data after subtracting the best-fit model. If the general relativistic model is accurate, the residuals (the remaining signal after subtracting the best-fit GR waveform from the detected GW150914 data) should primarily consist of instrumental noise, with no significant remaining gravitational wave signal. They combined data from multiple detectors (like LIGO Hanford, LIGO Livingston, and Virgo) for the single gravitational wave event GW150914 coherently to improve sensitivity and check for consistency across the network. They determined the maximum SNR value the residuals could have with 95% confidence, assuming the residuals are just noise. The observed residual SNR below this threshold means the residuals are consistent with instrument noise. No statistically significant evidence of any leftover signal could not be explained by noise.

8. The collaborations computed the Bayes factor between a damped-sinusoid waveform model (the oscillations of the black hole as it settles into a stable state after the merger, gradually losing energy) and Gaussian noise to test the consistency of the data with the predicted least-damped QNM of the remnant black hole. By comparing the damped-sinusoid waveform to Gaussian noise, LIGO could determine whether the observed signal contains the expected QNM oscillations or was just random noise.

9. Finally, they conducted a parametrized deviation analysis of the GW150914 waveform using a generalized inspiral-merger-ringdown (GIMR) waveform model to check for deviations from general relativity. This analysis involves modifying the theoretical waveform model by adding parameters representing possible deviations from general relativity. By fitting these modified models to the GW150914 data, the collaborations could see whether any deviations from general relativity were necessary to explain the data.

These tests aimed to detect deviations from predictions of general relativity and, within the limits set by LIGO's sensitivity and the nature of GW150914, found no statistically significant evidence against the hypothesis that GW150914 was emitted by two black holes merging to form a single, rotating Kerr black hole, per Einstein's vacuum field equations (1.288) and (1.289).

use probabilistic methods to predict where the bill might land, acknowledging the inherent uncertainty. Similarly, LIGO's analyses deal with uncertainties, approximations, and computational limitations. The detection and interpretation of gravitational wave signals require sophisticated algorithms that can sift through noisy data and match observed signals with theoretical models, all while acknowledging the limits of these models. Neurath and Cartwright's metaphor underscores the need for continuous refinement and adaptation of scientific models to better handle real-world complexities. This is mirrored in LIGO's ongoing efforts to improve gravitational wave detection and analysis. The collaboration continuously refines its techniques, updates its models, and incorporates more data to enhance the accuracy and reliability of its findings. Thus, the LIGO experiments represent an approach that acknowledges and addresses the complexities and uncertainties inherent in real-world data, moving beyond the simplified models to the practical understanding of gravitational phenomena in the real, *mottled* world.

Part II
Quantum Gravity

Chapter 4
Challenging General Relativity: Black Hole Thermodynamics

4.1 Introduction

I begin with Einstein's views on the Second Law, particularly his analogy of the perpetuum mobile of the second kind (closely related to Maxwell's Demon). He viewed thermodynamics as a principle theory akin to his theory of relativity. I show that Einstein recognized that principles of impossibility underpinned both theories. Thermodynamics was constrained by the impossibility of perpetual motion machines of both the first and second kind (machines that continuously extract heat and convert it into work without any other effect), while the impossibility of superluminal signals bounded relativity. These impossibilities were fundamental to the structure of physical laws. Einstein's involvement with thermodynamic principles had practical implications. His insights contributed to the understanding of statistical mechanics and quantum theory.

The subsequent topic addressed in this chapter, black hole thermodynamics, is closely related to the preceding discussion. By pondering the nature of black holes, John Archibald Wheeler contemplated an irreversible process where entropy increases, as exemplified by two tea cups of different temperatures reaching equilibrium. He suggested that dropping these cups into a black hole could seemingly violate the second law of thermodynamics. In response to Wheeler's tea cup thought experiment, Jacob Bekenstein suggested that a black hole should possess finite entropy proportional to its event horizon's surface area. This idea, foundational to black hole thermodynamics, led Bekenstein to formulate the *Generalized Second Law of Thermodynamics* (GSL). Bekenstein encountered a challenge when Robert Geroch proposed a thought experiment, known as *Geroch's engine*, that contradicted the GSL. This experiment involved a massless box filled with radiation, which, when lowered towards a Schwarzschild black hole and opened at a specific point, allowed the radiation to escape into the black hole. This process seemed to convert all radiation energy into work without changing the black hole's state, implying it had zero temperature. Bekenstein attempted to reconcile this by suggesting an immense entropy for black holes and providing a quantum-based explanation to prevent the box from

G. Weinstein, *Einstein's Legacy*, History of Physics,
https://doi.org/10.1007/978-3-031-73572-1_4

reaching the critical point. Hawking found this unconvincing, arguing that a black hole with zero effective temperature could indefinitely accumulate entropy without alteration. In this sense, a black hole transcends the second law of thermodynamics. Bekenstein countered by assigning a finite, non-zero temperature to black holes linked to their surface gravity. Hawking then argued that this implied black holes could thermally equilibrate with radiation at non-zero temperatures. However, this presents a paradox since black holes can absorb but not emit matter. Bekenstein's cumbersome explanation for why his GSL remained unviolated in these scenarios did not convince Hawking, who saw no feasible equilibrium for a black hole with thermal radiation at a temperature other than zero.

Faced with these paradoxes, Hawking suggested that the resolution lay in accepting that black holes emit thermal radiation, as per quantum mechanics. In 1974, he derived an Eq. (4.42) supporting this by using a semi-classical approximation combining quantum field theory with classical general relativity. By June 1975, he demonstrated that the emission spectrum of this *Hawking radiation* was thermal, linking it to various black hole properties like event horizon area and surface gravity. In August 1975, Hawking further substantiated his theory on Hawking radiation. Hawking revisited the challenge that posed a threat to the GSL and solved the problem of a black hole surrounded by black-body radiation by describing a black hole emitting Hawking radiation. This narrative underscores the intellectual tug-of-war between Bekenstein, who champions black hole thermodynamics, and Hawking, who advocates quantum field theory.

Hawking initially resisted Bekenstein's ideas, surprised to find his calculations suggesting particle emission from black holes [Haw-88]. In contrast, Bekenstein, despite acknowledging Hawking's contributions, noted the irony in Hawking's initial opposition to his thermodynamic theories. He suggested that while the entropy of black holes could bear both their names, the radiation should solely be attributed to Hawking, recognizing his unique contribution to its understanding [Cla]. While Bekenstein's work on black hole thermodynamics, specifically the GSL, encountered problems, these issues spurred further theoretical development. The effort to resolve the problems in black hole thermodynamics led to Hawking's discovery of Hawking radiation. This discovery was significant but did not overturn the existing theories of general relativity or quantum mechanics. The discovery of Hawking radiation introduced a new paradox—the black hole information paradox—which did not result in rejecting the existing theories but opened new areas of inquiry. Hawking radiation initially served as a conceptual tool to address the apparent contradictions and challenges in black hole thermodynamics, particularly concerning the GSL. The introduction of Hawking radiation simplified the complex problem of black hole entropy and temperature, providing a more intuitive understanding of how black holes could adhere to the laws of thermodynamics. However, Hawking radiation is not just a problem-solving tool but a fundamental theoretical contribution that significantly altered our understanding of black holes and quantum mechanics. Hawking's work on black hole radiation pioneered integrating concepts from quantum field theory and general relativity.

In this chapter, I explore how Hawking's discovery of Hawking radiation emerged within unresolved challenges and theoretical dilemmas. Bekenstein's struggles to address the quandaries posed by Geroch's thought experiment and other challenges in the field significantly contributed to the backdrop against which Hawking's research unfolded. This ultimately led to a deeper understanding of black holes, quantum mechanics, and general relativity, fundamentally altering our perceptions of these complex phenomena.

4.2 Revisiting Classical Relativity and Thermodynamics

4.2.1 The No-Hair Theorem and Hawking's Area Law

Brandon Carter (in 1971) and Werner Israel (in 1967) suggested *uniqueness theorems* in black hole solutions. Israel's theorem establishes the uniqueness of the Schwarzschild and the Reissner-Nordström solutions as the only static, asymptotically flat, vacuum solutions of Einstein's field equations with a regular event horizon [Isr]. Carter identified the Kerr and Kerr-Newman solutions as the unique stationary, axisymmetric solutions to Einstein's field equations in a vacuum that feature an absolute event horizon [Cart]: For a given mass M and angular momentum J, the Kerr metric is the unique, asymptotically flat, stationary, axisymmetric solution of the vacuum Einstein field equations.

In 1969, Roger Penrose visualized a scenario describing the collapse of a body to the size of its gravitational radius, followed by the formation of a trapped surface and an absolute event horizon. He suggested that the field would eventually settle into a Kerr solution or a Kerr-Newman solution if a nonzero net charge is present in the black hole [Pen-69]. Penrose pondered whether there was a connection between the existence of a trapped surface and a physical spacetime singularity, like the one at $r = 0$ in the Schwarzschild solution. In 1970, Penrose and Stephen Hawking answered this by demonstrating through a singularity theorem that a trapped surface invariably indicates a spacetime singularity [HP]. Hawking and Penrose's singularity theorem implies that a singularity must emerge when forming a stellar-mass black hole. Hawking referred to this as the Carter-Israel conjecture, stating that if an absolute event horizon develops in an asymptotically flat spacetime, the exterior solution will asymptotically approach a Kerr-Newman solution over time [Haw-71]. Additionally, Penrose noted that the Kerr and Kerr-Newman solutions are explicit asymptotically flat stationary solutions of the general relativity field equations, characterized by just three parameters: mass, angular momentum, and electric charge [Pen-69].

John Archibald Wheeler later added a distinct perspective to the Kerr metric, famously and metaphorically suggesting that black holes have "no hair." in this analogy, "hair" represents additional features that could differentiate one black hole from another, which, according to the *no-hair theorem*, do not exist. According to the no-hair theorem, all physical properties of matter falling into a black hole are obliterated, rendering two black holes indistinguishable if they share the same mass, angular momentum, and electric charge [Whe-81].

According to the Israel-Carter uniqueness theorems, a single black hole formed from the merger will settle into a Kerr black hole, defined by its mass and spin. The LIGO and Virgo collaborations analyzed the GW150914 gravitational wave signal's evolution in frequency and amplitude, concluding that the final object was consistent with a Kerr black hole. They identified the final black hole's mass as $M_{\text{BH}} = 65M_{\odot}$ (where $M_{\odot}$ is the mass of the sun) and a spin parameter $a_3 = 0.7$. The ringdown phase's damping time was approximately equal to the oscillation period, supporting the conclusion that the signal corresponded to a Kerr black hole. For a black hole with a spin of $a_3 = 0.7$, they calculated a ringdown frequency of approximately $260Hz$ and a damping time of 4 milliseconds. This value is proportional to the black hole's mass, meaning a more massive black hole would have a proportionally longer damping time. Thus:

$$\text{Ringdown frequency} \approx \frac{260\text{ Hz}}{\frac{M_{\text{BH}}}{65M_{\odot}}} = \frac{260\text{ Hz} \times 65M_{\odot}}{M_{\text{BH}}} \tag{4.1}$$

$$\text{Damping time} \approx 4\text{ ms}\left(\frac{M_{\text{BH}}}{65M_{\odot}}\right) \tag{4.2}$$

The final mass of $65M_{\odot}$ was consistent with the merger of two black holes of around $35M_{\odot}$ each, after accounting for the energy emitted as gravitational waves. The extracted parameters for the progenitor black holes showed slight differences in their masses and spins, reflecting the complexity of their formation and evolution. Specifically, the LIGO and Virgo collaborations extracted the following parameters for the masses and spins of the two progenitor black holes [LIG-1]:

$$M_{\text{BH}_1} = 36.2^{+5.2}_{-3.8}M_{\odot},\ a_1 = 0.3^{+0.5}_{-0.3},\ M_{\text{BH}_2} = 29.1^{+3.7}_{-4.4}M_{\odot},\ a_2 = 0.5^{+0.5}_{-0.4}, \tag{4.3}$$

and the measured mass and spin for the remnant produced by the binary coalescence is:

$$M_{\text{BH}_3} = 62.3^{+3.7}_{-3.1}M_{\odot},\ a_3 = 0.68^{+0.05}_{-0.06}, \tag{4.4}$$

This is the mass of the new, larger black hole formed after the merger. The mass of the remnant black hole is less than the sum of the masses of the two progenitors.

The above masses represent the masses in the source frame, not the detector frame.[1] The relation between the two frames involves the redshift z of the source[2]:

$$z = 0.09^{+0.03}_{-0.04}. \tag{4.6}$$

The mass of the black hole as it is observed in the detector frame M_f is related to the mass of the black hole in the source frame by[3]:

$$M_f = (1+z)M_{BH3}. \tag{4.7}$$

The collaboration detected a mass of the black hole in the detector frame of approximately $M_f \approx 67.9 M_\odot$. Assuming the remnant is a Kerr black hole, and using the fundamental mode $l = m = 2$ and its first overtone $n = 0$, the GW150914 ringdown was analyzed.[4]

In 2016, Hawking reflected on the observation of gravitational waves from the GW150914 event detected by LIGO's Hanford and Livingston observatories on September 14, 2015. Hawking expressed that this event marked two significant tests in astrophysics. First, the initial detection of a binary black hole system and the first-ever observation of black holes merging. Second, he noted that the characteristics of this system aligned with his theoretical predictions made in 1970 in Cambridge. Specifically, he highlighted that the area of the final black hole formed by the merger was larger than the combined areas of the two initial black holes, a phenomenon he had predicted in his area law [Haw-16].

---

[1] When gravitational wave signals are detected, the reported masses of black holes can be expressed either in the source frame or in the detector frame. The source frame refers to the actual physical masses of the objects as they would be measured locally in the vicinity of the source.

[2] Cosmological redshift z measures how much the universe's expansion has stretched the wavelength of gravitational waves (or light) as they travel from the source to Earth. The cosmological redshift is described by the redshift parameter z, where:

$$1 + z = \frac{\lambda_o}{\lambda_e} \tag{4.5}$$

with λ_o and λ_e being the observed and emitted wavelengths, respectively. The *cosmological redshift* is different from the *gravitational redshift* Einstein predicted in 1907 and 1911, which is related to time dilation in a gravitational field.

[3] M_f represents the detector frame or redshift mass, the mass of the black hole as observed in the detector frame, after being adjusted for the redshift effects.

[4] In the ringdown phase of a black hole merger the new black hole emits gravitational waves as it oscillates and settles into a final state. These oscillations can be described by quasi-normal modes, characterized by specific frequencies and decay rates determined by the black hole's mass and spin. The fundamental mode $l = m = 2$ and its first overtone $n = 0$ are particularly significant. The fundamental mode $l = m = 2$ is often the dominant mode observed in the ringdown signal because it tends to be the strongest and most observable. This signal's characteristic frequency and decay rate are functions of the black hole's mass and spin. Thus, this fundamental mode is significant for gravitational wave detectors like LIGO and Virgo, providing key information about the black hole's properties.

In 1971, Hawking proposed an area law stating that the area of a black hole's event horizon never decreases. Hawking considered a scenario where, initially, there were two black holes at a considerable distance apart. These black holes are assumed to have formed earlier due to gravitational collapses or amalgamation of smaller black holes. The two black holes, initially separate, merge to form a single black hole that eventually settles into a Kerr black hole.[5] The energy during this process is emitted as gravitational radiation (waves). Hawking used this scenario to explore the merger's theoretical limits on efficiency ϵ[6]:

$$\epsilon = \frac{M_{BH_1} + M_{BH_2} - M_{BH_3}}{M_{BH_1} + M_{BH_2}} \tag{4.8}$$

$$\epsilon = \frac{\text{ENERGY EMITTED AS GRAVITATIONAL WAVES}}{\text{TOTAL INITIAL MASS- ENERGY OF THE BLACK HOLES}}.$$

The limit on the efficiency ϵ depends on the specific dynamics of the merger, including the masses and spins of the initial black holes. The efficiency is influenced by the spin a of the black holes involved. When two non-spinning, equal-mass black holes merge, the resulting black hole is a Schwarzschild black hole characterized by no spin. Hawking found that the efficiency of gravitational wave emission in such cases is significant but not as high as in some mergers involving spinning black holes [Haw-71].

Hawking then discussed how the area of a black hole's event horizon behaves during black hole mergers and the absorption of matter and radiation by black holes. He articulated an *area law*. Hawking's area law states that the total area of the event horizons of the colliding black holes cannot decrease over time. The combined area of the event horizons of two merging black holes must be less than or equal to the area of the event horizon of the resulting black hole.

Hawking gave formulas involving the masses M_{BH} and specific angular momenta a of the initial and final black holes, indicating how the event horizon's area could be calculated and how it changed through the merger process. He gave the following equation for the area of the event horizon of the newly formed black hole [Haw-71][7]:

$$A_K = 8\pi \left(\frac{GM_{\mathrm{BH}}}{c^2}\right)^2 \left(1 + \sqrt{1 - a^2}\right), \tag{4.9}$$

[5] Hawking uses the term *coalescence* to describe the process of two black holes merging into one. This term refers to the phase where the black holes combine their masses and angular momenta into a single Kerr black hole.

[6] The numerator $(M_{BH_1} + M_{BH_2} - M_{BH_3})$ represents the energy emitted as gravitational waves during the merger, quantified in terms of the mass-energy equivalent (according to $E = mc^2$). The denominator $(M_{BH_1} + M_{BH_2})$ represents the total initial mass-energy of the merging black holes. This efficiency ϵ measures how much of the mass-energy of the black holes is radiated away as gravitational waves versus how much remains as the mass-energy of the remnant black hole.

[7] The term $1 + \sqrt{1 - a^2}$ accounts for the black hole's rotation effects on its event horizon area, showing that rotation increases the event horizon area for a given mass.

where he showed that the area A_K of the event horizon of the resulting black hole is greater than the sum of the areas of the event horizons around the original black holes[8]:

$$A_K \geq (A_1 + A_2). \tag{4.10}$$

Hawking substituted the formula for A_K from (4.10) for each of the black holes (initial and final). He obtained an inequality that compares the combined mass and spin parameters of the initial black holes (M_{BH_1}, a_1 and M_{BH_2}, a_2) with those of the final black hole (M_{BH_3}, a_3) (4.9), which is expressed as follows [Haw-71][9]:

$$M_{\mathrm{BH}_3}^2\left(1+\sqrt{1-a_3^2}\right) > M_{\mathrm{BH}_2}^2\left(1+\sqrt{1-a_2^2}\right) + M_{\mathrm{BH}_1}^2\left(1+\sqrt{1-a_1^2}\right). \tag{4.11}$$

Substituting Eqs. (4.3) and (4.4) into Eq. (4.11) leads to the conclusion that the area law is consistent with the GW150914 data. Equations (4.3), (4.4) and (4.9) give:

$$A_1 + A_2 = (2.2 \pm 0.2) \times 10^5 km^2,\ A_K = (3.8 \pm 1.3) \times 10^5 km^2. \tag{4.12}$$

Hawking's law (4.10) is shown to be consistent with the GW150914 data [GMU]. The confirmation of Hawking's law is specific to the GW150914 event.

4.2.2 The Second Law and Einstein's Relativity

Einstein frequently compared the foundational principles of relativity with those of thermodynamics [Gio]. He viewed both as theories of principle grounded in fundamental observations about nature. In discussing the Michelson-Morley experiment and Lorentz's and Minkowski's theories, Einstein referred to thermodynamics. He likened Lorentz's dynamical contraction (due to physical forces) and Minkowski's kinematical interpretation (independent of physical forces) to different ways of approaching the Second Law based on molecular motion versus a principle-based approach [CPAE3]. This perspective resonates throughout his work, from his *Autobiographical Notes* to various correspondences and lectures.

In his *Autobiographical Notes*, Einstein discussed relativity in the context of thermodynamics. Einstein recalls that before 1905, he was looking for a general principle for the electrodynamics of moving bodies, a principle of the kind one finds

[8] In 1973, Bardeen, Carter, and Hawking formulated the four laws of black hole mechanics, with the second law being Hawking's 1971 area law [BCH]. This law asserts the non-decreasing nature of the area of the event horizon of each black hole and its implications in the case of black hole coalescence.

[9] This equation extends the event horizon area conservation to express a relationship between the masses and spin parameters of the black holes before and after the merger, indicating that the mass-energy equivalent of the area of the final black hole's event horizon is greater than the sum of the mass-energy equivalents of the areas of the merging black holes' event horizons.

in thermodynamics, the Second Law. He was inspired by thermodynamics, based on a fundamental rule: the impossibility of creating a *perpetuum mobile* (of the first and second kind). How could one discover such a universal principle? After a decade of contemplation, he found an answer rooted in a paradox he encountered at sixteen. Imagine chasing a beam of light at its speed, the speed of light in a vacuum. The light beam would appear as a stationary oscillating electromagnetic field if possible. Yet, there seems to be no such a phenomenon in practical experience or according to Maxwell's equations [Ein-49-1]. In discussing relativity, Einstein often returned to thermodynamics as a foundational analogy. He saw the development of relativity, like thermodynamics, as driven by fundamental empirical observations. This viewpoint underscored his belief in the importance of principle theories, which are grounded in unassailable physical principles rather than the details of constructive models. For thermodynamics, this observation is the impossibility of perpetual motion (both of the first and second kind). For relativity, it is the constancy of the speed of light and the principle of relativity itself.

In 1919, Einstein emphasized that Thermodynamics aims to analytically derive the essential conditions that events must meet based on the widely accepted fact that perpetual motion is impossible [Ein-19]. In an undated letter to Maurice Solovine, Einstein states that the approach used in the theory of relativity is similar to that used in thermodynamics. Thermodynamics provides a structured response to the question of how natural laws must be designed to prevent the occurrence of perpetual motion [Ein-34]. He writes that the ideal scenario arises when new fundamental hypotheses are derived from real-world experiences. The hypothesis that perpetual motion is impossible, which underpins thermodynamics, is a prime example of such a fundamental hypothesis based on experiential evidence. Similarly, the foundational hypotheses of the theory of relativity fall into this same category [Ein-34].

From the aforementioned examples, it is clear that Einstein often distilled the core principle of the Second Law of Thermodynamics to the impossibility of constructing a perpetual motion machine, whether of the first or second kind. The latter is conceptually similar to Maxwell's demon.[10]

---

[10] When Peter Guthrie Tait was working on his book *Sketch of Thermodynamics* published in 1868, he sought James Clerk Maxwell's input for potential corrections. Maxwell replied in a letter dated December 11, 1867, introducing the idea of *Maxwell's Demon*, a term later coined by William Thomson (Lord Kelvin). According to Maxwell, a hypothetical being could manipulate individual gas molecules, potentially invalidating the second law of thermodynamics. Maxwell aimed to highlight a potential flaw in this law. He described this demon as a finite entity capable of observing and controlling gas molecules within two separate vessels, A and B, divided by a diaphragm. This being could selectively allow faster or slower molecules to pass through the diaphragm, thereby affecting the temperature of the vessels without performing any work, thus challenging the second law of thermodynamics [Kno]. Maxwell elaborated on this idea in 1870 in a letter to John William Strutt (Lord Rayleigh). He described a gas-filled vessel with two compartments and a small hole controlled by an intelligent, quick, and observant but finite doorkeeper. This being would allow only certain molecules to pass through based on their velocity, leading to a temperature imbalance between the two compartments without expending work [Har]. In 1871, Maxwell published a text further detailing this thought experiment. He imagined a being able to track every molecule in a gas-filled vessel divided into two parts, A and B. By controlling a small hole in the division, this

Einstein highly regarded Ludwig Boltzmann's work, particularly his statistical interpretation of thermodynamics. This admiration is evident in Einstein's correspondence and writings [CPAE1], [CPAE4], [Ein-49-1].

Boltzmann developed the H-theorem, correlating H inversely with entropy. This theorem suggested that a system's entropy would increase, making a return to an initial state impossible. Boltzmann thus argues that while Poincaré's recurrence theorem is mathematically correct, its practical implications for thermodynamics are limited due to the immense timescales involved.[11] Boltzmann emphasized that theorems of gas theory, including Maxwell's law of velocity distribution among gas molecules, are statistical truths. These are not derivable from ordinary mechanics alone but are highly probable for many molecules (see explanation in footnote 11). From a molec-

---

being could selectively allow faster or slower molecules to pass, thus heating one part and cooling the other, contradicting the second law of thermodynamics [Max-71]. In 1877, Maxwell clarified that this being could not harness natural energies but could track and manipulate molecular motion [Max-77].
In 1874, Thomson famously referred to this hypothetical entity as *Maxwell's Demon*, describing it as a creature that could prevent diffusion by selectively controlling molecular movement, thereby separating hot and cold parts of a system [Tho-74-1]; See [Dau]. Maxwell himself outlined the demon's characteristics in an undated letter: named by Thomson, they were lively but small beings capable only of manipulating valves without doing work, with the primary purpose of demonstrating the statistical nature of the second law of thermodynamics [Kno]. In 1879, Thomson further described the demon as a minute, agile being, likening its manipulation of atoms to a pianist playing a piano. This demon could sort molecules and separate gases within a container, again highlighting its role in challenging the conventional understanding of thermodynamics [Tho-79]. In Thomson's hands, Maxwell's demon was transformed into *Thomson's demon*, resembling an almost human master chemist [Can].

[11] In the 19th century, significant theoretical advancements in thermodynamics, energy conservation, kinetic gas theory, statistical mechanics, and electrodynamics led physicists to explore the foundational relationship among these theories and their connection to mechanics. This exploration raised questions about which theory underpinned the entire field of physics, resulting in a reevaluation of physics' foundations. Three main worldviews emerged: the electromagnetic, energetic, and mechanistic perspectives, each championed by different scientists. Henri Poincaré, in 1893, recognized challenges faced by proponents of the mechanistic worldview, particularly in reconciling mechanics with observable phenomena. The mechanistic view posited that all phenomena should be reversible, yet real-world observations showed many irreversible phenomena. Poincaré noted that molecules' small and numerous nature made them difficult to track, which could explain the apparent irreversibility. He referred to Maxwell's hypothetical demon capable of manipulating individual molecules as an entity that could negate this irreversibility. Poincaré proposed his recurrence theorem. This theorem states that certain systems will return to a state very close to their initial state after a sufficiently long but finite amount of time. The key assumptions for this theorem are that the system is isolated (no external energy or matter enters or leaves the system), has a finite amount of energy, and is in a finite volume. Under these conditions, since the system will naturally fluctuate and revisit all possible states, including low-entropy states, over an infinite timeline, the natural dynamics of the system, as described by Poincaré's theorem, would eventually accomplish what Maxwell's demon is hypothesized to do in a closed chamber [Poi]. While theoretically interesting, Poincaré's recurrence theorem operates under idealized conditions. In reality, the timescales involved in such recurrences, especially in complex systems, are often so vast that they exceed the practical and observable timescales of the universe. Additionally, the theorem does not contradict the second law of thermodynamics, which is a statistical law implying that while entropy might decrease locally or temporarily, the overall trend for a large system is toward increasing entropy.

ular viewpoint, Boltzmann also stresses that the second law of thermodynamics is statistical. It is not an absolute rule but rather a highly probable outcome. Boltzmann's H-theorem shows that for a stationary gas in a stationary container, a certain quantity H (a measure of deviation from Maxwell's distribution) tends to decrease over time, increasing entropy. This supports the second law of thermodynamics. Boltzmann explains that while the H-curve typically stays close to the axis (indicating a state close to Maxwell's distribution), there are occasional peaks (deviations from this state). However, the probability of high peaks is extremely low. While Boltzmann agrees that the recurrence of initial states is possible (as per Poincaré's theorem), he argues that such occurrences are extremely improbable, especially for systems with many molecules. Boltzmann notes that the timescales are so vast that such recurrence is practically unobservable. The systems we observe behave according to the second law of thermodynamics for all practical purposes [Bol-97].[12]

Johann Josef Loschmidt challenged Boltzmann's interpretation of the second law of thermodynamics, particularly the idea that entropy always increases. Suppose an isolated system transitions from an initial to a final state with lower H. There should also be a scenario where H increases over time based on time-reversible microscopic mechanical laws. This scenario would involve reversing the velocities of all molecules from the final state, creating a time-reversed evolution, which contradicts

[12] Ernest Zermelo used Poincaré's recurrence theorem to argue that a system could eventually return to its initial or a very similar state [*(Poincaré's) Zermelo's Wiederkehreinwand*], challenging Boltzmann's conclusion. Zermelo aimed to establish that all processes in a closed system of gas molecules are periodic from a mathematical viewpoint and, hence, not irreversible in the strict sense. Zermelo argues that probability theory cannot explain the direction of irreversible processes because it does not inherently include a direction of time. He challenges Boltzmann's use of probability in explaining the increase in entropy, suggesting that an increase should be as probable as a decrease. In statistical mechanics, the probabilities of different states of a system are considered, but these probabilities do not dictate a forward or backward direction in time. The fundamental laws of physics, including the laws of molecules in gas theory, are time-symmetric. This means that the equations governing these processes do not prefer a direction of time—they are just as valid if time runs backward. When probability theory is applied to these physical processes, it inherits this time symmetry. In other words, probability theory alone cannot impose a direction on time. If a process is likely to evolve from state A to state B over time, the reverse process (from state B back to state A) is equally probable regarding the system's fundamental laws. Irreversibility in thermodynamics, particularly the idea that entropy always increases in an isolated system (as stated in the second law of thermodynamics), implies a definite direction of time—from past to future. However, Zermelo points out that if the underlying mechanics are time-reversible and probability theory does not distinguish between past and future, then the basis for this irreversible increase in entropy is not adequately explained. Zermelo uses Poincaré's recurrence theorem to bolster his argument. The theorem suggests that a system will return to a state close to its initial state over a long enough period. This recurrence is at odds with the principle of continuous entropy increase, implying that entropy could decrease back to its initial state. Zermelo argues that if both an increase and a decrease in entropy are equally probable (as they should be in a time-symmetric framework), then observable thermodynamic processes should sometimes show a decrease in entropy. However, this is not what is observed; entropy seems to consistently increase in isolated systems, a fact that probability theory cannot explain. Consequently, Zermelo suggests that additional physical principles or assumptions must be invoked to account for the observed irreversibility and the directionality of time in thermodynamic processes. This challenges the sufficiency of statistical mechanics, as formulated by Boltzmann and others, to fully explain the second law of thermodynamics [Zer].

the second law of thermodynamics.[13] Boltzmann quickly responded to Loschmidt's ideas, understanding the implications of the reversibility paradox. In his reply, Boltzmann thoroughly discussed the paradox and its implications for the second law of thermodynamics.

Boltzmann suggests it is impossible to prove that a system of particles, like gas molecules in a container, will necessarily reach a uniform distribution of states after a certain time, regardless of its initial conditions. While a non-uniform distribution is improbable, it is not impossible. Boltzmann argues that there are many more ways for a system to be in a uniform state than a non-uniform one. Therefore, a system is more likely to be found in a uniform state over time. He suggests that the relative numbers of different state distributions could be used to calculate their probabilities, which might offer a new method for calculating thermal equilibrium. The transition from a non-uniform to a uniform state corresponds to an increase in entropy (and thus a negative $\Delta S = \int \frac{dQ}{T}$). It results from the higher probability of uniform distributions rather than a strict law. He highlights the significance of Loschmidt's theorem in showing the connection between the second law and probability theory. It implies that for every process where entropy seems to increase, there could be a very improbable initial condition where entropy decreases. While it is theoretically possible for molecules to arrange themselves non-uniformly, such an arrangement is exceedingly unlikely, especially in systems as large as those in typical thermodynamic scenarios, where the number of molecules is 10^{23} (Avogadro's number). Boltzmann concludes by suggesting that while reducing the second law to a matter of probability might make its universal application questionable, the principles of probability theory are consistently validated in laboratory experiments [Bol-77-1]. The specific initial conditions leading to non-uniform distributions, as considered by Loschmidt, are extremely improbable. Therefore, from Boltzmann's viewpoint, such scenarios are practically impossible [uff].

According to Boltzmann's statistical mechanic framework, non-equilibrium microstates (the gas clustered in one-half of the container) occupy a minuscule region of the system's phase space. In contrast, equilibrium states fill almost the entire phase-space volume. Therefore, the system's evolution from non-equilibrium to equilibrium, and consequently an increase in entropy, is statistically inevitable over time. The dynamics of such large systems are inherently chaotic and sensitive to initial conditions, making it practically impossible to revert to a previous state that

[13] Loschmidt challenged the Second Law. His idea is known as *Loschmidt's demon* or *Loschmidt's reversibility paradox* (*Umkehreinwand*). Loschmidt briefly mentioned the reversibility paradox in his article, "About the thermal equilibrium of a system of bodies subject to gravitational forces." Loschmidt's primary aim was to demonstrate that thermal equilibrium could be achieved without uniform temperature distribution, challenging the idea of the heat death of the universe. Loschmidt proposed that the second law of thermodynamics, traditionally seen as dictating an irreversible increase in entropy over time, could be formulated as a mechanical principle without referencing the sequence of events in time. He aimed to destroy the nimbus of the second law and provide a more optimistic outlook on energy resources and the universe's fate. Central to his argument was the reversibility principle. Loschmidt claimed that in any system if the velocities of all particles are reversed at a certain instant, the entire course of events would retrace itself. This implies that processes deemed irreversible under the second law could, in principle, be reversed [Loc], [Bru].

contradicts the second law of thermodynamics. Moreover, no real system is perfectly isolated; external influences, however minor, contribute to this chaotic behavior.[14]

Boltzmann further linked entropy to probability, stating that a system naturally evolves toward its most probable state, typically one of equilibrium [Bol-77-2], [uff]. Boltzmann related the entropy S of a macrostate to the number of microstates W corresponding to it. The famous formula $S = k \log W$ quantitatively links entropy with the logarithm of the number of microstates. Here, k is Boltzmann's constant, and W represents the number of ways the particles in a system can be arranged while still producing the same macroscopic properties. This equation implies that higher entropy states are those with more corresponding microstates.

In 1905, Einstein referred to the Boltzmann entropy formula as the *Boltzmann principle*, emphasizing its foundational role in connecting the microscopic and macroscopic worlds. Einstein's acknowledgment of this principle underscored its importance in understanding statistical mechanics and thermodynamics [Ein-05], [Pai].

Much later, Einstein articulated his understanding of Boltzmann's probabilistic approach to entropy in his *Autobiographical Notes*. In these notes, Einstein delved into the relationship between entropy and probability, building on Boltzmann's ideas. He explained how entropy, from the perspective of the kinetic theory of gases, is equivalent to the logarithm of the probability of a given state [Ein-49-1]. This explanation by Einstein in his *Autobiographical Notes* is significant because it highlights his insights into the foundations of statistical mechanics and thermodynamics. By connecting entropy with probability, Einstein clarified how macroscopic phenomena (like irreversibility) emerge from the microscopic behavior of individual particles. This idea was central to Boltzmann's statistical interpretation of thermodynamics.

Einstein's discussions suggest that he viewed the Second Law as inherently statistical, a viewpoint influenced by Boltzmann's work. He recognized that the law describes tendencies rather than absolute certainties, which aligns with his broader views on the nature of physical laws. By relating thermodynamics to relativity, Ein-

[14] This discussion centers around the macroscopic properties of thermodynamic systems, like temperature, pressure, and volume, which are determined by the collective behavior of microscopic particles (microstates). While indistinguishable at the macroscopic level, these microstates vary in terms of individual particle positions and velocities. The behavior becomes statistically predictable in systems as large as those found in thermodynamics, typically containing $N = 10^{23}$ of particles. For instance, a gas occupying half the volume of a container initially will, over time, distribute itself evenly throughout the entire container. This behavior is due to the overwhelming number of microstates that correspond to an even distribution compared to those that correspond to a more clustered arrangement. In modern statistical mechanics, the notion of *Typicality* has been developed as an extension of Boltzmann's ideas [All], [Gol], [Leb-93-1], [Leb-93-2], [Fri] [Laz]. Typicality argues that the behavior of microstates evolving towards equilibrium (occupying a larger phase space volume) is not just probable but typical. This concept suggests that the detailed dynamics of the system (described by Hamiltonian equations) might be less relevant than the overall structure of the phase space and the relative sizes of different regions corresponding to non-equilibrium and equilibrium states. However, this approach has been critiqued for oversimplifying the role of dynamics [Hag], [Pit]. Critics like Uffink argue that the evolution of a system must consider its dynamics and initial conditions, not just the statistical measure of different phase space regions [uff].

stein highlighted both fields' universal applicability and fundamental nature. He saw both as providing overarching principles that guide understanding physical phenomena, whether the behavior of heat and energy or the nature of space and time.

In 1949, Einstein explained that from the point of view of thermodynamics, sending a signal is an irreversible process associated with an increase in entropy. This notion contrasts with the understanding that fundamental physical processes at the microscopic level are reversible [Ein-49-1]. For Einstein, this implies that signals are inherently linked to thermodynamics: one cannot move backward in time because no signal transmission faster than light is possible. Similarly, the entropy of a system cannot decrease. The statement that sending a signal represents an irreversible process associated with increased entropy reflects the understanding that any real-world transmission of information (such as a signal) involves energy dissipation and an increase in entropy.

This is analogous to Léon Brillouin's lamp in Maxwell's demon thought experiment, where the lamp used by the demon to observe particles also constitutes an irreversible process that increases entropy. Einstein's assertion supports that acquiring information (such as observing particles or sending signals) incurs thermodynamic costs.[15]

As I show at the beginning of this section, Einstein often compared the foundational principles of relativity with those of thermodynamics. Both theories stemmed from fundamental empirical observations—the impossibility of perpetual motion in thermodynamics and the constancy of the speed of light in relativity. In various writings, Einstein highlighted that the principles of relativity, like those of thermodynamics, rule out the possibility of perpetual motion. This link reflects his belief in

[15] In 1949, Léon Brillouin considered using photons for Maxwell's demon to "see" the molecules. However, he noted the difficulty of distinguishing the photons from the background blackbody radiation within an enclosure at thermal equilibrium. The demon needs to see the molecules to sort them, but he is in a black body enclosure at a constant temperature. In such an environment, which is in equilibrium, it is impossible to see anything because the radiation is that of a black body, where all incident radiation is absorbed and equally re-radiated. To overcome this visibility issue, the demon could use a torchlight. The torchlight is a source of radiation that provides energy in the form of light photons. However, introducing a torchlight adds a complication. It introduces *negative entropy* into the system. The torchlight introduces radiation that is not in equilibrium with the system. It adds energy in a structured form rather than random thermal energy. To solve this problem, Brillouin introduced the notion of *negentropy*, i.e., *negative entropy*, which he associated with information [Bril-49]. The torchlight provides a coherent light source, which injects *negentropy* into the system. This *negentropy* is used as information by the demon to distinguish between fast and slow molecules. The demon uses this information to manipulate the gate, sorting the molecules and decreasing the entropy of the gas. Brillouin describes a cycle where *negentropy* (structured energy from the torch) is converted into information (knowledge about the molecules' speeds): 1. The torchlight introduces *negentropy* into the system, allowing the demon to gather information about the molecules' velocities. 2. The demon uses the information obtained from the light signals (*negentropy*) to operate the sorting mechanism separating fast (hot) and slow (cold) molecules, ostensibly reducing the system's entropy. 3. The decrease in the gas's entropy due to the sorting process is compensated by the increase in entropy elsewhere in the system. This increase occurs due to the energy required to gather information, process it, and operate the sorting mechanism. 4. The total entropy of the system, including the demon, the torchlight, and any auxiliary machinery, increases or remains constant, guaranteeing compliance with the second law of thermodynamics.

the importance of principle-based theories in physics [Ein-19], [Ein-54], [Ein-34]. Einstein sought a universal principle in electrodynamics, akin to the Second Law of thermodynamics, which led to the development of his theory of special relativity. His earlier contemplation of chasing a beam of light guided this search, questioning the nature of electromagnetic fields in different reference frames. While Einstein does not explicitly mention Maxwell's demon in these contexts, his discussions about perpetual motion of the second kind indirectly touch upon the same conceptual ground.[16]

4.3 Black Hole Thermodynamics

4.3.1 Bekenstein: Black Hole's Entropy and the GSL

In 1971, Wheeler contemplated the implications of objects, like tea cups, falling into a black hole and vanishing. He considered the impact of the second law of thermodynamics, which posits a universal trend towards uniform temperature and entropy increase. Wheeler hypothesized an irreversible process where combining a hot tea cup with a cold one results in a uniform temperature and increases the universe's entropy and information loss. During Jacob Bekenstein's graduate studies at Princeton, he visited Wheeler's office. Wheeler playfully suggested that he could

---

[16] In 1900, Karl Pearson presented a speculative idea in his 1900 book *The Grammar of Science*, about a relativistic version of Maxwell's demon and the notion of time reversal in the context of faster-than-light (FTL) travel. Pearson imagines a hypothetical scenario involving a being akin to Maxwell's demon. If this being were to travel away from Earth at a speed greater than that of light, it would perceive natural processes and historical events in reverse. For this observer, events would unfold backward, with effects preceding their causes. Pearson adds to this discussion by highlighting that understanding historical change and time could be considered relative motion problems. Louis Napoleon George Filon suggested this idea to him. Pearson emphasized that it is interesting from the perspective of the relativity of all phenomena [Pea]. Pearson's suggestion is an interesting precursor to some ideas later explored in the theory of relativity. However, it is important to contextualize Pearson's ideas within the scientific understanding of his time and how they relate to later developments in physics by Einstein. Pearson suggested that the irreversibility of natural processes is relative and depends on the observer's motion relative to these processes. This aligns with the broader relativistic idea, which posits that the observation of physical phenomena can vary depending on the observer's frame of reference. Pearson's imaginative scenario, where a being (akin to Maxwell's demon) travels away from Earth faster than light and observes history in reverse, challenges the conventional understanding of time. However, it is essential to note that this idea predates the full development of relativity theory and the modern understanding of the impossibility of FTL travel according to special relativity. Einstein read Pearson's *The Grammar of Science* before developing his special theory of relativity. While it is unclear how much this specific idea influenced him, Pearson's book likely contributed to Einstein's broader thinking about the nature of space, time, and relativity. In Einstein's special relativity, the flow of time is relative and depends on the observer's velocity. However, it does not support the notion of time reversal or FTL travel. Special relativity sets the speed of light as the ultimate speed limit, making FTL travel and the associated paradoxes (like time reversal) physically impossible. While Pearson's version of Maxwell's demon involves a speculative scenario unsupported by special relativity, the broader question of how the demon operates in a relativistic context is interesting.

conceal the evidence of this entropy-criminal act by dropping both tea cups into a black hole. Bekenstein later termed this hypothetical scenario *Wheeler's demon*, noting its apparent violation of the second law of thermodynamics, as the entropy would seemingly be destroyed inside the black hole. The no-hair theorem indicated that such an act would go unnoticed externally, as a black hole eradicates all distinct physical properties of the matter it absorbs. However, after pondering this problem, Bekenstein returned to Wheeler with a key insight. He argued that Wheeler's demon had not circumvented the increase in entropy; rather, it had merely relocated it. According to Bekenstein, the black hole possesses entropy, and by dropping the teacups into it, one would only add to this entropy [Bek-80], [Whe-08].[17]

In 1972, Bekenstein formally attributed entropy to black holes. In his Ph.D. thesis, he had already proposed that the entropy of a black hole is proportional to the surface area (A) of its event horizon [Bek-72][18]:

$$S_{BH} = \frac{\eta k A}{l_p^2}. \tag{4.14}$$

Bekenstein provided a rationale for his choice of equation (4.14). Influenced by Hawking's area theorem (see Sect. 4.2.1), Bekenstein was convinced that a critical characteristic of black hole entropy should be its tendency to increase. Equation (4.14) aligns with this principle and the area theorem: when two black holes merge to form a single Kerr black hole, the entropy of the resultant black hole increases. This suggests that the area of the black hole, a geometric quantity, exhibits properties remarkably

---

[17] Demetrios Christodoulou's work on more efficient reversible processes led Bekenstein to realize that black holes must adhere to the principles of thermodynamics. In 1970, Christodoulou, a graduate student under John Wheeler, introduced the concept of *irreducible mass* (m_{ir}) of a black hole. Christodoulou developed an equation to represent the total mass-energy (m^2) of a black hole:

$$m^2 = m_{ir}^2 + \frac{J^2}{4m_{ir}^2}. \tag{4.13}$$

Here, $G = c = 1$ and J is the angular momentum of the black hole. Equation (4.13) is equivalent to $E^2 = m^2 + p^2$. It shows that the black hole's total mass-energy is composed of its irreducible mass and a rotational energy term associated with its angular momentum. Christodoulou explored two types of transformations a black hole can undergo: reversible and irreversible. Reversible transformations can change the black hole's rotational energy without altering its irreducible mass m_{ir}. Reversible processes can increase or decrease the rotational energy term $\frac{J^2}{4m_{ir}^2}$ in equation (4.13) for the total mass-energy of the black hole. Irreversible transformations involve an increase in the black hole's irreducible mass. Any process that increases m_{ir} is considered irreversible. Christodoulou stated that no process can decrease the irreducible mass of a black hole. He identified that reversible changes in black holes are the most efficient processes, in contrast to irreversible transformations, which are less efficient [Chri]. Bekenstein and Christodoulou noted that for Schwarzschild black holes, the irreducible mass squared is proportional to A, the event horizon's area: $m_{ir}^2 \propto \frac{A}{16\Pi}$ [Chri], [Bek-73], [Bek-80].

[18] Following Wheeler's recommendation, Bekenstein chose the proportionality constant to be on the order of $\frac{1}{l_p^2}$, where l_p represents the Planck length [see Eq. 4.61] and η is a constant [Bek-01].

similar to entropy.[19] Bekenstein conjectured that the loss of information about a black hole's initial conditions would manifest as an increase in S_{BH}. Consistent with Hawking's area theorem, S_{BH} is predicted to increase monotonically as the black hole evolves [Bek-72], [Bek-73], [Bek-01].

Bekenstein adopted the perspective that information is related to entropy. His suggestion mentioned above aligns with information theory, where entropy quantifies uncertainty and lack of knowledge. In this context, S_{BH} represents our ignorance about the internal state of the black hole. An observer outside the black hole (at infinity) cannot access information about its interior, thus increasing the perceived entropy. Bekenstein's perspective on black hole entropy aligns with Claude Shannon's view of information entropy. Shannon defined entropy as a measure of uncertainty about a system's state. Similarly, Bekenstein viewed S_{BH} as quantifying our lack of knowledge about the black hole's internal configuration.[20]

[19] As a black hole approaches equilibrium, it "loses its hair," leaving only mass M, angular momentum J, and charge Q as its defining parameters (see Sect. 4.2.1). Bekenstein suggested that the entropy of a black hole, S_{BH}, could be viewed as a measure of the information hidden within the black hole.

[20] In 1948, Shannon defined entropy as a measure of the amount of information generated by a source per symbol or second. In information theory, entropy measures the randomness in a message source. If a source emits a random sequence of symbols, entropy measures how much information we can expect to gain from each symbol. It is calculated using a logarithmic function (base 2, resulting in measurements in bits), reflecting the number of binary choices that would be needed, on average, to represent a symbol from the source. Entropy is also a measure of unpredictability. A source with high entropy produces symbols more unpredictably, i.e., each new symbol provides more information. Conversely, a predictable source (low entropy) provides less new information with each symbol because each new symbol is more expected. In communication, a higher entropy source requires a higher capacity channel to transmit the information efficiently. Shannon defines the conditional entropy of the message. The conditional entropy represents the amount of information needed to describe the outcome of a random variable, given that the outcome of another related variable is known; the uncertainty about a message given the signal received. This uncertainty arises due to noise and other distortions in the communication channel. Given the received signal, the conditional entropy of a message quantifies the missing information—what is lost or unaccounted for in the transmission process. It measures how much additional information one would need to thoroughly understand the original message, considering what has been received and the distortions introduced during transmission [Shan]. In 1961, Rolf Landauer advanced the understanding of the thermodynamics of information, particularly focusing on the relationship between information processing and energy dissipation. His principles and subsequent elaborations by Charles Bennett significantly impacted the field of information theory and thermodynamics. Landauer argued that information is physically stored and can be quantified using the formula $kT\ln(2)$ for the smallest unit of information, a bit. Here, k is Boltzmann's constant, and T is the temperature. Referring to Brillouin's work (see footnote 15), Landauer pointed out that measuring a system to obtain information (like the state of a molecule) requires energy dissipation. Specifically, an energy amounting to $kT\ln(2)$ must be dissipated to acquire one bit of information. However, Landauer suggested that it is not the measurement but the erasure of information from memory that requires energy dissipation. Erasing memory is an irreversible process, meaning it discards information about the previous logical state of a device. This erasure generates an equivalent amount of entropy in the environment, equal to the discarded information. Erasing one bit of information, therefore, results in the dissipation of $kT\ln(2)$ energy as heat into the environment [Lan], [Ben-88]. In quantum

Léon Brillouin introduced the notion of *negentropy* (negative entropy), which he related to information. In this view, gaining information about a system reduces its entropy, increasing its negentropy [Bril-59] (see footnote 15 in Sect. 4.2.2). Bekenstein suggested that higher entropy (or lower negentropy) correlates with less knowledge about a system. In the context of black holes, Bekenstein proposed that the entropy of a black hole, which is proportional to the surface area of its event horizon, corresponds to the amount of information hidden within the black hole.

4.3.2 Penrose Process and Geroch's Heat Engine

In 1969, Roger Penrose asked whether extracting energy from a black hole is possible. He proposed a mechanism by which rotational energy could theoretically be extracted from a rotating black hole (a Kerr black hole). Imagine, wrote Penrose, a civilization that could build a stabilized structure S surrounding a rotating black hole. They lower a mass on a light, inextensible, unbreakable rope until it reaches the stationary limit L (see explanation in footnote 22). The mass can be bartered at this point for its energy content. Another structure, S^*, is built, somewhat rotating with the black hole. This structure further lowers the mass beyond the stationary limit L (into what would in 1971 be termed the ergosphere). The mass is then dropped through H (the event horizon, see footnote 23) into the black hole so that its energy content, as measured from S, is negative. This negative energy contribution results in energy extraction from the black hole itself.[21] By carefully managing the energy content of

computing, error correction is critical due to the susceptibility of quantum states to environmental noise. Quantum error correction involves detecting and correcting errors in quantum states (qubits) without measuring the actual quantum information. This is achieved by creating and measuring ancillary quantum states (syndrome measurement) and applying corrective quantum operations based on these measurements. The quantum error correction process can decrease the entropy of a quantum system. This decrease happens because the error-correction algorithms restore qubits to their intended states, counteracting the effects of environmental decoherence and noise that increase entropy. This process can be likened to the actions of Maxwell's demon on the quantum scale. The demon (in this case, the error-correction algorithm) observes and manipulates the system at a microscopic level to decrease its entropy, akin to how Maxwell's demon hypothetically sorts molecules to decrease entropy. The first stages involve syndrome measurement (identifying errors without disturbing the quantum information) and applying corrective operations. These steps can be designed to be thermodynamically reversible. The key stage is erasing the quantum information in error detection and correction. According to Landauer's principle, erasing each bit of information from a quantum computer's memory increases the environment's entropy by at least $kT \ln(2)$. The increase in environmental entropy due to memory erasure compensates for the decrease in the entropy of the quantum system achieved during error correction. This assures that the overall entropy of the combined system (quantum system plus environment) does not decrease, thus adhering to the second law of thermodynamics [Nei-10], [Nei-98], [83Ved-00], [Ved-06].

[21] When the mass is lowered slowly using a tether, the potential energy lost by the mass can be converted into work done by the tether. This means energy is being extracted from the gravitational field of the black hole. The work done by the tether in lowering the mass can be thought of as

the mass, the civilization can extract more than the mass's initial energy, extracting the rotational energy of the black hole. This process also reduces the black hole's angular momentum [Pen-69].

Two years later, Penrose and R. M. Floyd elaborated on this idea in a paper discussing the extraction of rotational energy from black holes, known today as the *Penrose process*. Penrose and Floyd first identified two distinct regions outside a Kerr black hole's horizon: The first is the *stationary limit*.[22] This limit lies outside the *event horizon*.[23] Between these two boundaries is another region. The same year, Remo Ruffini and John Wheeler designated this region as the *ergosphere* [Ruf-Whe]. Penrose (and Floyd) did not specifically use the term ergosphere in their descriptions.[24]

Penrose and Floyd explored the possibility of tapping into the mass-energy of a rotating Kerr black hole as a source of available energy. This inquiry led them to a specific method for extracting rotational energy from a Kerr black hole. They focused on the Killing vector $K = \frac{\partial}{\partial t}$, which becomes null at the stationary limit and space-like within the ergosphere. They used this vector to define a particle's energy with four-momentum p_0 as measured from infinity: $E = -K p_0$. Inside the ergosphere, where K is space-like, the energy E can be negative even though the four-momentum remains time-like. Consider a particle with four-momentum p_a entering the ergosphere and splitting into two. According to momentum conservation, $p_{a0} =$

---

extracting energy from the black hole's rotational energy, which is facilitated by the frame-dragging effect. Using S^*, the mass is dropped beyond L (into the ergosphere). As the mass is lowered, the strong, ideally massless tether held by an observer far away from the black hole gains kinetic energy due to the gravitational pull of the black hole but loses gravitational potential energy. The key is to balance these energy changes such that the total energy of the mass (including contributions from the frame-dragging effect) becomes negative. This involves lowering the mass slowly and carefully, using the tether to control its descent, and assuring it does not gain too much kinetic energy. Due to frame-dragging, any object in the ergosphere must move in the direction of the black hole's rotation. However, the observer at S^* can somewhat counteract this motion with the tether. By carefully controlling the descent, the mass can move so that its total energy E (as seen from infinity) decreases and eventually becomes negative. The energy E becomes negative when the potential energy term (which is negative and grows stronger as the object gets closer to the black hole) outweighs the rest mass and kinetic energy terms.

[22] The stationary limit is a surface outside the event horizon of a rotating black hole. At the stationary limit, a particle would need to move at the speed of light to appear stationary to a distant observer. Only massless particles, such as photons, can travel at the speed of light. These particles can orbit the black hole at the stationary limit, but massive objects cannot.

[23] The event horizon is the black hole's effective boundary, beyond which no information or matter can escape. There is no ergosphere in a non-rotating Schwarzschild black hole, and the stationary limit coincides with the event horizon. These two surfaces are distinct for rotating (Kerr) black holes, creating the ergosphere between them.

[24] In 1971, Ruffini and Wheeler wrote that according to Penrose, a particle approaching a black hole and entering its surrounding region, the "ergosphere," can extract energy from the black hole [Ruf-Whe]. In the ergosphere, the frame-dragging effect caused by the black hole's rotation is so strong that all objects, including light, are forced to rotate in the direction of the black hole's spin. Due to the intense frame-dragging effect, particles can enter and exit the ergosphere but cannot remain stationary relative to a distant observer. Within the ergosphere, an object can split into two parts: positive energy and negative energy relative to an observer at infinity. The part with negative energy can fall into the black hole, while the part with positive energy can escape to infinity.

$p_{a1} + p_{a2}$, and energy conservation implies $E_{a0} = E_{a1} + E_{a2}$. In this scenario, the first particle, with negative mass energy measured from infinity $E_{a1} < 0$, crosses the event horizon and falls into the black hole. Meanwhile, the second particle, possessing positive energy $E_{a2} > 0$ as measured from infinity, escapes back to infinity, carrying more mass-energy than the original particle: $E_{a2} > E_{a0}$ [PF].

The location of the event horizon (the outer horizon radius) of a Kerr-Newman black hole is given by [Haw-71][25]:

$$r_+ = M_{BH} + \sqrt{M_{BH}{}^2 - \left(\frac{J}{M_{BH}}\right)^2 - Q^2}. \tag{4.15}$$

Another equation for the area of the event horizon that implies that any changes in M_{BH} and r_+ resulting from particle absorption:

THE AREA OF THE EVENT HORIZON:

$$\boxed{A = 8\pi M_{BH} r_+,} \tag{4.16}$$

must warrant Eq. (4.39).

Penrose and Floyd derived an inequality equation for the particle absorbed by the black hole, formulated as[26]:

THE PENROSE FLOYD INEQUALITY:

$$\boxed{2M_{BH} r_+ E_{a1} - \Omega\Lambda_1 \geq 0.} \tag{4.17}$$

As a result of this interaction, the mass of the black hole decreases:

THE MASS CHANGE:

$$\boxed{dM_{BH} = -E_{a1} \quad \text{(mass change).}} \tag{4.18}$$

The energy E_{a1} carried by the particle reduces the mass of the black hole.

The black hole's angular momentum also changes due to the extracted rotational energy:

$$dJ_1 = -\Lambda_1, \quad \text{(angular momentum change).} \tag{4.19}$$

This indicates that the angular momentum $-\Lambda_1$ carried by the particle decreases the black hole's angular momentum. If the change in angular momentum is given by Eq. (4.19), then the change in the event horizon radius dr_+ is given by:

[25] r_+ is the location of the surface of the horizon, M_{BH} is the mass of the black hole, J represents the angular momentum. For a Kerr black hole, the charge $Q = 0$.

[26] Here Ω represents the angular velocity of the black hole, and Λ_1 is the angular momentum of the first particle that falls into the black hole.

EVENT HORIZON RADIUS CHANGE:

$$\boxed{M_{BH} dr_{+} = \Omega \Lambda_1.} \tag{4.20}$$

Consequently, the area of the event horizon of the black hole does not decrease, satisfying the condition $dA \geq 0$ [as indicated in Eq. (4.39)].[27] This observation reaffirms the area theorem. Penrose and Floyd observed that, under general conditions, the surface area of a black hole's event horizon naturally tends to increase over time in any non-stationary situation. They proposed that the ideal scenario of maximum efficiency is achieved when this increase in surface area is minimized, corresponding to a reversible process [PF].[28]

---

[27] Given Eqs. (4.16), (4.17), we substitute Eqs. (4.18) and (4.20) into the differentials of equation (4.16):

$$dA = 8\pi (dM_{BH} r_{+} + M_{BH} dr_{+}), \quad \text{and obtain:} \tag{4.21}$$

$$\frac{dA}{8\pi} = -E_{a1} r_{+} + \Omega \Lambda_1. \tag{4.22}$$

Rearranging, we get:

$$dA = 8\pi (\Omega \Lambda_1 - E_{a1} r_{+}). \tag{4.23}$$

We rearrange to isolate terms involving Λ_1 using the Penrose inequality (4.17):

$$\Omega \Lambda_1 \leq 2 M_{BH} r_{+} E_{a1}. \tag{4.24}$$

Substituting the inequality (4.24) into Eq. (4.23), we get:

$$dA \geq 8\pi (2 M_{BH} r_{+} E_{a1} - E_{a1} r_{+}). \quad \text{Rearranging, gives:} \tag{4.25}$$

$$dA \geq 8\pi (2 M_{BH} - 1) E_{a1} r_{+}. \tag{4.26}$$

E_{a1} is the energy of the particle falling into the black hole during the Penrose process. It is negative. Now, we need to determine the conditions under which the right-hand side of the inequality (4.26) is non-negative. If E_{a1} is negative, and we want that $dA \geq 0$, then the mass of the black hole $2M_{BH} \geq 1$. Under these conditions, a black hole warrants the Penrose process, consistent with the area theorem (4.39).

[28] In footnote 17 in Sect. 4.3.1, I have elaborated on Christodoulou's contributions related to the concept of *irreducible mass* m_{ir}. Christodoulou, in his Eq. (4.13), identifies that the total mass-energy m^2 of a black hole comprises two components: the irreducible mass m_{ir} and the rotational energy term $\frac{J^2}{4m_{ir}^2}$. Christodoulou shows that his formulation helps understand the dynamics of black holes, particularly in energy extraction processes like the Penrose process [Chri]. In the Penrose process, the escaping particle can carry away more energy than the original particle, extracting energy from the black hole. Christodoulou's formula is consistent with the Penrose process in that it accounts for the rotational energy of the black hole. The extraction of energy, as described by Penrose, would reduce the rotational energy $\frac{J^2}{4m_{ir}^2}$ of equation (4.13), thus affecting the total mass-energy m^2 of the black hole [Chri]. Christodoulou's reversible and irreversible transformations relate to the efficiency of energy extraction processes. The Penrose process describes the rotational energy extraction from a Kerr black hole. Christodoulou demonstrates that this process can be seen as a reversible transformation under his framework, which involves altering the black hole's rotational

In concluding their paper, Penrose and Floyd thanked R. P. Geroch for his valuable discussions. The paper was submitted in December 1970 [PF]. In December 1971, during a colloquium at Princeton University, Robert Geroch proposed a thought experiment known as the *Geroch heat engine* involving a heat engine that uses a Schwarzschild black hole as an energy sink.

Geroch's heat engine posed a potential challenge to Bekenstein's equation (4.14). Bekenstein only wrote the black hole's entropy (4.14). However, a potential contradiction arose because he only considered the entropy (4.14). While entropy increases within a black hole, it simultaneously decreases in the surrounding region, contradicting Hawking's area theorem. According to Bekenstein's entropy formula (4.14), the entropy of a black hole S_{BH} is proportional to the area of its event horizon. Hawking's area theorem (4.39) states that the area of a black hole's event horizon cannot decrease, implying that the entropy of a black hole should also not decrease $dS_{BH} \geq 0$. Bekenstein addressed a potential paradox: a contradiction arose if he only considered one entropy separately, the black hole or the external entropy. Bekenstein resolved this by showing that even if the black hole's entropy increases at the expense of the external matter's entropy, the total entropy dS_g would still satisfy the second law of thermodynamics. This law states that the total entropy of an isolated system can never decrease over time. Thus, Bekenstein addressed the paradox by proposing that the combined change in entropy dS_g—encompassing the black hole dS_{BH} and the matter outside it dS_M—should always be non-decreasing ($dS_g \geq 0$):

$$dSg = dS_{BH} + dS_M \geq 0. \tag{4.27}$$

He named this principle the *Generalized Second Law of Thermodynamics* (GSL). The GSL extends beyond the requirement of Hawking's area theorem, which states that the area A of a black hole's event horizon should not decrease. The GSL further stipulates that if entropy S_M is absorbed by black holes, leading to a loss, then there must be a sufficient increase in A to assure that the resulting increase in S_{BH} compensates for the loss in S_M [Bek-72], [Bek-73], [Bek-75], [Bek-81].

Let us explore the Geroch heat engine thought experiment. To do this, we should start by considering the Schwarzschild-Droste metric [Eq. (2.94) in Sect. 5.3.2]:

$$ds^2 = \left(1 - \frac{2R_g}{r}\right) c^2 dt^2 - \frac{dr^2}{1 - \frac{2R_g}{r}} - r^2 \left(d\theta^2 \sin^2 \theta d\phi^2\right), \tag{4.28}$$

where R_g the gravitational radius is given by Eq. (2.93) (refer to Sect. 2.2.3).

Here, I present a detailed explanation of the steps involved in the Geroch heat engine process, along with subsequent theoretical discussions and challenges:

---

energy. Still, the irreducible mass m_{ir} remains unchanged, making it an efficient way of extracting energy. In the Penrose process, the black hole's angular momentum J decreases when energy is extracted. According to Christodoulou's formula, a decrease in J would result in a decrease in the second term of the Eq. (4.13) $\frac{J^2}{4m_{ir}^2}$, which represents the rotational energy. However, the process is reversible since the irreducible mass m_{ir} remains constant [Chri].

(1) A box filled with radiation of energy $E = Mc^2$, temperature T, and high entropy S is slowly lowered towards the horizon of a Schwarzschild black hole. For argument's sake, assume the box and the rope used for lowering are weightless. The goal is to extract work from the Geroch engine.

(2) The box is lowered as close to the black hole until $r \to 2R_g$, where r is the distance from the black hole. The box does not reach $2R_g$, but it is opened when it is very close, allowing the radiation to escape into the black hole. The gravitational field performs work on the box during this process, given by:

$$W = \frac{2R_g}{r} Mc^2. \tag{4.29}$$

The gravitational potential is: $U = -\frac{2R_g}{r} Mc^2$, and the redshift factor is:

$$\alpha \approx \frac{2R_g}{r}. \tag{4.30}$$

Since the radiation is redshifted significantly near the event horizon, the actual increase in the black hole's mass dM is negligible, meaning $dM \approx 0$. [29]

(3) The box is then pulled back up, requiring almost no work, and can be refilled with thermal radiation from the reservoir.

With $r \to 2R_g$, the work extracted from the Geroch heat engine is $W = Mc^2$. The black hole returns to its initial state, and the energy of the radiation can be converted entirely to work in the laboratory [Bek-72].

Geroch concluded that black holes are systems at absolute zero temperature ($T_C = 0$). This is based on the idea that if you could convert all the mass-energy of infalling matter into work, the black hole must be at absolute zero to achieve perfect efficiency.[30] This would violate Bekenstein's GSL, defined by the Eq. (4.27).

[29] α is the gravitational redshift factor. It quantifies how much the energy of the radiation is reduced due to the gravitational field of the black hole. The gravitational redshift becomes highly significant as the box containing high-entropy radiation is lowered very close to a black hole's event horizon. As r approaches $2R_g$, α [Eq. (4.30)] increases. Gravitational redshift causes the energy of the radiation to decrease, as seen from a distant observer. The closer the box reaches the event horizon, the more extreme this effect becomes. When the box is near $2R_g$, the energy required to keep it stationary becomes enormous due to the intense gravitational field, leading to a very high redshift of the radiation it emits. Because of the gravitational redshift, the energy of the radiation that eventually crosses the event horizon is diminished to nearly zero from the perspective of a distant observer. Thus, the effective contribution to the black hole's mass from the infalling radiation becomes negligible, leading to $dM \approx 0$, where dM is the change in the black hole's mass due to the infalling radiation.

[30] The Carnot efficiency of the Geroch heat engine is: $\eta_{TH} \leq \left(1 - \frac{T_C}{T_O}\right)$, where T_O is the temperature of the radiation entering the black hole and T_C is the black hole's temperature. If all mass energy is converted to work, the efficiency will approach 1, $\eta_{TH} \leq 1$, implying that T_C (the black hole's temperature) would have to be 0.

In the process of lowering a box towards a black hole to nearly $2R_g$ and releasing high-entropy radiation into it, the GSL could be violated since the black hole's mass would not increase.[31]

Bekenstein resorted to Christodoulou's work on the Penrose process (see explanation in footnote 28) and resolved the apparent violation of the GSL. He suggested that when a particle falls into a black hole, it radiates gravitational waves with energy equivalent to its rest mass, leading to an increase in the black hole's area and, consequently, S_{BH}, in line with the GSL [Bek-72]. To address the Geroch engine, Bekenstein proposed:

$$S_{BH} = \frac{\eta k A}{l_p^2 \hbar}, \quad \text{where:} \quad \hbar = \frac{h}{2\pi}. \tag{4.31}$$

This implies that a black hole's entropy is significantly larger than ordinary matter of similar size and energy.[32]

Hawking objected to Bekenstein's entropy formula (4.14), with most black hole experts, except for Bekenstein's mentor Wheeler, aligning with Hawking. Wheeler encouraged Bekenstein, finding his ideas intriguingly unconventional.[33]

Bekenstein grappled with the intricacies of the Geroch paradox concerning his GSL. The core issue was that even if black holes possess entropy S_{BH} and additional mass dM is added to them, the resultant increase in the horizon area might not be sufficient to offset the loss of entropy, potentially leading to a violation of the GSL.

---

[31] Lowering the box happens close to the event horizon (near $2R_g$). Thus, the mass of the black hole does not increase ($dM_{BH} = 0$). See explanation in footnote 29. The event horizon's area is: $A = 4\pi R_g^2$ [R_g is defined by Eq. (2.93)]. Therefore, the area (and thus the entropy) depends on the square of the mass. If the mass of the black hole does not increase, then the Schwarzschild radius R_g remains unchanged. Consequently, the area A of the event horizon would remain unchanged; thus, the entropy of the black hole dS_{BH} also remains unchanged. As the box of radiation (with entropy S) is lowered near the event horizon and the radiation falls into the black hole, the entropy outside the black hole decreases by S. Thus, $dS_M = -S$. The radiation falling into the black hole is highly redshifted, leading to a negligible increase in the black hole's mass and, thus, its entropy. We can consider $dS_{BH} = 0$ for practical purposes. The total change in generalized entropy would then be $dS_g = dS_{BH} - S = 0 - S < 0$, which implies a decrease in the generalized entropy. This violates the GSL. This is because before the box is dropped, the radiation's entropy S contributes to the total entropy outside the black hole. Once the radiation has fallen into the black hole, this entropy is no longer present outside the black hole. Therefore, the entropy of the outside region decreases by S. This results in a decrease in the generalized entropy.

[32] Bekenstein initially argued this from a quantum perspective, suggesting quantum effects prevent the box from reaching $2R_g$. However, he later admitted these arguments were forced [Bek-80], [Wal], [Un-01].

[33] Bekenstein maintained that black holes have enormous entropy [Eq. (4.31)] and interpreted his Eq. (4.14) as the black hole's internal configuration. However, leading physicists, including Hawking and Kip Thorne, disagreed, citing the presence of a singularity inside black holes rather than atoms or molecules [Thor-94].

4.3.3 Bekenstein's Challenge: A Black Hole in a Cavity

Bekenstein's GSL encountered another challenge from a thought experiment involving a Kerr black hole of mass M_{BH} placed in a cavity filled with black body radiation at temperature T. The setup assumed that the cavity did not rotate, meaning the radiation possessed zero angular momentum and the accretion process had reached a steady state. For radiation of energy E (as measured from infinity) falling into the black hole, a corresponding change in the black hole's entropy was given by:

$$dS_{BH} = \frac{E}{T_{BH}}. \tag{4.32}$$

Bekenstein pointed out that since the accretion occurred in a steady state, alongside the energy E, the entropy of the radiation:

$$S = \frac{E}{T}, \tag{4.33}$$

also entered the black hole. Therefore, for $T \geq T_{BH}$, the equation:

$$dS_g = dS_{BH} - S \geq 0, \tag{4.34}$$

holds, meaning the GSL remains valid.[34] However, for $T < T_{BH}$, the GSL appeared to be violated.[35]

To address this issue (anomaly), Bekenstein proposed that the wavelength in the radiation $\frac{\hbar}{T}$ becomes larger than the size of the black hole at very low temperatures, corresponding to very long wavelengths. Under these conditions, radiation does not flow continuously into the black hole but must tunnel inside discretely and individually. He argued that this tunneling rate is sensitive to wavelength: shorter wavelengths have a higher tunneling rate. Therefore, for $T \geq T_{BH}$, the rate is high enough to maintain the GSL, but for $T < T_{BH}$, the rate becomes so slow that radiation hardly enters the black hole. For the problematic case where $T < T_{BH}$, Bekenstein introduced an *ad hoc* hypothesis suggesting that irreversible processes within the cavity material or the radiation itself would generate additional entropy. This additional entropy would

---

[34] Substituting the expressions (4.32) and (4.33) into the GSL (4.34), we get:

$$dS_g = \frac{E}{T_{BH}} - \frac{E}{T} \geq 0. \tag{4.35}$$

For $T \geq T_{BH}$, we know that: $\frac{1}{T} \leq \frac{1}{T_{BH}}$. This means: $\frac{E}{T_{BH}} - \frac{E}{T} \geq 0$.

[35] Again, using the same expressions (4.32) and (4.33), $\frac{1}{T} > \frac{1}{T_{BH}}$. This means: $\frac{E}{T_{BH}} - \frac{E}{T} \leq 0$. Hence, $dS_g < 0$. In this case, the violation of the GSL suggests that the process of accreting radiation with a temperature lower than that of the black hole reduces the overall entropy of the system. This is because the black hole's entropy increase is insufficient to compensate for the entropy carried by the radiation. The black hole, acting as a hotter reservoir, loses entropy faster than the colder radiation gains it, leading to a net decrease in total entropy.

be sufficient to uphold the GSL in situations where $T < T_{BH}$ [Bek-74]. However, this explanation appeared to be somewhat forced and did not align well with the principle of Occam's razor, which favors simpler explanations.

On thermodynamic grounds, Bekenstein proposed that a black hole should possess a finite entropy S_{BH} proportional to the area of its event horizon A as indicated in Eq. (4.14). He also suggested that the temperature of the black hole should be considered as being proportional to its surface gravity κ as detailed in Eqs. (4.41) and (4.37). Hawking further reasoned that if a black hole has a finite entropy proportional to the area of its event horizon A [Haw-74]:

$$A = 4\pi \left(2R_g\right)^2 = 16\pi \frac{(GM_{BH})^2}{c^4}, \tag{4.36}$$

then it must also have a finite temperature T_{BH} that is proportional to the black hole's surface gravity κ:

$$\kappa = \frac{GM_{BH}}{\left(2R_g\right)^2} = \frac{c^4}{4GM_{BH}}. \tag{4.37}$$

R_g is defined by Eq. (2.93).

4.3.4 *The Laws of Black Hole Mechanics*

In 1973, James Bardeen, Brandon Carter, and Hawking formulated the four laws of black hole mechanics analogous to the laws of thermodynamics. According to the no-hair theorem, a black hole in a stationary state is characterized only by its mass (M_{BH}), angular momentum (J), and electric charge (Q). The four laws of black hole mechanics are as follows [BCH]:

(0) The surface gravity κ is constant over the event horizon of a stationary black hole.

(1) The first law relates the changes in mass, angular momentum, and charge of the black hole to the change in its entropy:

$$d\left(M_{BH}c^2\right) = T_{BH}dS_{BH} - \Omega dJ - \Phi dQ. \tag{4.38}$$

Here $d\left(M_{BH}c^2\right)$ is the change in the black hole energy, Ω is the black hole's angular velocity, and Φ represents the electric potential. The terms ΩdJ and ΦdQ are analogous to the work on the black hole, similar to PdV in thermodynamics.

(2) Hawking's 1971 area theorem (see Sect. 4.2.1):

$$dA \geq 0. \tag{4.39}$$

This implies that the area A and, consequently, dS_{BH} can only increase.

(3) It is impossible to reduce κ to zero in a finite number of steps.

Robert Wald noted that black holes obeying these laws supported Bekenstein's thermodynamic ideas [Wal].

Bekenstein responded to Bardeen, Carter, and Hawking's paper by assigning a finite, non-zero temperature to the black hole. He reformulated Eq. (4.38) as follows:

$$d\left(M_{BH}\right)=\frac{\kappa c^{3}}{8\pi G}dA-\Omega dJ-\Phi dQ, \tag{4.40}$$

where κ is given by Eq. (4.37).

Hawking also acknowledged this formulation (4.40) [Haw-73]. According to Eqs. (4.38) and (4.40), if $G = c = 1$, $T_{BH}dS_{BH} = \left(\frac{\kappa}{8\pi}\right)dA$. Using this form of the first law and Eq. (4.14), Bekenstein derived [Bek-73]:

$$T_{BH}=\frac{\kappa l_{p}^{2}}{\eta k}. \tag{4.41}$$

Bekenstein later reflected that this equation established the black hole temperature, although its operational meaning was unclear at the time [Bek-01].

In 1975, Hawking acknowledged that Bekenstein was the first to suggest that a multiple of κ could represent the temperature of black holes and that the relation expressed in Eq. (4.40) was valid [Haw-76-1].

4.4 Hawking Radiation

4.4.1 Hawking Radiation and Evaporation

In 1973, Hawking, Bardeen, and Carter viewed Bekenstein's resolution to the paradox of a black hole in a thermal radiation bath as implausible. They argued that a black hole cannot be in equilibrium with black body radiation at a non-zero temperature. This is because a black hole cannot emit radiation, whereas it will always absorb some radiation falling into it. If the radiation has a very long wavelength, corresponding to a low black body temperature, as Bekenstein suggested, the absorption rate would be very slow. However, true equilibrium would only be achievable without radiation, i.e., if the external black body radiation temperature was zero [BCH]. Hawking eventually argued that if a black hole has a finite temperature proportional to its surface gravity, it suggests that it could be in thermal equilibrium with radiation at a temperature other than zero. While this is a standard principle for classical bodies, it seemed counterintuitive for black holes, known for their ability to absorb matter but not emit anything [Haw-77].

Hawking was puzzled by Bekenstein's efforts to preserve his GSL. He believed that equilibrium between a black hole and thermal radiation at a temperature other than zero was not feasible, as the black hole would absorb any incident thermal radiation without being able to emit anything in return [Haw-77]. The dilemma Hawking faced was: either the temperature of the black hole T_{BH} is zero, making its entropy S_{BH} infinite and rendering the idea of a black hole entropy meaningless, or black holes must emit thermal radiation at some finite non-zero temperature [Haw-76-2].

Around 1974, Hawking resolved the paradox by proposing that black holes emit particles constantly, with a characteristically thermal emission spectrum. He suggested that black holes generate and emit particles and radiation as if they were typical hot bodies with a temperature (T) related to their surface gravity (κ) and inversely proportional to their mass (M_{BH}) [Haw-74], [Haw-77].[36]

Hawking used the equation $E = kT$, where k is the Boltzmann constant, to demonstrate that black holes emit radiation that is thermally equivalent to that of a black body, with the temperature:

$$T = \frac{\hbar\kappa}{2\pi ck}. \tag{4.42}$$

From an outside observer's perspective, a Schwarzschild black hole appears to emit black body radiation at a temperature T that is proportional to the surface gravity κ of the black hole and inversely proportional to its mass M_{BH}:

$$T = \frac{c^3\hbar}{8\pi GM_{BH}k} \approx \frac{6.2 \times 10^{-8} K}{m}, \tag{4.43}$$

[36] Back in 1974, the common knowledge was that black holes absorb any radiation falling into them but are not expected to emit anything. This suggested a black hole in a radiation bath should keep absorbing radiation, increasing its mass and entropy. If the black hole is in a cavity filled with black body radiation at temperature T, for equilibrium, T should equal the black hole temperature T_{BH}. However, it was believed that black holes could not emit radiation, making proper equilibrium impossible unless $T - 0$. Hawking resolved this paradox in 1974 by showing that black holes emit radiation. This radiation has a thermal spectrum. Hawking corrected Eq. (4.41), showing that T_{BH} is inversely proportional to the black hole's mass M_{BH} [see Eq. (4.43)]. Black holes emit radiation as if they were hot bodies. This emission balances radiation absorption from the surrounding environment, allowing for a possible thermal equilibrium at a non-zero temperature. The realization that black holes emit Hawking radiation resolves the apparent violation of the GSL. When a black hole absorbs radiation, it gains energy E, and its entropy increases by an amount (4.32). The radiation that falls into the black hole has entropy given by Eq. (4.33). The emitted Hawking radiation carries away energy and entropy. This radiation warrants that the total entropy of the black hole and its surroundings does not decrease, maintaining the validity of the GSL. The change in total entropy considering Hawking radiation becomes: $\frac{E}{T_{BH}} - \frac{E}{T} + \frac{E_H}{T_H} \geq 0$. For $T < T_{BH}$, without Hawking radiation, the term $\frac{E}{T}$ would dominate, making $dS_g < 0$. However, with Hawking radiation, the term $\frac{E_H}{T_H}$ adds a positive entropy change, leading to $dS_g \geq 0$. Thus, Hawking's discovery that black holes emit thermal radiation reconciles the GSL with the presence of a black hole in a thermal radiation bath. It shows that black holes can achieve thermal equilibrium with their surroundings, preserving the second law of thermodynamics (see [Haw-76-1]).

where m represents the black hole mass expressed in terms of solar masses $\approx 2 \times 10^{30} kg$:

$$m = \frac{M_{BH}}{M_{\odot}}. \tag{4.44}$$

Bekenstein noted that, to Hawking's surprise, Hawking's research led him to discover that black holes are hot [Bek-01].

Utilizing Eq. (4.36), Hawking expressed the entropy of a black hole in the following manner [Haw-76-1]:

BEKENSTEIN- HAWKING ENTROPY

$$\boxed{S_{BH} = \frac{kc^3}{G\hbar}\frac{A}{4},} \tag{4.45}$$

where S_{BH} is called the *Bekenstein-Hawking entropy*.

This equation corresponds to Bekenstein's equation (4.14) or Eq. (4.31). Consequently, Hawking refined Bekenstein's constant of proportionality, setting $\eta = \frac{1}{4}$. This calculation brought the matter to a satisfactory conclusion, leading Wheeler to remark that Hawking ultimately acknowledged Bekenstein's correctness [Whe-08].

In a lecture on his 60th birthday, Hawking recounted how he discovered Hawking radiation. By the end of 1973, his work on singularity theorems with Penrose had shown that general relativity broke down at singularities. This realization led him to consider the integration of general relativity with quantum theory. Despite lacking a background in quantum theory, he started with an investigation into the behavior of quantum fields and particles near a black hole. To his surprise, he found that black holes seemed to emit radiation. Initially doubting this result, Hawking became convinced of its validity when he realized that the emission perfectly aligned with what was needed to equate the area of the horizon with the black hole's entropy [Eq. (4.45)]. Hawking expressed his desire for this simple formula to be inscribed on his tombstone [Haw-03].

In his 1975 paper, Hawking highlighted two vital aspects of S_{BH}: firstly, it should always increase when a black hole absorbs matter or radiation; secondly, the entropy of a new black hole formed from the merger of two black holes should be greater than the sum of the entropies of the original black holes [Haw-76-1]. Bekenstein himself acknowledged that he had somewhat conjectured the form of the function in equation (4.14), $S_{BH} = f(A)$ [Bek-01]. Hawking noted in 1975 that this entropy function was unique [Haw-76-1].

Bekenstein sought to integrate thermodynamics with classical general relativity, whereas Hawking aimed to merge quantum mechanics with classical general relativity. Hawking used a semi-classical approach to derive equation (4.45). Hawking made the groundbreaking discovery that black holes emit thermal radiation, underpinned by quantum effects causing the radiation. He utilized a semi-classical approximation, treating matter fields quantum mechanically within a classical spacetime framework, leading him to Eq. (4.42) [Haw-78]. In 1974, Hawking showed that the number of particles emitted from a black hole is proportional to the number of particles that would have been absorbed from past null infinity[37]: This relationship is precisely what one would expect from a black hole emitting thermal radiation at a temperature given by equation (4.42). Thus, Hawking provided a theoretical justification for the geometric reasoning used in black hole thermodynamics [Haw-74].

In a June 1975 paper, "Black Holes and Thermodynamics," Hawking showed that steady-rate black hole emission has a thermal spectrum. His 1974 semi-classical approximation led to the emergence of several key quantities: the area A of the event horizon [Eq. (4.36)], the surface gravity κ [Eq. (4.37)], the black hole's angular frequency of rotation Ω, and the potential of the event horizon Φ. Both Ω and Φ are inversely proportional to the area A as indicated in Eq. (4.38), which describes the first law of black hole mechanics (see Sect. 4.3.4). The most significant outcome of Hawking's approximation was the definition of the entropy of the black hole [Eq. (4.45)] [Haw-76-1],

Two months after his initial findings, Hawking submitted a paper elaborating more explicitly on his theory. Hawking's paper outlined how quantum field theory suggests the continuous creation and recombination of particle-antiparticle pairs (like electrons and positrons) in the vacuum. Near a black hole's event horizon, the black hole might capture one particle of a pair. At the same time, the other, having no partner to annihilate with, could either fall into the black hole or escape to infinity, appearing as radiation emitted by the black hole. Hawking described this phenomenon as the spontaneous creation of particle pairs near the black hole: one with negative energy,

[37] I will begin by discussing Hawking's original 1974 derivation. That year, Hawking calculated the number of particles produced and emitted to infinity during gravitational collapse. He used the example of a spherically symmetric collapsing star with the external field described by the Schwarzschild solution. He divided the asymptotically flat spacetime into three regions: Future null infinity ($\mathcal{I}^+$): The region where outgoing light rays (null geodesics) reach observers at infinity. The event horizon ($\mathcal{H}^+$): The boundary beyond which light cannot escape the black hole. Past null infinity ($\mathcal{I}^-$): The region where incoming light rays originate from infinity. Hawking considered a wave with a positive frequency at $\mathcal{I}^+$ propagating backward in time through spacetime. This wave does not cross the event horizon. Part of the wave scatters off the curvature of the Schwarzschild metric outside the black hole and reaches $\mathcal{I}^-$ with the same frequency. Another part travels backward into the black hole, through the origin, and again to $\mathcal{I}^-$, undergoing significant blue-shift. These blue-shifted waves can reach $\mathcal{I}^-$ with an asymptotic waveform for times $v < v_0$, where v_0 is the last advanced time allowing escape to $\mathcal{I}^-$. The black hole absorbs these waves. For $v > v_0$, the waves can pass through the origin and escape to $\mathcal{I}^+$, experiencing red-shift. Hawking's calculations (involving Fourier transforms of the asymptotic waveform) indicated that the total number of outgoing particles or wave packets in a specific frequency range is infinite. He concluded that this infinite number corresponds to a steady-rate emission, signifying that the black hole will continuously emit thermal radiation until it evaporates. Hawking established a relationship between absorption and emission by comparing incoming waves that escape to $\mathcal{I}^+$ with outgoing ones arriving at $\mathcal{I}^-$. He discovered that the number of particles emitted from a black hole in a particular wave packet mode is proportional to the number of particles that would have been absorbed by the black hole from a similar wave packet coming from $\mathcal{I}^-$.

falling into the black hole, and the other with positive energy, escaping to infinity, thereby contributing to the thermal emission observed from black holes [Haw-76-2]. A paradox lies in the fact that Hawking radiation, seemingly emanating from within the black hole, is created outside it. Still, it appears to observers at infinity as if the black hole emits radiation [Un-18].[38] Hawking suggested an alternative interpretation of the radiation: quantum mechanics permits particles to tunnel through spacelike or past-directed world lines, allowing them to escape black holes through the event horizon. Hawking reimagined the particles falling into a black hole as traveling backward in time. This perspective suggests that radiation might originate from the singularity and quantum tunnel out of the black hole [Haw-77], [Haw-78].[39] Hawking interpretations leverage quantum mechanics to explain the radiation from a black hole. From an observer's viewpoint at infinity, it appears as though the black hole is creating particle pairs, with one falling into the black hole and the other escaping [Haw-76-2].[40]

As the black hole emits particles through Hawking radiation, it loses energy, decreasing its mass according to the mass-energy equivalence principle [Eq. (4.49)]. Consequently, the temperature T of the black hole rises because the emission of Hawking radiation is inversely proportional to its mass. This rise in temperature

---

[38] It might seem counterintuitive to say the black hole emits radiation if the particles are created just outside its event horizon. However, the essential aspect of Hawking's explanation is that the black hole's gravitational field is necessary for separating these particles and the energy dynamics involved in Hawking radiation. This radiation should be viewed as a process where the black hole's gravitational field facilitates the conversion of virtual particles into real particles, leading to observable radiation. Thus, Hawking radiation is fundamentally driven by the black hole's gravitational field, making it a form of radiation from the black hole itself.

[39] Hawking suggested that particles could tunnel through spacelike (outside the causal structure of the spacetime) or past-directed (seemingly traveling backward in time) world lines. Normally, once particles cross the event horizon of a black hole, they cannot escape because the escape velocity exceeds the speed of light. However, due to quantum tunneling, particles can probabilistically escape from inside the event horizon. This implies that the event horizon is not an absolute boundary for particles at the quantum level. Hawking reimagined the particles falling into a black hole as traveling backward in time. This is a metaphorical way to describe the mathematical framework where certain solutions to the equations governing particle behavior near black holes allow for such interpretations. In this view, particles that appear to be falling into the black hole can be considered their anti-particles escaping from it, moving backward in time from our perspective. By considering these particles as originating from the singularity and quantum tunneling out of the black hole, Hawking provided a new angle on how black hole radiation could be understood. This perspective suggests that the black hole's singularity is a source of these particles.

[40] This description aligns with Penrose and Floyd's process involving particles in the ergosphere, where the black hole absorbs one part of a split particle with negative mass energy while the other part escapes. While Hawking radiation and the Penrose process involve particle interactions leading to energy extraction from black holes, they differ significantly in their underlying mechanisms, locations of occurrence, and the types of energy they extract. The Penrose process is based on classical physics and applies specifically to rotating black holes, whereas Hawking radiation is a quantum mechanical process applicable to all black holes. While the Penrose process upholds the area theorem (4.39), Hawking's radiation seems to violate it. Nevertheless, neither process contravenes the GSL (as discussed in Sect. 4.4.2.

causes an increase in the rate of Hawking radiation, which further reduces the black hole's mass M_{BH}. As the black hole becomes smaller, its temperature rises, accelerating the mass loss and increasing the radiation rate, leading to a runaway process that continues until the black hole evaporates completely.[41] Thus, the smaller the black hole, the hotter it becomes and the faster it loses mass. When the black hole reduces to the size of a micro black hole with a Planck mass:

[41] Let us calculate the power P emitted by a Schwarzschild black hole with mass M_{BH}. However, there is a complication in determining P, as Hawking noted in 1974. As the black hole emits thermal radiation at a temperature of approximately $\approx 10^{-6}$ [refer to Eqs. (4.42) and (4.43)], it is expected to lose mass. This mass loss would, in turn, increase the surface gravity, thereby accelerating the emission rate. Consequently, the black hole would have a finite lifespan, estimated by Hawking to be around: $10^{71}\left(\frac{M_\odot}{M}\right)^{-3}s$. For a black hole with the sun's mass, this duration is significantly longer than the current age of the Universe [Haw-74]. The Stefan-Boltzmann law, which pertains to black body radiation, states that the power P radiated per unit area of a black body is given by:

$$P = \sigma \epsilon A T^4, \tag{4.46}$$

where σT^4 represents the flux at which a black body emits, $\epsilon = 1$ because a black hole is a perfect black body and σ represents the Stefan-Boltzmann constant:

$$\sigma = \frac{\pi^2 k^4}{60 c^4 \hbar^3}. \tag{4.47}$$

To determine the rate at which a black hole emits energy across its horizon surface, as perceived by an external observer, we integrate Eqs. (4.46) and (4.47) along with Eqs. (4.36) and (4.43):

$$P = -\frac{dE}{dt} = \frac{\hbar c^6}{15360\pi G^2 {M_{BH}}^2} = \frac{9\times 10^{-22}}{m^2}\frac{erg}{s}. \tag{4.48}$$

m is defined in Eq. (4.44). The emitted energy is remarkably low. By inserting the equation:

$$E = M_{BH}c^2, \tag{4.49}$$

into Eq. (4.48), we can express the left-hand side as:

$$-\frac{dE}{dT} = -c^2\frac{dM_{BH}}{dt}. \tag{4.50}$$

Rearranging the terms in Eq. (4.48) gives us a new equation that details the rate at which the black hole's mass decreases over time due to Hawking radiation:

$$-\frac{dM_{BH}}{dt} = \frac{\hbar c^4}{15360\pi G^2 {M_{BH}}^2}, \tag{4.51}$$

From this, we can calculate the evaporation time t for the black hole, as observed from outside:

$$t = \frac{5120\pi G^2 {M_{BH}}^3}{\hbar c^4} \approx m^3 \times 6.67\times 10^{74} s, \tag{4.52}$$

where m is specified in Eq. (4.44). This duration is extraordinarily long.

$$m_p = \sqrt{\frac{\hbar c}{G}} = 2.176 \times 10^{-8} kg, \tag{4.53}$$

it almost completely evaporates. At this point, its temperature is so high that it will explode in a burst of energy. This process implies a violation of Hawking's area theorem (4.39), as the area of the black hole decreases while it radiates Hawking radiation.

Thus, *naked singularities* significantly challenge Hawking's area law. The essence of the problem lies in the disappearance of the black hole horizon. When the horizon vanishes, the area of the horizon diminishes, leading to a violation of the area law. This issue is intricately linked to the solutions of the field equations of general relativity that describe naked singularities not concealed by an event horizon. These scenarios pose severe problems to the Israel-Carter uniqueness theorems, as they stipulate that the Kerr black hole is a unique solution only if event horizons envelop all singularities.

In 1969, Penrose contemplated the existence of a hypothetical cosmic censor. This principle would prevent the formation of naked singularities because they are always enveloped within an event horizon. Penrose expressed uncertainty about whether naked singularities could emerge from a collapse that began from a non-singular initial state [Pen-69]. In 1975, Hawking formulated the *cosmic censorship* hypothesis, succinctly stating that nature abhors a naked singularity. This hypothesis suggests that an event horizon invariably conceals any singularities resulting from gravitational collapse for distant observers. However, Hawking also noted that the evaporation of black holes posed a challenge to this classical view of cosmic censorship. He explained that if we attempt to describe the process of a black hole losing mass and eventually vanishing through a classical spacetime metric, we are confronted with an inevitable naked singularity when the black hole ceases to exist [Haw-76-2]. Thus, in response to Penrose's remark that the singularity is completely exposed and visible to the external world [Pen-69], Hawking countered that an observer positioned infinitely far away is still able to witness events occurring at the singularity [Haw-76-2], implying that cosmic censorship might not be an absolute principle in black hole physics.

Thus, Hawking radiation and black hole evaporation challenge the Hawking area law. The ultimate resolution to this problem likely awaits a more complete understanding of quantum gravity.

4.4.2 Bekenstein Revisits the GSL

In 1975, Bekenstein revisited the GSL, acknowledging that without additional entropy generated by Hawking radiation, the GSL would be compromised. He initially suggested a complex mechanism for this but later recognized the simplicity and effectiveness of Hawking radiation in resolving this issue. This led Bekenstein to demonstrate that the GSL is maintained, on average, once the black hole's radiated energy integrates into the external environment (see footnote 36). Bekenstein re-evaluated the same theoretical setup in 1975, considering a black hole within a black body radiation bath inside a cavity at a temperature T_M significantly lower than the black hole's temperature T_{BH}. The entropy of the external radiation and matter, S_M, became a focal point in his analysis. Bekenstein theorized that transferring radiation from the cavity to the black hole (from a lower to a higher temperature) would infringe the GSL unless a Hawking radiation entropy-generating process existed. While he initially proposed a cumbersome and illogical mechanism in 1974 (see Sect. 4.3.3), by 1975, he recognized Hawking radiation as a more straightforward solution. Bekenstein thus showed that the GSL is respected in an averaged sense: the total entropy, S_{BH} plus S_M, increases after the assimilation of the black hole's emitted radiation into the ambient bath. Although Hawking radiation reduces the black hole's surface area A over time, the average entropy change (4.27) remains positive for all temperatures T and only vanishes when $T_M = T_{BH}$ [Bek-75].

Bekenstein later emphasized that Hawking radiation must carry sufficient entropy to more than compensate for any loss [Bek-01]. The Kerr black hole's area might decrease as the radiation progresses, but the radiation entropy increases. Bekenstein compared a black hole to a hot body, suggesting that the combined entropy of the black hole and its radiation will likely increase [Bek-75]. He clarified that the GSL was intended to supplant the conventional second law of thermodynamics. Unlike Hawking's area theorem, which requires that the area A does not diminish (see Sect. 4.3.1), the GSL makes a more potent assertion. Bekenstein stated that it could be expected to perform better as a fundamental quantum rule. Bekenstein noted that the quantum phenomenon of a Kerr black hole spontaneously emitting radiation, as discovered by Hawking, conspicuously violates the area law. However, the increase in external entropy because of this radiation is expected to be enough to uphold the GSL. While not strictly a quantum law, the GSL, according to Bekenstein, gains confirmation from Hawking's findings, highlighting the GSL's versatility and the significance of black-hole entropy [Bek-75].

In his 1975 paper, Hawking noted Bekenstein's earlier inconsistencies due to his attempt to merge the idea of finite entropy with classical theory [Haw-76-1]. However, Bekenstein interpreted this differently. He asserted that his 1975 argument was implicitly present in Hawking's affirmation of the GSL in black hole radiation processes. He acknowledged Hawking's straightforward proof that $dSg = dS_{BH} + dS_M$ is greater than before, thus validating Eq. (4.27) and rescuing the GSL from disproof in scenarios where $T_M < T_{BH}$ (see Sect. 4.4.1). Bekenstein expressed satisfaction, noting that Hawking, initially a critic of the GSL, ultimately contributed to its consistency with theoretical models [Bek-80].

In 1981, Bekenstein suggested that as long as the entropy S does not exceed $2\pi ER$, the GSL is not violated, leading to the formulation of what became known as the *Bekenstein bound* [Bek-74], [Bek-81], [Bou-18], [Bai][42]

$$S \leq 2\pi ER. \tag{4.59}$$

William Unruh and Robert Wald challenged this conclusion [Un-Wal-81].[43]

---

[42] Bekenstein simplified the problem by setting constants G, c, k, and $\hbar$ to 1. For a system of mass M, energy E, effective radius R, and entropy S dropped into a black hole, the mass increase of the black hole is given as:

$$dM_{BH} = \frac{ER}{4M_{BH}}. \tag{4.54}$$

This equation includes the gravitational redshift factor α defined by Eq. (4.30).
The increase in mass given by Eq. (4.54) leads to an increase in the surface area of the black hole's horizon, as shown by the equation:

$$dA = 32\pi M_{BH} dM_{BH}. \tag{4.55}$$

Here, A is defined by Eq. (4.36). Taking the differential of (4.36) gives Eq. (4.55). The factor 32 arises from differentiating (4.36). Substituting Eq. (4.54) into Eq. (4.55) gives:

$$dA = \frac{32\pi M_{BH} ER}{4M_{BH}} = 8\pi ER. \tag{4.56}$$

Since, the GSL requires that $dS_g \geq 0$ and $dS_{BH} \geq -dS_M = S$, Eq. (4.45) leads to: $\frac{dA}{4} \geq S$. Dividing Eq. (4.56) by 4 yields:

$$\frac{dA}{4} = \frac{8\pi ER}{4}, \tag{4.57}$$

and the black hole entropy increases by: $2\pi ER$. Dividing Eq. (4.16) by 4 yields:

$$\frac{dA}{4} = \frac{8\pi M_{BH} r_+}{4}, \quad \text{where: } r_+ = 2M_{BH}. \tag{4.58}$$

This gives the Bekenstein-Hawking entropy, consistent with the Bekenstein bound Eq. (4.59).

[43] Unruh and Wald argued that a buoyancy effect, caused by the acceleration radiation near a black hole, creates a lower bound on the energy delivered to the black hole. This complies with the GSL without the need for Bekenstein's bound. Bekenstein had reservations about this buoyancy effect, maintaining that his entropy bound was still necessary for the GSL. Despite Bekenstein's skepticism, Unruh and Wald's findings suggested that the Bekenstein bound was not essential for the GSL. Yet, Bekenstein supported the bound, citing direct statistical arguments for its validity. Unruh and Wald further elaborated on this approach. They argued that when attempting to lower a box (the Geroch box) containing rest energy E and entropy S into a black hole, an effective buoyancy force, resulting from acceleration radiation experienced by the box near the black hole, limits the energy delivered to the black hole. This causes a minimal increase in the black hole's area, sufficient for the GSL to be satisfied. Therefore, the rope performs less work in lowering the box, and more energy is transferred to the black hole than in classical scenarios. This is analogous to the Archimedes principle of buoyancy, where the energy of the box is equal to the energy of the displaced acceleration radiation. Unruh and Wald communicated this finding to Bekenstein, but he had reservations about the buoyancy effect. Bekenstein provided examples showing that the buoyancy effect alone might not be sufficient to uphold the GSL. He maintained that his entropy bound on the matter, as described in Eq. (4.59), was still necessary for the GSL's validity. In 2002, Bekenstein acknowledged that the original derivation of his bound, which did not account for the

4.4.3 Detecting Hawking Radiation

Hawking radiation, though a widely accepted concept, still lacks empirical validation through direct observational evidence, as per Hawking's original predictions. According to Eq. (4.43)), the temperature of Hawking radiation emitted by a stellar-mass black hole is extremely low, falling six to eight orders of magnitude below the cosmic microwave background radiation. This significant discrepancy makes detecting Hawking radiation from stellar mass, and even more so from supermassive black holes, unfeasible. Furthermore, Hawking acknowledged in 1974 that the evaporation time of a black hole [referenced in equation (4.52) in footnote 41] is directly proportional to the cube of its mass ${M_{BH}}^3$. Consequently, a black hole with a mass equivalent to one solar mass ($M_\odot \approx 2 \times 10^{30} g$ grams) would have an evaporation time exceeding $t \approx 2 \times 10^{67}$ years. This duration far surpasses the current estimated age of the universe, around 10^{10} years. The evaporation timeline extends even further for larger black holes; for instance, a stellar-mass black hole with a minimum mass of $3.3M_\odot$ and the supermassive black hole at the center of our galaxy have estimated evaporation times of approximately 10^{94} years. This extended duration is partly due to black holes continuously absorbing matter while gradually decreasing size.

Given these considerations, the possibility arises of detecting Hawking radiation from micro black holes. However, as Eq. (4.53) indicates, a black hole's mass cannot fall below the Planck mass.[44] In 1974, Hawking speculated on the existence of

---

buoyancy of the Geroch box, had sparked controversy. He conceded that accounting for buoyancy might only slightly increase the coefficient in his bound Eq. (4.59) but did not fundamentally challenge its validity [Un-Wal-81], [Un-01], [Wal], [Bek-83], [Bek-94], [Bek-02].

[44] The rationale is that if the mass were lower, the black hole would be smaller than its own Compton wavelength:

$$\lambda_c = \frac{\hbar}{m_p c}, \tag{4.60}$$

equal to its Schwarzschild radius. The Schwarzschild radius, $2l_p$ [$= 2R_g$ and defined by Eq. (2.93)], is thus of the same order of magnitude as the Planck length:

$$l_p = \sqrt{\frac{G\hbar}{c^3}} \approx 1.616 \times 10^{-35} m. \tag{4.61}$$

This implies that a black hole with a mass below the Planck mass would defy the fundamental principles governing its Schwarzschild radius. Consequently, a black hole smaller than this would not exhibit the defining characteristic of a black hole: the event horizon. Without an event horizon, it ceases to possess the essential properties that classify it as a black hole, thereby challenging the notion of detecting Hawking radiation in such circumstances [Bek-04]. The Planck time, a fundamental unit, is defined as follows:

$$\frac{G^2 {m_p}^3}{\hbar c^4} = t_p = \sqrt{\frac{G\hbar}{c^5}} \approx 5.39 \times 10^{-44} s, \tag{4.62}$$

Furthermore, as per Eq. (4.52) (see footnote 41), the evaporation time t for a black hole with a mass equivalent to the Planck mass can be derived:

$$t = \frac{5120\pi G^2 {m_p}^3}{\hbar c^4} = 5120\pi t_p \approx 8 \times 10^{-40} s. \tag{4.63}$$

significantly small black holes, possibly formed from fluctuations in the early Universe. He postulated that any such black hole with a mass less than $10^{15}g$ grams would have already evaporated by now. As these black holes approached the end of their lifespan, he theorized that their emission rates would dramatically increase. Hawking estimated that in the final 0.1 seconds of a black hole's life, its emission rate could reach 10^{30} erg. Though this might seem minor on an astronomical scale, Hawking likened the explosive force of such a black hole to approximately one million one-megaton hydrogen bombs [Haw-74]. Hawking also suggested the possibility of detecting Hawking radiation from hypothetical primordial mini black holes, comparable in size to a mountain on Earth, with masses less than or equal to $M_{BH} \leq 10^{12}g$. He proposed that a $10^{12}g$ primordial black hole formed in the early universe would radiate its mass over the universe's age and be in its final evaporation stages. The evaporation time for such a low-mass black hole is $t \approx 2.267 \times 10^9 y$. If these tiny black holes exist, we might observe them evaporating in a spectacular explosion. In 1975, Hawking introduced the idea that these mini-black holes could be analogous to hypothetical white holes. He described white holes as the temporal reverse of black holes. While black holes emit Hawking radiation and consume matter and radiation, white holes emit matter and radiation but prevent anything from entering their event horizons. According to Hawking, the emission rate from white holes is equivalent to that of black holes, assuming the emission of a white hole is independent of its surroundings. Thus, he concluded that to external observers, black and white holes appear identical. In a 1979 interview, Hawking summarized his view that he believed there was just a single entity (only holes). When large, these entities seem black; when small, they appear white [Over], [Haw-76-1].

Over the years, numerous proposals have been made regarding detecting Hawking radiation.[45]

This calculation reveals that a tiny black hole with a mass equivalent to the Planck mass would radiate away all its mass energy in a remarkably brief period, approximately $8 \times 10^{-40}s$. This rapid evaporation underscores the challenge in detecting Hawking radiation from such minuscule black holes, as their existence is fleeting and their radiation output, while theoretically significant, is practically ephemeral.

[45] In 1981, Unruh proposed an innovative laboratory experiment, an acoustic black hole, due to the challenge of detecting Hawking radiation. Unruh's idea is to simulate the phenomena of an astrophysical black hole through an artificial counterpart in the lab. The approach hinges on sound waves (phonons) moving in a fluid: waves co-moving with the fluid travel faster, while those moving against it slow down. When the fluid's velocity surpasses the wave's speed, a certain event horizon forms, beyond which no wave can travel upstream against the fluid flow. Acoustic black holes are used to investigate and potentially detect Hawking radiation analogs. In these systems, a phenomenon analogous to Hawking radiation is expected to occur, where pairs of phonons are emitted close to the region mimicking the event horizon, with one escaping and the other falling into the black hole [Un-81], [Leo]. However, the acoustic black hole model faces a significant limitation. It cannot replicate the Bekenstein-Hawking entropy (S_{BH}). This discrepancy arises because while Hawking radiation is a kinematic effect reliant solely on the presence of a specific metric and a type of horizon, S_{BH} is tied to dynamic factors, specifically the area of the event horizon. Einstein's field equations determine the area of a black hole's event horizon. For a Schwarzschild black hole, the radius of the event horizon (the Schwarzschild radius) is directly proportional to its mass. For more complex black holes, like those with rotation (Kerr black holes) or charge (Reissner-Nordström black holes), the calculations become more intricate but still fundamentally rely on solutions to Einstein's field equations [Vis].

4.4.4 Hawking's Black Hole Information Paradox

Typically, we present the information paradox in the following manner. A significant challenge to general relativity arises from the tension between the no-hair theorem and the principle of unitarity in quantum mechanics. Recall that the no-hair theorem in general relativity suggests that a black hole is entirely characterized by three externally observable parameters: mass, angular momentum, and electric charge. According to this theorem, all other information about the matter that formed the black hole or fell into it is lost to outside observers and erased (see Sect. 4.2.1). On the other hand, the principle of unitarity in quantum mechanics asserts that information is never lost in the quantum world. It implies that the evolution of a quantum system is deterministic and reversible, meaning that the system's entire history (or its wave function evolution) should always be inferable from its current state. This principle suggests that information about a system's initial state should always be preserved, even if the system undergoes extreme transformations, such as collapsing into a black hole.[46]

The discrepancy between these two principles—the apparent loss of information in a black hole (per the no-hair theorem) versus the preservation of information (per the unitarity principle)—is at the heart of the black hole information paradox. This paradox highlights a fundamental conflict between general relativity and quantum mechanics.[47]

However, Hawking described the information paradox by focusing specifically on the aspect of the black hole information paradox that arises from the conflict between Hawking radiation (which implies a non-unitary process and information loss) and the unitarity principle of quantum mechanics without involving the no-hair theorem. In this view, the paradox is centered around the nature of Hawking radiation. According to Hawking's calculations, this radiation is thermal (truly random)

[46] In quantum mechanics, unitarity warrants that the evolution of a closed quantum system is reversible. The wave function's evolution over time is governed by the Schrödinger equation, which is deterministic. Despite this deterministic evolution, quantum mechanics is inherently probabilistic regarding measurement. When we measure a quantum system, the wave function collapses to one of the possible eigenstates, and the square of the amplitude of the wave function in that state gives the probability of each outcome. Unitarity governs the evolution of the wave function, not the outcomes of measurements. Still, these processes are also described by unitary evolution in the context of the larger system, including the measuring device and observer. For instance, the EPR thought experiment's wave function is unitary before the particles are measured. If no measurement occurs, the system's wave function evolves deterministically and reversibly (see the derivation in Sect. 5.3.1).

[47] Bekenstein's perspective on the information paradox frames it as a conflict between the unitarity principle of quantum mechanics and the no-hair theorem of general relativity. Bekenstein observed that a black hole could transform a pure quantum state into a mixed one, which challenges the unitarity principle of quantum theory that asserts a pure state remains pure under unitary evolution. According to Bekenstein, this led Hawking to conclude that general relativity may violate quantum theory's unitarity principle. This conclusion is controversial: while general relativity researchers often accept information loss, particle physicists typically uphold the unitarity principle and traditional quantum theory. Bekenstein identified these opposing views as aspects of the black hole information loss paradox [Bek-93], [Bek-04].

(see Sect. 4.4.1). It does not contain information about the material that formed and fell into the black hole. This suggests that the process of black hole evaporation is non-unitary. In other words, the information about the initial state (the material that formed the black hole) is lost when the black hole evaporates. Such a non-unitary evolution contradicts the fundamental quantum mechanical principle of unitarity, which asserts that information is conserved in the quantum world and that the evolution of a quantum system is reversible.

This perspective on the paradox highlights the tension between the irreversible, information-erasing nature of black hole evaporation (as suggested by Hawking's radiation) and the principles of quantum mechanics, which insist on information conservation. While related to the broader discussion of black holes and information loss, the no-hair theorem is not directly involved in Hawking's specific formulation of the paradox.

The following is Hawking's formulation of the information paradox. Hawking initially suggested in 1975 that the process of black hole evaporation results in a loss of information about the state of particles that fell into the black hole, making the final state of the black hole and all the particles and radiation associated with it unpredictable if the black hole evaporated completely. If the black hole were to evaporate completely, the system's final state would be the state of all the radiation emitted, along with any remaining particles outside the black hole. Hawking's original semi-classical calculations suggested that the emitted radiation would be in a mixed state due to the apparent loss of information about the particles that fell into the black hole.[48] This implied that a non-unitary S-matrix (scattering matrix) would describe the system's evolution (from infalling particles to outgoing radiation), signifying a departure from standard quantum mechanics where a unitary S-matrix describes processes [Hawk1].[49]

---

[48] Unlike a pure state, described by a single vector in the Hilbert space, where we have complete knowledge about the system, a mixed state represents a lack of complete knowledge.

[49] Unitarity warrants that the total probability of all possible outcomes of any quantum event always sums to 1. This principle corresponds to the conservation of information, i.e., information is neither created nor destroyed in the evolution of a quantum system. In quantum field theory, the S-matrix describes how initial states (particles before an interaction) evolve into final states (particles after an interaction). A unitary S-matrix ensures that the probabilities of all possible outcomes of a quantum process are conserved. This means that the sum of the probabilities of all possible final states equals 1. If the S-matrix is non-unitary, it implies a violation of probability conservation in quantum processes. If the evolution of the quantum system (including the black hole and all associated particles and radiation) is governed by a non-unitary S-matrix, the system's evolution would deviate from the expected evolution in standard quantum mechanics and become unpredictable. This unpredictability occurs both during the black hole's evaporation and after it has completely evaporated, a significant issue known as the black hole information paradox.

Chapter 5
Addressing the Conflict Between Quantum Mechanics and General Relativity

5.1 Introduction

I start this chapter with black hole complementarity. In the previous chapter, I discussed Hawking's information paradox.[1]

In 1986, Douglas Macdonald, Richard Price, and Kip Thorne introduced the stretched horizon and the classical membrane paradigm. The stretched horizon is a surface model slightly outside the event horizon of a black hole, providing a conceptual and mathematical construct that simplifies the complex physics of black holes. It is a timelike surface, enabling well-defined geometry and finite redshift, making it more manageable using classical physics techniques.

In 1993, Leonard Susskind, Lárus Thorlacius, and John Uglum introduced the principle of black hole complementarity, asserting that observations by infalling and distant observers are complementary yet irreconcilable due to the event horizon's nature. Susskind and others suggested black hole complementarity to resolve the discrepancy between Hawking radiation's thermal nature and the unitarity principle. According to black hole complementarity, different observers (outside the black hole and those falling into it) have complementary but not contradictory descriptions of reality. No single observer can witness violations of the laws of physics.

Susskind et al.. used the stretched horizon to address the black hole information paradox. They utilized a 2D gravity model to describe the stretched horizon's behavior, kinematics, and statistical fluctuations. They developed four key postulates for black hole complementarity: 1. *Unitarity and Quantum Purity*: Black hole formation and evaporation, as seen by a distant observer, can be described by standard quantum

[1] Hawking's semi-classical calculations showed that black holes emit radiation due to quantum effects near the event horizon. This radiation causes black holes to lose mass and eventually evaporate. Hawking radiation is purely thermal. Thus, it carries no information about the matter that fell into the black hole. When the black hole evaporates completely, all that remains is the thermal radiation, which does not contain information about the matter initially consumed by the black hole. This apparent loss of information violates the principle of unitarity in quantum mechanics, which states that information must be preserved in a closed system.

G. Weinstein, *Einstein's Legacy*, History of Physics,
https://doi.org/10.1007/978-3-031-73572-1_5

theory. The process involves transforming infalling matter into Hawking radiation while maintaining the overall purity of the quantum state throughout its emission stages. 2. *Effective Field Theory*: Physical processes can be approximated outside the stretched horizon using semi-classical equations, which blend classical and quantum theories. This approximation is valid because the extreme effects of quantum gravity are not dominant in this region. 3. *Discrete Energy Levels and Quantum System*: A black hole, from the perspective of a distant observer, behaves like a quantum system with discrete energy levels. The number of independent quantum states for a black hole of mass M is given by the exponential of the Bekenstein entropy, implying a finite and well-defined number of quantum states. 4. *No Drama*: An infalling observer feels nothing unusual when crossing the event horizon, experiencing normal physics without singularities or discontinuities. This aligns with the equivalence principle, suggesting that locally, the effects of gravity are indistinguishable from acceleration in a non-gravitational field.

Susskind et al.. argue that a theory based on these postulates is needed so that an external observer does not reference events inside a black hole. They propose that the stretched horizon be viewed through a thermodynamic lens, assuming a microphysical foundation. Although the thermodynamic description is coarse-grained, they believe it has an underlying microphysical basis that explains the observed thermodynamic behaviors. The stretched horizon is conceptualized as a physical membrane with distinct mechanical, electrical, and thermal properties perceived by an external observer. It absorbs and scrambles information from objects approaching the black hole before crossing the event horizon and re-emits it. The infalling observer perceives a smooth crossing over the event horizon, while the outside observer sees the horizon as a dynamical membrane where information is scrambled and re-emitted. For an observer outside the black hole, the Hawking radiation appears thermal and does not contain information about the matter that fell into the black hole. However, the information is not truly lost but is scrambled so that an external observer finds it practically irretrievable. For the infalling observer, the experience remains smooth and uneventful at the horizon, maintaining the equivalence principle. These experiences cannot be reconciled due to the fundamental limitations imposed by the event horizon.

In 2013, the AMPS (Ahmed Almheiri, Donald Marolf, Joseph Polchinski, and James Sully) team critically analyzed the postulates proposed by Susskind and others regarding the behavior of black holes, particularly as they age and emit Hawking radiation. The focus was on black holes in the later stages of their evaporation, formed from a pure state and decaying over time by emitting radiation. The AMPS team's analysis revealed a fundamental contradiction between two of Susskind's postulates (Postulates 1 and 2) and another postulate (Postulate 4). AMPS restate the paradox by dividing Hawking radiation into early and late parts, focusing on the state after the black hole has emitted half of its initial Bekenstein-Hawking entropy. After the Page time, the early radiation's Hilbert space is much larger than that of the late radiation, implying near-random states for the late-time radiation. This aligns with Postulate 3, which states that a black hole appears as a quantum system with discrete energy levels to a distant observer. This is reflected in the large Hilbert space dimension and the high entanglement and complexity of the black hole's quantum states. According to black hole complementarity, the stretched horizon stores information about everything that has fallen into the black hole in a highly scrambled form, preserving quantum mechanics principles. Hawking radiation, emitted due to quantum effects near the event horizon, carries away this scrambled information, implying that the emitted radiation contains pieces of information stored in the stretched horizon. This information is gradually released as the black hole evaporates, maintaining quantum mechanics principles. However, this leads to complex entanglement structures challenging traditional notions, specifically the monogamy of entanglement. The apparent contradiction arises because Postulate 1's requirement for unitarity seems to conflict with Postulate 4's no-drama scenario. Postulate 1 emphasizes the unitarity of black hole processes, implying that Hawking radiation remains pure. Late Hawking radiation must be entangled with early radiation to preserve this purity. However, this creates a paradox, as late Hawking radiation must be fully entangled with particles inside the horizon, adhering to Postulate 4's no-drama scenario. This would imply that the late Hawking radiation is entangled with both early radiation and the modes inside the horizon, violating the monogamy of entanglement.

This scenario hints at a type of cloning that contradicts the strong subadditivity of entropy, a fundamental inequality related to quantum entropies. The AMPS paradox suggests a scenario where the late Hawking radiation is highly entangled with both early Hawking radiation and the black hole interior, potentially violating the strong subadditivity condition. AMPS conclude that if black hole evaporation is unitary and follows black hole complementarity postulates, a contradiction arises, leading to the unexpected result of à firewall at the event horizon. This firewall would be a highly energetic boundary that incinerates anything attempting to cross it, contradicting the smooth horizon expected from classical general relativity. For an infalling observer, encountering the firewall would be dramatic, leading to immediate destruction. For a distant observer, the infalling observer appears to be destroyed by the firewall as they asymptotically approach the horizon, consistent with the extreme gravitational time dilation effects. Thus, the firewall behaves similarly to the stretched horizon regarding its relative position to the black hole but is a destructive barrier rather than a benign region. This significantly challenges our understanding of black holes and quantum mechanics.

Rejecting the firewall, Susskind and Juan Maldacena proposed the ER = EPR conjecture, linking Einstein-Rosen (ER) bridges (wormholes) with Einstein-Podolsky-Rosen (EPR) pairs (entangled quantum particles). The ER = EPR conjecture offers insights into the firewall paradox. As a black hole emits Hawking radiation, the particles of this radiation are entangled with the remaining black hole. According to the ER = EPR conjecture, a pair of maximally entangled black holes or a black hole and its radiation, connected by an ER bridge, can maintain a smooth event horizon, contrasting with the firewall, which suggests a disruption of the horizon's smoothness.

The ER = EPR conjecture fits within the AdS/CFT framework for understanding gravity emerging from a lower-dimensional quantum theory. The AdS/CFT correspondence establishes a duality between a gravity theory in higher-dimensional

Anti-de Sitter (AdS) space and a lower-dimensional Conformal Field Theory (CFT) on its boundary. This correspondence implies that the boundary CFT contains all information about the gravitational phenomena in the AdS space. The ER = EPR conjecture proposes that entangled particles in an AdS space can be represented by ER bridges, linking the geometry of spacetime with quantum entanglement.

ER and EPR trace their origins to Einstein's pursuit of a unified field theory. Einstein's pursuit of a unified field theory was driven by his desire to integrate the laws of gravity (as described in general relativity) with electromagnetism, seeking a comprehensive framework. His approach was rooted in the geometric language of general relativity, aiming to extend it to encompass other forces of nature. However, he struggled to incorporate quantum mechanics into this framework. After Einstein's death, the focus of theoretical physics shifted. Instead of trying to fit quantum phenomena into the mold of general relativity, researchers began exploring how to formulate a theory of quantum gravity. This framework unifies the principles of quantum mechanics with general relativity. This shift represents a reversal of Einstein's approach, starting from the foundations of quantum mechanics and seeking to integrate gravitational phenomena rather than vice versa. The philosophical implications of this shift are significant. Einstein's realist worldview and the quest for a unified field theory under the umbrella of general relativity reflect a classical approach to physics. In contrast, the pursuit of quantum gravity, and often counterintuitive principles, embodies a shift in thought and experimental verification of quantum entanglement through experiments conducted after Einstein's death, further strengthening this change in perspective.

The next topic is the AdS/CFT correspondence. The AdS/CFT duality contested the notion of information loss in black hole processes by proposing that information is instead encoded in the quantum state of a CFT on the boundary of the AdS space. This shift in understanding has significantly influenced discussions on the black hole firewall paradox, underscoring the conservation of information and the feasibility of decoding Hawking radiation to recover details about the initial states of particles that entered the black hole. The focus on encoding and decoding has consequently integrated principles from quantum computation into quantum gravity. This chapter explores two famous decoding proposals addressing the firewall paradox. The first is Patrick Hayden and John Preskill's suggestion: old black holes that have radiated more than half of their initial entropy might reflect information via Hawking radiation rather than irretrievably destroy it. According to Hayden and Preskill, black holes past their Page time (when they have emitted half of their entropy) might act like information mirrors. Hayden and Preskill describe a thought experiment where Alice throws her diary into a black hole. Bob understanding the black hole's internal dynamics, can extract and decode the information from the Hawking radiation. Under Hayden and Preskill's proposal, information thrown into an old black hole can be recovered from Hawking radiation. This idea challenges the traditional view that black holes destroy information. It also touches on quantum principles like the no-cloning theorem, emphasizing that the scenario remains consistent with the principles of quantum mechanics without requiring a firewall, despite the challenges posed by the AMPS paper.

The second proposal is Daniel Harlow and Hayden's decoding task. Harlow and Heyden demonstrated that decoding the information contained in Hawking radiation is computationally so complex that it cannot be done within the black hole's evaporation time (a typical black hole's lifetime). The Harlow-Hayden protocol introduces an approach to addressing the AMPS firewall paradox by leveraging the principles of quantum computation and information theory. This innovative paradigm shift challenges traditional physics-based solutions and offers profound implications for the discourse surrounding the no-cloning theorem and the consistency of quantum mechanics. Historically, the firewall paradox posed a significant challenge to our understanding of black hole dynamics. Prior attempts to resolve this paradox primarily relied on modifications within theoretical physics, often invoking new principles to maintain consistency. In contrast, Harlow and Hayden's approach reframes the problem through the lens of computational complexity, providing a novel and rigorous framework for examining the feasibility of decoding Hawking radiation. Harlow and Hayden conceptualize the AMPS experiment as a quantum computational task, where the goal is to decode the information encoded in the Hawking radiation before the black hole evaporates completely. They employ a computational basis and define a unitary transformation U_R that scrambles the information. Alice's task is to apply the inverse transformation $U_R^{\dagger}$ to extract the encoded information. However, they demonstrate that the time complexity required for this decoding process scales exponentially with the number of qubits involved, rendering the task practically impossible within the black hole's lifespan. Introducing a computational complexity barrier shifts the focus from purely physical considerations to the intrinsic computational limits of quantum systems. By doing so, Harlow and Hayden effectively argue that while the information is not lost, as dictated by the principle of unitarity, the extraction of this information is computationally infeasible. This ensures that no observer can decode the information within a reasonable period, thus maintaining the integrity of the no-cloning theorem and the consistency of quantum mechanics without necessitating a physical firewall.

Scientists have explored an approach that leverages the principles of quantum mechanics and holography, called *quantum gravity in the lab*. Inspired by this possibility, researchers created a model for teleportation through a traversable wormhole. They introduced *teleportation-by-size*, where a qubit is scrambled on one side of the system and then unscrambled on the other. This idea is supported by the dual gravitational interpretation that a particle can pass through the holographic wormhole. I conclude this chapter by critically analyzing the wormhole experiment performed by Daniel Jafferis and his team, situating this discussion within the framework of Nancy Cartwright's *Tangle of Science*. By examining the wormhole experiment and the surrounding discourse through the lens of Cartwright's philosophy, I aim to elucidate the complex interplay between scientific models, experimental practice, and philosophical interpretation.

5.2 Black Hole Complementarity and the Firewall Paradox

5.2.1 *The Stretched Horizon and Black Hole Complementarity*

In 1986, Douglas A. Macdonald, Richard H. Price, and Kip S. Thorne introduced the notion of a *stretched horizon* and the classical *membrane paradigm*. In the membrane paradigm, the stretched horizon is an imaginary surface located just a tiny distance (an infinitesimal distance in theory) above the event horizon. The classical membrane paradigm, in general relativity, is a model that facilitates analysis rather than a literal physical entity [Pr-Th, Pr-Th-Mac] (see discussion in [Wall-mem.18, Wall-mem.19]). The stretched horizon idea facilitates the analysis of black hole dynamics and interactions, making the complex physics of black holes more tractable and comprehensible [Par-Wil][2]:

---

[2] The classical membrane paradigm conceptualizes black holes as having a two-dimensional surface or membrane. The membrane paradigm approach is a useful mathematical and conceptual tool. In this paradigm, the event horizon of a black hole is replaced by a two-dimensional membrane just outside the horizon, which has specific physical properties such as electrical conductivity, viscosity, and thermal conductivity. These properties are analogous to a physical membrane and allow the application of familiar principles from fluid dynamics and electrodynamics to studying black holes. The stretched horizon is only one Planck unit larger in area than the global event horizon, highlighting its proximity to the true horizon while still offering a practical analytical tool. Even though black holes do not have a physical surface, the stretched horizon helps simplify complex issues in black hole physics:

(1) *Coordinate singularities and metrics*: In Schwarzschild coordinates, the metric is singular at the event horizon, leading to a coordinate singularity. Alternative coordinates, such as Kruskal-Szekeres and Eddington-Finkelstein, are smooth through the event horizon and eliminate this problem. Considering a surface slightly outside the true horizon can help avoid these coordinate singularities without switching coordinates. The stretched horizon serves as this conceptual surface.

(2) *Infinite redshift and gravitational time dilation*: The event horizon is the location of infinite redshift: light emitted from just outside the horizon will be infinitely redshifted, appearing to have zero energy to an observer at infinity. This highlights the extreme gravitational effects near the event horizon. Gravitational redshift is closely related to time dilation; the closer an observer is to the event horizon, the slower their time runs relative to a distant observer. At the event horizon, this time dilation becomes infinite. The stretched horizon is a timelike surface chosen to be just outside the event horizon, with a well-defined geometry free from the singularities associated with the null event horizon. On this timelike surface, a family of time-like observers, fiducial observers, can hover just outside the true horizon. These fiducial observers experience a large but finite redshift and gravitational time dilation relative to distant observers. Unlike an observer at the true event horizon who would experience infinite gravitational time dilation, these fiducial observers have a well-defined time experience. This means they can experience time in a manageable way, allowing for practical analysis and interpretation of physical phenomena.

(3) *Tidal gravitational forces*: Tidal gravitational forces can become significant as one approaches the event horizon. These forces can remain small at the event horizon for sufficiently massive black holes and only diverge near the singularity at $r = 0$. Thus, an observer falling into a massive black hole might not experience severe tidal forces until they approach the singularity. The stretched horizon does not significantly alter these tidal forces but provides a manageable surface for studying the effects close to the event horizon.

In 1993, Leonard Susskind, Lárus Thorlacius, and John Uglum introduced the principle of black hole complementarity [STU]. By considering a 2D model of gravity, Susskind et al. outline the behavior, kinematics, and statistical fluctuations of the stretched horizon, emphasizing its importance in understanding black hole formation and evaporation while considering quantum mechanical principles. According to Susskind and colleagues, the stretched horizon paradigm addresses the dynamics of black holes at the quantum level by providing a framework to analyze the quantum gravitational effects near the black hole horizon.

Susskind et al.. initially develop four postulates to establish the foundation of black hole complementarity [STU]:

Postulate 1. Unitarity and Quantum Purity: The formation and evaporation of a black hole, as observed by a distant observer, can be entirely described within the framework of standard quantum theory. This includes the transformation of infalling matter into outgoing Hawking radiation. A unitary S-matrix describes the evolution from the infalling matter (the matter that forms the black hole) to the outgoing Hawking-like radiation (the radiation emitted as the black hole evaporates). The postulate implies that Hawking radiation's overall state remains pure throughout the various emission stages, both early and late.[3]

Postulate 2. Effective Field Theory: Outside the stretched horizon of a black hole, the physical processes can be approximated using semi-classical equations, a blend of classical and quantum theories. This approximation is valid because, in this region, the extreme effects of quantum gravity are not dominant, and classical theories like general relativity still provide a good description of gravitational phenomena. However, these equations are semi-classical because they also include some quantum corrections. The postulate implies that the modes of fields outside the stretched horizon of a black hole evolve predictably, as described by the semi-classical equations.

Postulate 3: Discrete Energy Levels and Quantum System: To a distant observer, a black hole behaves like a quantum system with discrete energy levels. This is a significant departure from the classical picture (the no-hair theorem), where a black

(4) *Jacobian matrix and coordinate transformations*: In Schwarzschild coordinates, the Jacobian matrix becomes singular at the event horizon, causing the transformation between coordinate systems to degenerate. Switching to a more suitable coordinate system, like Kruskal-Szekeres, eliminates this coordinate singularity and maintains a well-behaved metric tensor. In these coordinates, the Jacobians remain non-singular at the event horizon. The stretched horizon offers a practical approach to handling the Jacobian singularity without switching coordinate systems. Since the stretched horizon is a timelike surface, it does not suffer from the same coordinate singularities as the null event horizon. This timelike nature allows for a more stable and manageable framework for analysis. The metric remains well-behaved and non-degenerate on the stretched horizon, avoiding the singularities present at the event horizon in Schwarzschild coordinates. This allows meaningful physical quantities to be calculated without the complications of singular Jacobians. By conceptualizing a surface outside the true event horizon, physicists can work within a framework where the Jacobian matrix and the metric tensor remain well-behaved. This circumvents the problems associated with coordinate singularities and allows for applying standard techniques from classical physics.

[3] A pure state in quantum mechanics is a state that is fully described by a single wave function and is not a mixture of different states. This means Hawking radiation maintains full information about its quantum state throughout its evolution.

hole is described purely by its continuous properties: mass, charge, and angular momentum. The number of independent quantum states (or the dimension of the Hilbert space) that can describe the black hole of M is given by the exponential of the Bekenstein entropy $\exp^{S_{BH}(M)}$. This implies that the number of possible quantum states (the dimension of the subspace of states for a black hole of mass M) is extremely large and grows exponentially with the entropy. The entropy and the corresponding number of quantum states are finite and well-defined.[4]

While Susskind et al.. outline the above three postulates focusing on the perspective of an outside observer, they also bring up an additional assumption concerning the experience of an infalling observer as per the equivalence principle. This assumption is a form of a fourth postulate, although not explicitly labeled as such in the text. It contrasts with Postulate 1, as it hints at a divergence in the observed phenomena between an external observer and an infalling observer, particularly around the event horizon of a black hole:

Postulate 4. No drama: A freely falling observer feels nothing unusual when crossing the event horizon. Locally, an infalling observer experiences no dramatic changes and normal physics; the usual laws of physics hold without any singularities or discontinuities. According to *no drama*, an observer in free fall into a black hole would not perceive any dramatic signs at the event horizon. The transition should appear smooth, and they would continue to fall inward without feeling sudden forces or disruptions, especially in a massive black hole with a weak tidal gradient at the horizon. This scenario aligns with the equivalence principle, which posits that, locally, the effects of gravity are indistinguishable from acceleration in a non-gravitational field. Thus, a freely falling observer approaching and crossing a large black hole's event horizon would experience a free-floating sensation.

Susskind and colleagues suggest that if we want an external observer not to reference events inside a black hole, a theory based on Postulates 1–3 is needed. They state that Postulates 1–3 are best implemented by viewing the stretched horizon through a thermodynamic lens, which is assumed to have a microphysical foundation. Despite the coarse-grained thermodynamic description (which overlooks the minute details and focuses on broader, averaged properties), Susskind et al.. believe there is an underlying microphysical basis, i.e., a finer-grained level of description of the stretched horizon based on microscopic physical properties. This finer description would be more detailed and might explain the thermodynamic behaviors observed at the coarse-grained level regarding underlying quantum mechanical processes [STU]. In the finer-grained level of description of the stretched horizon, when information hits the stretched horizon, it gets scrambled-becomes highly entangled and mixed up with other information-but it is not lost. The scrambling is so effective that it appears to an outside observer that the information has been destroyed, although, in principle, it has been conserved.

[4] The idea that the number of states is finite and exponentially large ties into the holographic principle, which suggests that all the information contained in a volume of space can be represented as a theory on the boundary of that space with one fewer dimension.

The semi-classical stretched horizon of a black hole is conceptualized as a physical membrane with distinct mechanical, electrical, and thermal properties, as perceived by an external observer. The gravitational dynamics near a black hole render this stretched horizon extraordinarily complex and chaotic. It exhibits characteristics similar to those of a viscous fluid, and its surface area determines its thermal behavior.

According to black hole complementarity, the stretched horizon is the region where information about everything that has fallen into the black hole is stored in a highly scrambled form. From the perspective of a distant observer, the information never actually crosses the event horizon but is instead stored in the stretched horizon in a way that respects the principles of quantum mechanics. Hawking radiation is emitted due to quantum effects near the event horizon. Pairs of virtual particles are created; one falls into the black hole, and the other escapes, reducing the mass of the black hole over time. The escaping particle (Hawking radiation) is entangled with the particle that falls in. This entanglement implies that the escaping radiation carries information about the internal state of the black hole. The stretched horizon is where the information about the black hole's contents is stored in a scrambled form. When Hawking radiation is emitted, it is thought to carry away this scrambled information gradually. This means that the Hawking radiation we observe far from the black hole contains pieces of information stored in the stretched horizon. The black hole acts as an efficient scrambler of information, meaning any information that falls into it gets mixed and spread throughout the stretched horizon. Scrambling makes it extremely difficult to decipher the original information from the emitted Hawking radiation, but it is preserved in principle. As the black hole evaporates, it emits more Hawking radiation, slowly releasing the scrambled information stored in the stretched horizon. By the end of the evaporation process, the information about everything that fell into the black hole should, in theory, be contained in the emitted Hawking radiation, preserving the principles of quantum mechanics. However, this process results in intricate entanglement structures that manifest as apparent information loss.

When an object approaches the black hole, the membrane-like stretched horizon absorbs and scrambles the information from the object before it crosses the event horizon and re-emits it. Scrambling and re-emitting information before reaching the event horizon challenges the *no-drama* scenario. Moreover, the notion of a stretched horizon interacting dramatically with incoming matter contradicts the smooth crossing over the event horizon. However, the membrane paradigm maintains that the physics outside the black hole should remain unaffected by the dynamics inside it. This conceptual framework clarifies a black hole's internal and external regions, focusing on observable and measurable phenomena outside the event horizon.

In 1993, Susskind et al. demonstrated that black hole complementarity reconciles two seemingly contradictory observations: the infalling observer and the outside observer. They posited that the experience of an observer falling into a black hole (no drama) and that of an observer remaining outside (who perceives the horizon as a dynamical membrane or stretched horizon and sees information scrambled and re-emitted) are both valid. However, these experiences cannot be reconciled due to the fundamental limitations imposed by the event horizon. In other words, the nature of the event horizon creates a fundamental division between the experiences of inside and outside observers.

Niels Bohr introduced the idea of complementarity to reconcile quantum mechanics' wave-particle duality. Susskind et al.. invoke the complementarity principle to understand the nature of the stretched horizon and the quantum behaviors surrounding black holes. Susskind et al. suggest that different experimental setups reveal different aspects of quantum systems. Though seemingly contradictory, these aspects are complementary and collectively contribute to fully understanding the black hole information paradox. Like the wave-particle duality, they suggest that depending on the experiment conducted, one may detect a quantum membrane (stretched horizon) around a black hole. Both experiments provide valid but complementary information about the black hole's nature. They hint at a scenario where different experiments, akin to the different setups used to observe particle or wave behavior, will reveal either the quantum membrane aspect or some other aspect of black holes. One type of experiment can identify a quantum membrane. In contrast, a different type of experiment fails to do so. However, no observer can ascertain the results of both types of experiments simultaneously. The membrane is not an invariant that all observers can agree upon [STU, LPSTU].

5.2.2 The Firewall Paradox

In 2013, the AMPS team, comprising Ahmed Almheiri, Donald Marolf, Joseph Polchinski, and James Sully, critically examined the postulates proposed by Susskind et al. in the context of an aging black hole—a black hole in the later stages of its evaporation. This old black hole is conceptualized as having formed from a pure state and decayed over time, emitting Hawking radiation. AMPS's examination revealed that Postulates 1 and 2, posited by Susskind et al., contradicted Postulate 4. They utilize Eddington-Finkelstein coordinates to illustrate the contradiction from Susskind et al.'s postulates. They focus on an outgoing Hawking mode in the later part of the radiation.[5] This aligns with Postulate 1, which implies a well-defined state for

[5] Recall from Sect. 4.4.1 that Hawking radiation originates from the vicinity of the event horizon. Particle-antiparticle pairs are created near the event horizon, where one falls into the black hole and the other escapes as radiation. The particle that escapes is what the outgoing Hawking mode is. It moves away from the black hole, radiating into space. This Hawking mode is described as a localized packet with a width of order r_s and a superposition of frequencies. The order of magnitude of these frequencies is represented by $0(r_s^{-1})$. In other words, the mode of Hawking radiation is spatially concentrated around the black hole with a size roughly equivalent to the Schwarzschild radius r_s. The mode consists of a range of frequency components, where the typical frequencies are on the order of $f = \frac{c}{r_s}$.

the Hawking mode. According to Postulate 2, we can assign a unique, observer-independent lowering operator b to this mode. This operator describes the quantum state of a field mode, where acting with b on a state reduces the number of particles in that mode by one. The definition of this operator and the associated quantum state are consistent across different observers, at least outside the stretched horizon. We can project onto eigenspaces of the number operator $b^\dagger b$, allowing us to measure the number of photons in this mode. In other words, an observer making measurements on the early radiation can determine the number of photons present in a given mode of the late radiation. Following Postulate 2, we can relate this Hawking mode to one at earlier times, before it was emitted, as long as we stay outside the stretched horizon. As the Hawking mode approaches the black hole, its wavelength shortens significantly due to gravitational blue-shift effects near the event horizon, implying higher frequency and increased energy. This guarantees that the semi-classical approximation remains valid as long as we stay outside the stretched horizon and that the quantum state of the mode can be consistently traced back, providing a coherent description of its evolution.[6] According to Postulate 2, the evolution of the mode remains predictable outside the stretched horizon, leading to high-energy modes near the event horizon. This implies that an infalling observer would encounter these high-energy modes, which contradicts the *no drama* postulate. Postulate 4 suggests that an infalling observer should not experience anything unusual at the event horizon. This means that in the infalling observer's frame, the field should appear in the vacuum or ground state where no particles are detected. Encountering such high-energy modes would be a dramatic event, deviating from the smooth crossing anticipated by Postulate 4 [AMPS].[7]

Subsequently, AMPS reformulate the above paradox. Consider a black hole that forms from the collapse of some pure state and subsequently decays. According to the AMPS argument, the total state of the Hawking radiation can be divided into an early part and a late part after the Page time, when the black hole has emitted half of its initial Bekenstein-Hawking entropy.[8] This division leads to the following state of the Hawking radiation:

---

[6] Blue-shifting refers to a wave's increase in frequency (and therefore energy) as it moves towards a stronger gravitational field.

[7] For an infalling observer, Postulate 4 suggests finding the field in the ground state when measured using the a operator, where $a_\omega^\dagger a_\omega = 0$. This leads to a contradiction, as the field cannot simultaneously be in both a well-defined eigenstate of $b^\dagger b$ and the ground state of $a_\omega^\dagger a_\omega$, highlighting the fundamental tension between the postulates when considering the experiences of different observers.

[8] The trajectory that describes the rise and fall of the entropy of the radiation as a black hole emits Hawking radiation and eventually evaporates completely is called the *Page curve*. It shows that the entropy of Hawking radiation increases to a maximum at the *Page time* and then decreases so that information is not lost but redistributed between the black hole and its radiation:

1. *Initial stage*: At the beginning, the black hole is in a pure quantum state with zero *entanglement entropy of the radiation*. Entanglement entropy of the radiation quantifies the information or correlation shared between the subsystems, black hole + radiation, i.e., how entangled the black hole is with the radiation. As the black hole emits Hawking radiation, the entanglement entropy increases from zero. At this stage, the black hole is losing mass and emitting radiation, but the radiation itself contains very little information about the initial pure state of the black hole.

$$|\Psi\rangle = \sum_i |\psi_i^E\rangle \otimes |i^L\rangle. \tag{5.1}$$

$|i^L\rangle$ is an arbitrary complete basis for the late radiation, and $|\psi_i^E\rangle$ are the corresponding states in the early radiation.

After the Page time, the Hilbert space of the early radiation, which has accumulated more quanta over a longer period, becomes much larger than that of the late radiation. For typical states, $|\Psi\rangle$, the reduced density matrix ρ_L for the late radiation, obtained by tracing out the early radiation, is given by:

$$\rho_L = \mathrm{Tr}_E(|\Psi\rangle\langle\Psi|) \approx \frac{I}{d_L}. \tag{5.2}$$

I is the identity matrix, and d_L is the dimension of the Hilbert space of the late radiation. This indicates that the late radiation is nearly maximally mixed, appearing near-random and carrying little independent information about the black hole's initial state.

This is related to Postulate 3, which states that to a distant observer, a black hole appears to be a quantum system with discrete energy levels. Recall that according to Postulate 3, the dimension of the subspace of states describing a black hole of mass M is the exponential of the Bekenstein entropy $S(M)$. The large Hilbert space dimension reflects the vast number of possible quantum states, supporting the idea

2. *Growth stage*: As more radiation is emitted, the entropy of the radiation continues to increase. This is because each additional quantum of radiation is entangled with the black hole and carries away some of its energy. This growing entanglement between the emitted radiation and the black hole increases the entanglement entropy of the radiation. While the radiation is initially highly entangled with the black hole, it appears thermal, and information about the initial state remains hidden.
3. *Page time*: Around the midpoint of the black hole's evaporation process, the radiation's entanglement entropy reaches its maximum value. This point is known as the *Page time*. At this point, the information about the initial state of the black hole is equally distributed between the black hole and the radiation. The black hole has shrunk significantly, and much information has been transferred to the radiation.
4. *After the Page time*: The remaining black hole becomes smaller and less capable of maintaining high entanglement with new radiation emissions. After the Page time, the entropy of the radiation begins to decrease. This is because the remaining black hole continues to emit radiation. Still, due to the monogamy of entanglement, the new radiation must be correlated more with the already-emitted radiation than with the remaining black hole. Since the black hole is already highly entangled with the earlier emitted radiation, it cannot significantly entangle with the new radiation. The new radiation states are thus more entangled with the previously emitted radiation. As a result, the entanglement entropy of the entire radiation system starts to decrease. The newly emitted radiation carries away information already stored in the previously emitted radiation, decreasing the entanglement between the black hole and the radiation.
5. *Final stage*: When the black hole completely evaporates, the entanglement entropy of the radiation theoretically returns to zero, assuming the black hole evaporation process is unitary and no information is lost. All information about the initial state of the black hole is now contained in the Hawking radiation. This assumes that the black hole evaporation process is unitary, meaning no information is lost.

that the late-time radiation is in near-random states due to the high entanglement and complexity of the black hole's quantum states.

AMPS emphasize that a contradiction arises because Postulate 1's requirement for unitarity (that quantum information is conserved) seems to conflict with Postulate 4's no-drama scenario (an infalling observer experiences nothing unusual at the event horizon). Postulate 1 emphasizes the unitarity of black hole processes, implying that the state of Hawking radiation is pure. To preserve this purity, the late Hawking radiation must be entangled with the early Hawking radiation. The entire radiation emission process maintains quantum entanglement and coherence for Hawking radiation. According to quantum mechanics, a quantum system can be entangled with only one other system at a time. The paradox-often termed the *AMPS paradox* or the *Firewall Paradox*-arises when considering that, to preserve quantum purity (as per Postulate 1), the late Hawking radiation must be entangled with the early radiation. However, if we also consider the perspective of an infalling observer, the particles of the late Hawking radiation would also need to be fully entangled with the particles just inside the horizon, adhering to the *no drama* scenario of Postulate 4. This would imply that the late Hawking radiation is entangled with two systems simultaneously (both the early radiation and the modes inside the horizon), which violates the monogamy of entanglement.

This scenario hints at a type of cloning that contradicts the strong subadditivity of entropy. Strong subadditivity is a fundamental inequality related to quantum entropies. It states that for any three quantum systems A, B, and C, the total entropy satisfies the inequality:

$$S_{AB} + S_{BC} \geq S_B + S_{ABC}. \tag{5.3}$$

After the Page time, a paradox arises when considering the entanglement of the late Hawking radiation B with both the early Hawking radiation (system A) and the black hole interior (system C). The AMPS paradox suggests a scenario where the late Hawking radiation (system B) is highly entangled with the early Hawking radiation (system A) to preserve the quantum purity of the total Hawking radiation. Simultaneously, the late Hawking radiation (system B) is also highly entangled with the black hole interior (system C), in line with the no-drama hypothesis for an infalling observer. If both these high levels of entanglement are true, it would imply that the entropy of the combined system B (late radiation) and C (black hole interior) is much lower than expected, potentially violating the strong subadditivity condition. The high entanglement between B and C would typically reduce the joint entropy S_{BC}, making the inequality less likely to hold.

The main conclusion of AMPS' paper is that if one assumes that the process of black hole evaporation is unitary and follows the postulates of black hole complementarity, a contradiction arises. This leads to the unexpected result of a *firewall*, a barrier to extracting information from black holes at the event horizon. Recall that following Postulates 1 and 2, we can relate the Hawking mode to one at earlier times, before it was emitted, as long as we stay outside the stretched horizon. As the Hawking mode approaches the black hole, its wavelength shortens significantly due to gravitational blue-shift effects near the event horizon, implying higher frequency and increased energy. The blue-shifting effect means that the energy of the Hawking modes near the event horizon can become extremely high. According to the AMPS argument, this leads to a situation where these high-energy modes form a firewall. This firewall would be a highly energetic boundary that would incinerate anything that attempts to cross it, contradicting the idea of a smooth horizon expected from classical general relativity. Therefore, the firewall would very much be a drama for Alice, an infalling observer. She would encounter it as she approached the event horizon and be immediately destroyed. Bob, who stays safely outside the black hole and never falls in, will never see Alice cross the event horizon. Due to the extreme gravitational time dilation effects, as Alice gets closer and closer to the horizon, she appears to move more slowly. From Bob's viewpoint, Alice will asymptotically approach the horizon but never quite cross it. Alice would appear thermalized to Bob in the stretched-horizon hot Planckian layer just above the event horizon. She seems absorbed and radiated away as part of the black hole's Hawking radiation. From Bob's perspective, Alice appears to be destroyed by the firewall as she asymptotically approaches the horizon. The firewall, therefore, behaves much like the stretched horizon in terms of its position relative to the black hole. However, while the stretched horizon is a benign region that an observer can pass through without noticing anything unusual, the firewall is a highly energetic and destructive barrier that an observer cannot pass through unscathed.

5.3 AdS/CFT and ER = EPR

5.3.1 EPR and the Heisenberg Uncertainty Principle

Complementarity in quantum mechanics illustrates that a complete description of reality may require multiple perspectives that are individually limited and seemingly contradictory. The key aspect of black hole complementarity is that no single observer can verify both realities simultaneously, much like in the quantum complementarity principle, where we cannot simultaneously measure both the wave and particle aspects of light. The Heisenberg uncertainty principle states that certain physical properties, like position and momentum, cannot be simultaneously known to arbitrary precision. This principle illustrates quantum complementarity by showing that position and momentum are complementary variables. Thus, this principle manifests the broader model of quantum complementarity, which posits that objects can exhibit seemingly contradictory properties depending on the experimental context. For example, light can exhibit wave-like and particle-like characteristics, but these aspects cannot be observed simultaneously. The Heisenberg uncertainty principle and black hole complementarity embody the idea that our understanding of reality might require accepting multiple, seemingly contradictory perspectives, depending on the observational context.

After Einstein emigrated from Germany to the United States, he collaborated with his assistant Nathan Rosen and his colleague Boris Podolsky to publish a paper titled "Can Quantum-Mechanical Description of Physical Reality Be Considered Complete?" This paper was submitted in March 1935 and published in *Physical Review*. The paper introduced the Einstein-Podolsky-Rosen (EPR) paradox, challenging foundational principles of quantum mechanics, particularly the Heisenberg uncertainty principle.

Heisenberg, in 1927, introduced an equation illustrating that momentum P and position Q operators are non-commuting [Hei]:

$$PQ - QP = \frac{h}{2\pi i}, \tag{5.4}$$

where h is Planck's constant. This equation underpins the uncertainty principle, suggesting that a particle's exact momentum and position cannot be simultaneously known.

Einstein, Rosen, and Podolsky (EPR) argued against this through their thought experiment. They considered a scenario in which a quantum mechanical system comprising two particles, A and B, interacts for only a limited time.[9]

After the interaction, the particle A is separated from B.[10]

---

[9] Before interaction, the wave function is separated, which implies it takes a form similar to:

BEFORE THE INTERACTION:

$$\Psi(p_1, p_2) = \frac{1}{\sqrt{2\pi}} e^{\frac{2\pi i}{h} p_1 x_1} \cdot \frac{1}{\sqrt{2\pi}} e^{\frac{2\pi i}{h} p_2 x_2}, \tag{5.5}$$

where p_1 and p_2 are the momenta of particles A and B, respectively.

BEFORE THE INTERACTION:

$$\Psi(x_1, x_2) = \psi_A(x_1)\psi_B(x_2), \tag{5.6}$$

where $\psi_A(x_1)$ and $\psi_B(x_2)$ are the individual wave functions of particles A and B.

[10] The state of the system after the particles have interacted and become entangled but before any measurement is made is represented by:

AFTER THE INTERACTION:

$$\phi(x_2) = h\delta(x_1 - x_2 + x_0), \tag{5.7}$$

x_0 is a constant reflecting the fixed spatial relationship between the particles A and B. The right-hand side term indicates that the wave function has a significant value only when $x_1 - x_2 + x_0 = 0$. The delta function δ enforces the condition that the positions of particles A and B are correlated with a fixed separation x_0.

After the interaction, the wave function of the entangled particles in terms of momenta is represented as:

AFTER THE INTERACTION:

$$\psi(x_2) = \delta(p_1 + p_2). \tag{5.8}$$

From Eq. (5.8) (see footnote 10), which EPR referred to as the *conservation of momentum*, they inferred that the momentum of particle B is $-p$, achieved without disturbing particle B. EPR then declared momentum an *element of reality* for particle B, expressed in the equation for $\psi(x_2)$, the eigenfunction of P [EPR]:

$$P = \frac{h}{2\pi i} \frac{\partial \psi(x_2)}{\partial(x_2)}. \tag{5.9}$$

Similarly, knowing the separation between A and B, EPR deduced from Eq. (5.7) (see footnote 10), the position of B, making position another *element of reality*. This is given by the eigenfunction $\phi(x_2)$ of Q [EPR]:

$$Q = x_2. \tag{5.10}$$

EPR argued that if these elements of reality (momentum and position of particle B) could be determined simultaneously, it would contradict the Heisenberg uncertainty principle, which states that momentum and position, represented by non-commuting operators, cannot be precisely known simultaneously. EPR concluded that quantum mechanics is incomplete, adhering to local realism. They argued that no real change could occur in particle B due to measurements on particle A.[11]

The EPR thought experiment, while an ingenious idea, does not constitute a violation of the Heisenberg uncertainty principle. The Heisenberg uncertainty principle

---

The expression (5.8) indicates that the momenta of the two particles are perfectly anti-correlated. This means that if we know the momentum of one particle, we can determine the momentum of the other particle. The delta function on the right-hand side of the equation enforces the condition that $p_1 + p_2 = 0$. This implies that if the momentum of particle A is p_1, then the momentum of particle B must be $-p_1$ to satisfy the condition imposed by the delta function.

[11] The EPR thought experiment is founded on the principles of realism, locality, and causality:

1. *Realism*: The EPR argument is based on the notion that if the value of a physical quantity can be predicted with certainty without disturbing (measuring) the system, then there is an element of reality corresponding to this physical quantity.

2. *Locality*: An action at one point in space cannot instantaneously affect a distant object (a principle deeply rooted in classical physics). EPR objected to non-locality, where an action on one particle instantaneously affects its entangled partner, regardless of distance. However, EPR's thought experiment suggested that actions on one particle could instantaneously affect its entangled counterpart, regardless of the distance separating them. This instantaneous correlation between entangled particles was initially counterintuitive and contradicted classical ideas of locality and reality. This phenomenon was experimentally verified after Einstein's death, showcasing that quantum mechanics operates beyond the confines of classical locality and reality.

Causality: A realistic interpretation seems to suggest that a measurement of particle A affects the properties of the other particle faster than the speed of light would allow. However, using quantum entanglement to transmit information faster than the speed of light is impossible. Any measurement performed on one particle will only provide unpredictable results. We can, therefore, not use the EPR pair to transmit information faster than the speed of light. The two particles A and B are entangled, meaning a single, combined wave function describes their quantum states. This entangled state persists until a measurement is made on one of the particles. Hence, causality is not violated because a measurement performed on particle A collapses that particle's quantum state, causing it to take on a definite value.

embodies the broader principle of complementarity in quantum mechanics, stating that specific physical properties, such as position and momentum, cannot be simultaneously measured with arbitrary precision. This limitation arises not from deficiencies in measurement apparatus but from the inherent nature of these quantities within the quantum framework. The EPR paradox challenges the completeness of quantum mechanics by invoking classical assumptions about reality and measurement but does not refute the uncertainty principle. The EPR argument aims to determine position and momentum without disturbing the system. However, in quantum mechanics, the non-commutativity of operators, as represented in Heisenberg's principle, signifies an intrinsic uncertainty independent of measurement disturbance and is a fundamental characteristic of quantum states. Entanglement implies that a shared quantum state describes the particles. Measurement of one particle instantaneously influences the other, but this does not imply a direct measurement of both properties of the same particle. While the EPR setup allows for predicting one particle's properties based on the measurement of another, it does not provide simultaneous knowledge of the position and momentum of a single particle beyond the uncertainty limit. The non-commutativity of operators (position and momentum) means that measuring one property affects the outcome of the other. The critical point is that the EPR argument assumes classical realism, where properties such as position and momentum exist with definite values even when not measured. Quantum mechanics, however, dictates that these properties do not have definite values until they are measured.

After the publication of the EPR paper, Einstein significantly sharpened the formulation of the EPR argument. He argued against the notion that measuring particle A can have an immediate and direct effect on the state of particle B. Einstein formulated a *Separation Hypothesis*. The *Separation Hypothesis* implies that once particles A and B no longer interact, their subsequent states become independent. The real state of particle B should not depend on the specific measurement of particle A. The *Separation Hypothesis* challenges the idea that the measurement of particle A could have an immediate and direct influence on the physical state of particle B. Therefore, Einstein perceived the EPR thought experiment as highly problematic. If the physical state of particle B were determined by measuring particle A, then two different wave functions would be associated with the same physical state of particle B. Einstein argued that a complete description of a physical state should be unambiguous, meaning there should be a unique and definitive description of the state [Pop59].[12]

[12] According to the Copenhagen interpretation, when two particles (A and B) become entangled, their combined state is described by a single wave function (ψ-function) that encompasses both particles. The state of each particle is not independent but correlated with the other particle's state. When a measurement is performed on particle A, the wave function of the entire entangled system collapses into a state consistent with the measurement outcome. This collapse instantaneously affects the state of particle B, even if particle B is spatially separated from particle A. The type of measurement performed on A (e.g., position or momentum) determines the specific way the wave function collapses. Different types of measurements on A will collapse the wave function into different states, leading to different conditional states (ψ-functions) for B.

In his *Autobiographical Notes*, Einstein presented another iteration of the EPR thought experiment. He contended that resolving the EPR paradox would necessitate invoking either metaphorical telepathy (the instantaneous correlation between the states of the two entangled particles) or a breach of the separability principle—two approaches he found implausible. Einstein thus deduced that the EPR paradox was fraught with inconsistencies, further underscoring his belief that quantum mechanics was incomplete [Ein-49-1].

5.3.2 *The Einstein-Rosen Bridge and the Schwarzschild Singularity*

A few days before the EPR paper was published, Einstein and his assistant Rosen submitted the manuscript of the ER bridge paper, "The Particle Problem in the General Theory of Relativity," to *Physical Review*, and it was published on July 1, 1935, [ER]. In this paper, Einstein and Rosen tried to exclude singularities from the field theory.

Unlike his earlier work (1915–1916), Einstein's later work in theoretical physics included a prolonged effort to develop a unified field theory. Einstein's evolving views on his unified field theory were prominent from approximately 1919 until the end of his life in 1955. During this period, he grappled with the implications of singularities in the context of general relativity and his pursuit of a unified field theory. A key aspect of Einstein's vision for this theory was the exclusion of singularities. Einstein believed that a complete and satisfactory theory should not have such points, as they were seen as imperfections and breakdowns in the laws of physics.

Einstein distinguished between two types of singularities: singularities representing material particles and those not representing material particles. He believed that treating particles as singularities meant giving up on a real understanding of the structure of matter. However, by the late 1920s, Einstein had begun to reconsider his stance due to difficulties explaining elementary particles purely through field theory. He acknowledged the possibility of representing elementary particles as singularities within the field. This change of mind was partly due to the challenges Einstein faced in developing a field theory that could fully account for the properties and behaviors

---

If the ψ-function for B depends on the choice of measurement performed on A, this implies that B's state is influenced by a measurement performed on a spatially separated system. According to Einstein's view of locality, this is problematic. For B to have a well-defined physical state, that state should not depend on the measurement chosen for A, especially when A and B are no longer interacting. However, the Copenhagen interpretation of quantum mechanics suggests that B's ψ-function can vary based on these measurements. Einstein objected to this interpretation because it implies an ambiguity in describing B's state solely by its ψ-function. A complete description of a physical system should provide an unambiguous and objective account of its state. According to Einstein, if different measurements on A yield different ψ-functions for B, the ψ-function alone cannot fully describe B's physical reality, i.e., it cannot be the complete description of B's physical state.

of elementary particles. Thus, as a last resource, Einstein represented material particles as singularities in the field theory. However, in 1935, Einstein and his assistant Rosen found a way to describe elementary particles without relying on the presence of singularities in the field theory. They decided to introduce bridges, which were intended to represent particles. They aimed to develop a model free from the mathematical and conceptual difficulties associated with singularities using bridges instead of singularities.

Einstein's field equations (1.426) (see Sect. 1.7.2 in Chap. 1) relate the Ricci curvature tensor $R_{\mu\nu}$, the metric tensor $g_{\mu\nu}$, and the Ricci scalar R to the stress-energy tensor $T_{\mu\nu}$. In the absence of matter and energy, the stress-energy tensor $T_{\mu\nu}$ is zero, and the field equations reduce to the vacuum field equations[13]:

$$R_{\mu\nu} = 0. \tag{5.11}$$

Einstein and Rosen modified the vacuum field equations (5.11) by multiplying them by g^2, the square of the determinant of the metric tensor[14]:

$$g^2 R_{\mu\nu} = 0. \tag{5.13}$$

The Droste (Schwarzschild) solution [see Eq. (4.28) in Sect. 4.3.2 and Eq. (2.94) in Sect. 2.2.3 in Chap. 2) describes the spacetime geometry around a static, uncharged, spherically symmetric mass.[15] By multiplying the field equations (5.11) by g^2, Ein-

[13] In this simplified form, the vacuum field equations state that the Ricci tensor $R_{\mu\nu}$ is zero everywhere in a vacuum. This conveys the absence of curvature due to mass and energy in a vacuum. Equation (1.426) become $R_{\mu\nu} - \frac{1}{2} g_{\mu\nu} R = 0$. However, in a vacuum, the Ricci scalar R is also zero because it is the trace of the Ricci tensor. So, this equation reduces to the simplified form (5.11).

[14] Multiplying the vacuum field equations by g^2 guarantees that the field equations do not contain denominators that could potentially lead to singularities. This step helps redefine the equations to avoid points where g vanishes, thereby sidestepping singularities associated with these points. A zero determinant of the metric tensor indicates a point where spacetime geometry becomes degenerate or undefined, signifying infinite curvature and a breakdown of spacetime structure. Conversely, if the determinant is non-zero, the spacetime described by the metric tensor remains well-defined and non-singular:

$$g = |g_{\mu\nu}| \neq 0. \tag{5.12}$$

Einstein and Rosen's approach to avoiding singularities is focused on those that occur due to the vanishing of g. Modifying the field equations aims to create a mathematical framework where such singularities are excluded, redefining the theory to bypass these problematic points.

[15] During Einstein's lifetime, the singularity in the Schwarzschild solution to Einstein's field equations at $r = 2m$ (what we call today a coordinate singularity) was referred to as the *Schwarzschild singularity*. After Einstein died in 1958, David Finkelstein clarified that the Schwarzschild surface $r = 2m$ is not a singularity within the spacetime structure but acts as a perfect unidirectional membrane. This interpretation means that causal influences (such as light or any form of information or matter) can cross this boundary in only one direction. Finkelstein showed that changing the coordinate system could extend the solution through the Schwarzschild surface $r = 2m$ (the event horizon) smoothly and continuously. This extended solution made it clear that the Schwarzschild surface $r = 2m$ was not a barrier to spacetime but a one-way membrane [Fin].

stein and Rosen modified the equations so that the solutions remain well-defined even when $g = 0$. This approach redefined the field equations to avoid the infinite curvature associated with the singularity at $r = 0$ in the Schwarzschild solution. The modified field equations ensure that any singularity arising from $g = 0$ is mathematically excluded. This implies that the physical interpretation of the spacetime described by these modified equations does not include the problematic singularity at $r = 0$. The $g = 0$ marks the bridge.[16]

Einstein and Rosen aimed to avoid the singularity at $r = 0$ by formulating the modified field equations (5.13) so that they remain well-defined when $g = 0$. They used a coordinate transformation to reinterpret and avoid the Schwarzschild singularity at $r = 2m$, leading to the Einstein-Rosen bridge. This was done in two stages. First, Einstein and Rosen introduced a new transformation to eliminate the region containing the Schwarzschild singularity at $r = 2m$. Thus, they reinterpreted the Schwarzschild solution without the problematic singularity at $r = 2m$, presenting a more regular solution without infinite components. They introduced a new variable:

$$u^2 = r + 2m, \tag{5.14}$$

which redefines the Schwarzschild radial coordinate r in terms of a new coordinate u. The new solution represents a non-singular, spherically symmetric spacetime that connects two regions via a bridge:

THE BRIDGE SOLUTION REPRESENTING A NON- CHARGED PARTICLE:

$$ds^2 = -4(u^2 + 2m)du^2 - (u^2 + 2m)^2(d\theta^2 + \sin^2\theta d\phi^2) + \frac{u^2}{u^2 + 2m}dt^2. \tag{5.15}$$

The bridge is a finite, smooth, non-singular structure connecting two asymptotically flat regions of spacetime, meaning that the metric remains well-defined and free from

[16] In the Schwarzschild solution, the singularity at $r = 2m$ is not due to the determinant of the metric tensor g being zero but due to the individual components of the metric tensor becoming singular. The determinant of the metric tensor g for the Schwarzschild solution is given by Eq. (2.52) (refer to Sect. 2.2.2 in Chap. 2). This expression shows that the determinant does not explicitly depend on the mass m. Instead, it depends on the radial coordinate r and the angular coordinate θ. The term that leads to the $r = 2m$ singularity in the Schwarzschild metric is not the determinant g itself but the individual components of the metric tensor, specifically the g_{rr} and g_{tt} components. The singularity at $r = 2m$ occurs because the g_{tt} component, $-\left(1 - \frac{2m}{r}\right)$, approaches zero as r approaches $2m$. The g_{rr} component, $\frac{1}{1-\frac{2m}{r}}$, goes to infinity as r approaches $2m$. Thus, the singularity at $r = 2m$ in the Schwarzschild solution is due to these metric tensor components becoming singular (one going to zero and the other going to infinity).
Multiplying the field equations by g^2 [Eq. (5.13)], Einstein and Rosen aimed to create a framework where the equations remain well-defined and free from singularities arising from the vanishing determinant g. This approach does not directly address the specific singular behavior of the Schwarzschild metric components at $r = 2m$. According to Eq. (2.52), the determinant does not vanish at $r = 2m$. However, Einstein and Rosen's modification of the field equations (5.13) does address the singularity at $r = 0$ in the Schwarzschild solution. At $r = 0$, the determinant (2.52) goes to zero. The metric tensor becomes singular at this point, indicating a breakdown of the spacetime structure.

singularities at $u = 0$. When expressed in terms of u, the transformed metric does not exhibit any infinite curvature or undefined behavior at $u = 0$.[17] Einstein and Rosen tried to replace an elementary particle with mass but no charge with a topological structure, an Einstein-Rosen (ER) bridge of finite length in four-dimensional spacetime. The spatially finite bridge was identified as the neutron, an elementary particle with mass but no electric charge.

Einstein and Rosen then tried replacing an elementary charged particle with a spatially finite bridge. To obtain a bridge representing a charged particle, they considered the negative of the Maxwell stress-energy tensor in a vacuum and set the mass equal to zero. They obtained a solution free from singularities for all finite points in the space of two sheets, and a bridge between the two sheets again represented the charged particle. The ER bridge represented an elementary electric particle with electric charge but without mass (and negative energy density). Specifically, Einstein and Rosen defined the negative of the stress-energy tensor. Recall that in his 1916 review article, Einstein derived the stress-energy tensor for an electromagnetic field, $T_{\mu\nu}$ (1.386) (see Sect. 1.7.1) from the electromagnetic field tensor $F_{\mu\nu}$ [E1-16].[18] Einstein and Rosen introduced the stress-energy tensor $T_{\mu\nu}$ into the modified field equations (5.13)[19]:

$$g^2(R_{\mu\nu} - T_{\mu\nu}) = 0. \tag{5.16}$$

They considered the Reissner-Nordström metric (3.170) (see Sect. 3.3.3 in Chap. 3), the spacetime geometry around a charged, static, spherically symmetric mass.[20] By setting the mass $m = 0$ in Eq. (3.170), Einstein and Rosen got:

$$ds^2 = -\frac{1}{1 - \frac{e^2}{2r^2}}dr^2 - r^2(d\theta^2 + \sin^2\theta d\phi^2) + \left(1 - \frac{e^2}{2r^2}\right)dt^2 \text{ (with } m = 0), \tag{5.17}$$

---

[17] The solution (5.15) avoids the singularity at $r = 0$ and reinterprets the Schwarzschild solution in a new coordinate system u. The metric is regular and well-defined for all values of u. This is a bridge solution representing a non-charged particle. The bridge structure is described as follows: the new coordinate u ranges from $-\infty$ to $+\infty$. The point $u = 0$ corresponds to $r = 2m$, the location of the Schwarzschild singularity in the original Schwarzschild coordinates (2.94). At $u = 0$, the transformation (5.14) indicates that this point is the so-called "throat" of the bridge, where the two separate regions of spacetime are connected. These regions correspond to $u > 0$ and $u < 0$, each resembling a Schwarzschild-like solution without singularities.

[18] The negative stress-energy tensor (1.386) is necessary for a bridge solution representing a charged particle. This choice reflects the mathematical requirements to maintain a non-singular, continuous representation of spacetime in the presence of an electromagnetic field.

[19] Both Eqs. (5.13) and (5.16) are formulated under the assumption that the determinant of the metric tensor does not vanish, (5.12).

[20] Refer to Chap. 2, Sects. 2.3.1, 2.3.2, and 2.3.3 for a comprehensive discussion of the Reissner-Nordström metric. Recall that the Reissner-Nordström metric is a solution to Einstein's field equations in the presence of an electrically charged source. It modifies the Schwarzschild solution to account for the additional factor of electric charge, which also contributes to the curvature of spacetime. The term $\frac{e^2}{2r^2}$ adds the electric charge.

and by performing a coordinate transformation:

$$u^2 = r^2 - \frac{e^2}{r^2}, \tag{5.18}$$

Einstein and Rosen transformed Eq. (5.17) into a bridge structure at $u = 0$:

THE BRIDGE SOLUTION REPRESENTING A CHARGED PARTICLE:

$$ds^2 = -du^2 - (u^2 + \frac{e^2}{2})(d\theta^2 + \sin^2\theta d\phi^2) + \frac{2u^2}{2u^2 + e^2}dt^2. \tag{5.19}$$

This transformation leads to a solution free from singularities and represents an electric particle as a bridge between two non-singular spacetime sheets.

However, the ER bridge theory was not without its challenges, and soon enough, issues began to arise, casting a shadow over its initial promise. The electron, proton, and positron (anti-electron having the same mass and opposite charge to an electron) all have mass and electric charges. Einstein and Rosen concluded that each particle would be represented by two bridges between the congruent, identical flat sheets. They further expected that processes in which several elementary particles take part correspond to regular solutions of the field equations with several bridges connected to the two identical flat sheets [ER]. In this situation, the two flat sheets would become somewhat congested with parallel ER bridges, creating overcrowding. In collaboration with Rosen and Podolsky, Einstein persistently endeavored to tackle this many-body problem within the framework of the Einstein-Rosen bridge theory.[21] However, they encountered difficulties mathematically constructing a solution to the field equations of general relativity that accommodated multiple ER bridges. They were uncertain about regular solutions with more than one bridge [ER].

Einstein made many fruitless efforts to find a field representation of matter free from singularities based on a unified field theory program. Einstein and his assistants attempted to find a precise solution to the gravitational field equations by considering the electromagnetic field and avoiding singularities. They based their investigation on ER bridges but ultimately concluded that no such solution exists [EB].

[21] In a letter to his closest friend Michele Besso, Einstein mentioned his collaboration with Podolsky, referring to him as a "young colleague" and a "Russian Jew," and expressed their relentless struggle in addressing the many-body problem in the ER bridge theory (Einstein to Besso, February 16, 1936, in [Bes, Wei].

5.3.3 *ER Bridges and EPR Particles*

The ER = EPR conjecture suggests that an ER bridge (wormhole) between two black holes is created by EPR-like correlations (entanglement) between their microstates.[22] When these microstates are entangled, an ER bridge forms between the black holes, representing this entanglement geometrically. Not all entangled states result in ER bridges. The ER = EPR conjecture identifies a particular class of entangled states that correspond to the formation of ER bridges. These states are characterized by specific correlations among the black holes' microstates [Mald-Suss].[23]

The description of two black holes being connected by an ER bridge (wormhole) in the context of the ER = EPR conjecture can model two different scenarios [Mald-Suss]:

1. *Eternal black holes*: In this model, the two black holes are part of a solution that describes eternal black holes. In this case, the state of the eternal black hole can be interpreted as two black holes in disconnected spaces, highly entangled, with an ER bridge connecting them (this scenario is discussed in Sect. 5.3.4).

2. *Black hole and Hawking radiation*: In this model, the entanglement can be understood in the context of black hole evaporation and the firewall paradox. When a black hole emits Hawking radiation, the particles of this radiation are entangled with the remaining black hole. According to the ER = EPR conjecture, a pair of maximally entangled black holes (or a black hole and its radiation) connected by an ER bridge can maintain a smooth event horizon. The ER bridge does not introduce any high-energy phenomena at the horizon, thus preserving the equivalence principle. This is in contrast to the firewall paradox, which argues that such entanglement with external systems (like Hawking radiation) should lead to a firewall, disrupting the smoothness of the horizon. Maldacena and Susskind believe the firewall proposal is incorrect, and the challenging conjecture of ER = EPR pushes them to thoroughly explore the nexus between quantum information science and quantum gravity [Mald-Suss]. Susskind mentions that if there are firewalls at the horizons of black holes, they could straightforwardly resolve the firewall paradox without necessitating a rethinking of quantum mechanics. However, the mechanism through which these firewalls would form is yet to be identified or understood. This proposition preserves the standard interpretations of quantum mechanics but at the cost of introducing a radical new phenomenon at the event horizons of black holes. On the other hand, the ER = EPR conjecture leads to a more profound and potentially revolutionary understanding of quantum mechanics, tying together quantum entanglement and geometric connectivity [Suss16]. Yet, this idea also hinges on wormholes, theoretical constructs that have not been observed.

[22] Microstates refer to the various possible quantum states of the particles constituting the black holes.

[23] A concrete example is the creation of black hole pairs in a magnetic field. This process naturally results in a specific entangled state where an ER bridge connects the black holes. The magnetic field induces conditions that lead to the black holes being maximally entangled, thus forming an ER bridge.

Adam Brown and Leonard Susskind's paper, "A holographic wormhole traversed in a quantum computer," discusses the evolution and interconnectedness of two notions in theoretical physics: wormholes and quantum entanglement.[24] The paper starts with Einstein and Rosen by tracing the idea of a wormhole from 1935. In the same year, according to Brown and Susskind, Einstein, along with Rosen and Podolsky, wrote a separate paper that examined quantum mechanics without gravity and identified the phenomenon of quantum entanglement, which Einstein famously referred to as *spooky action at a distance* [BS]. Brown and Susskind's work takes a modern perspective that significantly differs from Einstein's original views and motivations. They delve into the ER=EPR conjecture, a modern theoretical framework connecting two concepts. This conjecture proposes that Einstein-Rosen bridges (wormholes) and Einstein-Podolsky-Rosen (EPR) pairs (entangled particles) are fundamentally linked. In Juan Maldacena's paper, "Black Holes, Wormholes and the Secretes of Quantum Spacetime," Maldacena points out a historical connection between the concepts of quantum entanglement and wormholes, which trace their origins to the two papers by Einstein and his colleagues in 1935. Although these papers addressed seemingly unrelated topics, Maldacena suggests these two distinct phenomena could be interconnected. This implies that the foundational work done by Einstein on these subjects, which initially appeared to cover separate aspects of physics, may have more profound and unexpected links, a possibility that Einstein might not have envisioned at the time [Mal-16].

Susskind and Maldacena's contemporary ideas diverge significantly from Einstein's original thinking and motivations. Einstein and Rosen's original bridge represents an elementary particle. It does not inherently imply any correlation or entanglement between the elementary particles. The ER bridges are separate structures, each representing one particle and two separate bridges representing the electron/proton. Thus, the particles the ER bridges represent behave according to their properties. This is most reasonable because Einstein adhered to the belief of local realism. From Einstein's standpoint, duality does not revolve around the relationship between entanglement and ER bridges. Instead, the purpose of the bridges is to reconcile the atomistic theory of matter with the principles of general relativity.

In 1935, Einstein did not consider the ER bridge a wormhole. Nor did he associate it with two black holes or mention "black holes" in his discussions. The modern interpretation of the ER bridge as a wormhole is attributed to John Wheeler, who introduced the wormholes after Einstein's death. In a paper presented in 1957 at a conference in Chapel Hill, North Carolina, Wheeler included a drawing of a two-mouth wormhole and wrote that in a two-dimensional space, a wormhole allows for the connection of two regions, P_1 and P_2, enabling an ant traveling from P_1 to instantly appear at P_2 without leaving the initial two-dimensional surface [Whe-57]. However, it is plausible to suggest that Wheeler drew inspiration from Einstein and Rosen's 1935 topological framework.

---

[24] The traversable wormhole experiment conducted using a quantum processor is examined in detail in Sect. 5.5.

It could be argued, with some justification, that the ER bridges theory does not explicitly tackle quantum phenomena. Moreover, in the context of the EPR argument, Einstein consistently referred to systems A and B rather than individual particles A and B (see, for instance, [Ein-49-1, Ein-36-1]). One could argue convincingly that this suggests Einstein intended to clearly distinguish between the individual particles in his ER bridge theory and the systems involved in the EPR paradox. However, I suggest that Einstein held a different perspective. In the ER bridge paper, Einstein expressed dissatisfaction with the incompleteness of general relativity. He argued that one of the limitations of general relativity is its incompleteness as a field theory, as it necessitates the introduction of an independent postulate concerning the motion of particles defined by the geodesic equation. On the other hand, a complete field theory exclusively describes phenomena using the language of fields, and material particles are solutions of the field equations with no singularities at all [ER]. Einstein aimed to elucidate the quantum characteristics of fundamental particles by revising general relativity. His objective was to reintroduce the notions of local realism, separability, causality, and determinism into this modified framework of general relativity and field theory without relying on the principles of the newly developed quantum mechanics, which entailed randomness and probabilistic outcomes. He proposed the idea of representing elementary particles using ER bridges but ultimately faced challenges in his endeavors with Podolsky.

Einstein ardently pursued a unified field theory. However, during his time, the advanced tools and concepts needed to discern a profound link between quantum entanglement, as illustrated by the EPR thought experiment, and the spacetime topology, as later explored in the ER = EPR correspondence, were still in their infancy. Notwithstanding this, Einstein's ideas on the EPR paradox and ER bridges have been instrumental in shaping the contemporary conjecture ER = EPR. Even though Einstein did not directly conceive these modern concepts, his work provided the essential groundwork from which they emerged.

5.3.4 AdS/CFT

Einstein spent a significant portion of his later life attempting to formulate a unified field theory in which he aimed to unify gravity with electromagnetism within the framework of general relativity. His approach was largely deterministic and geometrical, attempting to encompass quantum mechanics within a general relativistic framework. Modern efforts have primarily aimed at incorporating gravity into the quantum mechanical framework. This is a stark shift from Einstein's geometrical, classical approach towards a probabilistic, quantum approach. One promising pathway is the AdS/CFT correspondence, a part of the more significant holographic principles and string theory framework. AdS/CFT suggests an emergent nature of gravity and spacetime geometry from a lower-dimensional quantum theory, contrasting with Einstein's view of gravity as a fundamental force described geometrically.

This reverses Einstein's approach, focusing on deriving gravitational phenomena from quantum mechanics rather than vice versa.

Bekenstein and Hawking explored the idea that black holes can be associated with entropy, a measure of a system's disorder. Bekenstein calculated the amount of information a black hole could contain and expressed it in terms of its area, which led to the formulation of black hole entropy. A black hole's information content can be quantified and linked to its entropy. In other words, if we describe mathematically how information is stored in qubits, the information of all the material that forms the black hole gets encoded (represented) on the black hole's event horizon (the boundary).

According to Gerard 't Hooft, all phenomena within a volume can be described by degrees of freedom residing on the surface bounding that volume. This implies the world is a two-dimensional lattice of bits, with the bounding surface acting as a viewing screen onto which three-dimensional events are projected. Susskind proposed the holographic principle based on 't Hooft's ideas, positing that all the information contained within a volume of space can be encoded on a boundary surface surrounding that volume, much like a two-dimensional hologram can represent a three-dimensional object [Suss95]. The Bekenstein-Hawking formulation suggests this idea by demonstrating that a black hole's entropy (and hence information) scales with its surface area rather than its volume. The holographic principle was formulated before discovering the AdS/CFT correspondence. The AdS/CFT correspondence later provided a concrete realization of the holographic principle within the context of string theory by establishing a precise duality between a gravity theory in a higher-dimensional space [Anti-de Sitter (AdS) space] and a lower-dimensional quantum Conformal Field Theory (CFT) defined on its boundary. The boundary CFT encompasses all the information about the gravitational phenomena in the AdS space with a negative cosmological constant. The boundary of the AdS space is not within the same spacetime as the AdS but is an abstract mathematical framework. The correspondence implies that physical observables and correlations in the CFT are directly related to properties of fields and gravitational fluctuations in the AdS space. Thus, understanding the dynamics of CFT on the boundary helps comprehend the gravitational theory in the AdS space and vice versa.

Suppose we create maximally entangled pairs of particles, specifically EPR pairs of particles. We can describe this non-local correlation from the point of view of a CFT. According to the ER = EPR conjecture, if we describe the entangled particles in an AdS space, an ER bridge provides a geometric representation of the entanglement between the EPR particles. On the other hand, an ER bridge between two black holes implies that they are entangled; entanglement between two black holes implies that an ER bridge connects them. When we describe two distant black holes as entangled in the gravity picture, it does not imply that their quantum states are correlated. Instead, this statement suggests that an ER bridge (or a wormhole) connects the two black holes through their interior regions [Mal-13] (see discussion at the beginning of Sect. 5.3.3).

The entanglement of the CFT subsystems is not just a quantum mechanical phenomenon but also has a geometric counterpart in the bulk AdS space-a wormhole connecting the two regions associated with the entangled subsystems. According to Susskind and Maldacena, an eternal black hole is a stable black hole that does not lose mass or energy over time. It exists indefinitely and has two asymptotic regions, i.e., it does not emit Hawking radiation like an evaporating black hole does. A *thermofield double* TFD state describes the quantum state of two entangled black holes or two sides of an eternal black hole.[25]

The firewall paradox arises primarily in the context of evaporating black holes. Since an eternal black hole does not evaporate, the conditions that lead to the firewall paradox do not naturally arise. The entanglement between the two sides of the eternal black hole (as described by the TFD state) is stable, and there is no need to worry about the black hole losing mass or energy over time, which is the scenario that leads to the formation of a firewall. Even though the firewall paradox is more directly applicable to evaporating black holes, Susskind explores similar constructs in the context of eternal black holes. The TFD state in an eternal black hole provides a controlled environment to study how quantum entanglement, scrambling, and perturbations could lead to the appearance of firewalls, even if the black hole itself is not evaporating.

Susskind asks: what happens when the TFD state is perturbed, leading to potential instabilities like firewalls? He considers the following thought experiment. Consider a TFD state describing an eternal black hole with two asymptotic regions-one associated with Alice and the other with Bob. The interior of this black hole is connected by a wormhole (an ER bridge) between these two regions. Suppose Alice perturbs one side of the TFD state (the left asymptotic boundary of the eternal black hole). She introduces some disturbance to the system at time $t_L = 0$ on the left-hand side of the TFD. This perturbation could be as simple as throwing in a single photon or applying an operator that modifies the state of the left-hand side. This action changes

[25] Two black holes or the eternal black hole's left and right external bulk regions are linked through a wormhole. They are dual to two identical copies of the original CFT in the *thermofield double* (TFD) state, as proposed by Maldacena [Mal-18]. The TFD state represents an entangled pure state between these two copies of the quantum system (CFT):

$$\left|\mathrm{TFD}_\beta\right\rangle = \frac{1}{\sqrt{Z_\beta}} \sum_n e^{-\frac{\beta}{2}(E_{n,L}+E_{n,R})} \left|n\right\rangle_L \otimes \left|n\right\rangle_R, \tag{5.20}$$

The energies $E_{n,L}$ and $E_{n,R}$ are the energies associated with the left and right subsystems, respectively. The term $\frac{1}{\sqrt{Z_\beta}}$ normalizes the state (Z_β the partition function is the sum over all possible states, where $\beta = \frac{1}{k_B T}$). The states $|n\rangle_L$ and $|n\rangle_R$ represent the energy eigenstates of the left and the right systems, respectively. System R is introduced to mirror system L so the two systems are maximally entangled. This construction allows us to treat what would normally be a mixed thermal state as a pure state by working with the TFD state instead of the thermal density matrix. The thermofield double formalism enlarges the Hilbert space, incorporating both L and R and introducing an entangled state that represents the statistical properties of the thermal state. Specifically, the TFD state $|\mathrm{TFD}\rangle$ represents the thermal properties of system L by using this entanglement. When we focus solely on subsystem L, the state $|\mathrm{TFD}\rangle$ is indistinguishable from a thermal state described by the thermal density matrix $e^{-\frac{\beta}{2}E_L}$. This similarity arises because the reduced density matrix of subsystem L, obtained by tracing out subsystem R, resembles a mixed thermal state described by $e^{-\frac{\beta}{2}E_L}$.

the quantum state of the TFD slightly. A thermal photon would slightly disturb the equilibrium state, introducing a small shift in the system's configuration. Bob falls into the right asymptotic boundary of the eternal black hole long after Alice has perturbed the TFD state. Thus, the system has had time to evolve, and the effects of Alice's perturbation have spread and grown in complexity. Susskind emphasizes that Bob jumps into the black hole from the right-hand side very late. The significance of the late time is that in systems governed by quantum chaos, complex dynamics, and scrambling small perturbations can lead to significant changes as time progresses. Recall that quantum chaos is related to scrambling (see explanation in Sect. 5.2.1). When Alice throws a photon into the left-hand side of the black hole (her side), this perturbation affects the entire interior of the eternal black hole. Due to the complex dynamics of the black hole's interior, this perturbation could, over time, become highly concentrated. It might evolve into a significant, localized, highly energetic disturbance moving through the spacetime around the black hole, creating a shock wave or a firewall. When Bob jumps in, especially very late in the system's complexity, he might encounter this firewall.[26] [27]

According to Susskind, the system's specific quantum state (e.g., the exact TFD state or a perturbed TFD state) determines whether the horizon remains smooth or if a firewall or shock wave could form. If the system remains close to the TFD state with no significant perturbations, the horizon should be smooth, and Bob should not encounter a firewall. The timing of Bob's jump (early or late) and the timing of Alice's perturbation are significant because they determine how the system evolves. If Bob jumps in soon after Alice's perturbation, he might miss the fully developed shock wave. But if he jumps in very late, after the perturbation has had time to grow in complexity, he might encounter a shock wave that could act as a firewall. The combined effect of Alice's perturbation and the timing of Bob's entry into the black hole creates a scenario where the initially small perturbation introduced by Alice can lead to a significant disturbance at the event horizon if enough time has passed [Suss20].

The crux of the thought experiment is that the TFD state, which is highly symmetric and finely tuned, has a special property called *boost symmetry*.[28] Due to this

---

[26] According to the two entangled black holes connected by a wormhole, the ER = EPR scenario, the TFD state represents two entangled black holes, with a non-traversable wormhole connecting their interiors. The perturbation Alice introduces into her black hole can have effects that are felt on the other side of the wormhole due to the entanglement between the black holes. When Alice throws a photon into the first black hole (her black hole), this perturbation can travel through the wormhole, affecting the geometry of the spacetime connecting the two black holes. The wormhole's dynamics and the black holes' interiors determine how this perturbation evolves. If Bob jumps into the second black hole (the one on his side), especially very late, he might encounter the disturbance that originated from Alice's side. This disturbance could manifest as a shock wave or firewall in the wormhole region or near the event horizon of the black hole he enters.

[27] The ER bridge is non-traversable, i.e., no signal can pass through the wormhole in a way that would allow an observer in one black hole to send a message to an observer in the other black hole.

[28] Boost symmetry is one of the components of the larger Lorentz symmetry group. The Lorentz group includes transformations that preserve the spacetime interval between events, which means it includes both rotations in space and boosts. If boost symmetry is broken, then certain states that would be benign or non-physical (like a vacuum) in a perfectly symmetric scenario can become

symmetry, certain operations on the state, like those involving time shifts of the left and right sides of the TFD, do not change its physical properties. Susskind explains that when Bob jumps into the black hole very late, this is equivalent to him jumping in at an earlier time, thanks to this boost symmetry. Therefore, in the exact TFD state, even a very late entry by Bob should not affect the smoothness of the horizon. However, the situation changes if this boost symmetry is slightly broken (e.g., by a small perturbation from Alice). At sufficiently late times, this could result in a dangerous shock wave (or firewall) that Bob would encounter when crossing the horizon. Although the TFD state is robust against small perturbations, tiny perturbations can have significant effects when considering timescales as large as the quantum recurrence time.[29] Specifically, the perturbation can lead to the formation of a high-energy shock wave that acts like a firewall. This scenario provides an example where Bob, expecting a smooth horizon, could instead encounter a firewall due to the breakdown of boost symmetry over extremely long timescales [Suss20].

In Susskind's argument, the key idea is that when the system reaches a time close to the quantum recurrence time, the effects of a small perturbation (like a thermal photon introduced by Alice) can become significant. Susskind considers that a firewall could theoretically emerge under specific and extreme conditions involving quantum recurrences. However, he does not assert that Bob would encounter a firewall in any realistic situation.

Susskind's discussion about firewalls in eternal black holes explores how small changes and perturbations in a highly complex system might result in unexpected and dramatic consequences. The potential firewall in an eternal black hole scenario is about how the system responds to perturbations and the growth of quantum complexity over time. In contrast, the firewall in the context of an evaporating black hole is tied to the challenge of maintaining unitarity and dealing with Hawking radiation. In the case of an evaporating black hole, Susskind argues that wormholes (as part

---

highly non-trivial and potentially dangerous, as in the case of a firewall. This breaking could be due to quantum effects or specific configurations of matter or energy near the black hole. For instance, an operator (a photon) insertion at a specific time in the thermofield double state might correspond to an event that disturbs the symmetry, creating high-energy shock waves or firewalls.

[29] At very late times, the system may undergo quantum recurrence, where its state returns close to its initial condition after a sufficiently long time (because it is a closed quantum system). This idea is closely related to the Poincaré recurrence theorem in classical mechanics (see discussion in Sect. 4.2.2). Suppose Bob jumps into the black hole after the quantum recurrence has occurred. In that case, he would likely experience a situation similar to the system's initial conditions. The quantum state would have returned to a configuration close to its original TFD state. Bob might again encounter a smooth horizon, assuming the initial state was smooth and unperturbed. The shock wave or perturbation that had evolved into a dangerous structure before the recurrence would no longer exist in its harmful form after the recurrence, as the system's configuration would have effectively reset. However, the time it takes for quantum recurrence to occur is extraordinarily long and far beyond any realistic observational period, practically making it an unreachable and irrelevant event. Given the enormous timescales involved, the likelihood of observing such a recurrence is negligible within the universe's lifespan. Therefore, while quantum recurrence is theoretically guaranteed, it does not have practical implications for observers within the periods we consider. So, given the sheer magnitude of this time scale, it is practically impossible for any observer like Bob to encounter a firewall in any realistic context within the universe's lifespan.

of the ER = EPR conjecture) could prevent a firewall by maintaining smoothness across the horizon. Susskind argues that in the eternal black hole scenario, the firewall might appear due to perturbations and the breakdown of the boost symmetry after long timescales. However, in this context, Susskind's discussion of firewalls explores theoretical possibilities in highly speculative scenarios. In realistic observational periods, the emergence of a firewall remains an open question, with Susskind leaning towards the idea that while it might be theoretically possible under extreme conditions, it is unlikely to be encountered within the practical confines of our universe.

While Hawking initially proposed that information gets lost in black hole processes, the AdS/CFT correspondence suggests otherwise. All the information about the state of particles thrown into the black hole is not lost but is encoded in the quantum state of a CFT living on the boundary of the AdS space in which the black hole resides. Even though the details about the particles that fell into the black hole seem lost due to the black hole evaporation, this original form of the information is not truly lost; instead, it is encoded and transferred into the Hawking radiation. Extracting the information from the radiation would involve a process of decoding. The decoding process would hypothetically allow one to extract information about the initial state of the particles that fell into the black hole by analyzing the characteristics of the outgoing Hawking radiation.

Black holes in AdS space might not evaporate completely in the same way as in asymptotically flat spacetimes, i.e., like those often considered in general relativity, due to the reflection and reabsorption of Hawking radiation. While the principles of encoding and decoding information originated and were rigorously formulated within AdS/CFT, the broader ideas have influenced thinking about black holes in other contexts, including those not directly described by AdS/CFT. For example, discussions about the possibility of decoding Hawking radiation emitted by black holes in asymptotically flat spacetimes are inspired by the encoding/decoding notions from AdS/CFT, even though the direct application of AdS/CFT to these cases is not straightforward. The shift in understanding, primarily brought about by the AdS/CFT correspondence, moved the discourse from information loss to the conservation of information and the notions of encoding and decoding the information in the context of black holes. However, despite this shift, resolving the information paradox remains difficult. The paradox now hinges on understanding how the information is encoded in the Hawking radiation and how, in theory, it could be decoded. The practicality and exact mechanism of the decoding are among the complex issues intertwined with the black hole information or firewall paradox.

5.4 Decoding Hawking Radiation

5.4.1 Hayden and Preskill's Mirror

Consider black holes that have passed their Page time (see explanation in footnote 8 in Sect. 5.2.2). As black holes near the end of their Hawking radiation process, they would be considered old. In 2007, Patrick Hayden and John Preskill suggested that such old black holes behave more like an *information mirror* rather than a black abyss that swallows and destroys information irretrievably. They outlined a scenario in which Alice throws her diary into the black hole; instead of being lost, the information is reflected in Hawking radiation. This is akin to how a mirror reflects light.

First, Hayden and Preskill imagine a classical scenario where Alice, seeking to hide her sensitive diary, throws it into a large black hole. Bob, an expert with advanced recovery skills, aims to extract the information from Hawking radiation. In this deterministic model, the black hole's internal dynamics are predictable. When Alice's diary enters the black hole, it is transformed into a series of bits combined with the black hole's internal bit string. This combined string is then subjected to a known permutation, which is deterministic and known to Bob. With his advanced knowledge of black hole dynamics and the known permutation, Bob observes the bits emitted from the black hole via Hawking radiation. Once Bob receives a few more bits than the original size of the diary, he can decode the entire diary. The additional bits Bob needs to receive provide enough information, alongside his understanding of the permutation and the black hole's internal state, to deduce the diary's content. Initially, the black hole was considered a perfect medium to conceal and destroy the information in Alice's diary. However, Hayden and Preskill suggest that with enough understanding of the black hole's internal dynamics and by collecting slightly more data, Alice's diary can be decoded and recovered, challenging the idea of secure destruction of information[Hay-Pres].[30]

Subsequently, Hayden and Preskill consider two different quantum scenarios to explore the dynamics of information extracted from a black hole [Hay-Pres]:

1. In the first scenario, Alice encodes quantum information in a k-qubit quantum memory M (with the quantum state represented in a Hilbert space of dimension 2^k) and throws it into the black hole while she stays outside. If the black hole has evaporated past its halfway point, it has radiated away more than half of its initial

---

[30] When Alice throws her diary into the black hole, the information becomes scrambled within the black hole's degrees of freedom. This scrambling means the information is not lost but distributed complexly throughout the black hole's internal state. The Hawking radiation emitted after the Page time contains information correlated with the previously emitted radiation (see footnote 8 in Sect. 5.2.2). Bob collects the Hawking radiation emitted after the Page time. This new radiation is correlated with the earlier radiation that Bob has already collected. Bob can infer information about the black hole's internal state by analyzing the correlations between the newly emitted and earlier radiation. Specifically, he can decode the information scrambled within the black hole, including Alice's diary.

entropy. Hayden and Preskill introduce a third party, Charlie, who holds a reference system N that is maximally entangled with Alice's memory M. The black hole's complex dynamics would scramble the information in M.[31] Bob collects a subsystem from the Hawking radiation emitted by the black hole after Alice throws M into it. Suppose the radiation subsystem that Bob extracts is maximally entangled with Charlie's reference system N. This entanglement implies that the quantum information initially in M has been transferred to the Hawking radiation. In other words, the above entanglement is a channel that allows the transfer of quantum information from M out of the black hole. By utilizing this entanglement between the radiation and N, Bob can recover the quantum information initially in M.

2. In the second scenario, Alice carries the quantum memory M as she falls into the old black hole. The principle of black hole complementarity is invoked, suggesting that Alice and Bob have fundamentally different, non-overlapping experiences. From Alice's perspective, she falls into the black hole with M. From Bob's perspective, he extracts the information from the Hawking radiation outside the black hole. If the initial state of M was pure and the subsystem Bob extracts is maximally entangled with N, it again suggests that the information from M has been encoded in the Hawking radiation. The maximal entanglement with N allows Bob to recover the exact state in M from the radiation, extracting the original quantum information Alice tried to destroy. While Alice sees the information fall into the black hole, Bob obtains the information from the Hawking radiation, and according to black hole complementarity, neither can verify the other's experience.[32]

[31] Once Alice's information from M enters the black hole, it undergoes scrambling, wherein the information it carries becomes thoroughly mixed with the state of the black hole through a highly chaotic but deterministic evolution. This scrambling is hypothesized to happen extremely fast, making the information ostensibly lost, though it is conserved in a quantum mechanical sense. In the classical scenario, Alice throws a diary into the black hole, represented as a bit string. The black hole's internal state is represented as a bit string, and its dynamics are modeled as a known permutation of these bits. Bob, who understands the black hole's dynamics, observes the emitted Hawking radiation. Due to the deterministic scrambling, the information from Alice's diary is not destroyed but distributed within the emitted radiation. By collecting slightly more bits than the length of the diary, Bob can use his knowledge of the permutation to decode the diary's information. In the quantum scenario, Alice throws a quantum memory, represented by k-qubits, into the black hole. The black hole's internal dynamics are modeled as a unitary transformation that scrambles the information. Charlie holds a reference system N that is maximally entangled with Alice's memory M. After the black hole has evaporated past its halfway point, the emitted Hawking radiation becomes entangled with the previously emitted radiation. Due to the entanglement structure, Bob collects this radiation and can decode the information from M. The information is distributed non-locally within the Hawking radiation but not destroyed. Thus, if the black hole is a perfect scrambler in both classical and quantum scenarios, the information about Alice's memory M is not destroyed but is distributed within the emitted Hawking radiation. This aligns with the principles of quantum mechanics and challenges the traditional belief that information is permanently lost in black holes.

[32] According to black hole complementarity, the experiences of Alice and Bob cannot be directly compared or verified against each other. Alice sees the information falling into the black hole and believes it is lost to its interior. Meanwhile, Bob extracts the information from the Hawking radiation and can decode it using the entanglement structure. The complementarity principle warrants that no observer can witness a violation of the laws of quantum mechanics, maintaining consistency between different reference frames.

In principle, in both cases, the result is that Bob can recover the quantum information Alice attempted to destroy by extracting a subsystem from the Hawking radiation, which is maximally entangled with N. Bob, wanting to recover Alice's memory M, collects the Hawking radiation emitted by the black hole. Hayden and Preskill propose that by collecting the Hawking radiation and waiting for just a little more than k qubits to be emitted, Bob would then have enough information to begin decoding the state of Alice's qubits. Information of infalling qubits (i.e., Alice's memory M) gets scrambled with the black hole's state and is not permanently lost inside black holes. Hence, information that falls into a black hole is later emitted via Hawking radiation and can, in principle, be recovered, thereby aligning with the quantum mechanical principle of information conservation [Hay-Pres].

It seems Hayden and Preskill's scenarios violate the monogamy of entanglement. In scenario 1, Alice is outside the black hole, and Bob and Charlie's systems are outside the black hole. Scenario 2 is similar to Scenario 1, with Alice carrying M into the black hole. In the Hayden-Preskill scenario, Charlie's reference system N is initially maximally entangled with Alice's memory M. When M is thrown into the black hole, and Bob later extracts information from the Hawking radiation, it seems that the Hawking radiation is also entangled with N. This scenario suggests that N is simultaneously entangled with M (now part of the black hole) and the Hawking radiation (Bob's subsystem), which would violate the monogamy of entanglement. The key to resolving the apparent paradox lies in Charlie (N). The two scenarios do not violate the monogamy of entanglement because, at any given point, the maximal entanglement is between two parties only (either $M - N$ or Hawking radiation/Bob-N).[33]

Hayden and Preskill propose that information thrown into a black hole can be rapidly recovered from the Hawking radiation, provided the black hole has already emitted more than half of its initial entropy. If Bob can reconstruct the state outside the black hole while Alice still possesses it inside it, it seems the state has been cloned, violating the no-cloning theorem. In Hayden and Preskill's scenario, cloning refers to the possibility of having two copies of the same quantum information M: one inside the black hole (with Alice) and one outside (recovered by Bob from the Hawking

[33] Scenario 1: Charlie's reference system N is initially maximally entangled with Alice's quantum memory M. When Alice throws M into the black hole, the information from M becomes entangled with the internal degrees of freedom of the black hole. At this stage, the entanglement is between N and the black hole (including M). When Bob has collected enough radiation, his radiation subsystem becomes maximally entangled with system N. Scenario 2: Charlie's reference system N is initially maximally entangled with Alice's quantum memory M. As Alice falls into the black hole with M, the memory M becomes part of the black hole's internal state. The entanglement is now between N and the black hole (including M). Bob collects the Hawking radiation. When enough radiation is collected, this subsystem is maximally entangled with reference system N. The key element here is that the entanglement structure evolves. Initially, N is maximally entangled with M. When M enters the black hole, the entanglement shifts so that N is now entangled with the black hole's state (which includes M). At any given moment, the maximal entanglement is between two parties only. Initially, it is $M - N$. Once M enters the black hole, it is the black hole (including M) and N. After enough Hawking radiation is emitted and collected by Bob, the entanglement is between the subsystem of the radiation collected by Bob and N.

radiation). This raises concerns about the potential violation of the no-cloning theorem, which states that an unknown quantum state cannot be copied. The principle of black hole complementarity warrants that no single observer can witness the interior and exterior information simultaneously, thus preventing any direct verification of cloning. The information is either inside the black hole or outside in the Hawking radiation but never accessible simultaneously to any observer. The extreme conditions inside the black hole and the limited proper time before reaching the singularity prevent Alice and Bob from comparing their information effectively. Consequently, the no-cloning theorem remains unviolated because the potential for cloning is unverifiable, maintaining the consistency of quantum mechanics [Hay-Pres].[34]

In 2013, AMPS challenged black hole complementarity by pointing out a contradiction (see Sect. 5.2.2). Suppose one accepts the postulates used in their argument. In that case, it leads to the conclusion that there must be a firewall, a highly energetic barrier just outside the event horizon, which would burn anything falling into the black hole. This firewall would be at odds with the *no-drama* scenario, which predicts that free-falling observers should experience no special effects when crossing the event horizon of a large black hole. So, if one accepts the conclusions of the AMPS paper and still wants to maintain the consistency of quantum mechanics, then a firewall would seem to be required in the Hayden-Preskill scenario. This firewall would prevent Alice from entering the black hole intact, as she and her

[34] If Bob can reconstruct the quantum state M from the Hawking radiation while Alice still possesses the same information inside the black hole, it appears the quantum state has been cloned. Cloning of arbitrary quantum states is inconsistent with the linearity of quantum mechanics. Black hole complementarity was proposed to address the information loss or cloning quandary. According to this principle, Alice and Bob describe the same physical process in contrasting ways. Once inside the black hole, Alice's quantum information can be either retained within the black hole or emitted as Hawking radiation. No observer can simultaneously access the inside and outside perspectives to verify the cloning. Once Alice falls through the horizon, she cannot compare notes with Bob if he stays outside the black hole. To verify cloning, Bob would need to remain outside until he has extracted Alice's qubits from the Hawking radiation, then dive into the black hole to confirm that he and Alice have high-fidelity copies of the same quantum state. However, several factors make this verification impossible: the proper time Alice experiences between crossing the event horizon and reaching the singularity is extremely short, on the order of the Planck time, if Bob's entry into the black hole is delayed by a time comparable to the Schwarzschild radius (r_s). This short proper time limits Alice's ability to communicate with Bob about the status of her qubits. To send information to Bob after crossing the horizon, Alice would need to use super-Planckian frequencies, which are beyond our current understanding of strongly coupled quantum gravity. To prevent verifiable cloning, the black hole must retain Alice's information for a time interval on the order of $r_s \log r_s$. From this time scale, it follows that by the time Alice's information could be re-emitted as Hawking radiation, it would be impossible for Bob to verify the cloning by comparing notes with Alice inside the black hole. Even in the first scenario, where Alice throws her memory into the black hole but does not enter it herself, the no-cloning theorem is not violated. In this scenario, the potential concern is that if Bob can reconstruct M from the Hawking radiation, it might seem like the state has been cloned because Alice remains outside. However, after Alice throws M into the black hole, she can no longer access M. The information in M is now part of the black hole's internal state. Bob's reconstruction of M from the Hawking radiation is the only accessible version of M to any observer outside the black hole. In the first scenario, where Alice throws her memory into the black hole and stays outside, she is considered an outside observer like Bob.

memory M would be destroyed at the event horizon. However, as discussed earlier, the Hayden-Preskill scenario is consistent with the monogamy of entanglement and does not necessitate a firewall. While the AMPS paper raises significant questions about the nature of black hole horizons and challenges black hole complementarity, the Hayden-Preskill scenario remains consistent with quantum mechanical principles without requiring a firewall.

5.4.2 Standard and Strong Complementarity

Another approach attempting to address the feasibility of an infalling observer (Alice) being able to decode the Hawking radiation before the black hole completely evaporates is Daniel Harlow and Hayden's decoding task. Harlow and Hayden model an old black hole as a system of qubits and employ Page's theorem to represent its state as old.[35]

[35] Consider a bipartite quantum system with Hilbert space $\mathcal{H}_{AB} = \mathcal{H}_A \otimes \mathcal{H}_B$, where $\mathcal{H}_A$ and $\mathcal{H}_B$ are the Hilbert spaces of subsystems A and B, respectively. We assume that the dimension of A is less than or equal to that of B ($|A| \leq |B|$). The system is maximally entangled if the total state ρ_{AB} is pure. Still, the reduced density matrix ρ_A (obtained by tracing out subsystem B) is proportional to the identity operator on $\mathcal{H}_A$: $\rho_A = \frac{I_A}{|A|}$, where I_A is the identity operator on $\mathcal{H}_A$. Page's theorem asserts that a randomly chosen pure state in $\mathcal{H}_{AB}$ is likely to be very close to maximally entangled as long as $|A|/|B| \ll 1$. To choose a random pure state, we start with a particular state $|\psi_0\rangle$ and apply a random unitary transformation U: $|\psi(U)\rangle \equiv U|\psi_0\rangle$. The unitary U is chosen according to the Haar measure, a group-invariant probability measure on the unitary group. The reduced density matrix $\rho_A(U)$ is then obtained by tracing out subsystem B. The closeness of states is measured using the operator trace norm, also known as the L_1 norm. For any operator M, the trace norm $\|M\|_1$ is defined as: $\|M\|_1 = \mathrm{tr}\sqrt{M^\dagger M}$. The trace norm distance between two density matrices ρ and σ is $\|\rho - \sigma\|_1$. Another useful norm is the L_2 norm, defined as: $\|M\|_2 = \sqrt{\mathrm{tr}(M^\dagger M)}$. It is known that these norms satisfy: $\|M\|_2 \leq \|M\|_1 \leq \sqrt{N}\|M\|_2$, where N is the dimensionality of the Hilbert space. For any bipartite Hilbert space $\mathcal{H}_A \otimes \mathcal{H}_B$, Page's theorem states:

$$\int dU \|\rho_A(U) - \frac{I_A}{|A|}\|_1 \leq \sqrt{\frac{|A|^2 - 1}{|A||B| + 1}}. \tag{5.21}$$

This implies that as long as $|B|$ is significantly larger than $|A|$, the deviation of ρ_A from the maximally mixed state is very small. When $|B| \gg |A|$, the subsystem A will be almost maximally entangled with B. This means that ρ_A will be very close to $\frac{I_A}{|A|}$. Page's theorem helps justify the Page curve's form, which describes Hawking radiation's entanglement entropy over time (see footnote 8 in Sect. 5.2.2). As the black hole evaporates, the radiation becomes entangled with the remaining black hole and eventually with the earlier radiation, reflecting the behavior predicted by Page's theorem. The Page curve describes the entanglement entropy of the Hawking radiation as a function of time: The entanglement entropy S_{R} of the Hawking radiation increases as the black hole evaporates, reaching approximately half of the initial coarse-grained entropy S_{BH} (the black hole's initial entropy) at around half the evaporation time. At the Page time $t_{\mathrm{Page}} \approx 0.54 t_{\mathrm{evap}}$, the entropy $S_{\mathrm{R}} \approx 0.6 S_{\mathrm{BH}}$. After the Page time, the entropy decreases as the black hole emits radiation, eventually reaching zero when the black hole fully evaporates, confirming that the final state is pure. Thus, Page's theorem provides a rigorous foundation for understanding large quantum systems' entanglement properties and supports the Page curve's qualitative features [Har].

In this section, I will delve into the intricacies of the Harlow-Hayden protocol and elucidate how their approach has significantly broadened the scope of research in two particularly intriguing fields: black hole studies and complexity theory.

Harlow and Hayden begin by considering the quantum description of a Schwarzschild black hole in 3 + 1 dimensions from the points of view of an external observer, Charlie, and an infalling observer, Alice. The entropy of the black hole is proportional to M^2 in Planck units, where M is the mass of the black hole. The time it takes for the black hole to evaporate is proportional to M^3. Alice has a task to extract information from around $n \sim M^2$ bits of Hawking radiation within a time frame $T \sim n^{3/2}$ and jump into the black hole before it evaporates completely. Alice needs to apply a unitary transformation to the Hawking radiation to extract the desired information. This process unscrambles the desired information, making it accessible in a specific subfactor of the Hilbert space.[36] Charlie is positioned far from the black hole, at infinity. This position gives him ample time and memory resources to measure the Hawking radiation emitted by the black hole with great precision.

Harlow and Hayden start by considering Charlie's perspective. Charlie's description is based on the following three postulates [Har-Hay]:

(1) Charlie postulates that the black hole's formation and subsequent evaporation can be described as a *unitary process*.

(2) Furthermore, we can conceptualize the system as undergoing either continuous or discrete time evolution, wherein at any individual moment, it is in a pure quantum state $|\psi\rangle$, which lives in a specified Hilbert space labeled $\mathcal{H}_{\text{outside}}$. Charlie breaks down $\mathcal{H}_{\text{outside}}$ into *three subspaces*[37]:

---

[36] Alice performs the decoding task outside the black hole. She must collect the Hawking radiation emitted by the black hole and apply the necessary unitary transformation to decode the information. Decoding the information from the Hawking radiation is carried out in a region far from the event horizon of the black hole. Given the complexity and the sheer volume of data, it is highly unlikely that Alice can finish the decoding before the black hole evaporates completely. The computational difficulty of the decoding task is a significant obstacle, making it impractical for Alice to complete it within the available time frame. Since Alice is engaged in decoding outside the black hole and the decoding task is computationally infeasible to complete in the given time, she would not finish the decoding before the black hole evaporates. Consequently, at the end of the day (or, more accurately, at the end of the black hole's evaporation process), Alice cannot enter the black hole if she focuses on the decoding task.

[37] $\mathcal{H}_H$ represents the degrees of freedom inside or close to the black hole. Heuristically associated with the stretched horizon at a Schwarzschild coordinate radius given by $r = 2GM + \epsilon$, G is the gravitational constant, M is the black hole's mass, and ϵ is some ultraviolet cutoff.
$\mathcal{H}_B$ represents the field theory modes in the near-horizon region of the black hole. This region has a Schwarzschild coordinate radius in the range $2GM + \epsilon < r < 3GM$, indicating that it is close to the black hole but outside the stretched horizon described in $\mathcal{H}_H$. The geometry in this region is close to that of Rindler space. It will include modes with Schwarzschild energy less than the black hole temperature $T = \frac{1}{4\pi GM}$. Modes with higher energy are not confined to this near-horizon region and are considered part of $\mathcal{H}_R$.
$\mathcal{H}_R$ represents the Hawking radiation field, the modes of the radiation field outside the black hole, with Schwarzschild coordinate radius $r > 3GM$. $\mathcal{H}_R$ includes higher energy modes not confined to the near-horizon region described in $\mathcal{H}_B$.

$$\mathcal{H}_{\text{outside}} = \mathcal{H}_H \otimes \mathcal{H}_B \otimes \mathcal{H}_R. \tag{5.22}$$

We can relabel the different regions associated with a black hole and the surrounding environment as H, B, and R.[38] The dimensionalities of H and B are denoted by $|H|$ and $|B|$, which are related to the area of the black hole's horizon measured in Planck units.[39]

(3) The third postulate involves Page's theorem and the changes over time in the degrees of freedom or the size of $\mathcal{H}_H$ and $\mathcal{H}_B$.[40] These changes in size are proportional to the area of the black hole's horizon, and as the black hole evolves, these values change. In other words, the behavior and properties of the black hole differ depending on whether $|R|$ is larger or smaller than the product $|H||B|$, giving rise to the distinction between "young" and "old" black holes. As a black hole ages, the sizes of $|H|$ and $|B|$ decrease over time, illustrating the changes in the black hole dynamics. $\mathcal{H}_R$ starts restricted to a subset involved with the black hole dynamics. However, the size of $\mathcal{H}_R$ ($|R|$) increases over time, indicating that more states within the broader $\mathcal{H}_{\text{outside}}$ become relevant to the black hole's dynamics as time goes on. B and H are highly entangled when a black hole is young. As the black hole ages, B and H become a small part of a much larger system described in Hilbert space. After the Page time, the combined system of B and H reaches a state that can be described using a simple mathematical expression involving the identity operator, which means the system has become uniform [Har-Hay][41]:

---

[38] B is the zone just outside the event horizon of the black hole. B is where one can theoretically observe phenomena related to the near-horizon dynamics of the black hole, including the effects of Hawking radiation and other quantum gravitational effects. H refers to the black hole's horizon, and R represents the radiation field outside the black hole, which is farther from the black hole.

[39] The logarithms of these dimensionalities (i.e., $\log|H|$ and $\log|B|$) are proportional to the black hole's horizon area when we study the pure quantum state $|\psi\rangle$. The entropy of a system is related to the logarithm of the dimension of its Hilbert space. As the black hole evaporates, its horizon area decreases, corresponding to a decrease in entropy. This implies a reduction in the dimension of the Hilbert space $\mathcal{H}_H$ associated with the black hole.

[40] Consider a bipartite quantum system with Hilbert space $\mathcal{H}_{BH} = \mathcal{H}_B \otimes \mathcal{H}_H$. We assume that $|H| \leq |B|$. The system is maximally entangled if the total state ρ_{BH} is pure, i.e., the reduced density matrices ρ_B and ρ_H (obtained by tracing out the other subsystem) are maximally mixed. The density operator ρ_{BH} represents the combined system BH, with I_B and I_H being the identity operators on $\mathcal{H}_B$ and $\mathcal{H}_H$. Specifically, ρ_B (obtained by tracing out H) is proportional to I_B on $\mathcal{H}_B$: $\rho_B = \frac{I_B}{|B|}$. This state has maximum entropy, i.e., it contains no information about the specific quantum state of B alone. When the composite system BH is in a pure, maximally entangled state, ρ_B and ρ_H are maximally mixed. This means that B and H are fully entangled. Page's theorem suggests that for a randomly chosen pure state in $\mathcal{H}_{BH}$, ρ_B is likely to be very close to the maximally mixed state as long as the dimension ratio is much less than one ($\frac{|H|}{|B|} \ll 1$) (see footnote 35). Conversely, as $\frac{|H|}{|B|}$ increases, ρ_B becomes less mixed. After the Page time, ρ_B and ρ_H are less mixed than in earlier stages.

[41] Equation (5.23) implies that after the Page time, B and H become less entangled and more entangled with the remaining degrees of freedom in the $\mathcal{H}_{\text{outside}}$ [see Eq. (5.22)], which includes the radiation R. In other words, after the Page time, ρ_{BH} becomes less mixed (because the black hole is less entangled with its surroundings). This signifies that the subsystems B and H are no longer significantly entangled but are more entangled with the radiation subsystem R.

$$\rho_{BH} \approx \frac{1}{|B||H|} I_B \otimes I_H. \tag{5.23}$$

Harlow and Hayden decompose $\mathcal{H}_R$ into a direct sum of subspaces, $\mathcal{H}_{R_H}$ and $\mathcal{H}_{R_B}$, and another space termed as $\mathcal{H}_{\text{other}}$ (which does not contribute to the described entangled state):

$$\mathcal{H}_R = (\mathcal{H}_{R_H} \otimes \mathcal{H}_{R_B}) \otimes \mathcal{H}_{\text{other}}. \tag{5.24}$$

As time passes, the system reaches a state of maximal entanglement, where BH (the combination of B and H) is maximally entangled with R. $\mathcal{H}_{R_H} = |R_H|$ and $\mathcal{H}_{R_B} = |R_B|$ are their dimensions, which are set equal to $|H|$ and $|B|$ respectively, for the entanglement to hold [Har-Hay]. R_B represents a portion of the surrounding space farther away from the black hole, where the Hawking radiation emitted from the vicinity of the black hole (including B) can be observed. B and R_B are maximally entangled, describing a high degree of scrambling.

Let us keep in mind that we are still examining Charlie's perspective. Recall that the entire system (which encompasses both the black hole and the associated Hawking radiation, $\mathcal{H}_R$) is represented by a pure quantum state, $|\psi\rangle$, within $\mathcal{H}_{\text{outside}}$. $|\psi\rangle$ is in a superposition of product states from the H, B, R_H, and R_B Hilbert spaces[42]:

$$|\psi\rangle = \left(\frac{1}{\sqrt{|H|}} \sum_h |h\rangle_H |h\rangle_{R_H}\right) \otimes \left(\frac{1}{\sqrt{|B|}} \sum_b |b\rangle_B |b\rangle_{R_B}\right). \tag{5.25}$$

Harlow and Hayden consider the quantum state $|\psi\rangle$ of the black hole and its Hawking radiation. They utilize Schmidt decomposition, the quantum state of the black hole, and its radiation is represented with specific bases for two different

[42] The terms $|h\rangle_H$ and $|b\rangle_B$ represent basis states in the spaces associated with the black hole and B, respectively, while $|h\rangle_{R_H}$ and $|b\rangle_{R_B}$ represent basis states in the radiation Hilbert space. Equation (5.25) is factorized into two independent subsystems: one associated with H and one with B. The state in the Eq. (5.25) is constructed such that each basis state in H is entangled with a corresponding state in R_H and similarly for B and R_B. In other words, this structure reveals a complex entanglement pattern: subsystem H and R_H are entangled pairs. However, their entanglement is independent of the entanglement between subsystems B and R_B, another set of entangled pairs. By constructing the overall state $|\psi\rangle$ as a product of these two independent entangled pairs, the entanglement within H and R_H and also within B and R_B do not violate the monogamy of entanglement. Each pair of subsystems H with R_H and B with R_B is treated separately, preserving the principle of monogamy of entanglement. The above entanglement structure in each subsystem (between H and R_H, and B and R_B), which guarantees the state $|\psi\rangle$ remains pure, aligns with the Schmidt decomposition. The Schmidt decomposition expresses a bipartite pure state as a sum of product states of orthonormal basis vectors.

subsystems, B and H. The subsystems B, H, and R represent different parts of $|\psi\rangle$. The fundamental argument of Harlow and Hayden is grounded on Eq. (5.25).[43]

Now consider the perspective of Alice, and compare her observations with those of Charlie. Here are the key points and steps of her argument. Alice also perceives her surroundings through a description based on a Hilbert space, $\mathcal{H}_{\text{inside}}$, divided into several subspaces[44]:

$$\mathcal{H}_{\text{inside}} = \mathcal{H}_A \otimes \mathcal{H}_B \otimes \mathcal{H}_R \otimes \mathcal{H}_{H'}. \tag{5.26}$$

To avoid contradictions in Alice and Charlie's observations, an *overlap rule* warrants that both agree on the experiments' results they can communicate. Charlie and Alice, despite potentially having different perspectives, must come to the same conclusion regarding the density matrix of:

$$\mathcal{H}_B \otimes \mathcal{H}_R. \tag{5.27}$$

By observing outside the black hole, Charlie would have different observations and interpretations, particularly about regions like $\mathcal{H}_H$. Alice cannot directly access or verify the entanglement between B and R_B because these are external to her localized

[43] Harlow and Heyden write two equations: (5.23) and (5.25). The first equation sets the stage for understanding the entanglement structure after the Page time, and the Schmidt decomposition provides a concrete representation of that entanglement. Specifically, the first equation shows that BH as a combined subsystem is maximally entangled with the rest of the system R. It also indicates the relationship between the sizes of the subsystems B and H, which determines the entanglement properties. On the other hand, the Schmidt decomposition in Eq. (5.25) provides a precise mathematical form for the entangled state, showing how each subsystem is paired with its entangled counterpart: 1. How the entanglement is structured between the subsystems H and B and their corresponding purifications R_H and R_B. 2. The entanglement between BH and R is such that the subsystems H and R_H and B and R_B have equal dimensions.

[44] $\mathcal{H}_A$ represents the Hilbert space associated with A, which encompasses the field theory modes inside the black hole's event horizon from Alice's perspective. A is just inside the event horizon of the black hole with a range $GM < r < 2GM - \epsilon$. $\mathcal{H}_B$ and $\mathcal{H}_R$ are outside the black hole and accessible to Alice and Charlie; $\mathcal{H}_{H'}$ is related to Alice's horizon, distinct from the black hole's horizon. From Alice's perspective, $\mathcal{H}_H$ (the region $2GM - \epsilon < r < 2GM + \epsilon$) is absent. While $\mathcal{H}_H$ might be significant for an outside observer like Charlie, this region is crossed in an extremely short time for Alice. This rapid transit means she cannot make meaningful measurements in $\mathcal{H}_H$, rendering it operationally insignificant from her perspective. One may argue that there is a problem here with the perspectives of observers near and outside the event horizon of a black hole. Indeed, the calculation touches on a fundamental aspect of the interplay between different observers' perspectives. Alice's calculations involve theoretical constructs that consider the entanglement and information across the event horizon, even though she is physically outside. Although Alice is performing the decoding task outside the black hole, the theoretical framework of the calculation involves considering the entanglement structure and information flow across the event horizon. This is why the Hilbert space $\mathcal{H}_A$ (representing modes inside the event horizon) is considered in the theoretical setup, even though Alice cannot directly access it. In her perspective, the absence of $\mathcal{H}_H$ is due to the rapid crossing time, making it operationally insignificant. This does not imply that she is inside the black hole but that her calculations consider the full entanglement structure necessary for understanding the information flow.

perspective near the black hole. Alice, near the black hole, cannot directly access or verify R_B because it is distributed over a wide area outside the black hole and may not be practically accessible to her, being near the black hole. The above entanglement dynamics are part of Charlie's observation, who remains outside the black hole.

One way to see this more precisely, say Harlow and Hayden, is through strong subadditivity [Eq. (5.3), see Sect. 5.2.2 [Har-Hay]:

$$S_{AB} + S_{BR_B} \geq S_B + S_{ABR_B}, \tag{5.28}$$

along with the condition:

$$S_{BR_B} = 0, \tag{5.29}$$

indicating that for an old black hole, the combined system of the states B and R_B is in a pure state, i.e., that there is maximum entanglement between them. A pure state has zero entropy. Given Eqs. (5.29), (5.28) simplifies to:

$$S_{AB} \geq S_B + S_{ABR_B}. \tag{5.30}$$

Now we consider the total system ABR_B. The entropy of ABR_B is equivalent to that of A.[45] This can be written as:

$$S_{ABR_B} = S_A. \tag{5.31}$$

Given conditions (5.29) and (5.31), the inequality (5.30) translates to:

$$S_A + S_B \leq S_{AB}, \tag{5.32}$$

which leads to mutual information:

$$I_{AB} \equiv S_A + S_B - S_{AB} = 0. \tag{5.33}$$

However, the mutual information I_{AB} is zero. This indicates that no information is shared between A and B, implying no entanglement between these regions. This result contradicts the expectation that regions A and B should be entangled to maintain the no-drama hypothesis at the event horizon.[46]

[45] Since B and R_B form a pure state with zero entropy and A is not entangled with either B or R_B, the entropy of the whole system S_{ABR_B} should reflect the entropy of A. This is because the only non-zero entropy in the system comes from A.

[46] Mutual information I_{AB} quantifies the information shared between A and B. When two regions A and B are entangled, the mutual information I_{AB} is generally non-zero. Zero mutual information $I_{AB} = 0$ implies no correlation or shared information between A and B, meaning they are not entangled. For the no-drama hypothesis to be true, regions inside the horizon (A) and outside the horizon (B) must be entangled. This means the entropy of the combined system AB should reflect the entanglement. If $S_{AB} = 0$ (no drama), it implies that the total system AB is in a pure state,

Let us consider this from the opposite perspective. From Charlie's perspective, the entanglement between R_B and the states in front of the horizon B is expected. It fits with the standard quantum mechanical understanding of entanglement in Hawking radiation. To maintain this entanglement, the mutual information I_{AB} must be zero [Eq. (5.33)], and according to AMPS' suggestion, there must be a firewall of high-energy particles at the event horizon of an old black hole. This firewall breaks the entanglement between A and B. Charlie would see B as entangled with R_B from outside the black hole. As Alice approaches and attempts to cross the event horizon, from Charlie's perspective, instead of a smooth vacuum, there is a firewall—a region of high-energy particles—at the horizon of an old black hole, which annihilates her. The firewall forms at the Page time (when a black hole has emitted about half its particles and half the information) [AMPS, Har-Hay].

We can navigate the problem of the firewall in two directions [Har-Hay]:

(1) *Strong complementarity*: Alice's and Charlie's observations are treated separately. They must only agree on what is experimentally observable to both. Alice cannot access quantum states in the Hilbert space $\mathcal{H}_{R_B}$ associated with R_B. In Alice's theory, there are two potential solutions to address this problem:

1. Disentangle $\mathcal{H}_{R_B}$ from $\mathcal{H}_B$: This would involve creating a theory where the states in R_B are no longer entangled with those in B. This freeing up of B would entangle it with A, supporting Alice's expectation of a smooth horizon (no drama).

2. Remove $\mathcal{H}_{R_B}$ from her theory entirely. Alice's theory would ignore the states in R_B, treating them as irrelevant. This would again free up B to become entangled with A. This would grant Alice a smooth journey across the horizon. Thus, in Alice's frame, neglect of $\mathcal{H}_{R_B}$ is postulated to allow a self-consistent description of a smooth horizon, harmonizing the dual descriptions by Charlie and Alice.

(2) *Standard complementarity*: While described using different theoretical frameworks, Alice's and Charlie's observations can be compatible because Alice's theory can be seen as a subset embedded in Charlie's more encompassing theory. A theoretical framework is developed where Alice's interior operators to describe phenomena in $\mathcal{H}_A$ are formulated using the exterior operators that Charlie would use to describe $\mathcal{H}_{R_B}$. This attempt is made to avert the problem of firewalls and maintain the notion of a smooth space at the horizon [Bou-12]. Post the Page time, the modes (or quantum states) just behind the black hole horizon (A) can be identified with the radiation that has been emitted by the black hole (R_B). This is expressed as $A = R_B$ and is often called a *donkey map*. By developing a theory where Alice's interior description (on $\mathcal{H}_A$) is formulated using exterior operators that Charlie would use to describe $\mathcal{H}_{R_B}$,

---

suggesting perfect entanglement between A and B. If the mutual information I_{AB} is zero, it implies no entanglement between A and B:

$$I_{AB} \equiv S_A + S_B - S_{AB} = 0 \quad \rightarrow \quad S_A + S_B = S_{AB}. \tag{5.34}$$

This suggests that the entropy of the combined system AB is simply the sum of the individual entropies, indicating no correlation or entanglement between A and B. However, for the no-drama hypothesis to hold, there should be significant entanglement between A and B, leading to non-zero mutual information.

standard complementarity aims to avoid the problem of firewalls and uphold the idea of a smooth transition at the event horizon [Bou-13, STU, LPSTU, Suss12].[47]

However, identifying Alice's A with Charlie's R_B results in a paradox where Alice, while theoretically unable to access R_B, can do so according to this approach, thereby contravening the theory that posits complementarity.[48]

Harlow and Hayden solve this problem by introducing a novel perspective that incorporates the idea of computational complexity into the discussion about black holes. The donkey map attempts to map the information inside the black hole to the

---

[47] The donkey map is a mapping that equates the quantum states behind the horizon with those in the emitted radiation. It establishes a many-to-one correspondence between the state of the black hole's zone B and a vacuum state $|O\rangle_{B\tilde{B}}$ [Bou-13]:

$$\rho_B = \mathrm{tr}_{F-B}\left(|\psi\rangle_F\,{}_F\langle\psi|\right) = \mathrm{tr}_{\tilde{B}}\left(|O\rangle_{B\tilde{B}}\,{}_{B\tilde{B}}\langle O|\right). \tag{5.35}$$

This means that for the state $|\psi\rangle_F$ in the full Hilbert space F (which represents the state of the entire system outside the black hole), the donkey map $D(|\psi\rangle_F\,; t)$ can be realized between $\hat{B}(t)$ (the purified state associated with the Hawking radiation) and $\tilde{B}(t)$ (the corresponding state inside the black hole) [Bou-13]:

$$D(|\psi\rangle_F\,; t) : \hat{B}(t) \to \tilde{B}(t). \tag{5.36}$$

The donkey map primarily addresses the perspective of the infalling observer. It resolves the firewall paradox by redefining the relationship between the internal and external degrees of freedom in a black hole. Outside the black hole, various states of the Hawking radiation R_B and the near-horizon zone B can exist. Inside the black hole, for the infalling observer (Alice), there is an expectation of a smooth vacuum state consistent with the no-drama hypothesis. This state is represented as $|O\rangle_{B\tilde{B}}$, where B is the zone near the horizon and $\tilde{B}$ is the corresponding interior region. The donkey map defines a process where multiple different quantum states of the Hawking radiation R_B outside the black hole are mapped to the same infalling vacuum state $|O\rangle_{B\tilde{B}}$ inside the black hole. This maintains a smooth horizon. Consider different possible states of the Hawking radiation $|\psi\rangle_F$ outside the black hole. According to the donkey map $D(|\psi\rangle_F\,; t)$, regardless of the specific state of $|\psi\rangle_F$, the resulting state inside the black hole for an infalling observer is always the same $|O\rangle_{B\tilde{B}}$. Alice will always encounter the same vacuum state, regardless of the different possible configurations of the radiation outside. Since, according to the mapping, there is a consistent vacuum inside the black hole, it avoids creating a high-energy firewall that would violate the equivalence principle and the no-drama hypothesis. From the perspective of Alice, according to the donkey map, ρ_B, the state of B matches the thermal state. A thermal state is one where the near-horizon modes B are in thermal equilibrium at the Hawking temperature for black holes. This thermal equilibrium state implies that the degrees of freedom in B are mixed and not pure. So, this thermal state indicates that B is in a mixed state, consistent with the expected behavior of Hawking radiation. By matching ρ_B to the thermal state, B is not in a pure state that would be maximally entangled with R_B. This prevents the maximal entanglement needed to form a firewall.

[48] For the no-drama hypothesis to hold, A should be entangled with B, leading to a smooth experience at the event horizon. However, once the black hole is old, the principle of unitarity and scrambling require B and R_B to be maximally entangled. Charlie should see the entanglement between B and R_B, while Alice should experience a smooth transition through the event horizon without encountering a firewall, meaning A should be entangled with B. If A is identified with R_B, then Alice's observation of A would mean she is indirectly observing R_B, which is supposed to be outside her sight. This contradicts the principle of complementarity, which asserts that no observer should be able to access both the inside and the outside information simultaneously.

information outside (in Hawking radiation) while maintaining consistency with quantum mechanics. However, the donkey map becomes problematic if Alice attempts to decode R_B just outside the black hole. If Alice could decode R_B outside the black hole, she would reconstruct the quantum information that escaped the black hole as Hawking radiation. This means she would have a copy of the information inside the black hole. After decoding R_B, if Alice enters the black hole, she would have access to the original information still present inside the black hole. If Alice decodes R_B and then enters the black hole, she would possess two copies of the same quantum information: one from the Hawking radiation (R_B) and one from the interior of the black hole (A). This situation violates the no-cloning theorem because it results in two identical copies of the same quantum state existing simultaneously in different locations (outside as part of R_B and inside the black hole). Suppose a qubit of information is entangled between the black hole and the Hawking radiation. The radiation R_B carries away information entangled with the remaining black hole B. If Alice, while outside, applies a complex unitary transformation U_R to R_B, she decodes the information. This process extracts the original state that was inside the black hole. Upon entering the black hole, Alice would find the same quantum information entangled within its degrees of freedom. The presence of the same information inside the black hole and in Alice's possession outside (via R_B) creates a paradox. It implies the quantum state has been cloned, which the no-cloning theorem prohibits.

Harlow and Hayden address this potential paradox by introducing computational complexity. They argue that decoding the information from R_B is so computationally complex that it cannot be accomplished in any practical period, especially within the universe's lifetime. Therefore, while theoretically possible, it is practically impossible for Alice to decode R_B and then enter the black hole with the decoded information. Because Alice cannot realistically decode R_B, she cannot extract information about the black hole's interior before crossing the horizon. Thus, Harlow and Hayden introduce computational complexity to ensure that Alice cannot practically clone the quantum information. This prevents the paradox of observing the same quantum information in two places, thus avoiding the no-cloning problem.

Thus, Harlow and Hayden present a form of strong complementarity: from Charlie's perspective, subsystem B is entangled with R_B. This entanglement is required to preserve the unitarity of the overall quantum system. Charlie cannot access the information inside the black hole, represented by A. For Alice, the entanglement necessary for a smooth horizon and to avoid encountering a firewall implies that A is entangled with B (near the horizon). Alice cannot access R_B because decoding the information from the Hawking radiation is computationally infeasible within a practical period due to the high complexity of the required operations.

5.4.3 The Harlow-Hayden Decoding Task

Harlow and Heyden frame the AMPS (Almheiri-Marolf-Polchinski-Sully) experiment as a quantum computational problem [Har-Hay].[49] Initially, they phrase the discussion in terms of Charlie's Hilbert space. Since both Charlie and Alice are required to agree on the density matrix that describes the system in terms of $\mathcal{H}_B$ and $\mathcal{H}_R$, it implies that a consistent representation of $|\psi\rangle$ can be achieved from the perspective of Charlie's Hilbert space. Hence, despite potentially different vantage points, Alice and Charlie will converge on a shared description, abiding by the constraints of the overlap rule. In other words, Harlow and Hayden follow the tenets of standard complementarity, maintaining a coherent and consistent description of the black hole system from Alice and Charlie's respective viewpoints. However, they challenge this approach by considering the implications of computational complexity.

First, Harlow and Hayden specify that they will use a particular basis (*computational basis*) to describe the radiation field $\mathcal{H}_R$, which will be convenient for Alice to work with. They write an equation that denotes the specific computational basis state ($|bhr\rangle_R$) in $\mathcal{H}_R$ using n, k, and m that describe various aspects of that state.[50]

Second, Harlow and Hayden define U_R as a particular (scrambling) unitary transformation on $\mathcal{H}_R$ and write $|\psi\rangle$ [Eq. (5.25)] in the computational basis:

$$|\psi\rangle = \frac{1}{\sqrt{|B||H|}} \sum_{b,h} |b\rangle_B \, |h\rangle_H \, U_R \, |bh0\rangle_R \, . \tag{5.37}$$

U_R's exact form or structure depends on the details of the black hole's initial state and the quantum gravity, a yet-to-be-realized theory. The discussion here pertains

[49] The AMPS experiment analyzed using the principles and tools of quantum computation means extracting information from approximately n bits of Hawking radiation before the black hole evaporates. The goal is to extract the information encoded in the Hawking radiation to test if it matches the expected quantum information inside the black hole. If this decoding is possible, it would suggest a violation of the no-cloning theorem. By framing the AMPS experiment as a quantum computation, Harlow and Hayden leverage the mathematical and theoretical framework of quantum information theory to provide insights into the firewall paradox and the feasibility of the AMPS experiment. The decoding process is modeled as a quantum computation problem, and Harlow and Hayden explore the computational complexity of performing such a decoding task. Using principles from quantum error correction, they analyze how information is protected and how errors (erasures or loss of information) could be corrected. The entanglement properties between the black hole's interior and the Hawking radiation are central to this analysis, similar to how entanglement and error correction are vital in quantum computation.

[50] Working exclusively with this basis is intended to simplify calculations for Alice as she tries to work through the decoding problem with a basis where the mathematical expressions become more tractable. Alice uses n, k, and m to represent different quantities related to her problem of decoding information from the black hole's Hawking radiation: n represents the total number of qubits involved in the problem. This is related to the logarithm to base 2 of the dimension of R: $n \equiv \log_2 |R|$; k represents the number of qubits associated with $\mathcal{H}_B$; and m represents the number of qubits associated with $\mathcal{H}_H$. We can think of $k + m$ as the number of qubits remaining in the black hole.

to Schwarzschild black holes instead of AdS eternal black holes that almost do not evaporate and maintain a dynamic equilibrium with Hawking radiation. This dynamic equilibrium significantly changes the conditions and assumptions on which the Harlow and Hayden decoding task and the U_R are defined [Har-Hay].

Alice's challenge involves applying the inverse of the unitary transformation U_R, denoted $U_R^\dagger$, to the Hawking radiation.[51] By decoding the Hawking radiation, Alice aims to extract the information encoded in the entangled states between $\mathcal{H}_B$ and $\mathcal{H}_{R_B}$. This allows her to confirm the entanglement between these two components of the system without the need to access $\mathcal{H}_{R_B}$ directly. Alice can gain insights into the states and dynamics occurring in R_B through her operations on the Hawking radiation. Alice's decoding method allows her to gather information about R_B indirectly, maintaining the smooth horizon and upholding the principles of standard complementarity. The *donkey map* is used in the decoding task. The map suggests that the information content in the black hole's interior is mirrored or encoded in the radiation. The success of the decoding task depends on the validity of the donkey map assumption. Following the logic of the donkey map, if Alice has direct access to and can analyze or decode R_B, then in principle, she gains knowledge about the states in A without having to enter the black hole. This is because A and R_B are considered equivalent regarding their information content. Alice would not need to enter the black hole to observe its interior states in this framework. The donkey map suggests she can obtain the necessary information from the Hawking radiation alone. If Alice finds out she cannot perform the decoding task, this challenges the donkey map and its implications. It suggests that the simple equivalence posited by the donkey map might not be sufficient to fully explain the relationship between the black hole interior and the Hawking radiation.

However, the time Alice would need to decode the information is exponential and proportional to 2^{k+m+n}. This time complexity is extremely large, indicating a computational task of astronomical proportions. Since the black hole is old, it implies that n (representing the age parameter) is significantly larger than the sum of k and m. This condition allows us to derive a lower bound for the reduced time complexity expression: $n > k + m$. Using this condition in 2^{k+m+n}, we can establish a lower bound on the reduced time complexity as at least $2^{2(k+m)}$. But this expression also involves an exponential function and, therefore, would be extremely large for non-trivial values of k and m.

Thus, reversing U_R exactly would be practically impossible for Alice. Hence, a framework is required to quantify how close Alice can reach the perfect task. Harlow and Hayden use a *trace norm*, which allows them to define a notion of closeness, how close Alice needs to get to test the entanglement accurately [Har-Hay]. In other words, they aim to find a unitary operation close enough to the ideal operation $U_R^\dagger$ such that the error in the subsequent measurements is within acceptable bounds. Defining a

[51] This process reverses the unitary transformation carried out by U_R, thereby *decoding* the information encoded in the Hawking radiation through U_R. Applying the inverse unitary transformation $U_R^\dagger$ reverses the encoding process that scrambled the information, extracting it from the Hawking radiation.

threshold for the closeness using trace norm would define how accurately Alice needs to perform the unitary transformation to test the entanglement reliably. It establishes a criterion for the allowable error in Alice's operation, such that she can still validate the entanglement through her measurements. By working within this framework, Alice aims to construct a unitary transformation that, while not the same as the ideal $U_R^\dagger$, is close enough according to this defined metric to allow her to verify the entanglement adequately. This approach acknowledges the practical difficulties and uncertainties in constructing the exact unitary transformation. It provides a pathway to achieve Alice's goal within tolerable error margins.[52]

Alice uses a quantum computer to perform the transformation $U_R^\dagger$. Applying $U_R^\dagger$ to the radiation R, Alice aims to distill (isolate and extract) the specific information about R_B, the part of the Hawking radiation entangled with B.[53] This process attempts to reverse the scrambling, disentangling the information and isolating the purification of subsystem B from the radiation R. Successfully performing this operation would allow Alice to extract and interpret the quantum information about R_B.

However, finding the initial state $|\psi\rangle_C$ for the computer to start the computation is extremely challenging.[54] Even if Alice finds an initial state for her quantum computer, the computation involved is highly non-trivial due to the vast amount of

---

[52] Allowing for a non-ideal transformation, where Alice tries to get close enough through trace norms, might reduce the complexity of the task. However, even when settling for a non-ideal transformation, Alice still faces an enormously complex quantum computational task. Though the requirement is somewhat relaxed compared to achieving the ideal transformation, the task remains a high-order polynomial time problem, implying a computational duration that vastly exceeds the black hole's lifespan.

[53] Distillation in the context of quantum entanglement is akin to purification. Both processes aim to obtain a set of purer or more highly entangled states from a set of less entangled or mixed states. The analogy with distillation or purification in chemistry is often made to help intuitively understand the process. Just like in a chemical process where you might distill a liquid to remove impurities and obtain a purer substance, in quantum information theory, you would distill (or purify) a mixed quantum state to obtain a state of higher entanglement, which is often more useful for various quantum information processing tasks. The terms *distillation* and *purification* convey the idea of extracting and isolating the most useful part of something from a less pure or mixed state.

[54] Alice uses a quantum computer whose initial state is defined in a new Hilbert space $\mathcal{H}_C$. This space is distinct from the radiation Hilbert space $\mathcal{H}_R$. The quantum computer interacts with $\mathcal{H}_R$, undergoing a unitary evolution described by U_{comp}. This evolution occurs in a larger Hilbert space, the tensor product of $\mathcal{H}_R$ and $\mathcal{H}_C$, denoted as $\mathcal{H}_R \otimes \mathcal{H}_C$. Alice aims to use U_{comp} akin to $U_R^\dagger$ to reverse the effects implemented by U_R and distill information about R_B. This is intended to decode the information entangled with B during the evolution governed by U_R. However, Alice's major challenge is finding the appropriate initial state $|\psi\rangle_C$ for her computer. The goal is to choose this state such that the unitary operation U_{comp} decodes the information. The following equation represents the decoding process:

$$U_{\text{comp}} : U_R\,|bh0\rangle_R \otimes |\psi\rangle_C \to |\text{something}\rangle \otimes |b\rangle_{\text{mem}}\,. \tag{5.38}$$

Here, Alice is seeking an initial state $|\psi\rangle_C$, such that after U_{comp}'s operation, the qubits representing $|b\rangle$ (associated with the B basis) are isolated in the computer's memory. They are stored in the first k qubits of her computer's memory, separated from the other parts of the system described by $|\text{something}\rangle$ [Har-Hay].

entanglement and scrambling of information. Theoretically, a Schwarzschild black hole, being non-rotating and uncharged, has a long lifespan determined by Hawking radiation emission. However, this lifespan is still finite. Astrophysical black holes, typically more massive, have even longer lifespans, but they are still not infinite. The computation required to decode the Hawking radiation is expected to take an extraordinarily long time, likely much longer than the remaining lifespan of any black hole. Given the complexity and the time constraints, the probability that Alice's quantum computer will complete the necessary computations within the black hole's lifespan is minuscule.[55] The process requires a time that scales exponentially with the number of qubits, which indicates extremely long computation times for complex operations involving many qubits. This makes Alice's task of implementing $U_R^\dagger$ within a reasonable time frame highly unlikely [Har-Hay].[56]

---

[55] The probability is exponentially small in terms of the dimensions of the Hilbert space. To estimate the likelihood of finding a successful initial state, Harlow and Hayden use the trace norm to create a set close to which every pure state in the Hilbert space can be found. The trace norm helps quantify how close typical states are to being maximally mixed. This insight shows that decoding the Hawking radiation to extract the original information is theoretically difficult and practically infeasible. The computational resources required to perform such decoding within the universe's lifespan are enormous. Even after considering various potential unitary evolutions, the probability of finding a successful initial state remains extraordinarily small. The time required to find a suitable U_{comp} by random chance is identified as the quantum recurrence time, the time scale over which a quantum system revisits a particular state it was in at some earlier time due to its natural dynamics. The recurrence time is immensely long, reaching up to $10^{10^{40}}$ years. The recurrence time in which a quantum system revisits near an initial state by pure chance is especially long for the complex system Alice is dealing with, involving a black hole and Hawking radiation. This time scale is doubly exponential in terms of the entropy of the whole system, i.e., it increases extremely fast as the system's complexity increases. Thus, the recurrence time is potentially longer than the lifespan of a Schwarzschild or an astrophysical black hole. Despite the immense challenge posed by the quantum recurrence time, Alice has another strategy up her sleeve. Instead of the brute force approach, waiting for the right U_{comp} to occur by chance, she aims to actively find it by exploiting patterns in how U_{comp} evolves. If there are any predictable structures in its evolution, she can use this information to guide her search, significantly reducing the time it would take to find the right U_{comp}. By leveraging the structures in U_{comp}'s evolution, Alice can potentially reduce the computational time from a double exponential dependency on the entropy of the radiation to a single exponential dependency. This is a very long time since single exponential growth is very fast. But it is much more manageable compared to a double exponential growth. So, this strategy offers a glimmer of hope. Despite the initially grim prognosis suggested by the quantum recurrence time, if Alice can understand and leverage her system's underlying physics and mathematical properties well enough, she might be able to achieve her goal in a reasonable amount of time, though still astronomically long from a human perspective.

[56] How many quantum gates would be necessary to implement a unitary transformation as complicated as $U_R^\dagger$? Harlow and Hayden show that the number of gates required scales with a single exponential function of the number of qubits (n), 2^{2n} times a logarithm of a parameter (ϵ): $2^{2n}\log(\frac{1}{\epsilon})$. This is a significant result because it means the computational time is a lot less than previously assumed based on a crude double exponential scaling (for instance, 2^{2^n}). This improvement is partly credited to the Solovay-Kitaev theorem, which states that any unitary operation can be approximated to an arbitrary degree of precision using a relatively short sequence of gates from a finite set of universal gates (CNOT, Hadamard, and the phase gate). Despite this improvement, reducing the computational time further seems unlikely, indicating that $U_R^\dagger$ is still a highly non-

However, we can speed up computations significantly compared to initial expectations. Harlow and Hayden investigate whether the dynamics of a black hole might constrain U_R in a manner that would allow Alice to implement it faster. They explore this idea but ultimately conclude that such a simplification does not occur. A transformation U_{dyn} is defined as a unitary transformation that describes the evolution of the black hole's state, including the interactions and entanglement between the black hole's internal degrees of freedom and the emitted Hawking radiation. This transformation represents how the black hole evolves and produces the state of the radiation and the remaining black hole matter. It is reasonable to assume that U_{dyn} can be generated by a polynomial number of gates, and this assumption is based on insights from AdS/CFT. The CFT can be simulated by a quantum circuit with a polynomial number of gates. The computational operation does not require an astronomical number of quantum gates. This makes the problem of simulating black hole evolution tractable in principle, though still potentially challenging in practice [Har-Hay].[57]

Harlow and Hayden define the state $|\psi\rangle$ as arising from the action of the polynomial size circuit U_{dyn}:

$$U_{\text{dyn}}|0\rangle_{BHR} = \frac{1}{\sqrt{|B||H|}} \sum_{bh} |b\rangle_B \, |h\rangle_H \, U_R \, |bh0\rangle_R \, . \tag{5.39}$$

Alice wants to determine if a small circuit for U_{dyn} implies a small circuit for U_R. If such a circuit exists, she could more easily decode R_B from the Hawking radiation. U_R is derived from U_{dyn}, which is sensitive to the system's initial state. For the sake of simplicity, an initial state is chosen with all bits set to zero. Harlow and Hayden write an expression for U_{dyn} acting on this initial state, intending to learn more about U_R:

$$U_{\text{dyn}} \, |00000\rangle_{\text{init}} \approx \frac{1}{\sqrt{|B||H|}} \sum_{b,h} |b\rangle_B \, |h\rangle_H \, U_R \, |bh0\rangle_R \, . \tag{5.40}$$

U_{dyn} can be expressed as a product of U_R and another operation called U_{mix}. U_{mix} is a unitary operation that creates entanglement between different subsystems, mixing or scrambling the information within them to create a highly entangled state. Here, U_{mix} is conceptualized as a straightforward circuit that entangles the initial four subfactors in the chosen initial state.

---

trivial task even under optimal conditions. Adjusting the model, such as using different gates or more complex quantum entities like qutrits (which have three basis states) instead of qubits, does not fundamentally change the 2^{2n} scaling of the problem [Har-Hay].

[57] AdS/CFT specifically describes black holes in AdS space. However, the principles and mathematical structures it reveals about quantum gravity can offer insights into the dynamics of black holes in a dS space. The assumption that U_{dyn} can be generated by a polynomial number of gates comes from the general expectation that physical processes governed by quantum mechanics should, in many cases, be simulatable by a quantum circuit with a reasonable (i.e., polynomial) number of operations.

Harlow and Hayden emphasize the simplicity of implementing U_{mix}, noting that a universal circuit (built from a universal gate sets that include the Hadamard gate, the phase gate, and the CNOT gate) would suffice:

$$U_{\text{mix}}\,|00000\rangle_{\text{init}} = \frac{1}{\sqrt{|B||H|}} \sum_{b,h} |b\rangle_B\,|h\rangle_H\,|bh0\rangle_R\,. \tag{5.41}$$

They then define a new operator $\tilde{U}_R$ as: $\tilde{U}_R = U_{\text{dyn}} U^{\dagger}_{\text{mix}}$, which has the property:

$$\tilde{U}_R \frac{1}{\sqrt{|B||H|}} \sum_{b,h} |b\rangle_B\,|h\rangle_H\,|bh0\rangle_R = \frac{1}{\sqrt{|B||H|}} \sum_{b,h} |b\rangle_B\,|h\rangle_H\,U_R\,|bh0\rangle_R\,, \tag{5.42}$$

stressing that a sequence of universal gates can efficiently approximate the unitary transformation $\tilde{U}_R$ (it undoes the effects of U_R when applied to a specific superposed state, a complex combination of states from B and H) [Har-Hay].[58]

Applying $\tilde{U}_R$ ideally would unscramble the quantum information, extracting it from the entangled state and converting it into a format that can be more straightforwardly accessed and analyzed. However, a complication arises; the above process is highly non-trivial because the new inverse operator $\tilde{U}_R$ is intricately dependent on the precise initial states of the qubits in B and H, which are part of an entangled system associated with a black hole. Although it seems that $\tilde{U}_R$ could be utilized to decode the information, it encounters a pivotal issue; the operation of $\tilde{U}_R$ involves all qubits, including those in B and H, to which Alice does not have access.[59] Hence,

[58] The transformations U_{dyn}, U_{mix}, and $\tilde{U}_R$ are complex unitary operations that need to be implemented on a quantum computer. The Solovay-Kitaev theorem (see explanation in footnote 56) provides a theoretical foundation for breaking down the complex unitary transformations, U_{dyn}, U_{mix}, and $\tilde{U}_R$ into sequences of simpler, universal gates. The theorem assures us that for any desired unitary operation U, a sequence of gates exists from a universal gate set that approximates U to within any desired accuracy ϵ. According to the theorem, the sequence length needed to achieve the approximation grows logarithmically with $\frac{1}{\epsilon}$. This means that achieving a high degree of precision does not require an impractically long sequence of gates (see footnote 56). While Harlow and Hayden do not provide the exact sequence of gates, they refer to the Solovay-Kitaev theorem, which guarantees that such a sequence exists and can be efficiently found. To translate such abstract unitary transformations into a sequence of quantum gates, we first determine the high-level structure of the unitary transformation and understand the overall unitary transformation needed, which parts involve simple operations like entanglement (e.g., U_{mix}) and which parts involve more complex dynamics (e.g., U_{dyn}). We then decompose the high-level operations into simpler, universal quantum gates. Universal gates include the Hadamard gate, the phase $Z_{\frac{1}{4}}$ gate, and the CNOT gate, which can be combined to create any unitary transformation. This process is guided by the (decomposition) Solovay-Kitaev theorem. However, the practical realization of even a small circuit is constrained by the limitations of the Noisy Intermediate-Scale Quantum (NISQ) era. Thus, we cannot use the identified sequences of gates to build the quantum circuit capable of performing the required transformations with current technology.

[59] This refers to Alice's practical accessibility. Even though Alice's theoretical calculations involve the modes inside the black hole (A), she does not have (and will never have) practical access to these modes if she remains outside the black hole. The operations and measurements required to

implementing $\tilde{U}_R$ would be infeasible. Alice, therefore, considers a strategy where she replaces the unavailable qubits from B and H with some ancillary qubits in a random state, hoping this would allow her to use $\tilde{U}_R$ to undo U_R. The difficulty remains as $\tilde{U}_R$ is fundamentally tied to the initial states of the B and H.

Ultimately, it is highly unlikely for Alice to find a simple, small circuit to implement U_R due to the constraints imposed by U_{dyn} and the complexity arising from $U_R^{\dagger}$. Because Alice cannot access all the qubits she needs, she cannot directly apply $U_R^{\dagger}$ to reverse the dynamics and extract the information encoded in R_B. The transformations she would like to apply, including $\tilde{U}_R$, are sensitive to the states of all the zones, including B and H, which she cannot access. Therefore, Alice cannot take a straightforward approach to reverse the dynamics in a manageable amount of time, specifically in a time that scales polynomially with the system's entropy. So, Alice is stuck with having to try a brute force strategy, where she uses a tremendous number of operations (2^{n+k+m} gates, which indicates a vast, potentially impractical number) to construct $U_R^{\dagger}$. Harlow and Hayden, therefore, suggest a kind of pessimism or skepticism that there would be an easy solution to Alice's problem [Har-Hay]. Thus, due to the extreme computational complexity of implementing the unitary transformation $U_R^{\dagger}$, doing so would take longer than the lifetime of the black hole itself. Alice will not have enough time to complete the necessary computations.[60]

decode $\tilde{U}_R$ involve subsystem H that is not directly manipulable from outside the event horizon (the horizon modes). However, in theoretical frameworks, Alice might include the modes inside the black hole A and H to understand the complete entanglement structure and information dynamics.

[60] While the black hole can produce certain states without requiring exponential time, distilling B from the radiation poses a significant computational challenge. The discrepancy between the time it takes for a black hole to encode information into Hawking radiation and the time required to distill specific information from the radiation is rooted in quantum processes and computational complexity. The emission of Hawking radiation is a natural consequence of the black hole's interaction with its surrounding quantum fields. The laws of quantum mechanics and general relativity govern this process. Black holes are believed to be among the fastest scramblers of information. This means that the information about the initial state of the matter that formed the black hole or fell into it quickly spreads throughout its degrees of freedom and then into the emitted Hawking radiation. One must reverse the scrambling process to decode the information from the Hawking radiation. Due to the chaotic nature of the scrambling, this inverse process involves extremely complex quantum operations that grow exponentially with the size of the system. This highlights a boundary in our understanding and manipulation of black hole information. The inability to decode Hawking radiation is a prominent issue in black hole physics because it implies a lack of a complete quantum mechanical description of its interior. The distilling transformation $U_R^{\dagger}$ is established, which could act solely on radiation. However, this is nonconstructive because it only proves there is such a transformation without providing a method to achieve it. This leaves a gap in understanding the complexity of the distillation process, which is fundamental for further analyses. The absence of a constructive method to achieve the transformation $U_R^{\dagger}$ leaves an important aspect of the black hole quantum mechanics—the complexity of the distillation process—unexplored. The complexity of this process is pivotal for understanding the practicality and feasibility of distilling and analyzing quantum information from black hole radiation [Har-Hay, Har].

5.4.4 The Harlow-Hayden Error-Correcting Code

Alice's lack of access to B and H is represented mathematically as an erasure of all qubits in this system. These erasures pose a significant challenge to Alice's task, as they take away information vital for her to successfully decode the information about the black hole's initial state. Alice uses the error-correcting code to recover information that has been erased and affected by the interaction of the black hole with its environment.

Up to this point in their discussion, Harlow and Hayden have focused on the errors that arise from the fact that Alice does not have direct access to all the necessary information from B and H to decode the information she seeks straightforwardly. However, they point out that this is not Alice's task's only source of potential errors. They bring attention to another substantial challenge. For instance, the black hole emits Hawking radiation, including hard-to-detect gravitons, let alone coherently manipulate. Thus, information about the black hole's state is carried away with that radiation, and some of it is lost. However, Harlow and Heyden highlight that losing all gravitons would not prevent successful error correction, emphasizing their model's resilience because even losses due to gravitons will not necessarily disrupt extracting information. In other words, they incorporate a robust error correction mechanism, showing that theoretically, their model upholds the principle of unitarity.[61]

Harlow and Hayden write Eq. (5.37) in an equivalent way[62]:

$$
\begin{aligned}
|\psi\rangle &= \frac{1}{\sqrt{|B|}} \sum |b\rangle_B \left|\overline{b}\right\rangle, \\
\left|\overline{b}\right\rangle &\equiv \frac{1}{\sqrt{|H|}} |h\rangle_H \, U_R \, |bh0\rangle_R .
\end{aligned}
\tag{5.43}
$$

[61] Harlow and Heyden define the following in their error correction scheme: n represents the total number of bits that can be theoretically lost while retaining the ability to decode the information successfully. n represents the entire set of bits in the encoding scheme. These bits include the information to be protected and the redundancy added for error correction. The total n bits are divided into various parts reflecting different aspects of the black hole model, $2k + m + n$: This formula represents the total number of bits in the encoding scheme. $2k$, m, and n correspond to different sets of bits within the encoding scheme. $2k$: The bits associated with the primary encoding. $k + m$ bits: These bits are already considered lost because Alice cannot access subsystems B and H. This represents the bits that are inherently inaccessible to Alice. Remaining Bits $\frac{n-k-m}{2}$: This represents the number of bits Alice can still afford to lose due to other errors and yet remain within the error correction capability (the system's built-in tolerance for errors, i.e., information that can still be extracted despite some loss). α denotes the fraction of the radiation consisting of gravitons. This fraction is significant because it accounts for less than half of the total radiation. Even if all gravitons are lost, Alice should theoretically still be able to extract the necessary entanglement information because she can afford to lose up to $\frac{n-k-m}{2}$ bits and still perform successful error correction. This demonstrates the robustness of the error correction mechanism [Har-Hay].

[62] The state $|\psi\rangle$ is expressed as a superposition of $|b\rangle_B$ and $\left|\overline{b}\right\rangle$, where $\left|\overline{b}\right\rangle$ is defined in terms of $|h\rangle_H$ and U_R. Harlow and Hayden are trying to simplify the problem by focusing on a smaller Hilbert space, where $\left|\overline{b}\right\rangle$ is a basis for a k dimensional subspace of $\mathcal{H}_H \otimes \mathcal{H}_R$. But they also emphasize the intertwined nature of the state $|\psi\rangle$ in the black hole and the Hawking radiation system, with its behavior determined by interactions between multiple and different Hilbert spaces.

Alice creates a quantum code using U_{enc}, which is built from U_R and $U_{\text{mix,H}}$: $U_{\text{enc}} \equiv U_R U_{\text{mix,H}}$.[63] When Alice applies U_{enc}, she identifies a subset of states (subspace) of the total Hilbert space within the larger Hilbert space.[64] When Alice wants to extract the information, she is looking to decode the information stored in this subspace. This decoding process reverses U_{enc} to extract the original information, or at least the pertinent part of it, even after parts of the system have been erased.[65] However, reversing U_{enc} to decode the information is not straightforward. Alice uses a set of quantum operations, a universal gate (CNOT gates and Hadamard transformations), to develop U_{enc}, and she tries to identify a suitable decoding process that can extract the original information from the encoded state.[66]

To facilitate the decoding process, Alice introduces an additional system, B', with the same number of qubits as B, and forms a new, more extensive set of encodings. Now, the major part of the decoding process involves a complex unscrambling of the entanglements between B' and R through another set of transformations U_{enc}, which is constructed from U_{dyn}. The complex encoding and subsequent decoding

---

[63] $U_{\text{mix,H}}$ is a simple entangling transformation analogous to U_{mix}, which is applied to a subset of the qubits representing a part of the radiation emitted. It affects the m qubits of H and the $n+1$ to $n+m^{th}$ qubits of R. This encoded information involves entangling pairs of qubits from H and R to produce new states.

[64] By focusing on this specific subspace, Alice reduces the computational complexity and resource requirements needed for her task. Operating within a smaller subspace allows a more efficient approach to extracting information, bypassing the need to manage the prohibitively large set of all possible quantum states in the full Hilbert space. This subspace, identified through U_{enc}, aids in encoding the information about the black hole state. The essential information can still be protected and extracted even if some information is lost or erased. The transformation creates a structure that supports robust error correction. In the context of black hole information, the full Hilbert space represents a vast universe of possible quantum states. Working within this space would be computationally infeasible. However, by applying the U_{enc} transformation, Alice narrows her focus to a more manageable set of states within a subspace where the black hole information is encoded. This simplification is necessary for making the decoding process feasible.

[65] Considering Alice's lack of access to subsystems B and H as erasures, the encoded information remains intact despite these erasures. U_{enc} is designed to distribute information across multiple parts of the black hole and its surrounding radiation. This is done so that even if some parts are inaccessible (i.e., have experienced errors), the original information can still be reconstructed according to unitarity. According to unitarity, no information is lost in the evolution of a closed quantum system, so proper decoding allows for recovering this information despite partial loss.

[66] To reverse the encoding operation U_{enc}, Alice applies $U_{\text{enc}}^{\dagger}$, which involves the following steps. First, she determines the sequence of CNOT and Hadamard gates used in the encoding. Second, she applies the inverse of each gate in the reverse order of their application during encoding. For example, if the encoding operation first applied a Hadamard gate to the state $|\psi\rangle$ followed by a CNOT gate applied to the resulting state $|\psi\rangle_{\text{enc}} = \text{CNOT} \cdot H \cdot |\psi\rangle$, the decoding would consist of the following (since both Hadamard and CNOT gates are their inverses, $H = \frac{1}{H}$ and $\text{CNOT} = \frac{1}{\text{CNOT}}$) $|\psi\rangle = H \cdot \text{CNOT} \cdot |\psi\rangle_{\text{enc}}$. First, Alice would apply the CNOT gate to $|\psi\rangle_{\text{enc}}$, and then she would apply the Hadamard gate H to the resulting state. Applying $H \cdot \text{CNOT}$ reverses the original encoding operation because applying the CNOT gate again disentangles the qubits, and applying the Hadamard gate again reverses the superposition. The actual sequence of operations is much more complex, involving many CNOT and Hadamard gates. Nevertheless, the general principle remains the same: Alice should identify the sequence of gates used in U_{enc} and then apply the inverse of each gate in the reverse order to achieve $U_{\text{enc}}^{\dagger}$.

processes should lead Alice to the original information being reconstructed. The transformation U_{enc}, which includes U_{dyn} and $U_{\text{mix},B'}$, creates a complex entanglement structure between B, B', H, and R. This entanglement protects the information. During decoding, the additional qubits in B' play a vital role. They provide the necessary degrees of freedom to unscramble the entanglement and reconstruct the original information, even if B and H are partially or wholly inaccessible. For instance, the encoding process involves applying $U_{\text{mix},B'}$ to entangle B with B', followed by U_{dyn}, which scrambles the information across B, B', H, and R. To decode, $U_R^\dagger$ is applied to reverse the scrambling effect of U_{dyn}. The entanglement created by $U_{\text{mix},B'}$ allows the recovery of the original state from the remaining accessible qubits.[67]

Harlow and Hayden argue that they present a robust theoretical framework for an error correction scheme, which does not prevent the AMPS experiment from being done. They show that what is encoded in Hawking radiation can theoretically be decoded using their proposed mechanisms. The main obstacle is the practical issue of computational complexity. This complexity increases with the black hole's size and the volume of information, making the decoding task infeasible within a reasonable period. Since the task must be completed before the black hole evaporates, this imposes a strict time limit. Though the process of black hole evaporation spans eons, in the context of the thought experiments, it feels like a race against a rapidly dwindling clock. Thus, in practice, Alice cannot access the necessary information within this limited period, which upholds the principles of quantum mechanics, including the no-cloning theorem and the consistency of unitary evolution. This prevents any observer from witnessing a paradox where quantum information appears

---

[67] Harlow and Hayden want to establish U_{dyn} as the encoding transformation U_{enc} and U_R as a correction operation. However, there is an issue with the dimensionality of the code space, which they address by introducing an additional system B' with the same number of qubits as B. They then apply a transformation $U_{\text{mix},B'}$ to entangle B' with B, thereby creating a k-qubit code subspace within a larger $2k+m+n$ qubit Hilbert space using U_{enc}. U_{enc} is defined by applying $U_{\text{mix},B'}$ followed by U_{dyn}. U_{enc} operates on initial states to generate a new state that incorporates B, B', H, and R, such that:

$$U_{\text{enc}} \left|b'\right\rangle_{B'} |0\rangle_{BHR} = \frac{1}{\sqrt{|B||H|}} \sum_{bh} U_{\text{mix},B'} \left|b'b\right\rangle_{B'B} |h\rangle_H \, U_R \, |bh0\rangle_R \,. \tag{5.44}$$

Harlow and Hayden acknowledge the presence of errors introduced by the environment, conceptualized as erasures affecting both B and H. To restore the initial state by unscrambling the entanglement between B' and R, one needs to implement $U_R^\dagger$. By applying $U_R^\dagger$ on the newly defined state, and subsequently acting with $U_{\text{mix},B'}$, the information b' is recovered. This process involves replacing the second element of each pair of qubits from the first k qubits of R instead of B, which is no longer accessible. This produces a state:

$$\left|b'\right\rangle \frac{1}{\sqrt{|B||H|}} \sum_{bh} |b\rangle_B \, |h\rangle_H \, |bh0\rangle_R \,. \tag{5.45}$$

Following the recovery of b', the initial state can be restored in polynomial time using ancillary qubits to reset BHR to the state $|0\rangle_{BHR}$ and then utilizing U_{enc} to revert to the desirable state formed by the encoding transformation (5.44) involving B' and BHR [Har-Hay].

cloned. Since Harlow and Heyden's decoding process is practically impossible, this makes the information not lost but remains hidden, maintaining unitary evolution.

Harlow and Hayden define a computational problem termed Error Correctability. The state of the system B, H, and R is transformed by U_{dyn} acting on an initial state, resulting in a new state $|\psi\rangle_{BHR}$, which exhibits maximal entanglement between B and HR:

$$|\psi\rangle_{BHR} = U_{\mathrm{dyn}}\,|000\rangle_{BHR}\,. \tag{5.46}$$

The crux of the Error Correctability problem is determining if maximal entanglement between B and HR can be decoded using information from R.[68]

Harlow and Hayden show that the Error Correctability problem has a QSZK proof by demonstrating that the problem can be verified using a quantum verifier with the help of an all-powerful prover. This setup fits within the QSZK class, illustrating that the Error Correctability problem is QSZK-complete. If the Harlow-Heyden decoding task could be done in polynomial time, it would imply QSZK = BQP. This would collapse the hierarchy of complexity classes, suggesting that a quantum computer could efficiently solve problems that require a prover, which is considered unlikely.[69] Scott Aaronson suggests against the above possibility. He

[68] From Alice's perspective, the subsystems B and H are considered erased (inaccessible). Alice cannot directly access or measure the qubits in B and H. The goal of the Error Correctability problem is to determine whether the entanglement information between B and H can be decoded using only the information from R, which represents the Hawking radiation that has escaped the black hole and is accessible to Alice. While Alice may not have full access to all of R immediately, the assumption is that with sufficient time (before the black hole completely evaporates), she would have access to a significant portion of R. However, this decoding must happen within the practical constraints of the black hole's evaporation time and the limits imposed by computational complexity.

[69] Suppose there exists an efficient (polynomial-time) error correction procedure for U_{dyn}. Alice can correct errors and decode the information encoded by U_{dyn} without requiring exponential time. This is significant because it suggests that certain complex quantum operations, which are generally thought to be computationally hard, can be performed efficiently. In QSZK (Quantum Statistical Zero-Knowledge) problems, a prover (someone with unlimited computational power) performs quantum error correction on R, and the verifier (someone with polynomial-time computational power) checks the result. If Error Correctability procedure for U_{dyn} is efficient, a quantum computer can perform these operations within polynomial time without needing an external prover. This implies that a quantum computer can verify the problem directly, reducing the need for the prover. Hence, it suggests that problems in QSZK could be solved within BQP (Bounded-error Quantum Polynomial time), collapsing the hierarchy between these complexity classes. Currently, QSZK is believed to be more powerful than BQP because it allows the verifier to use an all-powerful prover. If QSZK = BQP, it means that the additional power provided by the prover in QSZK does not add any computational advantage. This collapse in the hierarchy means that any problem that can be solved with the help of an all-powerful prover (in QSZK) can also be solved by a quantum computer alone within polynomial time (in BQP). Harlow and Heyden show that the Error Correctability problem has a QSZK proof, i.e., that it fits within the QSZK complexity class. The prover convinces the verifier that maximal entanglement can be decoded from R. The verifier verifies that only statistically negligible information is lost during the process. The prover implements Error Correctability on R and presents the result to the verifier. The verifier checks the result to confirm whether the entanglement between B and HR has been correctly decoded. If this problem can be solved efficiently, it suggests that the error correction required to decode the entanglement between B and HR (QSZK problems) could be solved on a quantum computer

proposes improvements to the complexity assumptions underlying Harlow-Heyden's decoding task. Aaronson's calculations reinforce the belief that QSZK and BQP are distinct, thus supporting the infeasibility of efficiently solving the Harlow-Heyden decoding Task. He argues that this task remains hard, even under stronger complexity assumptions [Aar].[70]

5.4.5 Noise Barrier and Causality Barrier

The Harlow-Hayden argument revolves around the idea that the task of decoding information from a black hole's Hawking radiation (a quantum system) is computationally impractically complex. This computational impracticability is rooted in the assumption that even with a perfectly closed quantum system—where no external noise or decoherence interferes—the task would require an astronomical amount of computational resources, far beyond what is physically possible within the universe's age. Theoretically, the Harlow-Hayden circuit can be designed using a universal gate set, which means the decoding could be formulated and simulated within a quantum system. The design of such a circuit is possible because the laws of quantum mechanics allow for such an arrangement of gates and operations, assuming ideal conditions and infinite computational resources. The theoretical aspect here is abstract and concerns what is allowed by the fundamental laws of quantum physics

---

without needing an all-powerful prover, and can be performed within polynomial time, thereby fitting into BQP.

[70] Aaronson uses injective one-way functions (OWFs) to strengthen the argument that solving the Harlow-Hayden decoding task efficiently is unlikely. An injective OWF is a function f that is easy to compute but hard to invert. Formally, for a polynomial-time computable function f: $f(x)$ can be computed in polynomial time. For any polynomial-time algorithm A, the probability that $A(f(x)) = x$ is negligible. The primary property of OWFs is that, given $f(x)$, it is computationally infeasible to find x. This property is essential for cryptographic security and underpins many cryptographic protocols. Consider a state $|\psi\rangle_{RBH}$ prepared using an injective OWF f:

$$|\psi\rangle_{RBH} = \frac{1}{\sqrt{2^{n+1}}} \sum_{x\in\{0,1\}^n} \left(|x,0\rangle_R\,|0\rangle_B\,|f(x)\rangle_H + |x,1\rangle_R\,|1\rangle_B\,|g(x)\rangle_H\right). \tag{5.47}$$

Here, f and g are injective functions. If the ranges of f and g are disjoint, the state $|\psi\rangle_{RBH}$ will not be entangled between R and B. If the ranges of f and g are equal, there is maximal entanglement between R and B. Aaronson demonstrates that distinguishing whether the ranges of f and g are equal or disjoint is as hard as inverting the OWF f (distinguishing whether these ranges are equal or disjoint is performed by a quantum algorithm A). This is because finding the preimage of $f(x)$ or $g(x)$ is inherently difficult if f is hard to invert. Consequently, if the Harlow-Heyden decoding task could be done in polynomial time, it would imply that one could efficiently invert f, contradicting the assumption that f is an OWF. The existence of OWFs implies that there are problems in QSZK that are hard to solve efficiently. By tying the Harlow-Heyden decoding task to the difficulty of inverting OWFs, Aaronson strengthens the argument that this task is hard and cannot be efficiently solved, thereby suggesting that QSZK and BQP are distinct [Aar].

without considering practical limitations. We can describe the process conceptually even if we can never do it.

Harlow and Hayden propose to use error-correcting codes in their thought experiment. However, these correcting codes are specifically aimed at mitigating the effects of erasures in the regions H and B, which is a different type of error compared to the noise and decoherence typically encountered in quantum computing. Therefore, to effectively implement the Harlow-Hayden circuit in a real quantum computer, we would need to employ fault-tolerant quantum error-correcting codes that can combat these additional errors.[71]

So, let us sprinkle in the realities of an open system where noise and errors abound. The Harlow-Hayden circuit involves a complex unitary transformation and error-correcting code, requiring deep quantum circuits. Each quantum gate operation in a universal set introduces potential errors due to decoherence and gate imperfections[72] Since Alice is attempting a complex decoding task, the cumulative effect of errors from multiple gates can significantly degrade the fidelity of the decoded information. Fault tolerance, which involves designing quantum circuits and error-correcting codes to detect and correct these errors, is a challenge and a fundamental requirement for any practical implementation of the Harlow-Hayden circuit in a noisy quantum system. It is essential to maintaining the fidelity of the computation. A deeper circuit would take more time to execute, which is problematic given that Alice has limited time before the information is lost due to quantum decoherence. Even in an ideal scenario, the deeper the circuit, the more challenging it becomes to complete the necessary computations within the allowed period (the evaporation time of the black hole). This period seems eternal. Yet, in the cosmic scale, even this vastness echoes the limits within infinity, where time is a fleeting breath in the face of insurmountable complexity.

This directly ties into Harlow and Heyden's assertion that decoding R_B is not feasible in practice due to the extensive time required. Without fault-tolerant mechanisms to manage errors, the cumulative effect can severely degrade the computation's accuracy, further complicating Alice's task. Thus, while the problem is already hard in a closed system, adding noise makes it impossible without fault tolerance. In other words, the noise and errors could prevent Alice from completing the decoding task, making the computational complexity barrier less of a concern than the task's physi-

[71] It is important to clarify the distinction here: erasure errors refer to the loss of information in specific qubits, which can be modeled and corrected within the thought experiment. In contrast, real quantum systems face additional challenges from noise, decoherence, and other operational errors, which require different and more complex error-correcting codes to manage effectively. In practice, this means that the quantum circuit would require two types of error correction: codes to mitigate erasures that address the issue of losing access to certain qubits (in H and B), which are modeled as erasures, and standard error-correcting codes designed to protect the quantum information from errors introduced by gate operations, qubit decoherence, and other noise sources inherent to the quantum hardware. However, adding such codes increases the circuit's complexity and depth, which can introduce more errors if not managed by fault-tolerant protocols.

[72] As in any quantum computation, each gate in the circuit introduces the potential for errors: gate fidelity, decoherence, calibration errors, crosstalk, measurement errors, etc. accumulating as the circuit depth increases.

cal realization. Even if the task were computationally feasible in a closed system, the imperfections of an open system would make it practically impossible without robust error management. Alice's challenge would shift towards overcoming these imperfections, and the complexity barrier would not be as critical in an open system with noise and errors. According to Harlow and Hayden, a firewall would appear if Alice could execute the Harlow-Hayden circuit. The appearance of a firewall is contingent on Alice's ability to decode R_B. However, due to the noise barrier (the degradation of quantum information due to interactions with the environment, which can make accurate decoding even more difficult and practically impossible), it is exceedingly unlikely that Alice could complete the decoding task, thus preventing the appearance of a firewall.

Leonard Susskind demonstrates that viewing the computational limitations proposed by Harlow and Hayden as a manifestation of more profound physical principles like causality is possible. From this perspective, the immense computational complexity required to decode R_B is not just a technical hurdle but a reflection of the underlying physical laws that enforce chronology protection. Specifically, allowing Alice to extract R_B and reintroduce it to the black hole A would create a closed causal loop, violating the principles of causality. It is as if A has traveled back in time as R_B, creating a closed causal loop. Hawking's chronology protection conjecture argues against the physical plausibility of closed timelike curves in nature. The introduction of a firewall can be seen as a mechanism to prevent this loop, thereby preserving the causality of spacetime and enforcing chronology protection [Suss12, Suss13, Suss20].

Susskind's suggestion can be interpreted as a complementarity principle where computational complexity and fundamental physical principles play important roles in preventing the scenarios envisaged in the AMPS paradox. Thus, the Harlow-Hayden protocol and Susskind's causality argument can be seen as complementary: the computational complexity practically acts as a barrier, and the physical principles of causality also serve as a barrier. Both perspectives are valid and complementary. The choice of which barrier to emphasize may depend on whether the observer comes from a physics or quantum computing background.

5.5 Quantum Gravity in the Lab

5.5.1 Teleportation Through a Traversable Wormhole

Adam Brown, Leonard Susskind, et al. have embarked on a groundbreaking endeavor to bring the elusive predictions of AdS/CFT and ER = EPR into an experimental science. Their work, called *quantum gravity in the lab*, aims to make the theoretical concepts of quantum gravity observable within a laboratory setting. Susskind et al. explore a model known as *teleportation-by-size*, where a qubit is scrambled on one side of a quantum system (such as a CFT) and then unscrambled on the other. This process is interpreted within the framework of holography, where the two CFTs are dual to a holographic wormhole in an AdS space. It requires two key conditions: the systems must be entangled before the message is sent and coupled afterward. Susskind and his team have introduced the principle of *size winding* to describe the signal's transmission through a holographic wormhole, with perfect size winding necessary for traversable wormhole behavior. Size winding is observed in the SYK model around the scrambling time, when information becomes thoroughly mixed throughout the system, indicating quantum chaos.[73] As a thermal operator is inserted into the system's left side, the operator's size grows, and its distribution exhibits a distinct winding pattern, typically in a clockwise direction. When a coupling between the left and right subsystems is introduced, this winding pattern reverses, mapping the operator from the left side to the right side. This behavior illustrates how information is scrambled and then unscrambled within the system, providing important insights into the dynamics of quantum chaos and the potential applications in holographic systems. *Perfect-size winding* occurs in the ground state at near-zero temperature, where the system closely resembles the TFD state, suggesting that systems with a holographic dual might exhibit this behavior (see footnote 25 in Sect. 5.3.4 for a definition of TFD state) [Bro-23, NL].

A wormhole links the entangled pairs in the CFT. Two copies of the CFT in the TFD state would be dual to the two-sided eternal black hole, and an eternal black hole can be thought of as having two separate black holes connected by a wormhole (see explanation in Sect. 5.3.4). This dual possibility made Ping Gao, Daniel Jafferis, and Aron C. Wall realize they could create a teleportation model through a traversable wormhole [GJ, GJW].[74]

---

[73] The Sachdev-Ye-Kitaev (SYK) model is a theoretical model in quantum mechanics that describes a system of interacting fermions with random couplings. The SYK model was initially proposed in condensed matter physics by Subir Sachdev, Jinwu Ye, and Alexei Kitaev. The model's large N limit exhibits emergent scale invariance, resembling conformal symmetry, due to many N interacting Majorana fermions. As the number of fermions increases, the system displays maximally chaotic behavior, characterized by chaotic dynamics and universal scaling properties of correlation functions. The SYK model demonstrates non-Fermi liquid behavior and possesses a low-temperature entropy that matches the Bekenstein-Hawking entropy of black holes. An important feature of the SYK model is the all-to-all interactions between Majorana fermions resulting from random and uncorrelated interaction terms. This feature accounts for the model's chaotic behavior and other unconventional phenomena observed in the system. The SYK model is particularly important in studying quantum chaos, holography, and the AdS/CFT correspondence. The scrambling time is the time scale over which the insertion of a qubit is spread throughout the entire system. One way to measure scrambling is through Out-of-Time-Order Correlators (OTOCs). OTOCs measure how the disturbance caused by a slight change in the initial conditions of a quantum system spreads over time. At the onset of quantum chaos, most of the OTOC functions tend to decay exponentially with time. The scrambling time is when the OTOC decays significantly, indicating that the information has been scrambled.

[74] Gao et al. propose that a wormhole could become traversable by leveraging the principles of ER = EPR and the AdS/CFT correspondence. Normally, wormholes require exotic matter that violates the null energy condition (NEC) to be traversable. This is generally forbidden in classical physics, but quantum field theory allows for NEC violations under certain conditions. The physicists consider

5.5.2 The Learned Hamiltonian and the Quantum Circuit

A team of scientists led by Jafferis developed and designed a quantum circuit to simulate this wormhole teleportation on Google's *Sycamore* quantum processor developed by Google's Quantum Artificial Intelligence division [Aru, GAI].[75] The quantum circuit emulates a holographic traversable wormhole using two SYK models. First, two

---

a setup involving two asymptotically AdS regions connected by a Schwarzschild wormhole (ER bridge). They suggest introducing a boundary interaction (a double trace deformation in the CFT) between the two boundaries of the AdS spacetime. Single-trace operators in the CFT correspond to bulk fields in the AdS space. A double trace deformation in the CFT corresponds to modifying the boundary conditions of the dual bulk fields in the AdS space. This can have significant effects on the dynamics of the bulk theory. The ER bridge can be opened by introducing a double trace deformation in the CFT, rendering the wormhole traversable. Thus, a qubit can be transmitted through the wormhole from one boundary to the other. The interaction causes the quantum matter stress tensor to have negative average null energy, essential for making the wormhole traversable. The process is akin to quantum teleportation, where information is not sent directly but instead via the entanglement between the two systems. The interaction between the boundaries acts like the classical communication step in quantum teleportation, making it possible for the qubit to traverse the wormhole. The two ends of the wormhole are connected by an ER bridge, which is analogous to a pair of entangled systems. The transmission of a qubit is mediated by the wormhole, which acts like a quantum channel between the two boundaries of the AdS space. The teleported qubit does not physically travel through space in the traditional sense. In quantum teleportation, the qubit's state is reconstructed on Bob's side based on the classical information given by Alice. The classical information (two bits) sent from Alice to Bob is essential for Bob to apply the correct operation to his qubit. Similarly, the quantum state is transmitted from one side of the wormhole to the other without a signal passing through spacetime in the usual sense. However, in the wormhole teleportation, the process is more direct. The wormhole is the physical conduit allowing the qubit to move from one boundary to another without additional classical information exchange. The traversable wormhole is open only briefly, determined by the duration of the interaction. After this, the wormhole closes, and the system returns to its non-traversable state.

[75] The traversable wormhole experiment was performed using a 72-qubit superconducting quantum processor. This version of the Sycamore processor includes improvements over the original 2019 version (a 53 qubits processor). The qubits in the original sycamore were laid out in a $6x9$ grid, i.e., 54 qubits in a complete grid. However, one qubit was intentionally removed to make the grid asymmetric, leaving it with 53 qubits [Aru]. The qubits were connected so that each qubit could interact with its nearest neighbors, facilitating the implementation of quantum gates between adjacent qubits. This arrangement helped reduce errors during quantum operations and allowed for efficient quantum circuit execution. The processor was expanded to 72 qubits, allowing for more sophisticated quantum error-correcting codes. Specifically, the Google Quantum AI team details the distance-5 surface code as a significant improvement in the Sycamore device's error correction capabilities. The distance-5 surface code is a quantum error-correcting code implemented on a grid of physical qubits that can detect and correct up to two errors, thereby enhancing the robustness of a logical qubit against physical errors in a quantum processor. The device also incorporates dynamical decoupling, a series of timed and oriented quantum gates applied to the qubits to mitigate decoherence. These gates average out the qubit's interactions with its environment over time, thereby reducing the impact of unwanted noise and preserving the qubit's quantum state. Dynamical decoupling is significant during qubit idling. During these periods, qubits are not actively participating in computations and are more susceptible to environmental noise and interactions that can cause decoherence. Applying dynamical decoupling during idling minimizes the impact of these unwanted interactions, helping to preserve the qubit's state over longer periods. These new enhancements to the device contribute to more accurate and reliable quantum error correction [GAI].

identical quantum systems are prepared and entangled in a TFD state, simulating two black holes connected by a wormhole. A message qubit is injected into the left subsystem, where it undergoes scrambling, spreading its information across the system's qubits. The left and right subsystems are then weakly coupled, creating a temporary interaction that allows the qubit to teleport through the wormhole to the right subsystem. Upon arrival, the qubit emerges unscrambled, transferring the message without additional decoding [JZ-1].[76]

The success of this quantum teleportation is verified by measuring the qubits and analyzing the mutual information.[77] This confirms that the process accurately

[76] Several components are involved in the described quantum circuit: registers P, Q, L, R, and T (collections of qubits that are treated as a single entity within the quantum circuit), SWAP gates, unitary operators, and interaction across the left and right subsystems with coupling operators. Register P holds a reference qubit entangled with a qubit q (the message) on register Q. Register L contains qubits that represent the left-hand side of the wormhole. The researchers applied a SWAP operation between registers Q and L to insert the qubit into the circuit. When a SWAP operation is applied between registers Q and L, the state of qubit q is transferred from Q to L, moving the qubit into the left side of the wormhole. The qubit q is now part of the wormhole's left subsystem, entangled with qubits in L. Subsequently, register L is evolved forward in time, i.e., the quantum state of the qubits in L are updated according to the sparsified Hamiltonian that governs their interactions (see footnote 79 for the sparse Hamiltonian). When the qubits in register L are evolved forward in time, the time-evolution operator ($\exp^{-iHt}$) is applied to those qubits. This evolution is approximated using Trotterization. The researchers used shallow Trotterization to simulate the interaction between Majorana fermions with fewer steps, thereby reducing computational overhead and error accumulation [JZ-1]. Trotterization (or Trotter-Suzuki decomposition) is a technique used to approximate the time evolution of a quantum system by breaking down a complex Hamiltonian into a sequence of simpler operations. Shallow Trotterization refers to using a minimal number of these steps, which reduces the depth of the quantum circuit and the number of gate operations required. The team implemented the Jordan-Wigner transformation (maps fermionic operators, Majorana fermions, onto spin operators, which can be directly implemented on a quantum processor) to optimize this process further, making the Trotterized steps more efficient to execute on the quantum processor so that the quantum circuit remains practical for current hardware. Applying the time-evolution operator to the quantum state, the entire state of the qubits in register L, including qubit q, evolves in time according to the Hamiltonian. In other words, the qubit q becomes rapidly scrambled and entangled with the carrier qubits in subsystem L due to the interactions dictated by the Hamiltonian. The next step involves weakly coupling side L to side R (employing the coupling operator $\exp^{i\mu V}$, V is the interaction term, and μ represents the coupling interaction). This coupling occurs suddenly, connecting all the qubits on side L with those on side R. After the qubits in L and R interact, the qubit q initially in L has its state transferred to R. To extract the qubit q from the wormhole, a SWAP operation is applied between register R (which now holds the state of q) and register T. This operation moves the state of q from R into T. After this SWAP operation, the qubit q is now in register T, extracted from the wormhole, and ready for measurements.

[77] Mutual information $I_{PT}(t)$ is a measure of the shared information between two subsystems, in our case, the qubits in registers P and T. It reflects how much information about one qubit (e.g., in P) can be learned by knowing the state of the other qubit (e.g., in T). A peak in the mutual information within a certain time window indicates the occurrence of quantum teleportation. Once the quantum circuit (see footnote 76) has completed its operations, the quantum chip measures the final state of the qubits. This process uses a readout mechanism that translates the quantum information (the superposition of states) into classical bits. The outcome of these measurements (the mutual information) is then passed to the classical computer, which calculates the mutual information values based on the measurement outcomes obtained from the quantum chip.

simulates the traversal of a wormhole in quantum gravity. Teleportation occurs via the wormhole at a near-scrambling time if the coupling interaction between the two sides of the system (e.g., through a sequence of CZ gates) is turned on. This coupling introduces correlations that mimic the geometry alteration caused by a negative energy shockwave in the bulk, modifying the wormhole's geometry and enabling traversability.

Due to the complexity of simulating the dense SYK model, the researchers simplified it, reducing the number of terms while preserving essential gravitational properties. This involved a machine-learning approach to generate a sparser Hamiltonian, which was then simulated on classical computers. The team explored the large-N SYK model, which is dual to nearly two-dimensional AdS (AdS2) space and is expected to exhibit perfect-size winding in the low-temperature limit. They sought to identify a straightforward Hamiltonian that preserves the gravitational physics of the dense SYK model and found that a model with $N = 10$ Majorana fermions and 210 terms could simulate behavior akin to a traversable wormhole, including size winding and teleportation phenomena. However, simulating this model on a quantum processor was not feasible due to the complexity and limitations of current quantum processors [JZ-1]. Instead, the team employed classical computational methods, including machine learning, to sparsify the Hamiltonian (i.e., reduce its complexity while preserving its essential properties). They reduced the original model to a sparsified version with $N = 7$ Majorana fermions and only five terms[78] They aimed to preserve the original model's essential gravitational properties while making it computationally manageable. This process involved training a neural network to minimize the number of terms while maintaining the model's accuracy, resulting in a reduced Hamiltonian capable of simulating the desired quantum dynamics with fewer resources [JZ-1].[79]

---

[78] The model:

$$\begin{aligned} H_{L,R} = &-0.36\psi^1\psi^2\psi^4\psi^5 + 0.19\psi^1\psi^3\psi^4\psi^7 \\ &- 0.71\psi^1\psi^3\psi^5\psi^6 + 0.22\psi^2\psi^3\psi^4\psi^6 + 0.49\psi^2\psi^3\psi^5\psi^7. \end{aligned} \tag{5.48}$$

ψ^i refer to the Majorana operators, where i denotes the index of the operator. Each ψ^i corresponds to a Majorana fermion in the system. The Hamiltonian $H_{L,R}$ describes the energy and dynamics of the system, specifically the interactions between the Majorana fermions represented by the Majorana operators. The first term on the right-hand side of Eq. (5.48) represents the interaction between the Majorana fermions associated with the operators $\psi^1\psi^2\psi^4$ and ψ^5. Similarly, the second term describes an interaction involving the Majorana fermions associated with $\psi^1\psi^3\psi^4$ and ψ^7, etc. The coefficients -0.36, $+0.19$, etc. in front of the products of the operators represent the strength (positive or negative) of these interactions. The 5 terms describe the interactions between the 7 Majorana fermions. Thus, only 5 specific interactions out of the potentially larger number of possible interactions are non-zero.

[79] Using machine learning techniques on a classical computer, the team trained an analog of a neural network to sparsify the SYK model with $N = 10$ and 210 terms, backpropagated over the Hamiltonian coefficients, and applied regularization, interpreting the Hamiltonian coefficients as neural network weights. Each term in the $N = 10$ SYK Hamiltonian was treated as a feature, and the corresponding coefficient was seen as the weight of that feature. Each data point in the dataset represented a specific configuration of the Hamiltonian terms. A neural network was constructed that

5.5.3 Setting Up the Simulation Environment

First, the full quantum circuit of the wormhole teleportation protocol was compiled. The team aimed to create a quantum circuit that uses fewer two-qubit gates. Two-qubit gates are typically more error-prone than single-qubit gates. So minimizing their number was important for the fidelity of the teleportation protocol. By carefully selecting how each Majorana fermion was represented, the circuit could be optimized to reduce the required gate operations, thus improving the overall efficiency and reducing potential errors in the simulation. The full circuit of the wormhole teleportation protocol was designed with a fixed sequence of operations (gates). The circuit included 164 CZ gates, and this structure did not change [JZ-1].[80] This is the step where the TFD state was prepared. The team used the *Variational Quantum Eigensolver* (VQE) algorithm to prepare the TFD state on the quantum hardware [JZ-1].[81]

---

mimics the structure of the SYK model. The network had an input layer with neurons corresponding to each term in the original Hamiltonian. These neurons serve as the inputs to the neural network. The training process involved adjusting the neural network's weights (Hamiltonian coefficients) to sparsify the model. This was done using backpropagation, where the errors were propagated from the output layer back to the input layer, and the weights were updated accordingly. The training objective was to find a set of weights that reduce the $N = 10$ SYK model to an $N = 7$ model with five terms. Weight regularization techniques were applied to encourage sparsity. These techniques introduced a penalty term, encouraging the model to have fewer non-zero weights. This helped select the most relevant terms and discard the less significant ones. Terms with low weights were removed. The Hamiltonian coefficients were interpreted as the neural network weights once the training process was complete. The non-zero coefficients indicate the terms selected in the reduced $N = 7$ model. The team employed a machine-learning approach to create another Hamiltonian featuring $N = 10$ and 8 terms. The specific objective of this Hamiltonian was to maximize the distinction in the teleportation signal between $-\mu$ (associated with wormhole teleportation) and $+\mu$ (related to scrambling teleportation) [JZ-1].

[80] CNOT (Controlled-NOT) and CZ (Controlled-Z) two-qubit gates were used to implement the interactions between qubits as part of the Trotterization process. The Trotterized time-evolution operator was constructed by applying a sequence of quantum gates corresponding to the Trotter-decomposed Hamiltonian terms.

[81] To determine the ground state of the learned Hamiltonian [Eq. (5.48) in footnote 78], the researchers employed a hybrid classical-quantum algorithm known as the VQE. VQE is an algorithm explicitly developed with NISQ devices in mind, aiming to get useful results even with noisy qubits. It involves running short quantum circuits multiple times and using classical computers to process and guide quantum computations. The algorithm starts with an ansatz, a parameterized wavefunction, approximating the ground state. The parameters of this ansatz are then iteratively optimized using a classical optimizer to minimize the energy expectation value. On the quantum processor, a circuit is constructed according to the ansatz. This circuit prepares a quantum state whose energy is then measured. The measured energy serves as feedback to a classical optimization algorithm, which adjusts the parameters of the quantum circuit to find the lowest possible energy state corresponding to the system's ground state. The team utilized the VQE algorithm to the following Hamiltonian: $H_{\mathrm{TFD}} = H_L + H_R + i\mu V$. V is the interaction term, and μ represents the coupling interaction. The result is that the ground state of H_{TFD} closely approximates the TFD state. Initializing the wormhole at $t = 0$ with $|\mathrm{TFD}\rangle$ using a VQE algorithm is done on the quantum chip by preparing the individual qubits in the state $|0\rangle$ and then entangling them appropriately using a sequence of controlled gates and single-qubit rotations.

Before running the main wormhole simulation (which involves the above quantum circuit with SWAP gates, time evolution, etc.), cross-entropy benchmarking (XEB) calibration was performed to verify that the quantum processor was optimally functioning. The quantum gates were calibrated using XEB. This process aimed to minimize errors so that the gates perform as close as possible to their intended behavior as specified by the quantum circuit. The team chose the same types of gates critical for the wormhole teleportation protocol; the generated random quantum circuit was executed on the quantum processor multiple times, and after each execution, the resulting quantum state was measured, and the output was recorded. In parallel, the exact theoretical output distribution of the random circuit was computed using classical simulation. The output bitstrings obtained from the quantum processor were then compared against this theoretical distribution using cross-entropy, a statistical measure that quantifies the difference between the two distributions. The feedback from XEB was used to fine-tune the control parameters of the quantum gates, improving the accuracy of their operations.[82] The team reported gate error rates, with the mean XEB error per cycle being approximately 0.5%, indicating high fidelity on the quantum chip [JZ-1].

After the calibration results were obtained, qubits were selected based on these results. The selection was done to minimize the errors in the quantum circuit. Specifically, qubits that showed the lowest errors for the necessary gate operations and had minimal idling errors were chosen.[83] The team first selected the best-performing qubits on the Sycamore chip. The simulation involved 7 qubits each for the left and right systems and 2 additional qubits for the teleportation process, totaling 16 (each

[82] XEB is a method used to evaluate the performance and fidelity of quantum gates on a quantum processor. A random quantum circuit is generated by applying a sequence of quantum gates to a set of qubits. Testing the quantum hardware's capabilities is performed by a complex random circuit. The comparison between the experimental (quantum processor) results and the theoretical (classically computed) results reveals how closely the quantum processor's output matches the ideal case. The closer the match, the lower the error rate of the quantum gates being tested. Repeating this process with different random circuits and gate combinations can comprehensively measure the quantum processor's performance. Cross entropy is used to quantify how closely the probability distribution of measured outcomes from a quantum circuit matches the expected distribution. The expected distribution comes from the quantum circuit. If the quantum circuit is ideal (with perfect gates), it should produce a specific probability distribution of outcomes over multiple runs. The measured distribution is obtained by running the quantum circuit on the actual quantum processor multiple times and recording the frequencies of different outcomes. By comparing the expected and measured distributions using the cross-entropy formula, one can measure how close the real circuit behavior is to the ideal behavior. The cross-entropy formula is logarithmic (derived from entropy, which uses logarithms to measure information content). So, in XEB, the cross-entropy score is used to quantify the fidelity of the quantum gates and circuits. A lower cross-entropy score (close to the theoretical minimum) indicates that the quantum processor is performing well, with high fidelity to the intended quantum operations.

[83] Idling errors occur because qubits are not actively participating in a computation (see explanation in footnote 75). Even if not active, they are still subject to various physical processes that can alter their quantum state. Qubits are often placed close to each other on a quantum processor. When a neighboring qubit is manipulated, the idle qubit might be inadvertently affected due to crosstalk, leading to an unintended change in its state. Idle qubits can also be influenced by gate operations performed on other qubits, especially if they are physically close.

Majorana fermion was represented by one qubit). The quantum processor utilizes a two-dimensional grid connecting each qubit to the nearest neighbors. This layout dictates which qubits can directly interact with one another. Since each qubit interacts primarily with its nearest neighbors, selecting a group of 16 qubits is vital where each qubit is as close as possible to others within the group. Some qubits may have higher error rates or less reliable connections, affecting their suitability for the circuit. Jafferis et al. identified a cluster of qubits physically close to the processor's lattice. They verified that this cluster could support the quantum operations required by their circuit. They avoided selecting qubits at the edges of the processor, as they have fewer neighbors and could limit interaction possibilities [JZ-1].

Finally, a depolarizing noise channel was implemented to simulate real-world noise. During the simulation of the quantum circuit (i.e., after the circuit is constructed and the gates are set), the depolarizing noise channel is applied on a gate-by-gate basis. The team simulated the circuit with their chosen qubits to verify that the layout worked as intended. The simulations were performed to assess the impact of noise on the quantum circuit under real-world conditions [JZ-1].[84]

5.5.4 *Simulating and Running the Teleportation Protocol*

After these steps, the quantum circuit for the wormhole experiment was run. This involved implementing the desired quantum operations, such as time evolution, SWAP gates, and the teleportation protocol.

Before moving to quantum simulations, Jafferis and his team checked that the simplified Hamiltonian they were using (with $N = 7$ Majorana fermions and five interaction terms) could accurately replicate the key behaviors predicted by the more complex $N = 10$ and 210 interacting terms SYK model. They used classical simulations to validate their sparsified model against the more complex $N = 10$ SYK model, showing that it could accurately represent the key dynamics of wormhole teleportation. On the classical computer, they investigated the effect of a negative coupling interaction μ in their model (see footnote 81 for a definition of μ). This negative sign induces a negative energy shockwave, which makes the wormhole traversable in the theoretical model. They verified the behavior of their sparsified learned Hamiltonian with $N = 7$ Majorana fermions and five interaction terms [Eq. (5.48) in footnote 78] aligned with the expected results from the $N = 10$ SYK model. This verification was

[84] XEB provides an error metric that reflects the fidelity of the quantum gate operations. This XEB error is then converted into a depolarization error e (a type of incoherent noise), representing the probability that any given gate would experience depolarizing noise. The result is a numerical value for e, which represents the likelihood of depolarization occurring during the execution of each gate in the circuit. The team simulated the effect of noise on the quantum circuit using this value of the depolarization error e. During the simulation, each quantum gate in the circuit is modeled to include the effect of the depolarizing noise channel. The outcome is a noisy simulation that provides a realistic model of how the quantum circuit would behave on a real quantum processor, accounting for the errors quantified by e [JZ-1].

done using classical computations. After establishing the behavior for $\mu < 0$, they explored the effect of reversing the sign of the coupling interaction to $\mu > 0$, which corresponds to generating a positive energy shockwave. In the theoretical framework, this is expected to disrupt the traversability of the wormhole, potentially preventing teleportation. The insights from classical simulations helped them understand the model's behavior and set the stage for the quantum simulations.

The team then moved to simulate the sparsified model on a quantum processor. For each value of μ, the quantum circuit was designed to either simulate a negative shockwave or a positive shockwave. Despite gate errors (inevitable in current quantum hardware), they observed a peak in the mutual information specifically for the negative coupling interaction $\mu < 0$ indicating that teleportation might have occurred. The results from the quantum processor showed qualitative agreement with the classical numerical simulations using the $N = 10$ SYK model. This comparison implied that the quantum simulation represented the $\mu < 0$ scenario, despite the noise and errors inherent in the quantum hardware [JZ-1].[85]

The team interlaced, i.e., alternated between the different quantum circuit configurations $\mu > 0$ and $\mu < 0$, to mitigate the impact of time-dependent coherent

[85] Susskind et al. and Jafferis et al. identified two distinct mechanisms through which the quantum circuit can accomplish teleportation [Bro-23, JZ-1]: 1. *Low-temperature teleportation, near the scrambling time.* Recall that the scrambling time is the characteristic timescale within the system during which information becomes highly mixed and spread throughout the system. This regime involves teleportation through the wormhole and applies to Hamiltonians with a holographic dual. There are two transmission mechanisms associated with low-temperature teleportation, with a strong dependence on the sign of μ: If $\mu < 0$, the qubit q experiences a time advance and reappears on side R, representing traversable wormhole teleportation. In this case, a peak in the signal $I_{PT}(t)$ (see explanation in footnote 77) is reached around t (t being approximately the scrambling time), indicating quantum teleportation within the time window when the wormhole is traversable. The LR coupling applied between the two sides of the wormhole allows for its traversability. If $\mu < 0$, the coupling operator generates a negative energy shockwave in the bulk, modifying the wormhole's geometry and enabling traversability. On the contrary, in the case where $\mu > 0$, the qubit becomes entangled with the qubits on side L, but it remains scrambled, indicating scrambling teleportation. This scenario reverses the sign of the coupling interaction, generating a positive energy shockwave. Consequently, the qubit is inevitably drawn to the singularity, preventing its emergence on the wormhole's right side R. 2. *High temperature teleportation, for $t >$ the scrambling time.* This mechanism, unexpected from gravity, does not involve signals traversing a geometric wormhole. Jafferis and his team investigated the low-temperature limit. They used classical simulations to show that the behavior of the learned Hamiltonian ($N = 7$ with 5 terms) [Eq. (5.48) in footnote 78] aligns with the original $N = 10$ SYK model for both $\mu = -12$ and $\mu = +12$. This correspondence was illustrated through the similarity in the curves of the two models. The similarity was evident in the peak positions of the $N = 10$ SYK model and the learned Hamiltonian. For a fixed injection time of $t_0 = 2.8$, it was anticipated that $I_{PT}(t)$ would reach its peak around the scrambling time t_* for $\mu = -12$ (the interaction) due to the negative energy shockwave making the wormhole traversable. Through the simulation carried out on the quantum processor with gate errors, a peak in $I_{PT}(t)$ was observed solely for $\mu = -12$, suggesting the occurrence of teleportation. The outcomes also revealed an asymmetry in $I_{PT}(t)$ when considering opposite signs of μ, specifically, a peak for $\mu = -12$ and a trough for $\mu = +12$ (corresponding to teleportation occurring using scrambling instead of through the wormhole, consistent with theoretical expectations). This observation demonstrated qualitative consistency with the numerical simulation performed on classical computers using the $N = 10$ SYK model.

errors.[86] They compared the effects of the different parameter values μ. Instead of running all the circuits for $\mu > 0$ first and then running all the circuits for $\mu < 0$ afterward, the researchers alternated between these two settings during data collection. For example, they could run the circuit with $\mu < 0$, immediately after, they could run the circuit with $\mu > 0$, and repeat this process many times (e.g., $10,000$ measurements for each setting, alternating back and forth) [JZ-1].[87]

5.5.5 Raising Concern About the Experiment

However, potential measurement errors could still influence the results. Experimental noise, decoherence, and gate errors could skew the results, particularly at the critical injection time $t_0 = 2.8$ (see footnote 85 for details). If the experiment is not perfectly calibrated or there are significant errors during the measurements, these could lead to artifacts in the mutual information that might be misinterpreted as a teleportation signal. And the way the interactions are implemented might influence the results in a way that does not accurately reflect the underlying physics. The results might not be as clear-cut as suggested. An alternative interpretation could be that the observed asymmetry in mutual information (i.e., observed peak in the mutual information specifically for $\mu < 0$) is not solely due to the intended gravitational dynamics (like the negative energy shockwave for $\mu < 0$) but might also be influenced by non-gravitational effects, such as those stemming from measurement errors. Another possibility is that the signal observed for $\mu < 0$ might be an artifact of the experimental setup rather than a true indication of successful teleportation through a wormhole. The difference between $\mu < 0$ and $\mu > 0$ might not be robust enough to definitively conclude that teleportation has occurred.

A team of physicists led by Norman Yao published a comment on the wormhole experiment, raising concerns about the simplified Hamiltonian's reliability and questioning whether it accurately represents the gravitational dynamics of the original SYK model ($N = 10$ with 210 terms). The team argued that the simplified model

---

[86] Coherent errors are systematic errors arising from consistent quantum operations deviations over time. These errors are particularly problematic because they can accumulate and significantly impact the quantum computation. Coherent errors might stem from imperfect calibration, drift in the control electronics, or environmental fluctuations (e.g., temperature changes). Coherent errors vary over time, i.e., the quantum processor's behavior might change slightly depending on when a particular circuit is run. This time dependence can introduce bias if not properly managed, especially when comparing different circuit outcomes.

[87] Interlacing between $\mu > 0$ and $\mu < 0$, any drift or noise that occurs over time is evenly distributed across both sets of measurements. This makes comparing the two settings more reliable, as both are subject to the same temporal conditions. The team mentions taking many measurements (e.g., $10,000$ measurements per configuration of the quantum circuit). They repeated the experimental procedure multiple times with the parameters $\mu > 0$ and $\mu < 0$. Each run involved tens of thousands of measurement outcomes. Since errors can vary over time (as discussed in footnote 86 with coherent errors), the team took multiple runs of the experiment, each consisting of the full set of required measurements.

might not represent all the essential features, especially regarding untrained fermions and fully commuting Hamiltonian terms. The model was trained to reproduce the teleportation signal for a particular pair of Majorana operators under negative coupling $\mu < 0$. They show that the successful observation of a single peak in the mutual information at $t_0 = 2.8$ for $\mu < 0$ was achieved only for the specific pair of operators involved in the machine-learning model training. When other (untrained) operators are used, the mutual information does not show the same consistent behavior. Instead, it exhibits large oscillations and multiple peaks, inconsistent with the expected dynamics of the original SYK model and gravitational teleportation through a wormhole. The team expressed concern that the observed teleportation signal (especially the peak in mutual information) might not be a general feature of the model but rather an artifact of the training process that only works well for the specific operators involved in training. This casts doubt on the broader applicability of the results, especially since the original SYK model (which is fully non-commuting and expected to represent holographic gravity more reliably) shows consistent teleportation signals across different operator pairs, unlike the learned Hamiltonian ($N = 7$ with 5 terms) [(5.48) in footnote 78] [Kor].

The criticism primarily revolves around sparsification and machine learning techniques, which are intrinsic to the classical computing side of the process rather than the quantum execution itself. The problems related to sparsification and machine learning are part of the classical setup and preparation of the model. These problems arise from how the Hamiltonian is constructed and optimized. While the quantum processor executes the quantum circuit based on the learned Hamiltonian, the concerns raised in the criticism do not pertain to the quantum hardware's performance but rather to the validity and generality of the model being used.

The learned Hamiltonian is a small-size, fully-commuting Hamiltonian that consists of only a few terms. The team argues in their criticism that fully-commuting Hamiltonians generally do not exhibit the same thermalization and scrambling behavior as non-commuting Hamiltonians like the SYK model.[88] They claim that the learned Hamiltonian does not truly thermalize. Instead, it exhibits strong oscillations in the correlation functions over time, indicating that the system does not lose the memory of its initial state (which would be expected in a thermalizing system). The decay observed in the experiment was likely an artifact of averaging over these oscil-

[88] Thermalization refers to the process by which a quantum system evolves towards a state that resembles thermal equilibrium, where correlation functions decay over time. A common type of correlation function is the two-point correlation function. The two-point function measures the correlation between two observables at different points in space or time in a quantum system. In thermalizing quantum systems, these correlations between observables typically decay over time as the system approaches thermal equilibrium, and the time it takes for the correlations to decay significantly is referred to as the thermalization time. The four-point correlation function probes the scrambling behavior in chaotic quantum systems. In these systems, the four-point function tracks how an initially localized perturbation spreads across the system. As the system evolves, the four-point function deviates significantly from its initial value, indicating the onset of quantum chaos. The rapid increase in the four-point function suggests the spread of entanglement across the system, and the time at which the four-point function reaches a characteristic value is known as the scrambling time.

lations. The physicists suggest that the experimental setup to generate the learned Hamiltonian may not represent the essential gravitational dynamics that the SYK model would exhibit. They point out that fully commuting models at small system sizes, like the learned Hamiltonian, might show some gravitational features (like size winding). Still, they do not thermalize or exhibit chaos, critical aspects of holographic duality [Kor].[89]

In response to the abovementioned criticisms, the experimenting team led by Jafferis published a rebuttal defending their learned Hamiltonian. They presented a counterargument related to the gravitational interpretation of thermalization. Jafferis and his team explained that the slower thermalization of lighter fermions, due to their more spread-out wave packets, leads to a slower decay of their two-point function. In their response, the team argued that the observed slow thermalization and delayed size winding of certain fermions are intrinsic to the physical dynamics of the system rather than being artifacts of computational methods like machine learning and sparsification, as suggested by the criticism. The team contended that the learned Hamiltonian is not biased to exhibit good-size winding only for the specific fermions trained during the experiment. Instead, they maintained that all fermions in the learned Hamiltonian display size winding behavior, albeit at different times. They pointed out that focusing on size winding at only one specific time point, such as $t_0 = 2.8$, might not represent the full range of dynamics across different fermions. Jafferis and his team further suggested that the criticism overlooked the broader behavior of the Hamiltonian over various timescales and across different fermions. They emphasized that size winding and teleportation signals were observed at various times, indicating that the teleportation behavior is not restricted to the trained operators. This broader perspective, they argued, supports the interpretation that the learned Hamiltonian represents the gravitational dynamics of the system, despite the concerns raised by the critics [JZ-2].[90]

The team then addresses the criticism that the commuting structure of the Hamiltonian was not conducive to realistic thermalization. They point out that size winding occurs before any significant revivals from the commuting structure of the Hamiltonian. Thus, for their specific goal—teleportation—the system must only exhibit sufficient scrambling to achieve the necessary quantum correlations. Full thermalization (where the entire system equilibrates) is not strictly necessary at the time scale of interest. The team focuses on a specific time scale at which the teleportation dynamics and the gravitational interpretation are observed. They argue that essential

---

[89] The oscillations and lack of decay in the correlation functions (two-point and four-point) suggest that the learned Hamiltonian does not properly thermalize, contrasting with the behavior expected from the SYK model or a system genuinely exhibiting holographic duality.

[90] The team argues that fermions exhibiting slower thermalization demonstrate significant size winding at later times. This behavior is attributed to the varying masses of the fermions: lighter fermions undergo slower thermalization and consequently take longer to traverse the wormhole. Specifically, the untrained fermions are the slowest to thermalize, leading to their achievement of size winding at slightly later times. This explanation is rooted in the physical behavior of the system, where the team observed that fermions with slower thermalization (specifically the untrained fermions) achieve size winding at later times, corresponding to different masses across fermions.

quantum phenomena-such as teleportation and the associated gravitational effects-occur at this time scale before the commuting Hamiltonian's potential negative effects (like lack of thermalization) become prominent. The system has undergone sufficient scrambling, and the necessary correlations have been established for the teleportation protocol to succeed. They assert these are the key features for interpreting the experiment within a gravitational framework. They argue that the early-time dynamics determine the primary gravitational effects, such as the successful teleportation signal and the associated size winding. The commuting nature of the Hamiltonian does not significantly hinder these dynamics because the interactions are designed to produce the required quantum effects (like entanglement and correlation) before thermalization (or the lack thereof) becomes a concern [JZ-2].

5.5.6 The Tangle of Science and Experiments

Nancy Cartwright introduces the model of the *Tangle of Science* to describe the interconnected network that constitutes scientific research. This tangle includes various components such as theories, models, methods, instruments, and practices, all of which are interdependent and collectively contribute to the overall success of scientific endeavors. According to Cartwright, in this complex environment, the focus should shift from merely determining the truth value of scientific claims to evaluating their reliability. Reliability, in this context, is a measure of the consistency, robustness, and practical effectiveness of scientific outputs—whether they are theories, models, or instruments—when applied in real-world scenarios. For Cartwright, reliability is not just about theoretical correctness but about how well scientific products withstand real-world scenarios' unpredictable and messy nature. By focusing on reliability, researchers can gain a deeper understanding of the effectiveness and applicability of scientific practices and their results in the environments where they are intended to be used. Furthermore, Cartwright emphasizes that criticism and rigorous testing of scientific models and experiments are essential within the *Tangle of Science*. The interplay between theory and experiment is a central aspect of this complex network, and the reliability of scientific outputs is contingent upon the methods, theories, and instruments used in research. In sum, Cartwright's approach shifts the focus of scientific evaluation from seeking universal truths to examining the reliability of scientific outputs within the complex and interconnected real world of scientific practice, where consistency, robustness, and practical effectiveness are paramount [Car-20, Car-22, CHMST, Car-23].

The traversable wormhole experiment is a prime example of the *Tangle of Science*, where various scientific components—such as the theoretical framework of holography and quantum gravity, machine learning for sparsification, and quantum processors—are deeply interconnected. The experiment simulated the dynamics of two entangled SYK models on a quantum processor. These two SYK models are dual to two sides of a traversable wormhole in an AdS space under the AdS/CFT correspondence. The setup involved creating a quantum circuit that mimicked the interaction between the two SYK systems, simulating the exchange of quantum information as it would theoretically occur through a traversable wormhole in the dual AdS space. This experiment provided a concrete realization of how information could be transferred between the two entangled systems within the holographic framework, analogous to a wormhole traversal.

Cartwright argues that the success of such an experiment depends on how well the various scientific components integrate and support each other. Therefore, the experiment's reliability hinges on the performance and interaction of these diverse elements within the complex tangle. Cartwright's approach shifts the focus from seeking an absolute truth to assessing the practical effectiveness of scientific tools and methods. For the wormhole experiment, this means evaluating how reliably the sparsified SYK model, implemented on a quantum processor, can simulate the desired quantum dynamics. The reliability of the quantum processor, including its error rates and noise levels, is essential for determining the overall success of the experiment. This aligns with Cartwright's view that scientific outputs should be judged by their consistency and accuracy in real-world applications. The experiment's reliance on a sparsified version of the SYK model (learned Hamiltonian) implemented on a NISQ (Noisy Intermediate-Scale Quantum) processor highlights the importance of practical effectiveness.

The critique of the physicists who published a comment about the traversable wormhole experiment is essential to Cartwright's process. As Cartwright highlights, this critical examination is essential for refining scientific practices' reliability and effectiveness. The criticism related to the commuting Hamiltonian and the behavior of untrained Majorana operators stems from the classical machine learning procedure used to derive the sparse Hamiltonian. This problem is intrinsic to the theoretical model used to approximate the system, which could introduce biases or limitations that do not accurately reflect the desired physical phenomena. On the other hand, issues like improper calibration, measurement errors, and the implementation of interactions during the experiment on the quantum processor could introduce artifacts into the results. These artifacts might lead to incorrect interpretations, such as mistakenly identifying noise or error effects as evidence of the intended gravitational dynamics (e.g., the peak in mutual information for $\mu < 0$). According to Cartwright's model, these two problems-experimental execution and classical model construction-are part of the *Tangle of Science* that determines the reliability of the scientific results.

The experimental results are influenced by the physical setup (quantum processor) and the theoretical underpinnings (machine learning-derived Hamiltonian). If the experimental execution is flawed or the theoretical model is inaccurate, the results may not reflect the underlying physics reliably. This interconnection means that both aspects must be carefully considered and validated to ensure the reliability of the conclusions. The experiment's success relies on the proper functioning of the quantum processor (how well the quantum gates and interactions are implemented) and the validity of the Hamiltonian model (how accurately the machine learning approach reproduces the relevant physics). Errors in one aspect can affect the overall reliability, i.e., the entire system—the tangle—must work together coherently. Cartwright's tangle suggests that scientific reliability is not just about getting the right results but also demonstrating that all scientific process components are robust and mutually supportive. This means addressing the technical execution and the theoretical modeling with equal rigor.

The debate over the learned Hamiltonian's reliability, particularly when implemented on a noisy quantum processor, underscores Cartwright's point that the effectiveness of scientific tools must be evaluated within the practical limitations of their intended use. NISQ processors, with their inherent noise and error rates, present significant challenges that can impact the reliability of the experimental results. Evaluating how these factors influence the Hamiltonian's performance is critical to understanding the experiment's success. While the simplified model shows promise, it has sparked debate within the scientific community about its validity and whether it truly reflects the complex dynamics of the original SYK model. This discussion emphasizes the challenges of using quantum circuits to simulate gravitational phenomena, particularly in noisy quantum environments.

References

[Aar] Aaronson, S. (2016). "The Complexity of Quantum States and Transformations: From Quantum Money to Black Holes." *arXiv*:1607.05256 [quant-ph]

[AN] Adamo, T. and Newman, E. (2016). "The Kerr-Newman metric: A Review." *Scholarpedia* 9, 31791.

[GAI] Acharya, R. et al. (2023). "Suppressing quantum errors by scaling a surface code logical qubit." *Nature* 614, pp. 676-681.

[Ale] Alexandrow, W., "Über den kugelsymmetrischen Vakuumvorgang in der Einsteinschen Gravitationstheorie." *Annalen der Physik* 72, pp. 141-152.

[All] Allori, V. (2020). "Some Reflections on the Statistical Postulate: Typicality, Probability and Explanation between Deterministic and Indeterministic Theories." In Allori, V. (ed.). *Statistical Mechanics: Scientific Explanation and Determinism, Indeterminism and Laws of Nature.* Singapore: World Scientific Publishing, pp. 65-111.

[AMPS] Almheiri, A., Marolf, D., Polchinski, J., and Sully, J. (2013). "Black Holes: Complementarity or Firewalls?" *Journal of High Energy Physics* 2013, pp. 1-19.

[AMPSS] Almheiri, A., Marolf, D., Polchinski, J., Stanford, D., and Sully, J. (2013). "An Apologia for Firewalls." *Journal of High Energy Physics* 2013, pp. 1-31.

[Aru] Arute, F. et al. (2019). "Quantum supremacy using a programmable superconducting processor." *Nature* 574, pp. 505-510.

[Bai] Bain, Jonathan (2018). "Black Hole Thermodynamics Part II." Lectures NYU.

[BCH] Bardeen, J. M., Carter, B., and Hawking, S. W. (1973). "The Four Laws of Black Hole Mechanics." *Communications in Mathematical Physics* 31, pp. 161-170.

[Bek-72] Bekenstein, J. D. (1972). "Black Holes and the Second Law." *Lettere al Nuovo Cimento* 4, pp. 737-740.

[Bek-73] Bekenstein, J. D. (1973). "Black Holes and Entropy." *Physical Review D* 7, pp. 2333-2346.

[Bek-74] Bekenstein, J. D. (1974). "Generalized Second Law of Thermodynamics in Black-Hole Physics." *Physical Review D* 9, pp. 3292-3300.

[Bek-75] Bekenstein, J. D. (1975). "Statistical Black-Hole Thermodynamics." *Physical Review D* 12, pp. 3077-3085.

[Bek-80] Bekenstein, J. D. (1980). "Black Hole Thermodynamics." *Physics Today*, January, pp. 24-31.

G. Weinstein, *Einstein's Legacy*, History of Physics,
https://doi.org/10.1007/978-3-031-73572-1

[Bek-81] Bekenstein, J. D. (1981). "Universal Upper Bound on the Entropy-to-Energy Ratio for Bounded Systems." *Physical Review D* 23, pp. 287-298.

[Bek-83] Bekenstein, J. D. (1983). "Entropy Bounds and the Second Law for Black Holes." *Physical Review D* 27, pp. 2262-2270.

[Bek-93] Bekenstein, J. D. (1993). "How Fast Does Information Leak Out from a Black Hole?" *Physical Review Letters* 70, pp. 3680-3683.

[Bek-94] Bekenstein, J. D. (1994). "Entropy Bounds and Black Hole Remnants." *Contemporary Physics* 49, pp. 1912-1921.

[Bek-01] Bekenstein, J. D. (2001). "The Limits of Information." *Studies in History and Philosophy of Modern Physics* 32, pp. 511-524.

[Bek-02] Bekenstein, J. D. (2002). "Quantum Information and Quantum Black Holes." In Bergmann P. G. and de Sabbata, V. (eds.) *Advances in the Interplay between Quantum and Gravity Physics*, Netherlands: Kluwer Academic Publishers, pp. 1-26.

[Bek-04] Bekenstein, J. D. (2004). "Black Holes and Information Theory." *Contemporary Physics* 45, pp. 31-43.

[Ben-88] Bennett, C. H. (1988). "Notes on the History of Reversible Computation." *IBM Journal of Research and Development* 32, pp. 16-23.

[Bir] Birkhoff, G. D. (1923). *Relativity and Modern Physics*. Cambridge: Harvard University Press (with the cooperation of R. E. Langer).

[Bol-77-1] Boltzmann, L. (1877). "On the Relation of a General Mechanical Theorem to the Second Law of Thermodynamics." In: Brush, S. G. (ed.) (1966). *Kinetic Theory, Vol. 2, Irreversible Processes*. London: Pergamon Press, pp. 188-193.

[Bol-77-2] Boltzmann, L. (1877). "Über die Beziehung zwischen dem zweiten Hauptsatze der mechanischen Wärmetheorie und der Wahrscheinlichkeitsrechnung, respective den Sätzen über das Wärmegleichgewicht." *Sitzungsberichte der Kaiserlichen Akademie der Wissenschaften. Wien* 76, pp. 373-435.

[Bol-97] Boltzmann, L. (1897). "On Zermelo's Paper 'On the Mechanical Explanation of Irreversible Processes'." In: Brush, S. G. (ed.) (1966). *Kinetic Theory, Vol. 2, Irreversible Processes*. London: Pergamon Press, pp. 238-245.

[Bou-12] Bousso, R. (2012). "Complementarity Is Not Enough." *Physical Review D* 87, 124023.

[Bou-13] Bousso, R. (2013). "Frozen Vacuum." *arXiv*:1308.3697v2 [hep-th].

[Bou-18] Bousso, R. (2018). "Black Hole Entropy and the Bekenstein Bound." In Brink, L., Mukhanov, V. F., Rabinovici, E., Phua, K. K. (eds.). *Jacob Bekenstein: The Conservative Revolutionary*. MA, US: World Scientific, pp. 139-158.

[Bo-Li] Boyer, R. H. and Lindquist, R. W. (1967). "Maximal Analytic Extension of the Kerr Metric." *Journal of Mathematical Physics* 8, pp. 265-281.

[Bril-49] Brillouin, L. (1949). "Life, Thermodynamics, and Cybernetics." *American Scientist* 37, pp. 554-568.

[Bril-59] Brillouin, L. (1959). *La science et la théorie de l'information*. Sceaux: Jacques Gabay Masson.

[BS] Brown, A. R., and Susskind, L. (2022). "A Holographic Wormhole Traversed in a Quantum Computer." *Nature* 612, pp. 41-42.

[Bro-23] Brown, A. R., Gharibyan, H., Leichenauer, S., Lin, H. W., Nezami, S., Salton, G., Susskind, L., Swingle B., and Walter, M. (2023). "Quantum Gravity in the Lab. I. Teleportation by Size and Traversable Wormholes." *PRX Quantum* 4, pp. 010320-1-010320-22.

[Bru] Brush, S. G. (ed.) (1966). *Kinetic Theory, Vol. 2, Irreversible Processes*. London: Pergamon Press.

[Cal] Calaprice, A. (ed.) (2002). *Dear Professor Einstein: Albert Einstein's Letters to and from Children*. New York: Prometheus Books.

[Can] Canales, J. (2020). *Bedeviled: A Shadow History of Demons in Science*. Princeton: Princeton University Press.

[Car-23] Cartan, E. J. (1923). "Sur les variétés à connexion affine et la théorie de la relativité généralisée (première partie)." *Annales scientifiques de l'École Normale Supérieure* 40, pp. 325-412.

[Car-24] Cartan, E. J. (1924). "Sur les variétés à connexion affine et la théorie de la relativité généralisée (première partie). (Suite)." *Annales scientifiques de l'École Normale Supérieure* 41, pp. 1-25.

[Cart] Carter, B. (1971). "Axisymmetric Black Hole Has Only Two Degrees of Freedom." *Physical Review Letters* 26, pp. 331-333.

[Car-99] Cartwright, N. (1999). *The Dappled World: A Study of the Boundaries of Science.* Oxford: Oxford University Press.

[Car-20] Cartwright, N. (2020). "X - Why Trust Science? Reliability, Particularity and the Tangle of Science." *Proceedings of the Aristotelian Society* cxx, pp. 237-252.

[Car-22] Cartwright, N. (2022). *A Philosopher Looks at Science.* Cambridge: Cambridge University Press

[Car-23] Cartwright, N. (2023). "Science In the Earthly Plane." Lecture delivered at the *University of Amsterdam. Institute for Advanced Study (IAS).*

[CHMST] Cartwright, N. Hardie, J. Montuschi, E. Soleiman, S. and Thresher, A. C. (2022). *The Tangle of Science: Reliability Beyond Method, Rigour, and Objectivity.* Oxford: Oxford University Press.

[Cha] Chandrasekhar, S. (1974). "The Black Hole in Astrophysics: the Origin of the Concept and its Role." *Contemporary Physics* 15, pp. 1-24.

[Chri] Christodoulou, D. (1970). "Reversible and Irreversible Transformations in Black-Hole Physics." *Physical Review Letters* 25, pp. 1596-1597.

[Chr] Christoffel, E. B. (1869). "Ueber die Transformation der homogenen Differentialausdrücke zweiten Grades." *Journal für die reine und angewandte Mathematik* 70, pp. 46-70.

[Cla] Clark, S. (2018). "A Brief History of Stephen Hawking: A Legacy of Paradox." *New Scientist,* March 14.

[CPAE1] *Collected Papers of Albert Einstein. Vol. 1: The Early Years, 1879-1902.* Stachel, J., Cassidy, D. C., and Schulmann, R. (eds.). Princeton: Princeton University Press, 1987.

[CPAE3] *Collected Papers of Albert Einstein. Vol. 3: The Swiss Years: Writings, 1909-1911.* Klein, M. J., Kox, A. J., Renn, J., and Schulmann, R. (eds.). Princeton: Princeton University Press, 1993.

[CPAE4] *Collected Papers of Albert Einstein. Vol. 4: The Swiss Years: Writings, 1912-1914.* Klein, M. J., Kox, A. J., Renn, J., and Schulmann, R. (eds.). Princeton: Princeton University Press, 1995.

[CPAE5] *The Collected Papers of Albert Einstein. Vol. 5: The Swiss Years: Correspondence, 1902-1914.* Klein, M. J., Kox, A. J., and Schulmann, R. (eds.). Princeton: Princeton University Press, 1995.

[CPAE6] *The Collected Papers of Albert Einstein. Vol. 6: The Berlin Years: Writings, 1914-1917.* Klein, M. J., Kox, A. J., and Schulmann, R. (eds.). Princeton: Princeton University Press, 1996.

[CPAE7] *The Collected Papers of Albert Einstein. Vol. 7: The Berlin Years: Writings, 1918-1921.* Janssen, M., Schulmann, R., Illy, J., Lehner, C., Buchwald Kormos, D. (eds.). Princeton: Princeton University Press, 1998.

[CPAE8] *The Collected Papers of Albert Einstein. Vol. 8: The Berlin Years: Correspondence, 1914-1918.* Schulmann, R., Kox, A.J., Janssen, M., Illy, J. (eds.). Princeton: Princeton University Press, 2002.

[CPAE15] *The Collected Papers of Albert Einstein. Vol. 15: Writings & Correspondence, June 1925-May 1927.* Kormos Buchwald, D., Illy, J., Kox, A. J., Lehmkuhl, D., Rosenkranz Z, Nollar James, J. (eds.). Princeton: Princeton University Press, 2018.

[Dar] Darrigol, O. (2022). *Relativity Principles and Theories from Galileo to Einstein.* Oxford: Oxford University Press.

[Dau] Daub, E. E. (1970). "Maxwell's Demon." *Studies in History and Philosophy of Science Part A*, pp. 213-227.

[Dro-16] Droste, J. (1916). *Het zwaartekrachtsveld van een of meer lichamen volgens de theorie van Einstein*. Leiden: E. J. Brill.

[Dro-17] Droste, J. (1917). "The field of a single center in Einstein's theory of gravitation and the motion of a particle in that field." *Koninklijke Nederlandse Akademie van Wetenschappen, Proceedings* 19, pp. 197-215.

[Ea-Jan] Earman, J. and Janssen, M. (1993). "Einstein's Explanation of the Motion of Mercury's Perihelion." In Earman, J., Janssen, M. and Norton, J. (eds.) (1993). *The Attraction of Gravitation: New Studies in the History of General Relativity*. Einstein Studies 5. Boston: Birkhäuser.

[Ein-05] Einstein, A. (1905). "Über einen die Erzeugung und Verwandlung des Lichtes betreffenden heuristischen Gesichtspunkt." *Annalen der Physik* 17, pp. 132-148.

[Ein-07] Einstein, A. (1907). "Über das Relativitätsprinzip und die aus demselben gezogenen Folgerungen." *Jahrbuch der Radioaktivität und Elektronik* 4, pp. 411-462.

[Ein-11] Einstein, A. (1911). "Über den Einfluß der Schwerkraft auf die Ausbreitung des Lichtes." *Annalen der Physik* 35, pp. 898-908.

[E1-12] Einstein, A. (1912). "Lichtgeschwindigkeit und Statik des Gravitationsfeldes." *Annalen der Physik* 38, pp. 355-369.

[E2-12] Einstein, A. (1912). "Zur Theorie des statischen Gravitationsfeldes." *Annalen der Physik* 38, pp. 443-458.

[E3-12] Einstein, A. (1912). "The Zurich Notebook Transcription and Facsimile." In Renn, J., Norton, J., Janssen, M. and Stachel, J. (eds.) (2007). *The Genesis of General Relativity*. Vol. 1. New York: Springer.

[E13] Einstein, A. (1913). "Zum gegenwärtigen Stande des Gravitationsproblems." *Physikalische Zeitschrift* 14, pp. 1249-1262.

[E2-14] Einstein, A. (1914). "Die formale Grundlage der allgemeinen Relativitätstheorie." *Königlich Preußische Akademie der Wissenschaften (Berlin). Sitzungsberichte*, pp. 1030-1085.

[E1-15] Einstein, A. (1915). "Zur allgemeinen Relativitätstheorie." *Königlich Preußische Akademie der Wissenschaften (Berlin). Sitzungsberichte*, pp. 778-786.

[E2-15] Einstein, A. (1915). "Zur allgemeinen Relativitätstheorie. (Nachtrag)." *Königlich Preußische Akademie der Wissenschaften (Berlin). Sitzungsberichte*, pp. 799-801.

[E3-15] Einstein, A. (1915). "Erklärung der Perihelbewegung des Merkur aus der allgemeinen Relativitätstheorie." *Königlich Preußische Akademie der Wissenschaften (Berlin). Sitzungsberichte*, pp. 831-839.

[E4-15] Einstein, A. (1915). "Die Feldgleichungen der Gravitation." *Königlich Preußische Akademie der Wissenschaften (Berlin). Sitzungsberichte*, pp. 844-847.

[E1-16] Einstein, A. (1916). "Die Grundlage der allgemeinen Relativitätstheorie." *Annalen der Physik* 49, pp. 769-822.

[E2-16] Einstein, A. (1916). "Eine neue formale Deutung der Maxwellschen Feldgleichungen der Elektrodynamik." *Sitzungsberichte der Königlich Preußischen Akademie der Wissenschaften (Berlin)*, pp. 184-188.

[E3-16] Einstein, A. (1916). "Näherungsweise Integration der Feldgleichungen der Gravitation." *Königlich Preußische Akademie der Wissenschaften (Berlin). Sitzungsberichte*, pp. 688-696.

[E4-16] Einstein, A. (1916). "Hamiltonsches Prinzip und allgemeine Relativitätstheorie." *Sitzungsberichte der Königlich Preußischen Akademie der Wissenschaften (Berlin)*, pp. 1111-1116.

[Ein-17] Einstein, A. (1917). *Über die spezielle und die allgemeine Relativitätstheorie, gemeinverständlich*. Braunschweig: Vieweg und Sohn.

[E1-18] Einstein, A. (1918). "Über Gravitationswellen." *Königlich Preußische Akademie der Wissenschaften (Berlin). Sitzungsberichte*, pp. 154-167.

[E2-18] Einstein, A. (1918). "Notiz zu E. Schrödingers Arbeit 'Die Energiekomponenten des Gravitationsfeldes'." *Physikalische Zeitschrift* 19, pp. 115-116.

[Ein-19] Einstein, A. (1919). "What is the Theory of Relativity." *Ideas and Opinions*. New York: Crown Publishers. pp. 227-232.

[E-22] Einstein, A. (1922). *The Meaning of Relativity: Four Lectures Delivered at Princeton University in May 1921*. Translated by Edwin Plimpton Adams. Princeton: Princeton University Press.

[E-23] Einstein, A. (1923). "Zur affinen Feldtheorie." *Sitzungsberichte der Königlich Preußischen Akademie der Wissenschaften (Berlin)*, pp. 137-140. A concise English version: "The Theory of the Affine Field." *Nature* 112, pp. 448-449.

[Ein-34] Einstein, A. (1934). *Letters to Solovine 1906 - 1955*. New York: Open Road Integrated Media.

[Ein-36-1] Einstein, A. (1936). "Physics and Reality." *Journal of the Franklin Institute* 221, pp. 349-382.

[Ein-49-1] Einstein, A. (1949). "Autobiographisches/Autobiographical Notes." In Schilpp, P. A. (ed.) *Albert Einstein: Philosopher-Scientist*. La Salle, IL: Open Court., pp. 1-95.

[Ein-49-2] Einstein, A. (1949). "Notes for an Autobiography." *The Saturday Review of Literature*, November 26, pp. 9-12.

[Ein-54] Einstein, A. (1954). "Physics and Reality." *Ideas and Opinions*. New York: Crown Publishers. pp. 290-323.

[EB] Einstein, A., and Bergmann, P. (1938). "On a Generalization of Kaluza's Theory of Electricity." *Annals of Mathematics* 39, pp. 683-701.

[EF] Einstein, A. und Fokker, A. D. (1914). "Die Nordströmsche Gravitationstheorie vom Standpunkt des absoluten Differentialkalküls." *Annalen der Physik* 44, pp. 321-328.

[EG13] Einstein, A., und Grossmann, M. (1913). *Entwurf einer verallgemeinerten Relativitätstheorie und einer Theorie der Gravitation. I. Physikalischer Teil von Albert Einstein. II. Mathematischer Teil von Marcel Grossmann*. Leipzig und Berlin: B. G. Teubner.

[EG14] Einstein, A., und Grossmann, M. (1914). "Kovarianzeigenschaften der Feldgleichungen der auf die verallgemeinerte Relativitätstheorie gegründeten Gravitationstheorie." *Zeitschrift für Mathematik und Physik* 63, pp. 215-225.

[EPR] Einstein, A., Podolsky, B., and Rosen, N. (1935). "Can Quantum-Mechanical Description of Physical Reality be Considered Complete?" *Physical Review* 47, pp. 777-780.

[ER] Einstein, A., and Rosen, N. (1935). "The Particle Problem in the General Theory of Relativity." *Physical Review* 48, pp. 73-77.

[Eis] Eisenstaedt, J. (2006). *The Curious History of Relativity: How Einstein's Theory of Gravity Was Lost and Found Again*. Princeton: Princeton University Press.

[Fer] Fernández, C. G. (1956). "My Tilt with Albert Einstein." *American Scientist* 44, pp. 204-211.

[Fin] Finkelstein, D. (1958). "Past-Future Asymmetry of the Gravitational Field of a Point Particle." *Physical Review* 110, pp. 965-967.

[Fri] Frigg, R. (2009). "Typicality and the Approach to Equilibrium in Boltzmannian Statistical Mechanics." *Philosophy of Science* 76, pp. 997-1008.

[GJW] P. Gao and D. L. Jafferis and A, C. Wall (2017). "Traversable wormholes via a double trace deformation." *Journal of High Energy Physics volume* 2017, pp. 1-24.

[GJ] Gao, P. and Jafferis, D. J. (2021). "A traversable wormhole teleportation protocol in the SYK model." *Journal of High Energy Physics* 2021, pp. 1-43.

[Gio] Giovanelli, M. (2020). "'Like Thermodynamics Before Boltzmann.' On the Emergence of Einstein's Distinction Between Constructive and Principle Theories." *Studies in History and Philosophy of Modern Physics* 71, pp. 118-157.

[Gio] Giovanelli, M. (2021). "Nothing but coincidences: the point-coincidence and Einstein's struggle with the meaning of coordinates in physics." *European Journal for Philosophy of Science* 11, pp. 1-64.

[GMU] Giudice, G. F., McCullough, M., and Urbano, A. (2016). "Hunting for Dark Particles with Gravitational Waves." *Journal of Cosmology and Astronautical Physics* 10, pp. 1-43.

[Gol] Goldstein, S. (2011). "Typicality and Notions of Probability in Physics." In Ben-Menahem, Y. and Hemmo, M. (eds.) *Probability in Physics*. Berlin: Springer, pp. 59-71.

[Hag] Hagar, A. (2017). "On the Tension Between Ontology and Epistemology in Quantum Probabilities." In Lombardi, O., Fortin, S., Holik, F., and López, C. (eds.). *What Is Quantum Information*. Cambridge: Cambridge University Press, pp. 147-178.

[Har] Harlow, D. (2016). "Jerusalem Lectures on Black Holes and Quantum Information." *Reviews of Modern Physics* 88, pp. 015002-1- 015002-58.

[Har-Hay] Harlow, D., and Hayden, P. (2013). "Quantum Computation vs. Firewalls." *Journal of High Energy Physics* 2013, pp. 1-56.

[Har] Harman, P. M. (ed.) (1990). *The Scientific Letters and Papers of James Clerk Maxwell. Volume II. 1862-1873*. Cambridge: Cambridge University Press.

[Hay-Pres] Hayden, P., and Preskill, J. (2007). "Black Holes as Mirrors: Quantum Information in Random Subsystems." *Journal of High Energy Physics* 2007, pp. 1-20.

[Haw-71] Hawking, S. W. (1971). "Gravitational Radiation from Colliding Black Holes." *Physical Review Letters* 26, pp. 1344-1346.

[Haw-73] Hawking, S. W. (1973). "The Analogy Between Black Hole Mechanics and Thermodynamics." *Annals of the New York Academy of Sciences* 224, pp. 268-271.

[Haw-74] Hawking, S. W. (1974). "Black Hole Explosions?" *Nature* 248, pp. 30-31.

[Haw-76-1] Hawking, S. W. (1976). "Black Holes and Thermodynamics." *Physical Review D* 13, pp. 191-197.

[Haw-76-2] Hawking, S. W. (1976). "Breakdown of Predictability in Gravitational Collapse." *Physical Review D* 14, pp. 2460-2473.

[Haw-77] Hawking, S. W. (1977). "The Quantum Mechanics of Black Holes." *Scientific American* 236, pp. 34-42.

[Haw-78] Hawking, S. W. (1978). "Black Holes and Unpredictability." *Physics Bulletin*, January, pp. 23-24.

[Haw-88] Hawking, S. W. (1988). *A Brief History of Time: From the Big Bang to Black Holes*. Introduction by Carl Sagan. New York: Bantam Books.

[Haw-03] Hawking, S. W. (2003). "Sixty Years in a Nutshell." In Gibbons, G. W., Shellard, E. P. S., and Rankin, S. J. (eds.) *The Future of Theoretical Physics and Cosmology: Celebrating Stephen Hawking's 60th Birthday*. Cambridge: Cambridge University Press, pp. 105-118.

[Hawk1] Hawking, S. W. (2005). "The Information Paradox for Black Holes." *arXiv*:1509.01147v1 [hep-th].

[Haw-16] Hawking, S. W. (2016). "Stephen Hawking Congratulates LIGO Team." *Courtesy of BBC News*, Feb 12.

[HP] Hawking, S. W., and Penrose, R. (1970). "The Singularities of Gravitational Collapse and Cosmology." *Proceedings of the Royal Society of London A* 314, pp. 529-548.

[Hei] Heisenberg, W. (1927). "Über den anschaulichen Inhalt der quantentheoretischen Kinematik und Mechanik." *Zeitschrift für Physik* 43, pp. 172-198.

[Hil15-1] Hilbert, D. (2015). "Die Grundlagen der Physik. (Erste Mitteilung.)" Presented at the session of November 20, 1915. A set of preserved proofs for this paper (SUB Göttingen, Cod. Ms.D. Hilbert 634). In Sauer, T. and Ulrich, M. *David Hilbert's Lectures on the Foundations of Physics 1915-1927*. Heidelberg: Springer.

[Hil15-2] Hilbert, D. (1915). "Die Grundlagen der Physik. (Erste Mitteilung)." *Nachrichten von der Gesellschaft der Wissenschaften zu Göttingen, Mathematisch-Physikalische Klasse*, pp. 395-407.

[Isr] Israel, W. (1967). "Event Horizons in Static Vacuum Space-Times." *Physical Review* 164, pp. 1776-1779.

[JZ-1] Jafferis, D., Zlokapa, A., Lykken, J. D., Kolchmeyer, D. K., Davis, S. I., Lauk, N., Neven, H., and Spiropulu, M. (2022). "Traversable wormhole dynamics on a quantum processor." *Nature* 612, pp. 51-55.

[JZ-2] Jafferis, D., Zlokapa, A., Lykken, J. D., Kolchmeyer, D. K., Davis, S. I., Lauk, N., Neven, H., and Spiropulu, M. (2023). "Comment on 'Comment on "Traversable wormhole dynamics on a quantum processor'."*arXiv*:2303.15423 [quant-ph], pp. 1-5.

[Jan-99] Janssen, M. (1999). "Rotation as the Nemesis of Einstein's *Entwurf* Theory." In Goenner, H., Renn, J., Ritter, J., and Sauer, T. (eds.) *The Expanding Worlds of General Relativity*. Boston: Birkhäuser.

[Jan-07] Janssen, M. and Renn, J. (2007). "Untying the Knot: How Einstein Found His Way Back to Field Equations Discarded in the Zurich Notebook." In Renn, J., Norton, J., Janssen, M. and Stachel, J. (eds.) (2007). *The Genesis of General Relativity*. Vol. 2. New York, Berlin: Springer.

[JRSNS] Janssen, M., Renn, J., Sauer, T., Norton, J. and Stachel, J., "The Zurich Notebook Transcription and Facsimile. Einstein's Zurich Notebook, Commentary, and Essays." In Renn, J., Norton, J., Janssen, M. and Stachel, J. (eds.) (2007). *The Genesis of General Relativity*. Vol. 1. New York, Berlin: Springer.

[Jan-21] Janssen, M. and Renn, J. (2022). *How Einstein Found His Field Equations: Sources and Interpretation*. Switzerland: Springer.

[Jeb] Jebsen, J. T. (1921). "Über die allgemeinen kugelsymmetrischen Lösungen der Einsteinschen Gravitationsgleichungen im Vakuum." *Arkiv för Matematik, Astronomi och Fysik* 15, pp. 1-9.

[Ken-07] Kennefick, D. (2007). *Traveling at the Speed of Thought: Einstein and the Quest for Gravitational Waves*. Princeton: Princeton University Press.

[Ker-63] Kerr, R. P. (1963). "Gravitational field of a spinning mass as an example of algebraically special metrics." *Physical Review Letters* 11, pp. 237-238.

[Ker-08] Kerr, R. P. "Discovering the Kerr and Kerr-Schild metrics." *Arxiv*:0706.1109v2.[gr-qc].

[Kno] Knott, C. G. (1911). *Life and Scientific Work of Peter Guthrie Tait*. Cambridge: Cambridge University Press.

[Kor] Kobrin, B., Schuster, T., and Yao, N. Y. (2023). "Comment on 'Traversable wormhole dynamics on a quantum processor'." *arXiv*:2302.07897 [quant-ph], pp. 1-9.

[Kox] Kox, A. J. (1988). "General relativity in the Netherlands, 1915-1920." In Eisenstaedt, L. and Kox, A. J., *Studies in the History of General Relativity*. Boston: Birkhäuser, pp. 39-56.

[Kre] Kretschmann, E. (1915). "Über den physikalischen Sinn der Relativitätspostulate. A. Einsteins neue und seine ursprüngliche Relativitätstheorie." *Annalen der Physik* 53, pp. 575-614.

[Lan] Landauer, R. (1961). "Dissipation and Heat Generation in the Computing Process." *IBM Journal of Research and Development* 5, pp. 183-191.

[Lau] Laue, M. (1911). *Das Relativitätsprinzip*. Braunschweig: Vieweg und Sohn.

[Laz] Lazarovici, D. and Reichert, P. (2015). "Typicality, Irreversibility and the Status of Macroscopic Laws." *Erkenntnis* 80, pp. 689-716.

[Leb-93-1] Lebowitz, J. (1993). "Macroscopic Laws, Microscopic Dynamics, Time's Arrow and Boltzmann's Entropy." *Physica A*, pp. 1-27.

[Leb-93-2] Lebowitz, J. (1993). "Boltzmann's Entropy and Time's Arrow." *Physics Today*, September, pp. 32-38.

[Leo] Leonhardt, U. and Philbin, T. G. (2008). "The Case for Artificial Black Holes." *Philosophical Transactions: Mathematical, Physical and Engineering Sciences* 366, pp. 2851-2857.

[LIG-1] LIGO Scientific and Virgo Collaborations (2016). "Properties of the Binary Black Hole Merger GW150914." *Physical Review Letters* 116, pp. 241102-1-241102-19.

[LIG-2] LIGO Scientific and Virgo Collaborations (2016). "Tests of General Relativity with GW150914." *Physical Review Letters* 116, pp. 221101-1-221101-19.

[LIG-3] LIGO Scientific and Virgo Collaborations (2016). "Directly comparing GW150914 with numerical solutions of Einstein's equations for binary black hole coalescence." *Physical Review D* 94, pp. 064035-1-064035-30.

[LMES] Lorentz, H. A., Minkowski, H., Einstein, A., und Sommerfeld, A. (1913). *Das Relativitätsprinzip. Eine Sammlung von Abhandlungen.* Berlin: Druck und Verlag von B. G. Teubner, Leipzig.

[Loc] Loschmidt, J. (1876). "Über den Zustand des Wärmegleichgewichtes eines Systems von Körpern mit Rücksicht auf die Schwerkraft." *Sitzungsberichte der Akademie der Wissenschaften, Wien* 73, pp. 128-142.

[LPSTU] Lowe, D. A., Polchinski, J., Susskind, L., Thorlacius, L., and Uglum, J. (1995). "Black Hole Complementarity versus Locality." *Physical Review D* 52, pp. 6997.

[Mal-13] Maldacena, J. (2013). "Entanglement and the Geometry of Spacetime." *Institute for Advanced Study.*

[Mal-16] Maldacena, J. (2016). "Black Holes, Wormholes and the Secrets of Quantum Spacetime." *Scientific American* 315, pp. 26-31.

[Mal-18] Maldacena, J. and Qi, X.-L. (2018) Eternal traversable wormhole, *arXiv*:1804.00491v3 [hep-th], pp. 1-74.

[Mald-Suss] Maldacena, J. and Susskind, L. (2013). "Cool Horizons for Entangled Black Holes." *Fortschritte der Physik* 61, pp. 781-811.

[Max-71] Maxwell, J. C. (1871). *The Theory of Heat.* London: Longmans, Green, and Co, 1902.

[Max-77] Maxwell, J. C. (1877). "Diffusion." *The Encyclopædia Britannica: A Dictionary of Arts, Sciences, and General Literature, 9th edition, Volume VII.* Edinburgh: Adam and Charles Black.

[Mie] Mie, G. (1912). "Grundlagen einer Theorie der Materie." *Annalen der Physik* 37, pp. 511-534.

[Neur21] Neurath, O. (1921). "Anti-Spengler." In Neurath, M. and Cohen, R. S. (eds.) (1973). *Empiricism and Sociology.* Dordrecht: D. Reidel Publishing Company, pp. 158-213.

[Neur32] Neurath, O. (1932). "Protokollsätze." *Erkenntnis* 3, pp. 204-214.

[Neur33] Neurath, O. (1933). "Unified Science and Psychology." In McGuinness, B. (1987). *Unified Science.* Dordrecht: D. Reidel Publishing Company, pp. 1-23.

[NCCEPT] Newman, E. T., Couch, E., Chinnapared, K., Exton, A., Prakash, A., and Torrence, R. (1965). "Metric of a Rotating, Charged Mass." *Journal of Mathematical Physics* 6, pp. 918-919.

[NJ] Newman, E. T. and Janis, A. I. (1965). "Note on the Kerr Spinning-Particle Metric." *Journal of Mathematical Physics* 6, pp. 915-917.

[NP1] Newman, E. T. and Penrose, R. (1963). "An Approach to Gravitational Radiation by a Method of Spin Coefficients." *Journal of Mathematical Physics* 3, pp. 566-578.

[NP2] Newman, E. T. and Penrose, R. (1963). "Errata: An Approach to Gravitational Radiation by a Method of Spin Coefficients." *Journal of Mathematical Physics* 4, p. 998.

[NUT] Newman, E. T., Tamburino, L., and Unti, T. (1963). "Empty Space Generalization of the Schwarzschild Metric." *Journal of Mathematical Physics* 4, pp. 915-923.

[NL] Nezami, S., Lin, H. W., Brown, A. R., Gharibyan, H., Leichenauer, S. Salton, G., Susskind, L., Swingle, B., and Walter, M. (2022). "Quantum Gravity in the Lab: Teleportation by Size and Traversable Wormholes, Part II." *arXiv*: 2102.01064v1 [quant-ph].

[Nei-98] Nielsen, M. A., Carlton, M., Caves, B., Schumacher, B., and Barnum, H. (1998). "Information-Theoretic Approach to Quantum Error Correction and Reversible Measurement." *Proceedings: Mathematical, Physical and Engineering Sciences* 454, pp. 277-304.

[Nei-10] Nielsen, M. A. and Chuang, I. L. (2010). *Quantum Computation and Quantum Information. 10th Anniversary Edition*. Cambridge: Cambridge University Press.

[Noe] Noether, E. (1918). "Invariante Variationsprobleme." *Nachrichten von der Gesellschaft der Wissenschaften zu Göttingen, Mathematisch-Physikalische Klasse* 1918, pp. 235-257.

[Nor-12] Nordström, G. (1912). "Relativitätsprinzip und Gravitation." *Physikalische Zeitschrift* 13, pp. 1126-1129.

[Nor-13-1] Nordström, G. (1913). "Träge und schwere Masse in der Relativitätsmechanik." *Annalen der Physik* 40, pp. 856-878.

[Nor-13-2] Nordström, G. (1913). "Zur Theorie der Gravitation vom Standpunkt des Relativitätsprinzips." *Annalen der Physik* 42, pp. 533-554.

[Nor-18-1] Nordström, G. (1918). "On the mass of a material system according to the gravitation theory of Einstein." *Verhandelingen der Koninklijke Akademie van Wetenschappen te Amsterdam* 20, pp. 1076-1091.

[Nor-18-2] Nordström, G. (1918). "On the energy of the gravitation field in Einstein's theory." *Verhandelingen der Koninklijke Akademie van Wetenschappen te Amsterdam* 20, pp. 1238-1245.

[Over] Overbye, D. (1979). "The Wizard of Space and Time." *Omni*, February, pp. 45-46, 104-107.

[Pai] Pais, A. (1982). *Subtle is the Lord: The Science and Life of Albert Einstein*. Oxford: Oxford University Press.

[Par-Wil] Parikh, M. K., and Wilczek, F. (1998). "An Action for Black Hole Membranes." *Physical Review D* 58, pp. 064011-1-064011-12.

[Pea] Pearson, K. (1900). *The Grammar of Science*. Third Edition. Revised and Enlarged. London: Adams and Charles Black, 1911.

[Pen-69] Penrose, R. (1969). "Gravitational Collapse: The Role of General Relativity." *Rivista del Nuovo Cimento, Numero Speziale* 1, pp. 252-275; *General Relativity and Gravitation* 34, 2002, pp. 1141-1165.

[PF] Penrose, R. and Floyd, R. M. (1971). "Extraction of Rotational Energy from a Black Hole." *Nature Physical Science* 229, pp. 177-179.

[Pet] Petrov, A. Z. (1954). "The classification of spaces defining gravitational fields." *Scientific Proceedings of Kazan State University* 114, pp. 55-69.

[Pir] Pirani, F. A. E. (1957). "Invariant formulation of gravitational radiation theory." *Physical Review* 105, pp. 1089-1099.

[Pit] Pitowsky, I. (2012). "Typicality and the Role of the Lebesgue Measure in Statistical Mechanics." In Ben-Menahem, Y. and Hemmo, M. (eds.). *Probability in Physics*. Berlin: Springer, pp. 41-58.

[Poi] Poincaré, H. J. (1893). "Le mécanisme et l'expérience." *Revue de Métaphysique et de Morale* 1, pp. 534-537.

[Pop59] Popper, K. R. (1959). *The Logic of Scientific Discovery*. London: Routledge, 2002.

[Pr-Th] Price, R. H. and Thorne, K. S. (1988). "The Membrane Paradigm for Black Holes." *Scientific American* 258, pp. 69-77.

[Reis] Reissner, H. (1916). "Über die Eigengravitation des elektrischen Feldes nach der Einsteinschen Theorie." *Annalen der Physik* 355, pp. 106-120.

[RT] Robinson, I. and Trautman, A. (1962). "Some Spherical Gravitational Waves in General Relativity." *Proceedings of the Royal Society of London. Series A, Mathematical and Physical Sciences* 265, pp. 463-473.

[Ruf-Whe] Ruffini, R. and Wheeler, J. A. (1971). "Introducing the Black Hole." *Physics Today* 24, pp. 30-41.

[Sch] Schrödinger, E. (1918). "Die Energiekomponenten des Gravitationsfeldes." *Physikalische Zeitschrift* 19, pp. 4-7.

[S-1] Schwarzschild, K. (1916). "Über das Gravitationsfeld eines Massenpunktes nach der Einsteinschen Theorie." *Königlich Preußische Akademie der Wissenschaften (Berlin). Sitzungsberichte*, pp. 189-196.

[S-2] Schwarzschild, K. (1916). "Über das Gravitationsfeld einer Kugel aus inkompressibler Flüssigkeit nach der Einsteinschen Theorie." *Königlich Preußische Akademie der Wissenschaften (Berlin). Sitzungsberichte*, pp. 424-434.

[Shan] Shannon, C. E. (1948). "A Mathematical Theory of Communication." *The Bell System Technical Journal* 27, pp. 379-423.

[Bes] Speziali, P. (1972). *Albert Einstein and Michele Besso: Correspondence 1903-1955*. Translated by P. Speziali, Paris: Hermann.

[Sta-02] Stachel. J. (2002). *Einstein from 'B' to 'Z'*. Washington D.C.: Birkhäuser.

[Suss95] Susskind, L. (1995). "The World as a Hologram." *Journal of Mathematical Physics* 36, pp. 6377-6396.

[Suss12] Susskind, L. (2012). "The Transfer of Entanglement: The Case for Firewalls." *arXiv*:1210.2098v1 [hep-th].

[Suss13] Susskind, L. (2013). "Black Hole Complementarity and the Harlow-Hayden Conjecture." *arXiv*:1301.4505v2 [hep-th].

[Suss16] Susskind, L. (2016). "ER = EPR, GHZ, and the Consistency of Quantum Measurements." *Fortschritte der Physik* 64, pp. 72-83.

[Suss20] Susskind, L. (2020). *Three Lectures on Complexity and Black Holes*. Switzerland: Springer.

[STU] Susskind, L., Thorlacius, L., and Uglum, J. (1993). "The Stretched Horizon and Black Hole Complementarity." *Physical Review D* 48, pp. 3743-3762.

[Syn] Synge, J. L. (1960). *Relativity: The General Theory*. Amsterdam: North-Holland Publishing Company.

[Thi] Thirring, H. (1918). "Über die Wirkung rotierender ferner Massen in der Einsteinschen Gravitationstheorie." *Physikalische Zeitschrift* 19, pp. 33-39.

[Thi-Len] Thirring, H. und Lense, J. (1918). "Über den Einfluß der Eigenrotation der Zentralkörper auf die Bewegung der Planeten und Monde nach der Einsteinschen Gravitationstheorie." *Physikalische Zeitschrift* 19, pp. 156-163.

[Tho-74-1] Thomson, W. (1874). "Kinetic Theory of the Dissipation of Energy." *Nature* April 9, pp. 441-444.

[Tho-79] Thomson, W. (1879). "The Sorting Demon of Maxwell." *Nature* June 5, p. 126.

[Thor-94] Thorne, K. S. (1994). *Black Holes and Time Warps: Einstein's Outrageous Legacy*. Foreword by S. Hawking. New York: W. W. Norton.

[Pr-Th-Mac] Thorne, K. S., Price, R. H., and Macdonald, D. A. (1986). *Black Holes: The Membrane Paradigm*. New Haven: Yale University Press.

[uff] Uffink, J. (2007). "Compendium to the Foundations of Classical Statistical Mechanics." In Butterfield, J. and Earman, L. (eds.), *Handbook for the Philosophy of Physics*. Amsterdam: Elsevier, pp. 924-1074.

[Un-81] Unruh, W. G. (1981). "Experimental Black Hole Evaporation?" *Physical Review Letters* 46, pp. 1351-1353.

[Un-01] Unruh, W. G. (2001). "Black Hole, Dumb Holes, and Entropy." In Callender, C. and Huggett, N. *Physics Meets Philosophy at the Planck Scale: Contemporary Theories in Quantum Gravity*. New York: Cambridge University Press, pp. 152-175.

[Un-18] Unruh, W. G. (2018). "Black Holes in the Laboratory." *Bekenstein Memorial Lecture*, Hebrew University of Jerusalem.

[Un-Wal-81] Unruh, W. G. and Wald, R. M. (1981). "Acceleration Radiation and the Generalized Second Law of Thermodynamics." *Physical Review D* 25, pp. 942-958.

[83Ved-00] Vedral, V. (2000). "Landauer's Erasure, Error Correction and Entanglement." *Proceedings: Mathematical, Physical and Engineering Sciences* 456, pp. 969-984.

[Ved-06] Vedral, V. (2006). *Introduction to Quantum Information Science*. Oxford: Oxford University Press.

[Vis] Visser, M. (2003). "Essential and Inessential Features of Hawking Radiation." *International Journal of Modern Physics D* 12, pp. 649-661.

[Wal] Wald, R. M. (2020). "Jacob Bekenstein and the Development of Black Hole Thermodynamics." In Brink, L., Mukhanov, V. F., Rabinovici, E., and Phua, K. K. (eds.). *Jacob Bekenstein: The Conservative Revolutionary*. MA, US: World Scientific, pp. 3-10.

[Wall-mem.18] Wallace, D. (2018). "The Case for Black Hole Thermodynamics Part I: Phenomenological Thermodynamics." *Studies in History and Philosophy of Modern Physics* 64, pp. 52-67.

[Wall-mem.19] Wallace, D. (2019). "The Case for Black Hole Thermodynamics Part II: Statistical Mechanics." *Studies in History and Philosophy of Modern Physics* 66, pp. 103-117.

[W17] Weinstein, G. (2017). *Einstein's Pathway to the Special Theory of Relativity, 2ED* (includes Einstein's pathway to general relativity). Newcastle, UK: Cambridge Scholars Publishing.

[W18] Weinstein, G. (2018). "Why did Einstein Reject the November Tensor in 1912-1913, only to Come Back to it in November 1915?" *Studies in History and Philosophy of Modern Physics* 62, pp. 98-122.

[Wei] Weinstein, G. (2020). "The Einstein-Rosen Bridge and the Einstein-Podolsky-Rosen Argument: Singularities and Separability." In Beisbart, C., Sauer, T., and Wüthrich, C. *Thinking About Space and Time: 100 Years of Applying and Interpreting General Relativity. Einstein Studies* 15, pp. 91-109.

[Whe-57] Wheeler, J. A. (1957). "The Present Position of Classical Relativity Theory and Some of its Problems." *The Role of Gravitation in Physics: Report from the 1957 Chapel Hill Conference*. C. M. DeWitt and D. Rickles. Max Planck Research Library, 2017, pp. 43-49.

[Whe-81] Wheeler, J. A. (1981). "The Lesson of the Black Hole." *Proceedings of the American Philosophical Society* 125, pp. 25-37.

[Whe-08] Wheeler, J. A. (2008). "Entropy of a Black Hole: Bekenstein, Stephen Hawking." *Web of Stories: Life Stories of Remarkable People*.

[Wey] Weyl, H. (1918). "Gravitation und Elektrizität." *Sitzungsberichte der Königlich Preußischen Akademie der Wissenschaften zu Berlin*, pp. 465-480.

[Zer] Zermelo, E. (1896). "On the Mechanical Explanation of Irreversible Processes." In Brush, S. G. (ed.) (1966). *Kinetic Theory, Vol. 2, Irreversible Processes*. London: Pergamon Press, pp. 229-237.

Index

G. Weinstein, *Einstein's Legacy*, History of Physics,
https://doi.org/10.1007/978-3-031-73572-1

Author Index

G. Weinstein, *Einstein's Legacy*, History of Physics,
https://doi.org/10.1007/978-3-031-73572-1

The manufacturer's authorised representative in the EU is Springer Nature Customer Service Centre GmbH, Europaplatz 3, 69115 Heidelberg, Germany. If you have any concerns regarding our products, please contact ProductSafety@springernature.com

Printed and bound by CPI Group (UK) Ltd, Croydon, CR0 4YY
01/10/2026
02225953-0019